Materials for Energy Conversion and Storage

Development of new energy-related materials is essential in addressing future energy demands. *Materials for Energy Conversion and Storage* focuses on the materials science related to energy conversion and energy storage technologies. It covers the principles of prospective energy technologies and their relationship to the performance of energy devices.

- Covers fundamental principles of energy conversion and storage
- Discusses materials selection, design, and performance tradeoffs
- Details electrochemical cell construction and testing methodologies
- Explores sustainable development of energy devices
- Features case studies

Aimed at readers in materials, electrical, and energy engineering, this book provides readers with a deep understanding of the role of materials in developing sustainable energy devices.

Hieng Kiat Jun obtained his MEngSci and PhD from the University of Malaya, Malaysia. He is an Associate Professor in the Department of Mechanical and Materials Engineering, Universiti Tunku Abdul Rahman, Malaysia. He conducts experimental research on nanosized energy devices in solar cells and electrochemical energy storage systems. His main topics of interest include dye- and quantum dot-sensitized solar cells, organic solar cells, perovskite solar cells, electrolyte characterization, radiative cooling, and thermoelectric devices.

Foo Wah Low is an Assistant Professor in the Department of Electrical and Electronics Engineering at Universiti Tunku Abdul Rahman, Malaysia. He received his PhD in nanomaterials and nanoscience from Universiti Malaya. His main research interests are in the areas of graphene nanomaterials incorporated with metal oxide for photovoltaic cell application, especially in dye-sensitized solar cells and perovskites for photovoltaic cells.

Emerging Materials and Technologies

Series Editor: Boris I. Kharissov

The *Emerging Materials and Technologies* series is devoted to highlighting publications centered on emerging advanced materials and novel technologies. Attention is paid to those newly discovered or applied materials with potential to solve pressing societal problems and improve quality of life, corresponding to environmental protection, medicine, communications, energy, transportation, advanced manufacturing, and related areas.

The series takes into account that, under present strong demands for energy, material, and cost savings, as well as heavy contamination problems and worldwide pandemic conditions, the area of emerging materials and related scalable technologies is a highly interdisciplinary field, with the need for researchers, professionals, and academics across the spectrum of engineering and technological disciplines. The main objective of this book series is to attract more attention to these materials and technologies and invite conversation among the international R&D community.

Smart Micro- and Nanomaterials for Pharmaceutical Applications
Edited by Ajit Behera, Arpan Kumar Nayak, Ranjan K. Mohapatra, and Ali Ahmed Rabaan

Friction Stir-Spot Welding
Metallurgical, Mechanical and Tribological Properties
Edited by Jeyaprakash Natarajan and K. Anton Savio Lewise

Phase Change Materials for Thermal Energy Management and Storage
Fundamentals and Applications
Edited by Hafiz Muhammad Ali

Nanofluids
Fundamentals, Applications, and Challenges
Shriram S. Sonawane and Parag P. Thakur

MXenes
From Research to Emerging Applications
Edited by Subhendu Chakroborty

Biodegradable Polymers, Blends and Biocomposites
Trends and Applications
Edited by A. Arun, Kunyu Zhang, Sudhakar Muniyasamy and Rathinam Raja

For more information about this series, please visit: www.routledge.com/Emerging-Materials-and-Technologies/book-series/CRCEMT

Materials for Energy Conversion and Storage

Edited by

Hieng Kiat Jun and Foo Wah Low

CRC Press
Taylor & Francis Group
Boca Raton London New York

CRC Press is an imprint of the
Taylor & Francis Group, an informa business

Designed cover image: © 2025 Hieng Kiat Jun

First edition published 2025
by CRC Press
2385 NW Executive Center Drive, Suite 320, Boca Raton FL 33431

and by CRC Press
4 Park Square, Milton Park, Abingdon, Oxon, OX14 4RN

CRC Press is an imprint of Taylor & Francis Group, LLC

ISBN: 978-1-032-32311-4 (hbk)
ISBN: 978-1-032-32312-1 (pbk)
ISBN: 978-1-003-31442-4 (ebk)

DOI: 10.1201/9781003314424

Typeset in Times
by codeMantra

Contents

Preface

Embarking on a profound exploration of *Materials for Energy Conversion and Storage*, this volume delves into various chapters, each dedicated to unraveling the complexities and advancements within specific domains of energy conversion technologies. In this 21st century, where rapid technological developments and the demands of our digitalized lifestyles rely heavily on electricity, the need for sustainable energy sources becomes increasingly imperative. Fossil fuels, our traditional mainstay, are finite and environmentally impactful, prompting a shift toward alternative, clean energy solutions.

This book's introductory chapter sets the stage by providing fundamental insights into the principles of energy conversion, in general and in photovoltaics, in the following chapter. Readers are guided through the intricate processes involved in harnessing solar energy for electricity, gaining a foundational understanding of various photovoltaic technologies, including silicon-based and thin films. The journey continues with the third chapter, which zooms into the captivating realm of graphene materials and their profound effects on perovskite solar cells (PSCs). Exploring the synergistic relationship between graphene and PSCs, this chapter illuminates the potential for pushing the efficiency boundaries of solar energy conversion.

Shifting the focus to fuel cells, the fourth chapter unfolds the fundamental principles and diverse applications of this technology. From transportation to stationary power generation, readers are introduced to the versatility of fuel cells and their role in providing clean and efficient energy solutions. Building upon this foundation, the subsequent chapter narrows its focus to delve into the intricate world of solid polymer electrolytes and their application in fuel cells. By unraveling the complexities of these materials, readers gain insights into the advancements driving fuel cell technology toward enhanced efficiency, durability, and real-world applicability.

The exploration extends to supercapacitors in the sixth chapter, with a specific emphasis on sustainably processed carbon quantum dots. This section addresses the pivotal role of these innovative materials in developing supercapacitors with improved performance and environmental sustainability. Continuing the journey into energy storage, the seventh chapter dives into the world of tin-based anodes, a crucial component in the development of next-generation lithium-ion batteries. Readers gain valuable insights into the potential of these anodes to enhance the energy density, lifespan, and safety of lithium-ion batteries.

The final chapter casts a wide net, encompassing sustainable materials for energy conversion systems. Synthesizing key learnings from preceding chapters, this section expands the discussion to address the broader landscape of materials contributing to sustainable energy conversion.

In essence, *Materials for Energy Conversion and Storage* presents a collective exploration of cutting-edge materials shaping the future of energy conversion. Each chapter not only delves into the technical intricacies of its respective fields but also emphasizes the broader goal of achieving efficiency, durability, and environmental sustainability in the quest for innovative energy solutions. As we navigate through these chapters, a holistic understanding of the intricate interplay between materials and energy conversion emerges, paving the way for a greener and more sustainable energy future.

Contributors

A. K. Arof
Universiti Malaya
Kuala Lumpur, Malaysia

Grishika Arora
Universiti Tunku Abdul Rahman
Kajang, Malaysia

M. H. Buraidah
Universiti Malaya
Kuala Lumpur, Malaysia

Mohammad Aminul Islam
Universiti Malaya
Kuala Lumpur, Malaysia

Hieng Kiat Jun
Universiti Tunku Abdul Rahman
Kajang, Malaysia

Muhammad Kashif
Tianjin University
Tianjin, China

Chiam-Wen Liew
Tunku Abdul Rahman University of Management and Technology (TARUMT)
Kuala Lumpur, Malaysia

Son Qian Liew
Universiti Tunku Abdul Rahman
Kajang, Malaysia

Foo Wah Low
Universiti Tunku Abdul Rahman
Kajang, Malaysia

Chai Yan Ng
Universiti Tunku Abdul Rahman
Kajang, Malaysia

Cheen Sean Oon
Monash University Malaysia
Bandar Sunway, Malaysia

Tejas Sharma
Universiti Tunku Abdul Rahman
Kajang, Malaysia

Pramod K. Singh
Sharda University
Greater Noida, India

L. P. Teo
Tunku Abdul Rahman University of Management and Technology (TARUMT)
Kuala Lumpur, Malaysia

Chun Hong Voon
University Malaysia Perlis
Perlis, Malaysia

1 Introduction
Energy and Materials

Hieng Kiat Jun

1.1 ENERGY ISSUE

Energy is the backbone of modern societies, powering everything from transportation and communication to agriculture and manufacturing. However, the world is facing an energy crisis due to various factors such as rising global population, urbanization, and the demand for better living standards. The issue of energy is multifaceted, and it encompasses various aspects of human activity, including environmental sustainability, economic development, and social justice (Butler et al., 2018). In this chapter, the energy issue and its impact on the world are briefly discussed.

One of the main challenges posed by the energy issue is the finite nature of fossil fuels, which currently supply the majority of the world's energy needs. Fossil fuels such as coal, oil, and gas are finite resources that will eventually run out, and their extraction and consumption contribute significantly to greenhouse gas emissions that cause climate change (Scovronick et al., 2019). This means that there is a pressing need to find alternative sources of energy that are sustainable and do not contribute to global warming (Figure 1.1).

Renewable energy sources such as solar, wind, hydropower, and geothermal are increasingly being viewed as viable alternatives to fossil fuels. These sources of energy are abundant, clean, and renewable. They have the potential to reduce greenhouse gas emissions while meeting the world's energy needs. However, the adoption of renewable energy sources is often hampered by the high cost of technology and the lack of infrastructure to support their use. Fortunately, the cost of renewable energy has been reducing from year to year due to policy and reduced manufacturing costs (IRENA, 2023).

Another aspect of the energy issue is the need for energy security. Many countries rely heavily on imported fossil fuels to meet their energy needs, which can lead to vulnerability and instability in the event of supply disruptions or price shocks. This has led to the exploration of domestic sources of energy such as shale gas, tar sands, and offshore oil and gas reserves. However, the extraction of these resources can be environmentally damaging, and there are concerns about the long-term sustainability of these sources of energy (Stevens, 2012).

The energy issue is also closely tied to economic development. The availability of affordable and reliable energy is essential for economic growth, but energy prices can also have a significant impact on economic stability. High energy prices can lead to inflation and slow economic growth, while low energy prices can discourage investment in renewable energy and reduce incentives for energy efficiency. This means

DOI: 10.1201/9781003314424-1

FIGURE 1.1 Reduction of greenhouse gas emission is paramount in reversing climate change.

that there is a delicate balance between energy prices and economic growth, and policymakers must carefully consider the impact of energy policies on the economy.

Besides, the energy issue also has social justice implications. Access to energy is a basic human right, and yet over one billion people worldwide lack access to electricity (Jenkins, 2018). This lack of access to energy can have a profound impact on health, education, and economic opportunities. Addressing energy poverty is therefore an important aspect of the energy issue, and it requires a concerted effort from governments, international organizations, and the private sector.

The energy issue is a complex and multifaceted problem that requires a comprehensive approach. The world needs to transition to a more sustainable energy system that relies on renewable energy sources and reduces greenhouse gas emissions. This transition will require significant investments in technology, infrastructure, and policy, as well as a shift in societal attitudes toward energy consumption and conservation. Addressing the energy issue will also require a commitment to social justice, ensuring that access to energy is available to all. One of the policy approaches is the energy transition policy (Figure 1.2).

1.2 ENERGY TRANSITION

Energy transition is the process of moving from traditional, non-renewable energy sources, such as fossil fuels, to cleaner and more sustainable sources of energy. The idea of energy transition is based on the need to reduce greenhouse gas emissions and mitigate the effects of climate change. What is more important is the impact of energy

FIGURE 1.2 Renewable energy like wind energy can be one of the solutions for energy issue.

transition that affects various aspects of society, including the economy, environment, and public health (Kabeyi and Olanrewaju, 2022). Meeting the needs of energy transition requires a concerted effort from governments, businesses, and individuals.

The main principle of energy transition is to shift from high-carbon energy sources to low-carbon or carbon-free energy sources (Khan et al., 2022). The primary goal is to reduce greenhouse gas emissions and mitigate the effects of climate change. This requires the use of renewable energy sources such as wind, solar, and hydroelectric power, as well as energy efficiency measures such as insulation and efficient appliances. Energy transition also involves the electrification of transportation, such as the use of electric vehicles, to reduce reliance on fossil fuels.

Energy transition offers numerous benefits. The most anticipated benefits are the reduction of greenhouse gas emissions and the improvement of air quality. These can have positive effects on public health. It can also stimulate economic growth and create new job opportunities in the renewable energy sector. However, energy transition can also induce negative impacts, such as job losses in the fossil fuel industry and higher costs of energy for consumers.

One of the main requirements for energy transition is the development and implementation of policies and regulations that support the transition to renewable energy sources (Steinbacher and Röhrkasten, 2019; Wang et al., 2021). Governments must provide incentives for the development and use of renewable energy, such as subsidies and tax credits. They can also impose regulations on greenhouse gas emissions and energy efficiency standards. Businesses also have a role to play, by investing in renewable energy technologies and adopting energy-efficient practices. Finally, individuals can contribute by making conscious choices in their energy consumption, such as using public transportation and reducing energy consumption at home.

FIGURE 1.3 Transitioning from fossil fuel–based energy to renewable energy is crucial for addressing climate change and pollution issues.

Another criterion for energy transition is the development of new and improved technologies for renewable energy production and energy storage. Renewable energy technologies, such as solar and wind power, have made significant progress in recent years but still face challenges such as intermittency and storage. Developing new technologies for energy storage, such as batteries and hydrogen fuel cells, can help overcome these challenges and increase the reliability and effectiveness of renewable energy sources. Furthermore, education and awareness are critical for achieving a successful energy transition. Educating the public about the benefits and challenges of renewable energy sources can help build support for energy transition policies and encourage individuals to make conscious choices in their energy consumption. Awareness campaigns can also help debunk myths and misconceptions about renewable energy, such as the belief that it is too expensive or unreliable. Understanding the principles of energy conversion is the beginning of the innovation of energy devices (Figure 1.3).

1.3 ENERGY CONVERSION

Energy conversion is the process of transforming one form of energy into another, such as converting solar energy into electricity or converting heat into mechanical work. The principle of energy conversion is based on the laws of thermodynamics, which govern the behavior of energy in physical systems (Lostaglio et al., 2014). The theory of energy conversion involves the study of the mechanisms and processes involved in energy transformation.

The First Law of thermodynamics states that energy cannot be created or destroyed but can only be transformed from one form to another. This law is also known as the law of conservation of energy. It signifies that the total amount of energy in a closed system remains constant and that energy cannot be created or destroyed within that system. Energy can only be transferred from one object or system to another or converted from one form to another.

The Second Law of thermodynamics states that the total entropy of a closed system always increases over time. Entropy is a measure of the disorder or randomness of a system. The Second Law implies that in any energy conversion process, some energy will inevitably be lost as unusable heat energy, and the overall efficiency of the process will be less than 100%. Therefore, energy conversion processes are always associated with some degree of inefficiency.

Energy conversion is an essential concept in physics and engineering and has numerous practical applications in various fields, including power generation, transportation, and electronics. There are various types of energy conversion processes, such as mechanical-to-electrical energy conversion, and photovoltaic energy conversion (Bejan, 2016).

In a mechanical-to-electrical energy conversion system, the process involves converting mechanical energy, such as the rotation of a wind turbine or the movement of a hydroelectric dam turbine, into electrical energy. This process involves the use of generators, which convert mechanical energy into electrical energy using the principle of electromagnetic induction. On the other hand, thermal-to-electrical energy conversion is a process of converting heat energy into electrical energy. An example of this system is a thermoelectric generator. This process involves the use of a thermocouple, which generates an electrical voltage when there is a temperature difference between two junctions.

For the conversion of chemical energy to electrical energy, the chemical energy stored in the system, such as batteries, is converted into electrical energy. This process involves a chemical reaction that generates an electrical potential difference between two electrodes. Electricity can also be generated from solar energy which is known as photovoltaic energy conversion. This process involves the conversion of solar energy into electrical energy, as found in a solar cell. Solar cell devices are usually made up of semiconductor materials, which absorb photons from sunlight, generating an electrical voltage in return.

The produced electricity is usually used to produce useful work or applications. In this case, electrical energy is converted to mechanical energy like in an electric motor. This process utilizes electromagnets to generate a magnetic field, inducing rotation in a conductor carrying an electric current through their interaction. In all the energy conversion systems, the amount of energy output produced determines the efficiency of the systems.

The efficiency of energy conversion processes is a crucial factor in determining their practical usefulness. It is defined as the ratio of the useful energy output to the energy input. For example, the efficiency of a wind turbine is defined as the ratio of the electrical power output to the wind power input. The efficiency of energy conversion processes is limited by various factors, including the laws of thermodynamics, the properties of the materials used in the conversion process, and the design of the conversion system (Nevskaya et al., 2023).

In recent years, there has been increasing interest in developing new and more efficient energy conversion technologies to address the growing demand for sustainable

FIGURE 1.4 Solar panel involves in converting solar energy to electric energy.

energy sources. One area of research is the development of materials with improved properties for energy conversion processes. For example, the development of new semiconductor materials with high efficiency and low cost is crucial for improving the efficiency of photovoltaic cells. Researchers are also exploring the use of nanomaterials and novel structures for improving the efficiency of energy conversion processes (Figure 1.4).

1.4 MATERIALS AND ENERGY

Materials science and energy are intricately linked, with materials playing a crucial role in the development of sustainable energy technologies. Materials science is a multidisciplinary field that encompasses the study of the structure, properties, and behavior of materials. It is a rapidly evolving field that has led to significant advancements in various areas like electronics, medicine, and energy.

We all acknowledge that energy is the driving force behind modern civilization. As the world continues to seek sustainable energy sources, the focus has shifted toward developing materials that can improve the efficiency, durability, and performance of energy systems. The use of suitable materials has enabled the development of new energy technologies in solar cells, wind turbines, and electric vehicles, to name a few. Indeed, materials science and engineering have played a vital role in the optimization and commercialization of these technologies.

One of the most significant challenges in the energy sector is to find sustainable and cost-effective solutions to meet the increasing energy demands of the world's population. The development of sustainable energy technologies is essential to address global challenges of climate change, energy security, and economic growth. To meet these challenges, there is a need for new and innovative materials that can enhance the performance of energy systems. In this case, materials research has contributed significantly to the development of renewable energy technologies. For example, the development of solar cells has been made possible by advances in the understanding of the properties and behavior of semiconductors. The use of new materials, such as perovskites, has led to significant improvements in the efficiency of solar cells (Zhang et al., 2022). Similarly, the development of wind turbines has been made possible by advances in the understanding of materials, such as high-strength alloys, that can withstand the harsh conditions of offshore wind farms (Tong, 2019). In another example, thermoelectric materials can convert heat into electricity and have potential applications in waste heat recovery (Yang et al., 2018).

Energy storage is another critical area where materials science has played a vital role. Particularly, energy storage is essential for renewable energy systems to address the variability of solar and wind energy. The development of materials for energy storage has led to the commercialization of lithium-ion batteries, which are widely used in electric vehicles and consumer electronics (Xu et al., 2023). However, there is a need for new materials that can store more energy (in terms of energy density) and be cost-effective and environmentally sustainable.

Hydrogen is another potential clean energy source that can be used for a variety of applications, from transportation to power generation. Materials science plays a crucial role in the development of materials for hydrogen storage and transportation. Researchers are exploring a range of materials, such as metal hydrides, which can store hydrogen safely and efficiently (Tarasov et al., 2021).

Carbon capture and storage technologies are essential for reducing carbon emissions from industrial processes and power plants. Materials science has contributed to the development of materials that can effectively capture carbon dioxide while being cost-effective and sustainable. Metal-organic frameworks, for example, are highly porous and can selectively capture carbon dioxide from flue gas streams (Pettinari and Tombesi, 2020).

Improving energy efficiency is one of the most effective ways to reduce energy consumption and greenhouse gas emissions. In this case, materials science has contributed significantly to the development of materials for energy-efficient systems. For example, light-emitting diodes (LEDs) are more energy-efficient than traditional lighting systems (Bessho and Shimizu, 2012). Another example is aerogel, which can improve the insulation properties of buildings, reducing energy consumption for heating and cooling (Zhang et al., 2023).

In summary, materials science plays a vital role in the development of sustainable energy technologies. The use of materials has enabled the optimization and commercialization of renewable energy technologies, as seen in solar cells and wind turbines. Materials science has also contributed to the development of materials for energy storage, hydrogen energy, carbon capture, and energy conversion technologies. Furthermore, improving energy efficiency is one of the most effective ways to reduce energy consumption and greenhouse gas emissions. The field has contributed significantly to the development of materials for energy-efficient systems, such as LEDs and aerogels, which have the potential to reduce energy consumption and

FIGURE 1.5 Progress of materials research in LED has enabled the development of efficient LED lighting with various color illumination.

mitigate the effects of climate change. The continued advancement of materials science will be crucial for meeting the energy challenges of the future and transitioning toward a sustainable energy future (Figure 1.5).

1.5 TRENDS IN MATERIALS RESEARCH

Without a doubt, energy is the driving force behind modern civilization. As the world continues to seek sustainable energy sources, the focus has shifted toward researching and developing materials that can improve the efficiency, durability, and performance of energy conversion systems. Some of the trends in materials for energy research that have emerged in recent years are briefly described in this section (Figure 1.6).

Energy storage is an essential part of renewable energy systems. The current challenge is to find materials that can store energy efficiently, while also being cost-effective and environmentally sustainable. One of the most promising materials

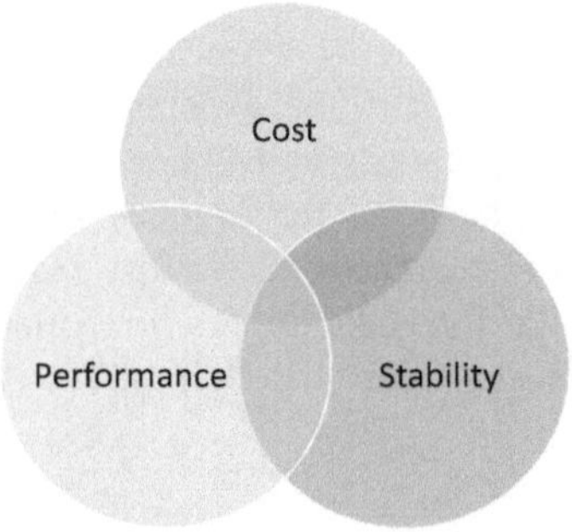

FIGURE 1.6 Trends in materials research for energy systems and devices.

in this field is graphene. Graphene-based batteries can store more energy and charge faster than traditional lithium-ion batteries, making them ideal for use in electric vehicles and other energy storage applications (Olabi et al., 2021). There has been an increasing trend in graphene research in general for the past few years. The application of graphene in energy storage application will see a significant uptake in the coming years.

Solar energy has also become an increasingly popular source of renewable energy. One of the main challenges in this field is to improve the efficiency of solar cells. Silicon-based solar cells have dominated the market for decades, but researchers are now looking at alternative materials, such as perovskites, that have the potential to be more efficient and cheaper. Perovskites are a type of crystal that can be made into thin films and have already shown promising results in laboratory settings (Chen et al., 2023). For scale-up and commercial application, the stability and cost of the solar cell need to be improved.

Another upcoming trend is hydrogen energy. Hydrogen has emerged as a potential clean energy source that can be used for a variety of applications, from transportation to power generation (Rosen and Koohi-Fayegh, 2016). The current challenge in this field is to find materials that can efficiently store and transport hydrogen. New materials that can store hydrogen safely and efficiently are needed. The current metal hydride materials require proper and extensive research for better storage capability (Tarasov et al., 2021).

As a way of promoting sustainable usage, organic materials have been the focus of energy devices. The challenge is to find suitable organic materials that can efficiently convert energy while also being durable and cost-effective. A wide range of organic materials that can be used in flexible and lightweight devices has been explored, especially wearable energy devices' applications (Zhang and Park, 2019).

In conclusion, materials research plays a crucial role in advancing sustainable energy technologies. New materials are explored, from graphene to perovskites to metal oxide frameworks in order to improve the efficiency, durability, and performance of energy systems or devices. With continued research and development, these materials have the potential to transform the way we generate, store, and use energy. Ultimately paving the way for a more sustainable future.

1.6 CONCLUSION

In summary, this chapter has covered crucial aspects of the energy landscape, addressing global challenges, exploring energy conversion principles, and highlighting the pivotal role of materials. Emphasizing the urgency of transitioning to sustainable energy, the emerging trends in materials for energy research have been discussed. This chapter serves as a guide for stakeholders for collaborative efforts to shape a cleaner and more resilient energy future through innovative materials solutions.

REFERENCES

Bejan, A. (2016). *Advanced Engineering Thermodynamics* (Wiley, New York).

Bessho, M., and Shimizu, K. (2012). Latest trends in LED lighting. *Electronics and Communications in Japan* 95, 1–7.

Butler, C., Parkhill, K.A., and Luzecka, P. (2018). Rethinking energy demand governance: Exploring impact beyond 'energy' policy. *Energy Research and Social Science* 36, 70–78.

Chen, Y., Zhang, M., Li, F., and Yang, Z. (2023). Recent progress in perovskite solar cells: Status and future. *Coatings* 13, 644.

IRENA.org. International Renewable Energy Agency (IRENA). (2023). Renewables Competitiveness Accelerates, Despite Cost Inflation. Available at https://www.irena.org/News/pressreleases/2023/Aug/Renewables-Competitiveness-Accelerates-Despite-Cost-Inflation (accessed November 2023).

Jenkins, K. (2018). Energy poverty and vulnerability: A global perspective. *Energy* 147, 388–389.

Kabeyi, M.J.B., and Olanrewaju, O.A. (2022). Sustainable energy transition for renewable and low carbon grid electricity generation and supply. *Frontiers in Energy Research* 9.

Khan, I., Zakari, A., Ahmad, M., Irfan, M., and Hou, F. (2022). Linking energy transitions, energy consumption, and environmental sustainability in OECD countries. *Gondwana Research* 103, 445–457.

Lostaglio, M., Jennings, D., and Rudolph, T. (2014). Thermodynamic laws beyond free energy relations. 11, arXiv preprint arXiv:1405.2188.

Nevskaya, M.A., Raikhlin, S.M., Vinogradova, V.V., Belyaev, V.V., and Khaikin, M.M. (2023). A study of factors affecting national energy efficiency. *Energies* 16, 5170.

Olabi, A.G., Abdelkareem, M.A., Wilberforce, T., and Sayed, E.T. (2021). Application of graphene in energy storage device: A review. *Renewable and Sustainable Energy Reviews* 135, 110026.

Pettinari, C., and Tombesi, A. (2020). Metal-organic frameworks for carbon dioxide capture. *MRS Energy and Sustainability* 7, 35.

Rosen, M.A., and Koohi-Fayegh, S. (2016). The prospects for hydrogen as an energy carrier: an overview of hydrogen energy and hydrogen energy systems. *Energy, Ecology and Environment* 1, 10–29.

Scovronick, N., Budolfson, M., Dennig, F., Errickson, F., Fleurbaey, M., Peng, W., Socolow, R.H., Spears, D., and Wagner, F. (2019). The impact of human health co-benefits on evaluations of global climate policy. *Nature Communications* 10, 2095.

Steinbacher, K., and Röhrkasten, S. (2019). An outlook on Germany's international energy transition policy in the years to come: Solid foundations and new challenges. *Energy Research and Social Science* 49, 204–208.

Stevens, P. (2012). The 'shale gas revolution': Developments and changes. Chatham House Briefing Paper 12.

Tarasov, B.P., Fursikov, P.V., Volodin, A.A., Bocharnikov, M.S., Shimkus, Y.Y., Kashin, A.M., Yartys, V.A., Chidziva, S., Pasupathi, S., and Lototskyy, M.V. (2021). Metal hydride hydrogen storage and compression systems for energy storage technologies. *International Journal of Hydrogen Energy* 46, 13647–13657.

Tong, C. (2019). Advanced materials enable renewable wind energy capture and generation. In: *Introduction to Materials for Advanced Energy Systems* (Springer, New York), pp. 379–444.

Wang, D., Liu, X., Yang, X., Zhang, Z., Wen, X., and Zhao, Y. (2021). China's energy transition policy expectation and its CO_2 emission reduction effect assessment. *Frontiers in Energy Research* 8.

Xu, J., Cai, X., Cai, S., Shao, Y., Hu, C., Lu, S., and Ding, S. (2023). High-energy lithium-ion batteries: recent progress and a promising future in applications. *Energy and Environmental Materials* 6, e12450.

Yang, L., Chen, Z.G., Dargusch, M.S., and Zou, J. (2018). High performance thermoelectric materials: progress and their applications. *Advanced Energy Materials* 8, 1701797.

Zhang, P., Li, M., and Chen, W.C. (2022). A perspective on perovskite solar cells: emergence, progress, and commercialization. *Frontiers in Chemistry* 10.

Zhang, X., Hu, H., Hou, Q., and Liu, W. (2023). Research progress of cellulose-based aerogel in heat preservation and insulation. *Chung-Kuo Tsao Chih/China Pulp and Paper* 42, 86–93.

Zhang, Y., and Park, S.J. (2019). Flexible organic thermoelectric materials and devices for wearable green energy harvesting. *Polymers* 11, 909.

2 Photovoltaic
Fundamental of Energy Conversion

Hieng Kiat Jun and Foo Wah Low

2.1 INTRODUCTION AND BRIEF HISTORY

A photovoltaic (PV) system, also known as a solar electric system, is a technology that converts sunlight into electricity. It is composed of solar panels that contain PV cells, which convert the energy from the sun into direct current (DC) electricity. This DC electricity can then be converted into alternating current (AC) electricity using an inverter, which allows it to be used in homes and businesses. PV systems can be used for a variety of applications, such as powering homes, businesses, and even vehicles. They are becoming increasingly popular as a source of renewable energy. This is because PV systems do not produce emissions and are considered a sustainable way to generate electricity.

A solar cell and a PV cell are essentially the same thing. A solar cell is a device that converts sunlight into electricity through the PV effect. The main difference between the two is that solar cell is a generic term that refers to any device that converts sunlight into electricity, while PV cell specifically refers to a type of solar cell that uses the PV effect to generate electricity. Solar cells can be made using a variety of materials, such as silicon (Si), cadmium telluride (CdTe), or copper indium gallium selenide.

The history of PV can be traced back to the 19th century, when scientists first discovered the PV effect. In 1839, French physicist Alexandre-Edmond Becquerel observed that certain materials would produce small amounts of electrical current when exposed to light (Palz, 2010). This phenomenon is known as the PV effect, and it forms the basis of modern PV technology.

In the early 20th century, researchers began experimenting with using the PV effect to generate electricity. In 1941, Russell Ohl discovered that silicon, a common semiconductor material, could be used to create a PV cell (Riordan and Hoddeson, 1997). However, the efficiency of these early PV cells was low and they were not cost-effective for widespread use at that time.

In the 1950s, Bell Labs developed the first practical silicon PV cell, which had an efficiency of 6%. This marked the beginning of the modern PV industry (Blandford and Watkins, 2009). Over the next few decades, researchers continued to improve the efficiency of PV cells and lower their cost. In the 1970s, the oil crisis led to increased interest in renewable energy sources, and PV technology began to be used

DOI: 10.1201/9781003314424-2

in a variety of applications. The increased interest was mainly attributed by oil companies venturing into solar firms (Wayback Machine, 2012).

By the early 21st century, PV technology had advanced significantly. The efficiency of PV cells has increased to over 20%, and the cost of PV systems has dropped significantly (Jaeger, 2021). As a result, PV systems began to be used for large-scale power generation and powering homes and businesses. Today, PV systems are becoming increasingly popular as a source of renewable energy. They are used in a variety of applications, such as powering homes, businesses, and even vehicles. They are also being integrated into the electricity grid as a source of distributed energy generation. In recent years, PV panel production has grown exponentially, with some countries like China and the US leading the way in terms of production and installation (IEA, 2015). With the cost of the solar panel dropping and the technology improving, the future of PV looks promising.

2.2 BASIC CHARACTERISTICS AND CHARACTERIZATION OF PHOTOVOLTAICS

The main characteristic of PV is the energy conversion capability. Specifically, PV cells convert light into electricity through the PV effect. In most cases, DC electricity is generated, which can then be converted into AC using an inverter. The performance of a PV cell is expressed in efficiency. The efficiency of a PV cell is a measure of how effectively it converts light into electricity. Typical commercial silicon-based PV cells have an efficiency of around 20% (Schultz et al., 2007).

Most commercial PV cells are durable and have a long lifespan. They are designed to withstand extreme weather conditions and have minimal maintenance requirements. Aside from that, PV systems are categorized as a clean and renewable source of energy. They do not produce emissions and have a low environmental impact.

One of the salient features of PV systems is their flexibility where they can be installed on a variety of surfaces, with most installations taking place on rooftops. Not only that, the systems can be scaled to meet the energy needs of a variety of applications, from powering a small device or single home to a large-scale power plant. In terms of cost, the cost of PV systems has dropped significantly in recent years, making them increasingly cost-effective compared to traditional sources of electricity derived from fossil fuels (Martin, 2016).

There are several methods used to characterize and evaluate the performance of PV cells and systems. The most frequently used method is current–voltage (I–V) measurement. In this method, the current and voltage responses of a PV cell or system are measured under standard conditions. The standard test condition entails a solar irradiance value of 1,000 W/m^2 with an exposure period of up to 2.74 hours a day. I–V characteristic can also be measured under varying light intensity and temperature. Such data can provide comprehensive data on the PV cell which is beneficial in designing integrated PV system. The output of the measurement is represented as I–V curve of J–V curve (J represents current density), which is used to determine the efficiency and maximum power point of a PV cell or system (Figure 2.1). From the same I–V or J–V curve, other parameters can be derived,

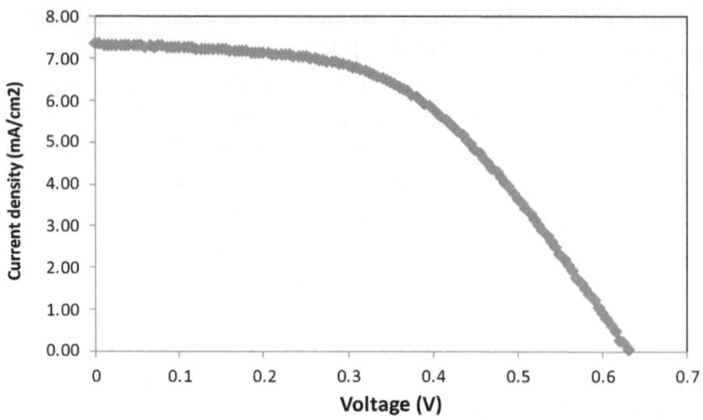

FIGURE 2.1 *J–V* curve of dye-sensitized solar cell using black dye as a sensitizer.

such as fill factor and maximum power point. Fill factor is a measure of quality of a solar cell while the maximum power point is the point where the product of *I* and *V* is the maximum.

In general, most of the PV cells do not respond to the entire spectrum of UV-visible regions. As such, it will be prudent to assess the spectral response of the PV active materials (i.e. materials responsible for the energy conversion). In this method, the electrical output of a PV cell or system illuminated or exposed to different wavelengths of light is measured. The output of the measurement can then be used to determine the cell's spectral response and identify the best wavelength for the highest efficiency.

Another less-discussed characterization of PV is temperature coefficient measurement. In this method, the performance of a PV cell or system is measured as a function of temperature. With increasing temperature, the efficiency of a PV cell usually will decrease. The data can be used to determine the temperature coefficient and identify the operating temperature range of the PV cell or system.

Other characterization or measurement methods include light-induced degradation measurement, current-matching measurement, electroluminescence imaging, and quantum efficiency (QE) measurement. The QE measurement is sometimes discussed in scientific journals pertaining to the PV materials' characteristics. QE denotes the proportion of photons that transform into electric current (i.e., gathered carriers) when the cell operates in short circuit conditions. In this method, the number of photocurrent generated by the PV cells per number of incident photons is determined. The data is then used to determine the internal QE of the cells.

2.3 PHOTOCURRENT GENERATION AND PHOTOVOLTAGE

The PV effect is the mechanism by which PV cells generate electricity. This effect occurs when photons of light are absorbed by a material, causing the release of electrons and the creation of an electrical current. In PV cells, the material that absorbs photons is typically a semiconductor, such as silicon. Semiconductors have a specific bandgap energy, which determines the energy required for an electron to be excited from the valence band to the conduction band. When a photon with energy greater than the bandgap energy is absorbed by the semiconductor, an electron is excited from the valence band to the conduction band, leaving a hole behind. These excited electrons and holes are referred to as free carriers, and they are able to move through the semiconductor. A PV cell is designed to collect these free carriers and separate them, creating an electrical current.

The positive and negative terminals of the PV cell are typically made of different semiconductor materials, known as p-type and n-type semiconductors. The p-type semiconductor has an excess of holes, and the n-type semiconductor has an excess of electrons. When the p-type and n-type semiconductors are brought into contact, a depletion region is formed at the interface between them. The depletion region is a region where the free electrons and holes are forced to recombine, creating an electric field. This electric field acts as a barrier, separating the free electrons and holes and creating a voltage across the terminals of the PV cell (Mishra, 2008).

In the *P–N* junction diode, four current components are present in equilibrium condition, namely, electron drift, electron diffusion, hole drift, and hole diffusion. In equilibrium condition, the net current is zero, which requires the drift and diffusion current of carriers to be equal and opposite. When PV is illuminating under light, it results large drift current with minority carriers (Figure 2.2), which flow from the n-type side to the p-type side. Since current flow is caused by the light, it is known as light-generated current, I_L. The generated photovoltage due to the light, biases the *P–N* junction in forward bias mode. Generated photovoltage reduces the junction's potential energy barrier and causes diffusion current flowing in the opposite direction to the I_L. But the magnitude of I_L is larger than the forward-biased diffusion

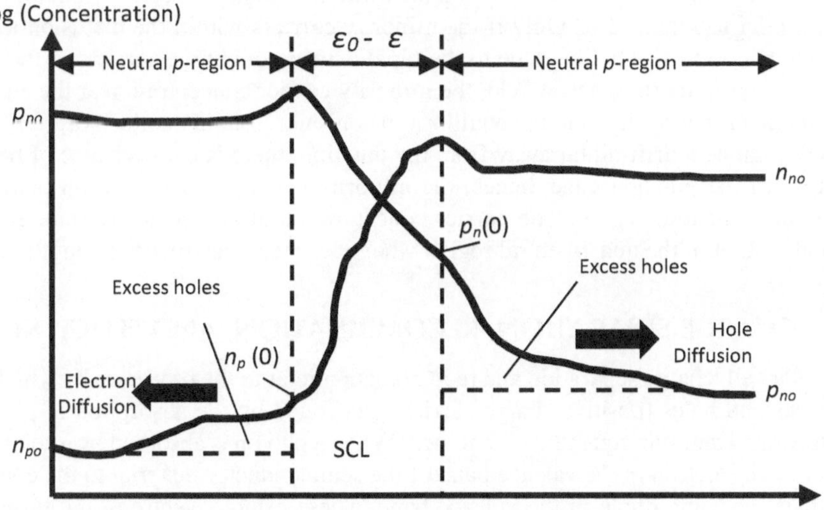

FIGURE 2.2 Minority and majority carriers profile for *P–N* junction.

current, and hence net current flows from the n-type side to the p-type side. Thus, when light shines on the PV cell, current flow is in the opposite direction respective to the generated voltage.

This voltage is known as the photovoltage, and the current generated by the flow of free carriers is known as the photocurrent. The combination of the photovoltage and photocurrent generates an electrical power output from the PV cell, which can be used to drive an electrical load. The photovoltage and photocurrent vary with the light intensity, temperature, and other environmental factors. Under uniform illumination condition, the generation of charge carrier will occur in the space charge region as well as the quasi-neutral region. The generated carriers in the space charge region will immediately swap due to the electric field. During efficient flow of electric field, the occurrence of recombination centers is quite less. Somehow, there are random motions of generated electron–hole pairs in quasi-neutral regions. Random motion of generated carriers eventually will be closer to the edge of the space charge region due to the force of the electric field. Only minority carriers will cross the junction as carriers always head down due to the energy force. The swapping of the *P–N* junction will induce a potential difference under light illumination. This phenomenon of photovoltage generation is known as PV effect.

There are seldom cases of random motion of carriers generated in the quasi-neutral region crossing the junction. In the event of this occurrence for these minority carriers, they will travel an average distance, L_n or L_p, before recombining with the opposite type of carrier. L_n and L_p are denoted as the diffusion length of electrons and holes, respectively. The minority carriers in the quasi-neutral region will be stuck at the edge of the *P–N* junction and will not contribute to the photovoltage. In rare cases, once the minority carriers manage to cross into the junction, the generated carriers within the width will be $(L_n + W + L_p)$. The rate of minority carriers depends on the intensity of light. In fact, the generation of carriers for a PV cell results in a significant increase in minority carrier concentration, while the majority carriers remain unaffected (refer to Figure 2.2). Only if the minority carriers within the diffusion length are able to cross the junction. Due to the rapid drifting of charge carriers in the *P–N* junction caused by the electric field, the minority carrier concentration at the edge of the junction is nearly equal to the equilibrium concentration. In another way, for those minority carriers drifting far away from the junction, there is a low chance of boarding toward the junction edge; hence, the minority carriers' concentration increases in the quasi-neutral region. The carrier concentration attains a steady state against recombination in the quasi-neutral region when it exceeds the diffusion length.

2.4 CHARGE SEPARATION, RECOMBINATION, AND EFFICIENCY

In a solar cell, charge separation and recombination refer to the processes by which the electrons and holes (positive charge carriers) generated by the absorption of photons are separated and subsequently recombined. When a photon is absorbed by a solar cell, it excites an electron in the valence band of the semiconductor material to the conduction band, leaving a hole in the valence band. These excited electrons and holes are known as free carriers, and they are able to move through the semiconductor material. The goal of a solar cell is to separate these free carriers and create an electrical current.

Charge separation is achieved by creating a built-in electric field at the junction between the p-type and n-type semiconductor materials. The p-type semiconductor has a surplus of holes, while the n-type semiconductor has a surplus of electrons. When the p-type and n-type semiconductors are brought into contact, a depletion region is formed at the interface between them. This depletion region acts as a barrier, separating the free electrons and holes and creating a voltage across the terminals of the solar cell (refer Figure 2.3).

Once the electrons and holes are separated, they flow in opposite directions to the terminals of the solar cell, generating an electrical current. However, this separation is not perfect and some of the free carriers will recombine before they reach the terminals of the cell. This recombination results in a loss of the electrical current and it reduces the solar cell efficiency.

Recombination can occur at various defects in the solar cell material such as grain boundaries, recombination centers, and surface states. When a working PV cell is under illumination, electrons and holes will be consequently generated, thus it will influence the equilibrium conditions of material. The process of recombination of charge carriers is in the opposite condition to the generation of charge carriers. In this comprehensive process, excited electrons will fall back from the conduction band into the valence band, reoccupying an empty energy state in the valence band. Henceforth, the electron–hole pairs are destructed. In other words, the recombination process can be considered as a restoring process of the carrier's mechanism. The recombination issue is not desirable in PV operation because it will be limiting the PV efficiency performance.

Assuming the concentration of electrons and holes in equilibrium condition in n_0 and p_0, when the light falls on the PV device, non-equilibrium concentration of electrons and holes will be induced with n and p. Therefore, the excess charge carrier concentration of electrons and holes under illumination condition, Δp and Δp, respectively, can be written as follows:

$$\Delta n = n - n_0 \tag{2.1}$$

$$\Delta p = p - p_0 \tag{2.2}$$

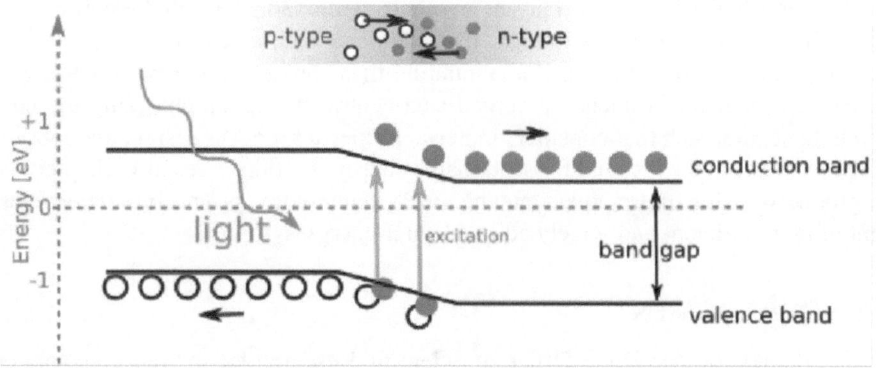

FIGURE 2.3 Electron–hole pairs generation under illuminated light.

Charge carrier lifetime is defined that the period for a charge carrier returned to the ground state from the excited state during the occurrence of recombination. Indeed, it should consider in average time taken for charge carriers to travel to the ground state, the symbol is tau, τ. τ_n and τ_p denoted for electrons and holes with time unit, respectively. The τ value could be as low as nanoseconds to milliseconds. The charge carrier lifetime depends on the excess carrier concentration and rate of recombination, R, which is given as follows:

$$\tau_n = \frac{\Delta n}{R} \tag{2.3}$$

$$\tau_p = \frac{\Delta p}{R} \tag{2.4}$$

In short, low structural defects and low undesirable impurity concentration are deserved to obtain a high carrier lifetime.

The efficiency of solar cell is affected by the recombination rate. The higher the recombination rate, the lower the efficiency of the solar cell. Therefore, one of the main goals of solar cell research is to reduce recombination and increase charge separation to improve solar cell efficiency. This can be achieved by using high-quality semiconductor materials, reducing defects in the solar cell material, and optimizing the design of the solar cell. To increase solar cell efficiency, researchers have developed various techniques such as passivating the surfaces of the solar cell, using doping techniques and multi-junction cells to increase the amount of energy that can be harvested from the sun.

2.5 CONCLUSION

This chapter covers the historical development, basic characteristics, and essential processes of photovoltaics. It began by tracing the evolution of PV technology, highlighting its pivotal role in solar energy utilization. The exploration of fundamental PV characteristics, including photocurrent generation and photovoltage, provides insights into the intricate interplay of photons, semiconductors, and electrical circuits. The discussion extends to crucial phenomena such as charge separation and recombination within PV cells. Understanding these processes is key to optimizing device performance. Efficiency emerged as a central theme, emphasizing the need for a holistic approach that considers material properties, device design, and fabrication processes. As the demand for sustainable energy solutions rises, this chapter lays the groundwork for future advancements in PV technology, promoting the development of more efficient and widely adopted solar energy systems.

ACKNOWLEDGMENT

HK Jun thanks the Air Force Office of Scientific Research for the research support under award number FA2386-21-1–4106.

REFERENCES

Blandford, R., and Watkins, M. (2009). Bell labs demonstrates the first practical silicon solar cell. *APS News* 18, 2.

IEA (2015). Snapshot of global PV 1992–2014. In: *Photovoltaic Power Systems Programme*, IEA, pp. 5–6.

Jaeger, J. (2021). *Explaining the Exponential Growth of Renewable Energy*. World Resources Institute, Washington, DC, pp. 1–9.

Martin, C. (2016). *Solar Panels Now So Cheap Manufacturers Probably Selling at Loss*. Bloomberg, London.

Mishra, U. (2008). *Semiconductor Device Physics and Design*. Springer, New York.

Palz, W. (2010). *Power for the World: The Emergence of Electricity from the Sun*. Pan Stanford Publishing, Belgium, p. 6.

Riordan, M., and Hoddeson, L. (1997). Origins of the pn junction. *IEEE Spectrum*, 34, 46–51.

Schultz, O., Mette, A., Preu, R., and Gluntz, S.W. (2007). Silicon solar cells with screen-printed front side metallization exceeding 19% efficiency. In: *Proceedings of the 22nd European Photovoltaic Solar Energy Conference and Exhibition*, Milano, pp. 6–9.

Wayback Machine. (2012). *The 1970s Energy Crisis Archived*, February 15, 2012. Available at https://web.archive.org/web/20120215190032/https://cr.middlebury.edu/es/altenergylife/70%27s.htm (accessed October 2023).

3 Photovoltaic

Graphene Materials and Their Effects in Perovskite Solar Cells

Foo Wah Low, Muhammad Kashif, Mohammad Aminul Islam, Hieng Kiat Jun, Cheen Sean Oon, and Chun Hong Voon

3.1 INTRODUCTION

In the 21st century, rapid technological development and the busy lifestyle of humans, often reliant on digitalization, have significantly increased the demand for electricity. Undeniably, fossil fuels remain our primary source for generating the essential energy needed for life. However, this reliance contributes to climate change through increased combustion, ultimately leading to incidents such as the devastating bushfires in Australia. Henceforth, the promotion of renewable energy as an alternative electricity source, such as solar energy, has been introduced for the betterment of human society. Compared to other renewable energy sources, solar energy stands out as the most reliable option, capable of converting sunlight into useful electrical energy—offering a clean and sustainable solution.

One solution is to transition our daily needs to a smarter lifestyle by incorporating technologies such as Artificial Intelligence with the Internet of Things, smart home appliances, and integrating solar PV or fuel cells into vehicles. However, these advancements are often considered "high-cost" compared to existing analog products. In the case of low-power products, there are still alternative ways, such as generating energy from solar cells. Solar energy is a redundant and renewable source that can be easily harnessed from the sun.

Currently, solid-state-based Perovskites (PSCs) PV technology has become a popular subject in the PV industry due to its high-efficiency performance and good light absorption characteristics. Despite the potential advantages, this technology is still in the preliminary stage and is not yet available in the market compared to other PV technologies. The most common electrode layer currently used is gold, which increases the overall cost of this technology and places a burden on customers. Moreover, PSCs are susceptible to rapid deterioration in moist environments, such as during rainy days, and the decay products may corrode the metal electrodes, leading to a reduction in the performance of the PSCs. To protect the solar cell, heavy encapsulation is added, contributing to increased production costs and weight. Considering

DOI: 10.1201/9781003314424-3

the interface and electron transporting layer (ETL), PSCs exhibit various carrier generation, electron transportation, and recombination blocking behaviors. The transfer rate of electrons at the ETL interface is expected to be slower than that at the hole transporting layer (HTL) in PSCs. This condition influences the hysteresis behavior in efficiency measurements and also affects the stability of PSCs, such as the migration of ions.

The photovoltaic (PV) effect was discovered by Becquerel, a French physicist, in 1839. Later, in 1876, British scientists Adams and other researchers found that semiconductors of selenium could generate electricity under sunlight (Blume, n.d.). In 1883, Fritts successfully invented a new semiconductor junction solar cell composed of a thin gold layer with germanium. However, the capability of his innovation was only about 1% (Fritts, 1883). A significant milestone in the PV industry occurred in 1954 when Pearson and other researchers from the US Bell Labs introduced the primary Crystalline Silicon Solar Cell (CSSC), achieving a power conversion efficiency (PCE) of up to 4.5% (Chapin et al., 1954). Nevertheless, these early-generation solar cells were highly costly and environmentally unfriendly.

Among PV generations, third-generation solar cells have been recently invented, with dye-sensitized solar cells (DSSCs) being the most popular. However, these PV cells have two major disadvantages in sunlight absorption. First, the manufacturing of DSSC involves a thickness that makes it challenging to absorb light using a thin absorption layer in solid-state solar cells. Second, light bleaching occurs in DSSC applications, making them less preferable for everyday use. To address these issues, Japanese scientists, led by Kojima and his team, discovered that organic metal halides have similar properties for absorbing sunlight, eliminating the need for dye functionalization (Kojima et al., 2009). The pioneering perovskite solar cells (PSCs), which operate entirely in a solid-state situation, achieved a PCE of 9.7%, as announced by Kim and his team in 2012 (Kim et al., 2016). PSCs have garnered global attention and have risen rapidly in popularity due to their lower cost compared to other PV generations and superior performance. The highest PCE of PSCs was successfully achieved in 2016, reaching up to 22.1%.

3.2 CRYSTALLINE SI-BASED PHOTOVOLTAIC

Silicon is the most commonly used material for PV applications. Silicon-based solar cells are capable of capturing photons from sunlight with an energy greater than 1.1 eV. However, CSSCs do not cover the full range of energy in the entire solar spectrum. The primary factor limiting the performance of silicon-based solar cells is the fixed bandgap of silicon material itself. Moreover, the fabrication of tandem solar cells requires materials with different bandgaps for each layer. Therefore, multijunction silicon-based solar cells show promise with the modification of silicon material by reducing the size of silicon crystals. The optical and electrical properties of semiconductor materials undergo changes when the crystalline size becomes smaller than the Bohr exciton radius (i.e., the Bohr exciton radius of silicon is <5 nm). Consequently, silicon is decorated with other materials exhibiting different properties. For example, silicon materials can be incorporated with high-bandgap metal oxides such as SiO_2 and Si_3N_4. The arrangement of silicon with high-bandgap metal

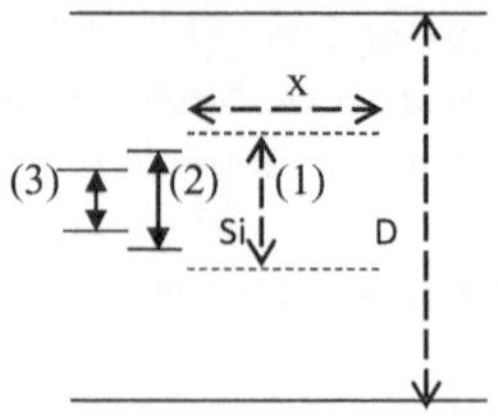

FIGURE 3.1 Schematic bandgap diagram of crystalline silicon in a dielectric matrix.

oxide materials tends to alter the respective bandgap. The energy bandgap diagram of silicon embedded with metal oxide material is illustrated in Figure 3.1. In this arrangement, the changes in electrical and optical properties are known as quantum confinement carriers (labeled "1"). As the size of silicon nanoparticles decreases, a lower Bohr exciton radius is required. In other words, the bandgap narrows between the lowest energy state in the conduction band and the higher occupied energy state in the valence band (labeled "2"). The bandgap becomes even narrower as the nanoparticle size decreases, promising a higher effective bandgap (labeled "3"). The equation describing the relationship between bandgap (E_g in eV) and nanocrystal size is summarized in Equation 3.1 (Conibeer et al., 2006).

$$E_g = E_g + \frac{k}{d^2} \tag{3.1}$$

where k = quantum confinement parameter, d = diameter of nanoparticles (nm), and E_g = bandgap of material.

The increased bandgap, influenced by small silicon nanocrystals and ultimately resulting in quantum confinement, is referred to as quantum dots (QDs). The narrower spacing of silicon QD bonds leads to a wider bandgap when incorporated with other semiconductor materials. By tuning the size of Si QD, the optimized bandgap is approximately 2.7 eV. In recent years, there has been a significant amount of research on silicon nanocrystals decorated with metal oxide due to their potential applications in PV.

3.3 III–V SEMICONDUCTORS MULTIJUNCTION-BASED PHOTOVOLTAIC

A III–V semiconductor is a compound, such as gallium nitride (GaN) or gallium arsenide (GaAs). Each of these compounds has three (III) and five (V) valence electrons, providing high electron mobility. Consequently, GaN or GaAs semiconductors are widely used in PV applications. GaAs, in particular, is notable for being the first multijunction PV cell that combines with single-junction solar cell technologies. In the context of multijunction PV structures, cells are stacked, resulting in a lower energy bandgap.

GaAs can absorb high-energy photons, but it can also transmit photons in the circuit loop with lower energies. The Shockley–Queisser limit refers to the maximum efficiency applicable for a single-material PV. Assuming that each energy gap gives

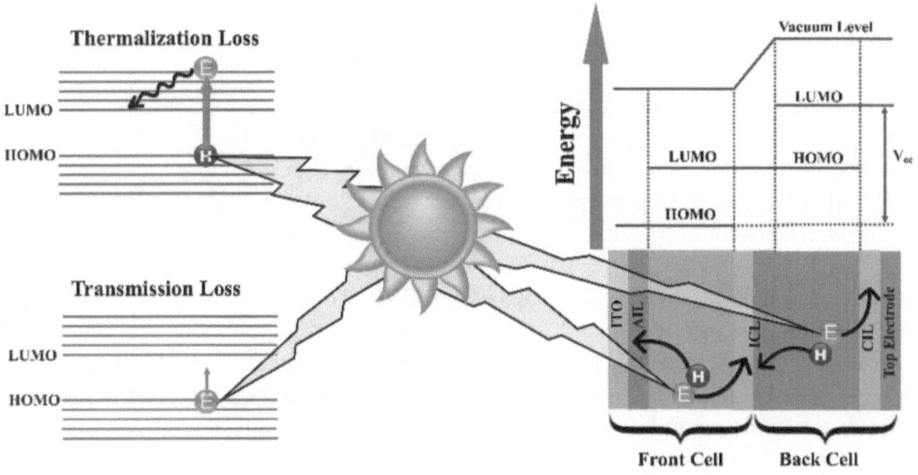

FIGURE 3.2 Schematic diagram of an electronic energy level of thermalization and transmission losses.

rise to only one electron–hole pair, the theoretical maximum PCE can be achieved up to 31%. In reality, there is a total loss of photon energy, approximately 56%, mainly attributed to thermalization and transmission losses (refer Figure 3.2).

Multijunction GaAs PV cells have achieved performances of up to 47.1% with a triple-junction cell (France et al., 2022). While high-efficiency GaAs PV cells are considered expensive, they are 40% more efficient than other types of PV.

3.4 THIN-FILM PHOTOVOLTAIC

Silicon-based solar technologies are generally categorized under first-generation technologies, while thin film–based PV is considered second-generation technology. The primary development objectives of second-generation thin film–based PV include reducing the fabrication cost of silicon wafer-based solar PV modules and achieving higher efficiency than silicon-based PV technologies. Thin-film PV technology involves the use of very thin layers of semiconductor material to convert sunlight into electricity. Materials commonly used in thin-film PV include amorphous silicon, cadmium telluride (CdTe), copper indium gallium selenide, and thin-film CSSC technologies. Another definition of thin film refers to the "random nucleation and growth process of individually condensing/reacting atomic/ionic/molecular species on a substrate." The structural, chemical, and physical properties of the specific film depend on the deposition parameters. In fact, the thickness of the film is a crucial parameter, ranging from a few nanometers to several tens of micrometers. Thin materials obtained through processes such as wafer thinning or depositing clusters of microscopic species, as in screen printing, are referred to as thick films.

In principle, any semiconductor material can be used for thin-film fabrication. However, only a few materials can achieve reasonable efficiency for light-to-electricity conversion. Ideally, the selection of semiconductor materials for thin-film fabrication must fulfill certain requirements to achieve high-efficiency conversion. These requirements are as follows:

- Bandgap between 1 and 1.5 eV.
- Absorption coefficient of the material should be high, 10^4–10^6 cm^{-1}.
- Recombination rate of generated charge carriers should be low.
- Diffusion length of generated charge carriers should be high.
- Materials should be widely available, reproducible, and non-toxic.

Basically, the operation of a solar cell is based on the generation of photo-voltage upon illumination. Semiconductor material is used to form a junction to facilitate the separation of charge carriers. There are two types of junction formation: homo junction (two similar materials) and hetero junction (two different materials). Thin-film technologies mostly apply inorganic materials for fabrication.

3.5 NANOCOMPOSITE SOLAR CELLS

Nanocomposite solar cells represent an innovative frontier in PV technology, leveraging the unique properties of nanomaterials to enhance solar energy conversion efficiency. By incorporating nanoscale structures or composite materials into the design, these solar cells aim to overcome traditional limitations, such as suboptimal light absorption or charge carrier mobility. The synergistic effects of combining different materials at the nanoscale allow for precise control over the solar cell's properties, leading to improved performance and promising advancements in the quest for more efficient and sustainable solar energy solutions. Examples of nanocomposite materials used in solar cells include perovskite and dye sensitizers.

3.5.1 Perovskites Solar Cells (PSCs)

PSCs are solid-state PV cells that comprise several materials, with the main component being a perovskite. Perovskite can refer to any material that forms a crystal structure similar to calcium titanium oxide. Consequently, PSCs exhibit higher efficiency in converting photons from sunlight into electricity. This superior efficiency in terms of PCE positions PSCs ahead of other solar energy technologies. Recently, PSCs have emerged as a prominent sector of research in addition to other PV technologies. Figure 3.3 shows the schematic layout for the PSCs structure.

There are two categories of PV technologies in the past: wafer-based PV and thin-film cell PV. Then, PSCs are categorized as emerging thin-film cell PV and are classified under the third generation in the PV industry. There are some other types of solar cells produced with PSCs, such as organic-based PV, quantum Dot PV, and DSSCs. Recent research on PSCs has shown that the efficiency gradually increased by 18.3% from 2009 until 2016 (Park et al., 2016). There are many advantages to PSCs compared to other PV technologies. The most attractive and essential advantage is their achievable highly efficient performances compared to other solar technologies. Remarkably, there is an increase from 3.8% to 22% from 2009 to 2016. Scientists and researchers believe that the PCE can be enhanced further as more

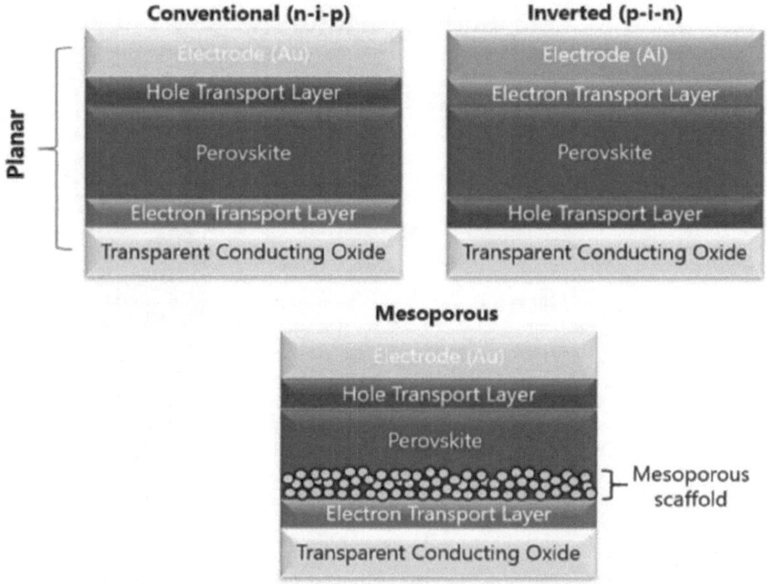

FIGURE 3.3 Basic PSCs structure (Valsalakumar et al., 2023).

studies focus on PSCs. In fact, PSCs have to be tolerant to defects, have a long carrier lifetime, long diffusion distance of photo-generated carriers, and a high absorption coefficient to obtain better PCE performance.

In fact, the cost of PSCs is also relatively low compared to other PV cells, as shown in Figure 3.4. This is because PSCs are made with man-made materials, unlike traditional solar cells that are made with crystalline silicon. Indeed, silicon material is extracted from the Earth and needs further processing before being made into a highly efficient solar cell, which makes traditional solar cells expensive. Unlike other traditional solar cells, PSCs are simply produced by a process called "solution processing," which follows the same procedures used for printing newspapers.

Furthermore, PSCs share many outstanding properties with other thin-film layer solar cells. Importantly, PSCs are more flexible and lightweight compared to traditional silicon solar cells. The special lightweight property of PSCs facilitates easier installation on roofs or walls, requiring less physical stress. Moreover, PSCs are also semi-transparent, aiding in more effective sunlight absorption. In the same vein, PSCs are easier to manufacture and excel in gathering sunlight. These properties make them more efficient and popular for consumers. Additionally, there are further advantages of PSCs in the future, and they represent a significant potential benefit for our daily lives, especially concerning the environmental impact related to the lead-based perovskite absorber. Researchers are studying alternative substitutions for lead, given its high toxicity to the environment. This effort is crucial for the future, aiming to reduce and eliminate the toxicity of lead to preserve our polluted environment for future generations. As such, PSCs have the potential to adopt a lead-free structure, fulfilling the basic aim of using solar cells to generate electricity.

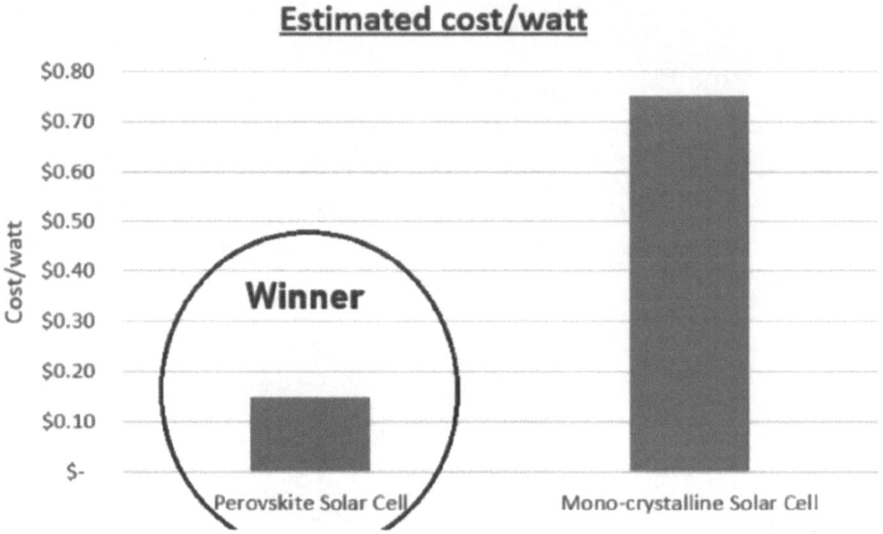

FIGURE 3.4 Cost of PSCs compared to silicon-based solar cell (Qian et al., 2019).

3.5.2 Structure of PSCs

Generally, perovskite is defined as $CaTiO_3$, with the molecular structure stated in the ABX_3 formula. Perovskite has attracted much attention due to its layered structures, characterized by a cubic lattice-nested octahedral shape, as well as its optical, unique thermal, and electrochemical properties. The perovskite used in solar cells is typically a compound of an organic metal halide and inorganic metal halide, such as $CH_3NH_3^+$, MA^+, or $CH(NH_2)_2^+$. FA^+ is positioned around the center of the cubic lattice, while the metal halide with Group 2 metals as cations, such as lead or tin, acts as the core of the octahedra. Meanwhile, halogens, such as chlorine or bromine, act as the apex of the octahedra. Figure 3.5 illustrates the structure of perovskite in solar cells.

3.5.3 Working Principle of PSCs

The structure of PSCs typically provides four features for the solar cell (refer Figure 3.6). First, the components maintain excellent photoelectric properties, with high optical absorption coefficients that can reach up to 10^4 cm^{-1}, and lower exciton binding energy. Second, solar cells that use perovskite as a layer to absorb light can increase efficiency in gathering sunlight. Besides that, electrons, holes, and dielectric constant can be obtained by the materials in a more effective way. Last, holes and electrons can be transported together, and the distance for transmission is approximately 100 nm and may go up to 1 μm.

These features contribute to the high open-circuit voltage (V_{OC}) and the short-circuit current density (J_{SC}). Initially, the perovskite layer absorbs photons, producing excitons (electron–hole pairs) when exposed to sunlight. Subsequently, free charge carriers can be formed by these electron–hole pairs, generating current due to the difference in

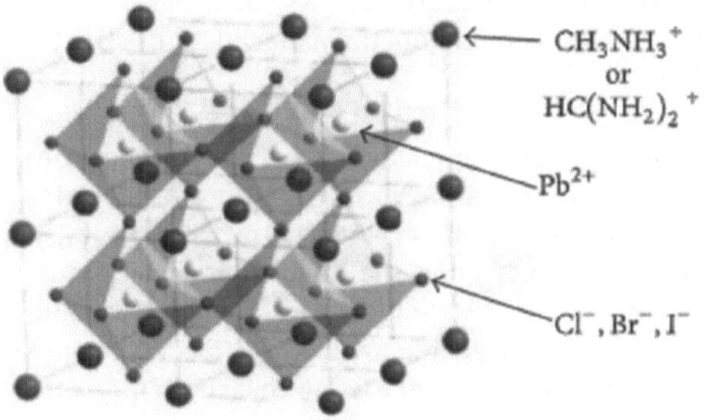

FIGURE 3.5 Structure of PSCs (Chen, 2023).

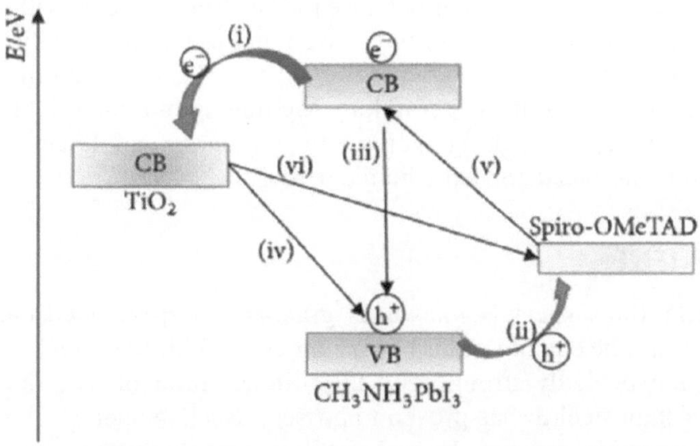

FIGURE 3.6 Mechanism of PSCs.

exciton energy levels for the perovskite material. The distance for diffusion and the duration of the carrier are long, as the probability of carrier recombination for $CH_3NH_3PbI_3$ ($MAPbI_3$) is low, coupled with the high mobility of the carrier. The lifetime and the distance of the carrier are directly proportional to the efficiency of the PSCs.

Next, the free holes and electrons are collected by the hole transport material (HTM) and electron transport material (ETM), respectively. Then, electrons are transferred from perovskite to the ETL. Afterward, the electrons are collected by fluorine doped tin oxide (FTO) from the ETL. Besides that, holes are carried to the HTL and then gathered by metal electrodes simultaneously. Finally, the FTO is connected to the metal electrodes to generate the photocurrent in the outer circuit.

3.5.4 Electron Transport Layer (ETL)

The electron transport layer (ETL) is a layer where its electrons exhibit high mobility and affinity. The primary purpose of the ETL in PSCs is to facilitate contact with the layer that harvests sunlight. This enhances the efficiency of photo-generated electrons to be extracted. As such, the ETL is a crucial element in PSCs. Different n-type semiconductor materials and structures have varying effects on the PCE.

When selecting the ETM for PSCs to operate more effectively, there are several aspects that take priority. First and foremost, n-type semiconductors with high carrier mobility offer advantages in providing higher efficiency for PSCs. Additionally, the ETM should allow penetration by sunlight, enabling PSCs to have a wide bandgap. Furthermore, the conditions for the preparation of PSCs should be gentle, and the ETM should be in a cold environment. Finally, the band structure should be compatible with PSCs to prevent errors and enhance performance. However, it's important to note that the ETM is not the main factor restricting efficiency.

Several aspects need investigation to strengthen the performance of the ETL, such as modifying the chemicals through doping and applying lamination to change the microstructure. These modifications enhance the electron transport within the PSCs. Laminated ETL improves the stability of PSCs and perovskite stability under UV illumination. For instance, the Mg-doping of a TiO_2 ETL improved the conduction band minimum alignment of the perovskite. Mg-doping of a SnO_2 ETL is ready to reduce the occurrence of cracking, increase the regularity and flatness of the ETL, and enhance the interfacial contact with the perovskite layer.

3.6 GRAPHENE

A 2D one-atom-thick sheet is known as graphene, a sp^2-bonded carbon material organized in a hexagonal pattern, serving as the building blocks of graphite. Graphene is unexpectedly strong, with the thinnest layer of material being 200 times stronger than steel. It has proven to possess excellent electrical, optical, and mechanical properties, along with good stability, contributing to the enhancement of the performance of PSCs. Graphene is also proficient at absorbing light and is a good conductor of heat and electricity. Much research has been conducted on graphene in the PV industry due to its unique characteristics, resulting in an increase in the efficiency of solar cells. The fusion of graphene into PSCs has been introduced, and extensive research has been carried out on this topic recently. This research covers various aspects, including holes and electrons transporting medium, electrodes, and so on, to ensure the devices become more stable. Figure 3.7 illustrates the structure of graphene.

Graphene is an extremely versatile material that can compound with other materials, such as gases and metals, to produce a different material with strong properties. Researchers worldwide are investigating its properties and applying them in many applications to benefit human life in this new era. The first application of graphene is its use as an essential material in batteries, such as solar cells. It is also employed in transistors, computer chips, antennas, water filters, supercapacitors, touchscreens for LCDs, and many more. Consequently, graphene has become a hotspot for researchers, especially Konstantin Novoselov and Andre Geim, who won the Nobel Prize in physics in 2010 as they were the first to isolate graphene in 2004 (Novoselov et al., 2004).

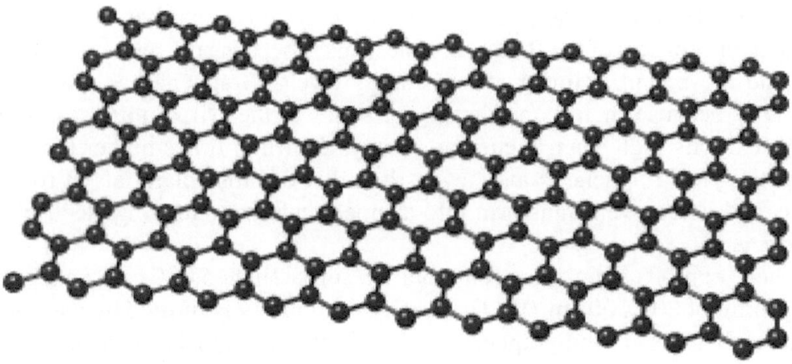

FIGURE 3.7 Structure of graphene.

3.6.1 Physical/Mechanical Properties of Graphene

Graphene is a material known as the thinnest in the world, measuring approximately 0.34 nm, and one of the toughest materials, being stronger than steel and diamonds in the same dimensions. Graphene can withstand more strength when stretched or pulled compared to other materials, as its tensile strength is even more than 1 TPa. Carbyne, a string of carbon, is the only material stronger than graphene. It also has a lightweight of only 0.77 mg/m^2. For comparison, 1 m^2 of paper is 1,000 times heavier than 1 sheet of graphene. Graphene also has the highest surface area since it exists in the 2D dimension. Graphene can stack to form the most stable structure of carbon called graphite. Moreover, graphene is water-resistant and cannot be penetrated by helium (Li et al., 2021).

Furthermore, graphene is excellent in conducting heat and has extraordinary potential for heat conduction based on its size. It has been found that the larger its size, the more heat it can conduct, as graphene has a stable bond pattern and a 2D structure. The thermal conductivity of graphene is expected to be around 3,000–5,000 W/mK at room temperature. Researchers believe that graphene's thermal conductivity is one of the highest compared to other materials. The high elasticity of graphene allows it to maintain its original size after being stretched. This is confirmed by atomic force microscopy for the spring constant with a thickness of 2–8 Nm, which is approximately 1–5 N/m. Its Young's modulus is approximately 0.5 TPa. This indicates that graphene can be stretched up to 20% of its starting size without damaging its structure.

3.6.2 Electronic Properties of Graphene

Graphene is a zero-overlapped semimetal, where its electrons and holes act as charge carriers. It is an excellent electrical conductor with high electrical conductivity. Carbon has six electrons, with two located in the s-orbit and the other four in the p-orbit. Only one of these electrons is freely available for electronic conduction while the atom is attached to the other three atoms. The high-mobility electrons are called pi electrons and contribute to enhancing the carbon bonds in graphene.

Holes and electrons in graphene do not have effective mass at the Dirac point, which is the six corners of the Brillouin zone. This is because the relationship between energy and movement is linear at low energy levels. Graphene has low electronic conductivity because there is no density of states at the Dirac point. Additionally, graphene exhibits high electric current density, a million times higher than copper. The resistivity of graphene is also lower than that of other materials at room temperature, and it can even transform into a superconductor, carrying electricity with 100% efficiency.

Graphene exhibits a high electron mobility, surpassing 15,000 cm^2/Vs and having an upper limit of 200,000 cm^2/Vs. However, this mobility is limited by the scattering of acoustic phonons around graphene. The mobility of graphene electrons is comparable to that of photons, as their mass is not significant enough to be measured. These charged carriers can travel sub-micrometers without scattering but are ultimately limited by the conditions of the substrate and the graphene itself. Despite being an excellent conductor, graphene is not suitable for making transistors because it lacks a bandgap.

Graphene is capable of harvesting 2.3% of white light, making it very suitable for manufacturing solar cells based on these properties. The unique characteristics of graphene are attributed to the high mobility of electrons. Researchers have demonstrated that the absorption of visible light is dependent on the Fine Structure Constant and not specific to the material. An interesting observation is that the amount of white light absorbed will be doubled if another layer of graphene is added to the first sheet. However, as the intensity of collected light reaches a threshold, the absorption of light will be reduced.

3.6.3 Graphene-Based Electron Transport Layer

As mentioned in Section 3.5, the ETL plays an essential role in PSCs. To enhance the PCE and stability for long-term performance, allowing efficient charge injection or collection at the electrodes, graphene-based or other 2D materials have been introduced into the structure of PSCs recently. Researchers have experimentally studied the effect of graphene-based ETLs in improving methylammonium lead iodide solar cells, leading to a significant enhancement in PSCs' efficiency.

The intersection between the ETL and perovskite is productively affected by the charge recombination process and material instability in PSCs. To slow down perovskite degradation and reduce nonradiative recombination, free charges are rapidly inserted from perovskite to ETL. Interface studies show that the conversion efficiency and stability are greatly enhanced when graphene flakes are inserted into the lithium-neutralized graphene oxide (GO-Li) and the mesoporous TiO_2 layer ($mTiO_2$) as an inter-layer at the perovskite and $mTiO_2$ interfaces. Hence, the effect of the substrate modified by mesoscopic graphene must be investigated in more detail to control the PSCs' performance in the PV industry.

Low-temperature graphene is often doped with tin oxide for planar PSCs. Doping graphene into the SnO_2 ETL, connected with the disabled charge fusion at the ETL and perovskite interfaces, has gradually enhanced efficiency. The graphene-SnO_2 device was found to display a higher efficiency of more than 18%. The device's J–V hysteresis is surprisingly attenuated, usually corresponding with the extraction of the electrons and enhanced mobility of electrons through the combination of the graphene conducted in two-dimensional electronics in the SnO_2 matrix. The graphene-SnO_2 improves stability, securing performance around 90% after exposure to air for 300 hours due to the graphene's hydrophobic nature.

There are major advantages to using graphene as a sandwich surface layer on the ETL to strengthen the intrinsic semiconducting properties of the ETL. It also transforms the surface and volume morphologies of the ETL. Through the utility of graphene with organic or inorganic molecules and chemical surface modification, the efficiency of graphene in the ETL of PSCs can be enhanced. As such, the surface chemistry and semiconducting properties can be measured so that it can adapt to different ETLs and PSCs. As graphene is stable in chemicals and possesses good physical properties, it enhances the stability of PSCs. Besides that, graphene can be integrated relatively easily into different conventional ETL materials.

3.6.4 Defects of Graphene

Scientists have extensively investigated graphene since its discovery and found that lattice imperfections in graphene cannot be prevented during its growth. The chemical and electronic properties of graphene are affected by the defects in its structure. Therefore, it is essential to understand the defects in graphene. In this project, the aspect of defects needs to be introduced, including the types of defects in graphene. Additionally, solutions for addressing these defects in graphene must be detailed to improve the PCE performance.

Graphene, with its honeycomb network structure consisting of two carbon atoms per cell, has two categories for ribbon edges: zigzag and armchair. These ribbon edges can impact the electronic properties of graphene, similar to carbon nanotubes. However, a weakness arises when graphene is used as the ETL of PSCs due to the absence of a semiconducting gap in graphene. This lack of a bandgap makes it challenging for a device produced with graphene, as it lacks the capability to switch off. It also poses a disadvantage in the low static power consumption of complementary metal-oxide semiconductor (CMOS) technology.

Structural defects degrade the efficiency of PSCs with graphene-based ETLs, as graphene's outstanding mechanical and electronic properties only manifest in highly perfect conditions. However, imperfections may have utility in certain conditions, as they can lead to new applications and tailor the local properties of graphene. Structural defects in graphene can be induced through chemical treatment or irradiation.

Rings with non-hexagonal shapes may enable graphene to remain flat when symmetry rules are satisfied, offering deflection in the graphene sheet. Sp^2-hybridized carbon atoms can align into different polygons to form various architectures. Unlike other massive crystal-like materials such as silicon, graphene utilizes these properties. The atomic network is reorganized to allow a consistent network defect, eliminating under-coordinated atoms. These reorganized defects increase the reactivity of the framework and enable the absorption of other atoms onto the graphene.

3.6.5 Type of Defects in Graphene

Recently, scientists have been studying the structural deficiencies in carbon nanotubes and graphite. Studies suggest that graphene should have deficiencies at the atomic level. However, accurately identifying all types of structural defects in graphene is challenging. High-resolution transmission electron microscopes (TEM) are employed to observe graphene at the atomic level, even down to single-layer graphene. Additionally, atomic force microscopes (AFM) and scanning electron microscopes (SEM) are experimental devices used to characterize nanomaterials and directly image configurations predicted theoretically.

There are two categories of defects in graphene: intrinsic defects and extrinsic defects. Intrinsic defects are characterized by non-sp^2 orbital hybridization in graphene. The existence of intrinsic defects is mainly due to non-hexagonal rings formed by surrounded hexagonal rings. The presence of non-carbon atoms disrupts the crystalline order in graphene. Studies on rebuilding the structure of carbon nanotubes under external energy disturbances indicate that the assumptions about defects may not always be stationary and can be random under certain conditions. They may move, governed by temperature and activation carriers.

Graphene has five categories of intrinsic defects: Stone–Wales defects, multiple-vacancy defects, single-vacancy defects, line defects, and carbon adatoms. Stone–Wales defects result from the rotation of a single pair of carbon atoms, causing adjacent carbon rings to become pentagons or heptagons. Rotation of carbon atom pairs creates Stone–Wales defects without carbon atoms being added or removed. The energy associated with these defects can reach about 5 eV. Stone–Wales defects can be induced by irradiation or rapid cooling in a high-temperature environment. Additionally, this defect may be formed by the impact of electrons. Figure 3.8 illustrates the atomic structure of Stone–Wales defects.

Graphene undergoes single-vacancy defects when one carbon atom is missing from the hexagonal rings. Additionally, graphene experiences Jahn–Teller distortion, causing two or three suspended bonds to connect at the missing atom to reduce its total energy. Geometry reasons may be attributed to the remaining pendant bonds. The energy produced by the vacancy defect is expected to reach 7.5 eV, demonstrating that higher energy than Stone–Wales defects is required to produce a vacancy defect with a pendant bond. Figure 3.9 illustrates the atomic structure of single vacancies in graphene.

The multiple-vacancy defect is also a structural defect in graphene, occurring when two or more carbon atoms are missing from the hexagonal rings (see Figure 3.10). Consequently, it can be inferred that the multiple-vacancy defect is based on the single-vacancy defect. For example, a double-vacancy defect is formed when two carbon atoms are missing from the ring. Double-vacancy defects result in the absence of pendant bonds for the rings formed by two pentagons and one octagon, instead of four hexagons. The energy required to form a double-vacancy defect is approximately 8 eV, and various numbered vacancy defects may form with different energy levels. This defect involves the rotation of carbon atoms, making the system more stable due to a reduction in total energy. Figure 3.10 illustrates the atomic structure of various multiple-vacancy defects.

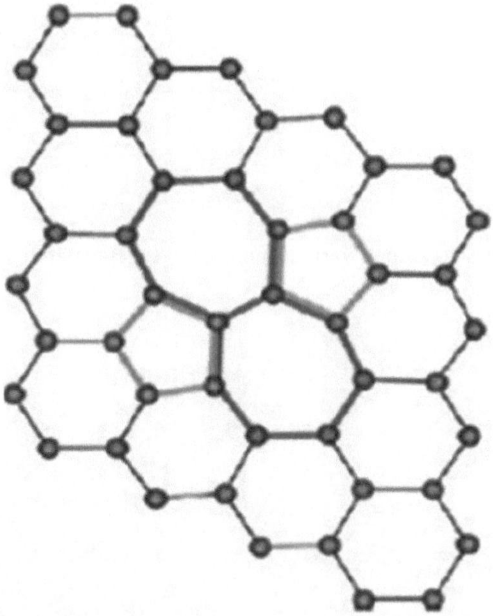

FIGURE 3.8 Atomic structure of Stone–Wales defects.

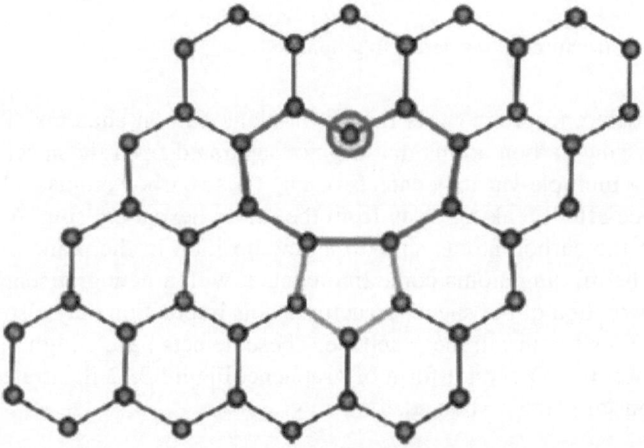

FIGURE 3.9 Atomic structure of single vacancy in graphene.

Another defect in graphene is the one-dimensional defect, also known as line defects, as shown in Figure 3.11. The line defects are slant borderlines dividing two regions of different networks, while the axis is perpendicular to the plane. Line defects can be considered as fixed-point defect lines that either have applied or not applied pendant bonds. One reason for the formation of these network regions is the mismatch of the lattice in graphene developed on the surface of Ni. These defects form due to octagons that divide the alternate lines of pentagon pairs.

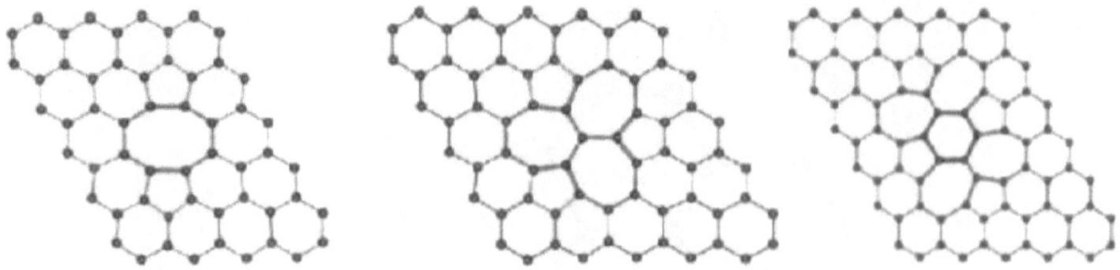

FIGURE 3.10 Atomic structure of various multiple-vacancy defect.

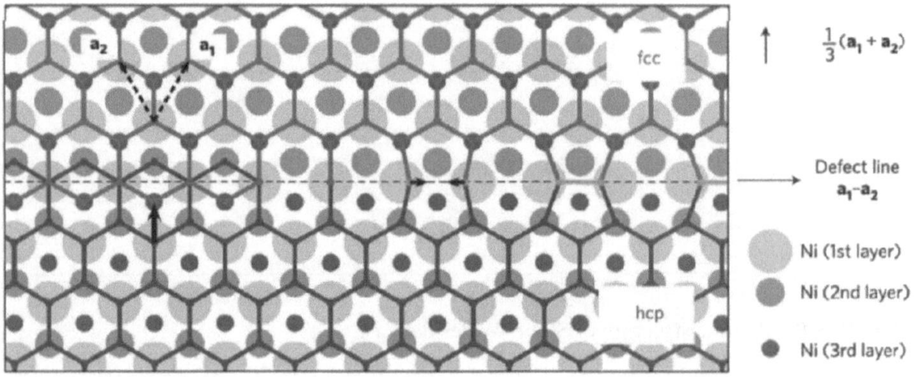

FIGURE 3.11 Structure of line defect in graphene.

The final defect in graphene is the out-of-plane carbon adatoms. This defect is caused by missing carbon atoms that are not separated from the sheet of graphene when single or multiple-vacancy defects occur. These carbon atoms shift on the graphene's surface after breaking away from the initial hexagonal ring. A new bond is created when the carbon atoms shift to a new position in the plane. A new defect will form if the missing atoms come into contact with a new graphene sheet, leading to the destruction of the sheet's structure. This interaction may also result in the creation of a 3D structure in the graphene. These defects have a high migration rate and can disrupt the 2D crystal form of graphene. Figure 3.12 illustrates the atomic structure of out-of-plane carbon adatoms in graphene.

3.6.6 Solution for Graphene Defects

Graphene defects are known for scattering phonons and charge carriers, reducing the length of the ballistic transmission path. The mobility of carriers and heat conductivity in graphene are negatively affected by the scattering of phonons and charge carriers. This negative impact of defects is particularly pronounced in graphene. For example, defects reduce the mobility of charge carriers produced by micromechanical cleavage. The defects are responsible for the properties of graphene films obtained by chemical methods, such as the chemical reduction of graphene oxide

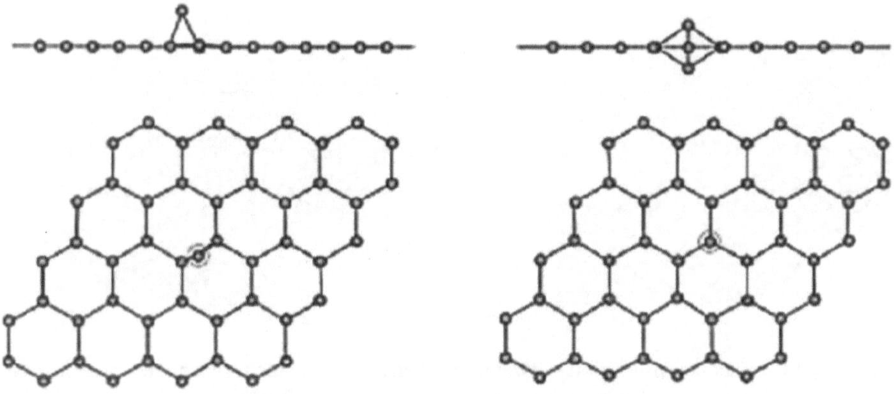

FIGURE 3.12 Defects and location of the carbon atoms (Tian et al., 2017).

(GO) and exfoliation. Therefore, solutions are needed to prevent these defects from producing high thermal and electrical conductivity nanostructures of carbon and, consequently, strengthen the mechanical strength.

First, thermal annealing is one of the solutions for defects by removing defects from the crystalline network. This process involves high temperatures in the presence of a hydrocarbon gas to restore the structure of graphite. Hydrocarbons break down to generate carbon atoms that can regenerate defective sites under certain conditions. López and his team recently introduced the processing of chemical vapor deposition (CVD) for chemically derived graphene films. They used ethylene carbon as a source to improve the film's conductivity by an order of magnitude ranging between 10 and 350 S/cm at room temperature. The enhancement of defective sites on the surface of graphitic material under conditions with hydrocarbon gas was observed and reported by Liu and his team. They investigated defect reactivity on the surface of highly oriented pyrolytic graphite (HOPG) at increasing temperatures using scanning tunneling microscopy (STM).

Self-healing is also a solution to remove defects in graphene. Tsetseris's team found that carbon adatom pairs and clusters of four or more self-interstitials remain hollow unless the system's temperature is elevated significantly. The clustering of three carbon adatoms causes the removal of hillock-like features, and the creation of mobile species causes the structure to self-heal. Liu's team demonstrated that most single-vacancy defects are likely to combine into larger vacancy holes, while the vacancies accumulate into a single hole in a layer through intra- and inter-layer migration in multi-layer graphene. Cooperation among defects simplifies the inter-layer migration of vacancies. Figure 3.13 illustrates the migration of vacancies between graphene layers, leading to the formation of vacancy holes in one layer and the self-healing of other layers.

Healing by absorption is also a method to reduce defects in graphene. Wang and his team introduced a method to control vacancy healing and N-doping in graphene. To heal the vacancy, CO and NO molecules need to be exposed sequentially. CO molecules are easily absorbed into the defective sites in graphene, while NO will

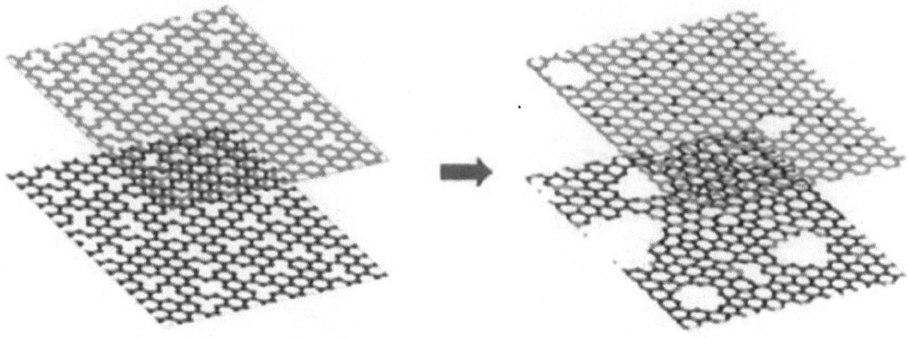

FIGURE 3.13 Vacancy migration between graphene layers leads to vacancies hole formation in one layer and the self-healing of other layers (Liu et al., 2014).

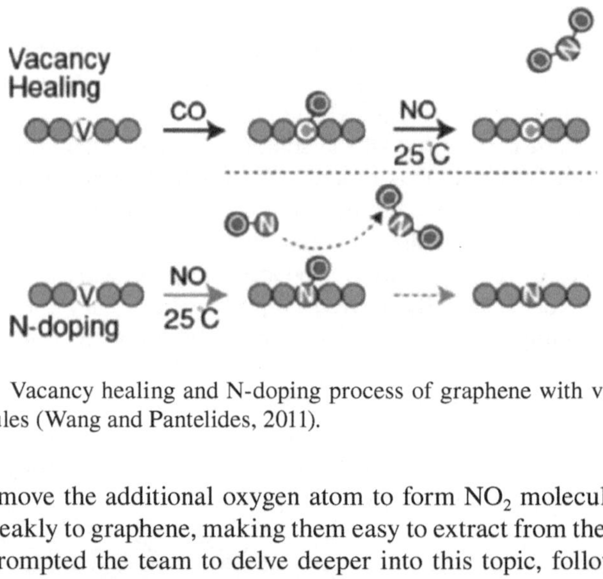

FIGURE 3.14 Vacancy healing and N-doping process of graphene with vacancies by CO and NO molecules (Wang and Pantelides, 2011).

chemically remove the additional oxygen atom to form NO_2 molecules. NO_2 molecules attach weakly to graphene, making them easy to extract from the material. This observation prompted the team to delve deeper into this topic, following the same steps until reaching an understanding of NO molecules. Therefore, the combination of CO and NO probably can facilitate synchronous healing and doping at a preferable ratio. This modulation is crucial for graphene in adjusting its bandgap. Figure 3.14 illustrates the vacancy healing and N-doping process of graphene with vacancies using CO and NO molecules.

Metal-assisted healing is also one of the solutions to remove defects in graphene. Karoui and his team investigated the healing of graphene developed on a metallic substrate using tight-binding Monte Carlo simulations. An isolated graphene sheet can become hardened, suggesting that its curing is thermally activated between 1,000 and 2,500 K. An excellent layer of graphene can be obtained when nickel is present as the substrate. The apparent structure indicates that the chemical bonds between nickel and carbon molecules keep breaking and reforming around the defective carbon zone. This process contributes to the primary communication necessary

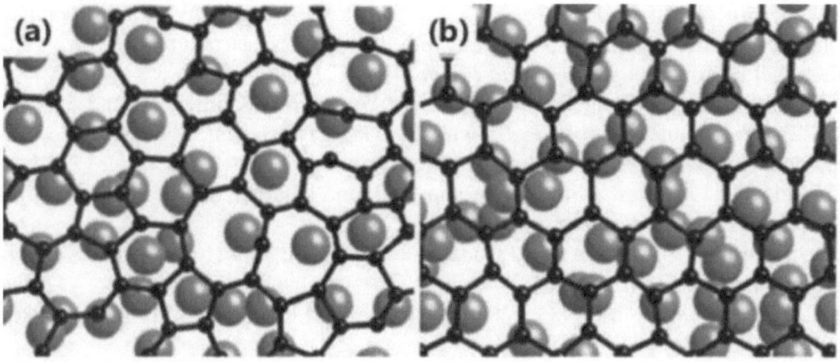

FIGURE 3.15 Defects at (a) 1000 K and (b) 2500 K (Karoui et al., 2010).

for healing. Significant defects are observed at a temperature of 1,000 K. Hexagonal and pentagonal rings are canceled out at high temperatures, leading to the healing of larger rings between 1,000 and 1,500 K. The graphene sheet is perfectly cured at 2,500 K. Figure 3.15 illustrates the defects at 1,000 and 2,500 K.

3.7 CONCLUSION

In conclusion, the defect density is unacceptable because it will influence the efficiency, performance, and stability of the PSCs. Additionally, heat dissipation is also a main factor for device thermal stabilization, with the help of a graphene layer. However, graphene has low conduction at accurate positions, and hence, it is suggested that graphene be doped with metallic ions. Furthermore, the interface of the graphene/perovskite layer must be set up with caution; otherwise, the recombination rate will be increased due to the huge network mismatch. The effect of the defect is also indicated as the transfer length (λ), in which the defect is expressed as the lengthening of the conductance channel. It is built upon the sheet resistance of graphene, and this amount contributes to a direct judgment for the PSC cell.

ACKNOWLEDGMENT

This research work was financially supported by the Fundamental Research Grant Scheme, FGRS (FRGS/1/2022/TK08/UTAR/02/39) from the Ministry of Higher Education and UTAR Research Fund, and UTARRF (IPSR/RMC/UTARRF/2022-C1/L12) under Universiti Tunku Abdul Rahman.

REFERENCES

Blume, E. (n.d.). Product review: PV technologies on the way to market. *Electric Perspectives,* 18:2.

Chapin, D.M., Fuller, C.S., and Pearson, G.L. (1954). A new silicon p-n junction photocell for converting solar radiation into electrical power. *Journal of Applied Physics* 25, 676–677.

Chen, C. (2023). Optimization strategies for electron transport layer aiming at high efficiency perovskite solar cells. *Journal of Physics: Conference Series* 2608, 012018.

Conibeer, G., Green, M., Corkish, R., Cho, Y., Cho, E.C., Jiang, C.W., Fangsuwannarak, T., Pink, E., Huang, Y., Puzzer, T., et al. (2006). Silicon nanostructures for third generation photovoltaic solar cells. *Thin Solid Films* 511–512, 654–662.

France, R.M., Geisz, J.F., Song, T., Olavarria, W., Young, M., Kibbler, A., and Steiner, M.A. (2022). Triple-junction solar cells with 39.5% terrestrial and 34.2% space efficiency enabled by thick quantum well superlattices. *Joule* 6, 1121–1135.

Fritts, C.E. (1883). On a new form of selenium cell, and some electrical discoveries made by its use. *American Journal of Sciences* 3, 465–472.

Karoui, S., Amara, H., Bichara, C., and Ducastelle, F. (2010). Nickel-assisted healing of defective graphene. *ACS Nano* 4, 6114–6120.

Kim, H.S., Seo, J.Y., and Park, N.G. (2016). Material and device stability in perovskite solar cells. *ChemSusChem* 9, 2528–2540.

Kojima, A., Teshima, K., Shirai, Y., and Miyasaka, T. (2009). Organometal halide perovskites as visible-light sensitizers for photovoltaic cells. *Journal of the American Chemical Society* 131, 6050–6051.

Li, X., Long, Y., Ma, L., Li, J., Yin, J., and Guo, W. (2021). Coating performance of hexagonal boron nitride and graphene layers. *2D Materials,* 3, 8.

Liu, L., Gao, J., Zhang, X., Yan, T., and Ding, F. (2014). Vacancy inter-layer migration in multi-layered graphene. *Nanoscale* 6, 5729–5734.

Novoselov, K.S., Geim, A.K., Morozov, S.V., Jiang, D., Zhang, Y., Dubonos, S.V., Grigorieva, I.V., and Firsov, A.A. (2004). Electric field in atomically thin carbon films. *Science* 306, 666–669.

Park, N.G., Grätzel, M., Miyasaka, T., Zhu, K., and Emery, K. (2016). Towards stable and commercially available perovskite solar cells. *Nature Energy* 1, 1–8.

Qian, J., Ernst, M., Wu, N., and Blakers, A. (2019). Impact of perovskite solar cell degradation on the lifetime energy yield and economic viability of perovskite/silicon tandem modules. *Sustainable Energy and Fuels* 3, 1439–1447.

Tian, W., Li, W., Yu, W., and Liu, X. (2017). A review on lattice defects in graphene: Types generation effects and regulation. *Micromachines* 8, 163.

Valsalakumar, S., Roy, A., Mallick, T.K., Hinshelwood, J., and Sundaram, S. (2023). An overview of current printing technologies for large-scale perovskite solar cell development. *Energies* 16, 1.

Wang, B., and Pantelides, S.T. (2011). Controllable healing of defects and nitrogen doping of graphene by CO and NO molecules. *Physical Review B: Condensed Matter and Materials Physics,* 24, 83.

4 Fuel Cells
Fundamental and Applications

Chiam-Wen Liew, Son Qian Liew, and Hieng Kiat Jun

4.1 INTRODUCTION TO FUEL CELL

Depletion of crude oil, natural gas, and coal catalyzes the development of the renewable energy source and planet-friendly technology (Viswanathan and Scibioh, 2007). Not only that, the concern about climate change exacerbates the development of alternative and sustainable energy sources. Among all electrochemical devices, fuel cell is a great choice to replace fossil fuels and oil because hydrogen can be an energy carrier source. Energy production using a clean technology becomes the core part of our daily life to solve the climate change issue. Fuel cell is an electrochemical device that converts the chemical energy of a fuel such as hydrogen to electrical energy through a redox reaction with minimized pollution (Wang et al., 2011; Kim et al., 2009). The concept of fuel cells was first introduced in 1839 by Sir William Grove, a Welsh judge and inventor, who discovered that electricity could be generated through the combination of hydrogen and oxygen. The first functioning fuel cell was built in 1843 by Sir Grove, who demonstrated that hydrogen and oxygen could be combined to produce electricity in a continuous and efficient manner. Fuel cell is a prominent electrochemical device as it exhibits a wide range of applications. Fuel cells have numerous advantages, such as being efficient, clean, safe, and reliable (Viswanathan and Scibioh, 2007). Among all the fuel cells, the proton exchange membrane fuel cell (also known as polymer electrolyte membrane fuel cell) (PEMFC) is a promising candidate. The PEMFC comprises of two porous gas diffusion electrodes (anode and cathode), proton conducting polymer electrolyte, anodic and cathodic catalyst layers (CLs), and current collectors. A great deal of effort has been made in the development of PEMFC as it has the potential to be an alternative energy source.

4.2 PRINCIPLES OF FUEL CELL OPERATION

4.2.1 Mechanism of PEMFCs

The working principle of a PEMFC is shown in Figure 4.8. The PEM utilized in PEMFCs is a membrane that separates the fuel (hydrogen) from the oxidant (air or oxygen), allowing the flow of the loaded molecules and ions (Giorgi et al., 2011). The catalyst used is mainly made of platinum and platinum-alloy nanoparticles, while the catalyst supports generally consist of high specific surface carbons such as volcanic

DOI: 10.1201/9781003314424-4

carbon, ordered porous carbon, hollow graphite particles, and more. Up to now, platinum with carbon-supported materials is the best candidate for PEMFC catalysts as Pt exhibits the most significant catalytic activity for the oxygen reduction reaction (ORR) to occur at the cathode and fuel oxidation at the anode side (Simya et al., 2018). The complex structure allows the electrochemical reactions to occur, where the transportation of reactant and product species, electrons, and protons from and to the active sites takes place. Moreover, the electrodes consist of anode and cathode, in which fuel oxidation and oxygen reduction occur, respectively. When hydrogen is used as fuel, protons and electrons are released due to oxidation at the anode. The protons generated are transported across the membrane, while the electrons are transported via the external circuit. At the cathode CL, the recombination of protons and electrons results in water generation (Weber et al., 2012). The electrochemical reactions involved in the operation of PEMFCs:

At anode: Fuel oxidation takes place.

- $H_2 \rightarrow 2H^+ + 2e^-$

At cathode: ORR takes place.

- $O_2 + 4H^+ + 4e^- \rightarrow 2H_2O$

Overall reaction

- $H_2 + 1/2O_2 \rightarrow H_2O$

Fuel cells operate in a continuous cycle, as long as hydrogen and oxygen are supplied to the electrodes. The basic working mechanism of a fuel cell can be summarized as follows:

1. First, hydrogen is supplied to the anode, where it is split into hydrogen ions and electrons.
2. The hydrogen ions pass through the proton exchange membrane to reach the cathode, while the electrons flow through an external circuit to generate an electrical current.
3. At the cathode, the hydrogen ions react with oxygen to form water, releasing heat in the process.
4. Finally, the electrical current generated by the flow of electrons can be used to power devices or be stored in a battery for later use.

4.2.2 Thermodynamics Principles of PEMFCs

Fuel cell is a type of electrochemical cell, where it converts chemical energy into electric energy. The chemical energy is usually supplied externally. During the energy conversion stage, water and heat are generated. The general layout of a fuel cell with its corresponding energy movement can be represented as in Figure 4.1.

In the case of a fuel cell system, it is an open system, where mass and energy flow through the system boundaries. As such, the work W is performed by the device on its surroundings. Thus, $-W$ over a period is to be expressed as

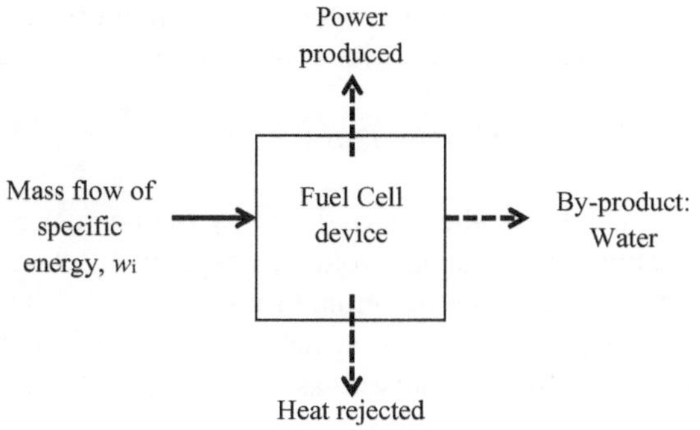

FIGURE 4.1 Schematic layout of a fuel cell with the respective energy movement.

$$-\Delta W = -\Delta W_{\text{electric}} + \int P\, dV \tag{4.1}$$

where W_{electric}, P, and V are electric work, pressure, and volume, respectively. If the fuel cell system is at a fixed volume, $\int P\, dV$ is zero. This leaves us a pure energy output in the form of electric energy. This electric work is supplied by the fuel cell device to another device (or load) via an external circuit. Alternatively, electric work can be expressed in terms of the electric potential difference between the two connecting electrodes or terminals. In this relationship, the number of electrons involved can be quantified as

$$-\Delta W_{\text{electric}} = n_e\, N_A\, e\, \Delta\phi_{\text{ext}} \tag{4.2a}$$

where n_e denotes the number of moles of electrons available, N_A is Avogadro's constant, and e represents the electron charge, thus $e\Delta\phi_{\text{ext}}$ becomes the energy difference for an electron. The product of N_A e is Faraday's constant, F. Thus, Equation 4.2 can be simplified as

$$-\Delta W_{\text{electric}} = n_e\, F\, \Delta\phi_{\text{ext}} \tag{4.2b}$$

Because fuel cell reaction involves a chemical reaction, Equation 4.2 can be rewritten to include the chemical potential difference, $\Delta\mu$ across the fuel cell device. This is the energy that can be absorbed or released during the reaction, and usually, it is made up by multiple components $\Delta\mu_i$:

$$-DW_{\text{electric}} = \sum_i n_{e,i} Dm_i \tag{4.3}$$

In general, the chemical potential of component i (in this case, it's the electrolyte in the fuel cell device) can be expressed as

$$\mu_i = \mu_i^{o} + RT \ln a_i \tag{4.4}$$

where R is the gas constant (8.3 J/K/mol), T is the temperature in Kelvin, a_i represents the activity coefficient in x_i of mole fraction, and μ_i^{o} is the standard potential of ith component.

From the definition of thermodynamics, the maximum amount of energy conversion by a fuel cell is represented by the Gibbs free energy.

$$G = U - T_{ref}S + P_{ref}V \tag{4.5}$$

where U, T_{ref}, S, P_{ref}, and V are the internal energy of the fuel cell system, absolute temperature, entropy, pressure, and volume, respectively. These conditions are defined as they are in any static state situation. From the first law of thermodynamics, increment of internal energy is defined as the net energy added to the system from the surroundings. This relationship is expressed as

$$\Delta U = \int dQ + \int dW + \int dM \tag{4.6}$$

where Q denotes the total net heat as supplied from the surroundings, W represents the net amount of work performed on the system by the environment, and M represents the net energy supplied to the fuel cell device. If we assume the fuel cell device to have a constant volume and the heat exchange with the surroundings is negligible, Equation 4.5 will become

$$-\Delta G = -\Delta W_{electric} \tag{4.7}$$

which means the electricity produced is correlated with the loss of free energy from the fuel cell. In this instance, we can deduce that Equation 4.7 provides the maximum work that is extracted or converted between the system and the environment. By definition, the fuel cell device is in equilibrium thermodynamically when its free energy is zero.

With reference to Figure 4.1, the energy efficiency of the fuel cell device at a given instant in time is

$$\eta = (dQ/dt + dm/dt\ \Delta w)/(Q_i + dm/dt\ w_i) \tag{4.8}$$

in which the energy flow rate is the result of the conversion of the mass flow rates by the specific energies ($\Delta w = w_i - w_o$). Ultimately, the device efficiency as defined by the first law of thermodynamics is simplified as

$$\eta = W/Q_i \tag{4.9a}$$

where W is the work done given by the ΔG. In this case, heat input is calculated based upon the higher heating value (HHV) of the fuel.

$$\eta = \Delta G/\mathrm{HHV} \tag{4.9b}$$

Thus, the maximum efficiency will occur at standard conditions, with the highest possible cell voltage. For a PEMFC with hydrogen and oxygen as the fuel, the maximum calculated efficiency is about 83%. Although the efficiency seems high, it is impractical due to no current drawn (at open circuit conditions). Therefore, the limitation of the efficiency can be addressed by using the second law of thermodynamics. Thus, the efficiency of providing work is

$$\eta = W/W_{max} \tag{4.10a}$$

$$\eta = \Delta G/\Delta G^* \tag{4.10b}$$

$$\eta = n_e\, F\, E/n_e\, F\, E^o = E/E^o \tag{4.10c}$$

where E is the system energy. The above equations are also known as voltage efficiency. It should be noted that the actual behavior of an energy conversion system device may not be accurately calculated due to various uncertainties with approximation. In any circumstances, direct application of the fundamental theories (as in this chapter) will serve as a guideline.

4.3 CATEGORY OF FUEL CELLS

Fuel cells are categorized by either their operating temperature (high temperature or low temperature) or the type of electrolyte used in the cell (for example molten carbonate salt is used in molten carbonate fuel cells – MCFCs). Fuel cells are categorized into five classes, i.e. alkaline fuel cells (AFCs), solid oxide fuel cells (SOFCs), phosphoric acid fuel cells (PAFCs), MCFCs, and PEMFCs, as listed in Figure 4.2.

4.3.1 Alkaline Fuel Cells (AFCs)

AFC is the firstly working fuel cell to generate electricity from hydrogen sources which was invented by Sir William Grove in the middle of the 19th century and developed into a workable system by Francis Bacon in the UK in the 1930s (Viswanathan and Scibioh, 2007). AFC is defined as the fuel cell that converts a controlled amount of hydrogen and oxygen into electricity using a circulating liquid alkaline electrolyte and two porous electrodes at a low operating temperature of 293–363 K. This cell usually consists of potassium hydroxide (30%–40% KOH) and two semi-porous electrodes with low platinum loadings of 0.3 mg/cm^2. The cell can

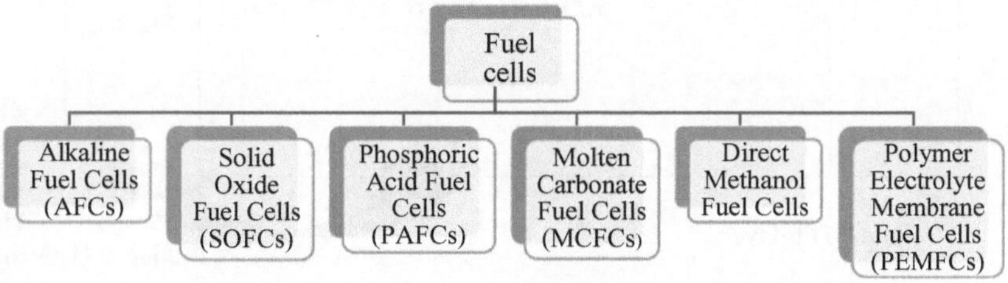

FIGURE 4.2 Types of fuel cells.

deliver power at ambient temperature. However, it just can produce full power at 343 K by applying electrical heating within 10 minutes of a cold start (Viswanathan and Scibioh, 2007). A schematic diagram of an AFC is shown in Figure 4.3.

An AFC cell is comprised of a potassium hydroxide (KOH) electrolyte and two porous electrodes. As shown in Figure 4.3, the oxygen from the air is supported to the positive electrode (cathode), whereas the hydrogen fuel is supplied to the negative electrode (anode). Anode and cathode are connected through an electric load. The voltage between these two electrodes of a single cell is between 0.9 and 0.5 V. There are two electrochemical redox reactions taking place at these respective electrodes. The reactions are illustrated as follows:

At cathode terminal: Oxygen is catalytically reduced to hydroxide anions.

- $\frac{1}{2}O_2\ (g) + 2e^- + H_2O\ (l) \rightarrow 2OH^-$

At anode terminal: Hydrogen is catalytically reduced to water.

- $H_2\ (g) + 2OH^- \rightarrow 2H_2O\ (l) + 2\ e^-$

Overall reaction

- $H_2\ (g) + \frac{1}{2}O_2\ (g) \rightarrow 2H_2O\ (l) +$ electrical energy + heat

The AFC was initially designed to produce electrical power for mission-critical services in space applications. Therefore, it can be used as reliable power sources with low volume and light weight. The benefits of an AFC are also high efficiency, low cost, insensitive to ambient humidity, good reliability, and long cycle life. Furthermore,

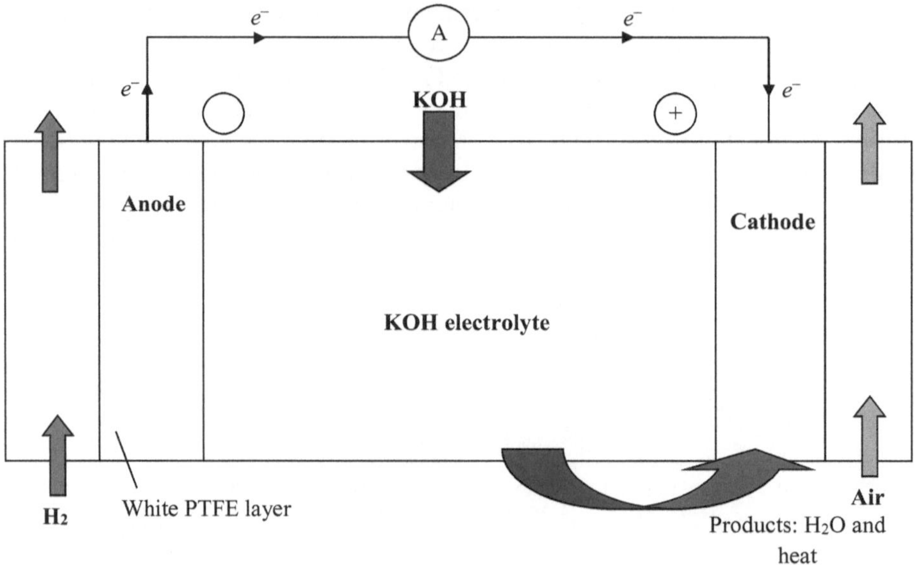

FIGURE 4.3 Schematic diagram of an AFC.

it does not rely on high-volume manufacture which can reduce the cost to an affordable level. Fuel cells have the potential to enable a fast-refueling process for a variety of commercially available battery-powered electric vehicles. Nevertheless, there are several drawbacks associated with the use of fuel cells, including uneven catalyst distribution on porous electrodes, challenges in creating a precise pore system for reactant transportation, corrosive electrolytes, high susceptibility to catalyst polarization, and hazardous materials such as the use of asbestos in separators that can harm consumer health. Furthermore, air-breathing AFCs are intolerant to carbon dioxide (CO_2).

Modern and non-noble metal catalysts based on nickel, silver, and cobalt can be used to replace platinum to reduce the cost of the fuel cell system and improve the electrochemical performance of AFCs. However, nickel and silver catalysts could be degraded easily in the electrolyte which reduces the life span of the catalyst. AFC cannot be operated, especially in mobile applications if there is any CO_2 in the cathode feed gas streams. The potential problems of electrolyte contamination by atmospheric CO_2 can be completely overcome by using an atmospheric scrubber or by changing the electrolyte at the service intervals.

4.3.2 Solid Oxide Fuel Cells (SOFCs)

SOFC is an electrochemical cell utilizing solid oxide or ceramics electrolyte, gaseous fuel, and oxidant to convert the chemical energy stored in fuel into electricity. SOFC is operated at high operating temperatures, ranging from 873 to 1,273 K which is much higher than AFC (Viswanathan and Scibioh, 2007). An SOFC is assembled with two electrodes (an anode and a cathode) and a ceramic electrolyte as a separator. The working principle of an SOFC is illustrated as follows (Figure 4.4):

Fuel (hydrogen, H_2 or methane, CH_4) is initially supplied to the anode. This fuel will react with the oxide ions from the electrolyte and, hence, release electrons (e^-) through an oxidation process. The released electrons are then transferred to the external circuit from the anode to the cathode. On the other side of the plate, oxidant (oxygen, O_2, or air) is fed to the cathode. As a result, the oxidants will accept the electrons from the external circuit and continue to supply the oxide ions (O^{2-}) for the electrolyte. Therefore, the oxygen ions conduction from the cathode and anode could maintain the overall electrical charge balance in the electrolyte. Hence, the

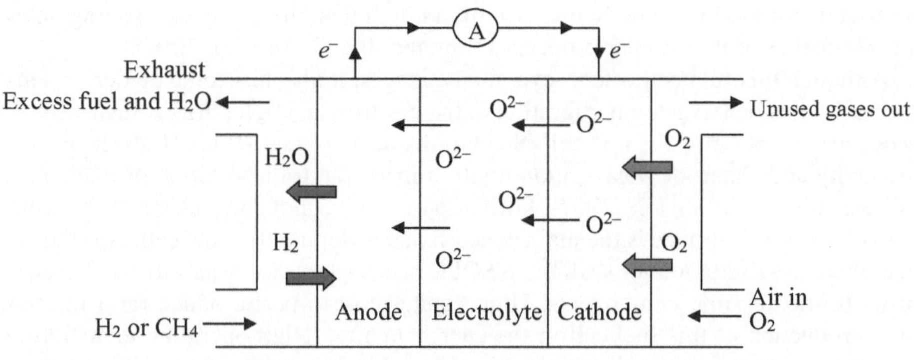

FIGURE 4.4 Schematic diagram of an SOFC based on oxygen-ion conductors.

electron flow and ion conduction will generate electricity from the combustion of the fuel and release the by-products of water and heat (Viswanathan and Scibioh, 2007; Stambouli and Traversa, 2002). The reaction at the anode and cathode of the cell is summarized as follows:

At anode terminal:

- $H_2\,(g) + O^{2-} \rightarrow H_2O\,(l) + 2e^-$ or
- $CH_4 + 4O^{2-} \rightarrow 2H_2O\,(l) + CO_2\,(g) + 2e^-$

At anode terminal:

- $CO\,(g) + O^{2-} \rightarrow CO_2\,(g) + 2e^-$

At cathode terminal:

- $O_2\,(g) + 4\,e^- \rightarrow 2O^{2-}$

SOFCs have some merits over other types of fuel cells such as high system efficiency, high power density, and simple design (Viswanathan and Scibioh, 2007). This fuel cell does not require an external reformer as the reforming natural gas; other hydrocarbon fuels to extract necessary hydrogen can be accomplished within the fuel cell (Stambouli and Traversa, 2002). This fuel cell employs green technology because of its pollution-free feature. Since there is no combustion in the system, there will be no emission of acidic gas (carbon dioxide gas) and/or solid which results from the combustion (Mehmeti et al., 2016a). Apart from that, the need of platinum electrocatalysts can be eliminated (Viswanathan and Scibioh, 2007). High operating temperature of this fuel cell enhances reactant activity, enables fast electrode kinetics, and thereby reduces activation polarization (Stambouli and Traversa, 2002). Other features are reduced oil consumption, low operating and maintenance cost, high efficiency for electricity production, flexibility in choosing the fuel, and operating quietly because of solid state construction and long life cycle (40,000–80,000 hours) due to high tolerance to impurities such as sulfur and low CO_2 emission. In addition, high conversion efficiency in SOFCs compensates high cost of fuel. Another attractive point is this SOFC can be used as water electrolyzers at elevated temperatures without major modification. SOFC also offers high flexibility and easy setting up as a consequence of its modular properties (Stambouli and Traversa, 2002).

Although this fuel cell offers several advantages, it also has some demerits. This fuel cell relies on oxygen ion migration in the electrolyte. High purity of hydrogen is necessary in operating the fuel cell (Stambouli and Traversa, 2002). High electrolyte resistivity and electrode polarization might impede the transportation of oxide ions (Viswanathan and Scibioh, 2007). Formation of low conducting phase at the cathode–electrolyte interface is the main concern in developing this fuel cell. Apart from that, there is a limitation for scaling up SOFC power plants to megawatt size because of its brittle ceramic components. High capital cost-to-performance ratio inhibits the introduction of this fuel cell in the energy market. High operating temperature of SOFCs is the main contributor to affect the cell performance. The high operating temperature of this cell requires a significant startup time. The drop in operating temperature will decrease the performance of the cell. This is due to the increased internal resistance to the flow of oxygen ions. Moreover, an extra thermal shielding is required for this fuel cell to protect personnel and to retain heat because of high operating temperature (Viswanathan and Scibioh, 2007).

Some of the aspects of this fuel cell must be improved in order to enhance the performance of the cells, for instance, electrolyte preparation and homogeneity of materials, and compact fuel cell designs. The efficiency, durability, and lifespan of an SOFC are also the aspects to be investigated for research and development (Mehmeti et al., 2016b). Lowering the operation temperature can be a solution to build up a low-cost SOFC by employing a cheaper electrolyte such as ferritic steels without protective coatings of lanthanum chromate ($LaCrO_3$) and structural components such as stainless steel (Viswanathan and Scibioh, 2007). The reduction in operating temperature of the cell can increase the overall efficiency of the system and thus reduce the thermal stresses in active ceramic structures. Hence, this will expand the expected lifetime of the cell.

4.3.3 Phosphoric Acid Fuel Cells (PAFCs)

PAFCs were the first fuel cells for the commercialization as electric power sources in small- and medium-scale applications (Viswanathan and Scibioh, 2007). PAFC is a fuel cell employing concentrated liquid phosphoric acid (85–100 wt.% H_3PO_4) as the electrolyte and carbon paper coated with a finely dispersed platinum (Pt) catalyst as electrodes (Strickland et al., 2018). Hydrogen is extracted from fuels and then a positively charged proton migrates through the electrolyte from anode to cathode. Electrons are stripped off by the catalyst at anode and thus migrated to cathode through an external circuit. Direct current electric power of these fuel cells is eventually produced. The material of anode is usually polytetrafluoroethylene (PTFE, also known as Teflon) bonded with Pt/carbon black, whereas the material of anode is PTFE bonded with silicon carbide (SiC).

The schematic diagram of the working mechanism in a PAFC is illustrated in Figure 4.5.

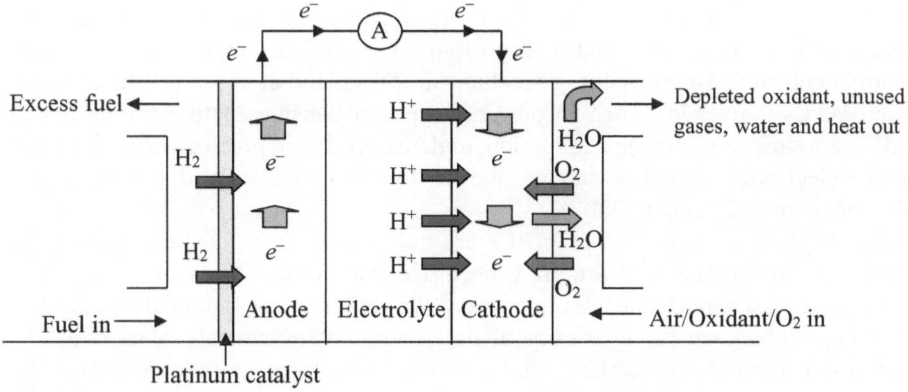

FIGURE 4.5 Schematic diagram of a PAFC.

The reaction occurred at both electrodes is listed as follows:

At anode:

- $H_2 \leftrightarrow 2H^+ + 2e^-$

At cathode:

- $\frac{1}{2}O_2 + 2H^+ + 2e^- \leftrightarrow H_2O$
- $O_2 + 2H^+ + 2e^- \leftrightarrow H_2O_2$
- $H_2O_2 + 2H^+ + 2e^- \leftrightarrow 2H_2O$
- $H_2O_2 \leftrightarrow 2H_2O + \frac{1}{2}O_2$

These PAFCs exhibit several advantages and disadvantages. Higher overall efficiency which comes from heat and power (which is around 80%) is an attractive feature for developing PAFCs. Simple manufacturing process and cell construction are some of the advantages of PAFCs as these cells are mainly comprised of carbon black/carbon, PTFE, and SiC. Simple cell construction could lead to fabrication of tailor-made PAFCs. PAFCs also portray high total energy with excellent overall cell efficiency. Phosphoric acid is a preferred candidate due to its lower volatility property as the PAFCs operate at higher temperatures which is around 423–473 K above the boiling point of water for a long cycle life. It is due to other acid electrolytes requiring water for conductivity which cannot be operated at high temperatures. Phosphoric acid shows superior thermal, chemical, and electrochemical stabilities, excellent capillary properties, and low vapor pressure, even though it is not a good ionic conductor. Other inorganic acids such as hydrochloric, hydrofluoric, sulfuric, and perchloric acids have lower thermochemical stability and higher vapor pressure. Therefore, these inorganic acids are not suitable to be used as electrolytes in fuel cells. It is important to take note that sulfur must be removed if the hydrocarbon fuel is petrol because it will damage the platinum catalyst. PAFCs exhibit high tolerance toward a large variety of fuel gases, good thermal integration, and no CO_2 contamination as it is naturally rejected from the electrolytes.

Platinum is used as catalyst at anode in PAFCs. However, rapid degradation of platinum is the main drawback of this type of FC. The degradation causes the loss in the surface area of the catalyst in the cell (Borup et al., 2007). The loss of surface area arises from dissolution of Pt at the electrode into the electrolyte at a cell voltage of more than 0.8 V and re-deposition of Pt to form larger particles via an electrochemical Ostwald-ripening mechanism (Strickland et al., 2018). In addition, these PAFCs deliver low specific power and power densities. Other limitations of PAFCs are slow rate of oxygen reduction, high material cost, precious metal required for the electrodes, and slow startup due to slightly higher operating temperature (Viswanathan and Scibioh, 2007).

Due to their expensive cost, PAFCs are not considered a favorable choice. As a result, various research efforts have been dedicated to their development, with a focus on several aspects such as improving the resistance of the anode catalyst to poisoning, enhancing the efficiency of pressurization instruments, stabilizing the relationship between the cathode and catalyst, improving the interfacial stability between the electrode and electrolyte, developing a more durable catalyst material that can overcome Pt dissolution, increasing power density through the application of suitable electrolytes, and creating a cost-effective PAFC with high power and energy densities. These aspects have been studied extensively, as indicated by research (Viswanathan and Scibioh, 2007).

4.3.4 Molten Carbonate Fuel Cells (MCFCs)

MCFCs are fuel cells operating at high temperatures of 873 K and above. MCFCs are developed for natural gas– and coal-based power plants for electrical utility, industrial, and military applications. An MCFC consists of molten carbonate salt as an electrolyte and nickel as electrode material. The electrolyte can be lithium carbonate (Li_2CO_3) or potassium carbonate (K_2CO_3). On the other hand, the anode is made of nickel alloy, for instance, nickel alloyed with chromium (Cr) or aluminum (Al), whereas cathode is made of in situ lithiated nickel oxide (Mehmenti et al., 2016a). Figure 4.6 shows the basic configuration of an MCFC.

However, there are several shortcomings of MCFCs. The anode is prone to creep and may experience a decrease in surface area due to sintering. Additionally, the carbonate melt may not fully wet the anode. The cathode is susceptible to dissolution in the electrolyte and may experience a short circuit through the metallic Ni matrix. The matrix may crack due to thermal cycling and may lose its capillary retention of electrolytes. The electrolyte may corrode cell components and may also experience evaporation (EG and Technical, 2016; Kulkarni and Giddey, 2012).

4.3.5 Direct Methanol Fuel Cells (DMFCs)

Direct methanol fuel cells (DMFCs) are electrochemical devices that utilize methanol and methanol solutions as fuel, operating at near room temperature. Electrons and protons are discharged as the methanol and water are oxidized in the anode CL during operations, as shown in Figure 4.6. The electrons are delivered to the cathode by an external circuit, while the protons enter through the electrolyte membrane.

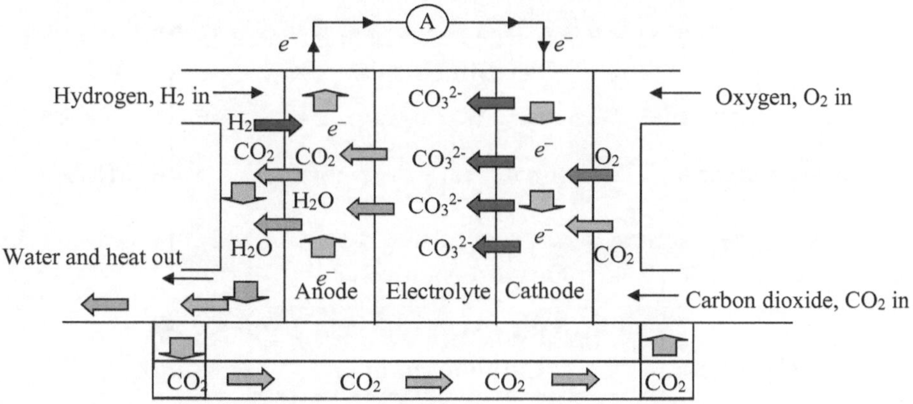

FIGURE 4.6 Schematic diagram of an MCFC.

When oxygen from the surrounding air reacts with the electrons and protons in the cathode CL, water is produced as a result (Li and Faghri, 2013) (Figure 4.7).

At anode CL: Methanol and water are oxidized to release electrons and protons.

- $CH_3OH + H_2O \rightarrow 6H^+ + 6e^-$

At cathode CL: Electrons and protons react with oxygen from atmospheric air, generating water.

- $6H^+ + 6e^- + 1.5O_2 \rightarrow 3H_2O$

Overall reaction

- $CH_3OH + 1.5O_2 \rightarrow CO_2 + 2H_2O$

DMFCs are suitable for portable devices with high consumption of power. This is mainly owing to the easy handling of methanol and its high theoretical energy density (which is 6.1 kWh/kg per day). DMFCs may be classified as active, semi-passive, or passive based on ways of supplying fuel and oxidants. Active DMFCs use pumps, fans, sensors, heaters, and humidity components to supply methanol and oxidizing solutions. Fuel and oxidant supply rates can be precisely controlled, and the best performance can be achieved in active DMFCs. Passive DMFCs do not need any auxiliary methanol, water, and oxygen components. Capillary forces transfer the supplies, gravity, and gradients passively from the methanol tank and ambient air to the membrane electrode assembly (MEA). Passive DMFCs are more compact, reliable, and straightforward, and are therefore more suited to portable and low-power micro/miniature applications

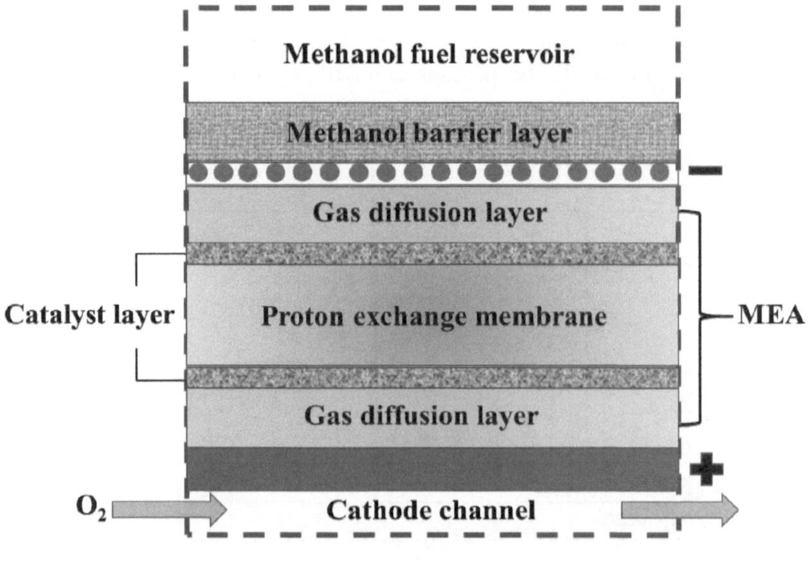

FIGURE 4.7 Semi-passive DMFC that is supplied with high concentration of methanol in the solutions. Adapted from Li and Faghri (2013).

(<10 W) (Verma, 2000; Liu et al., 2005; Faghri and Guo, 2008). In addition, semi-passive DMFCs have a passive and active anode or a passive cathode.

Nevertheless, passive DMFCs have specific barrier problems for commercialization purpose: methanol crossover (MCO), sluggish kinetic reaction of methanol, heat management, stability, durability, water management, cost, etc. (Munjewar et al., 2017).

4.3.6 Polymer Electrolyte Membrane Fuel Cells (PEMFCs)

PEMFCs, also known as proton exchange membrane fuel cells, have an anode, a cathode, and an electrolyte membrane. Hydrogen is oxidized at the anode, whereas oxygen is reduced at the cathode. An example of this process involves the transfer of protons through an electrolyte membrane from the anode to the cathode, while the electrons travel through external circuit loads. At the cathode, oxygen reacts with protons and electrons, resulting in the production of heat and water as byproducts (Spiegel, 2007). The whole PEMFC process is shown in Figure 4.8. In addition, depending on the working temperature, two distinct PEMFC types could be distinguished, where the first type is Low-Temperature Proton Exchange Membrane Fuel Cell (LT-PEMFCs), which works in the temperature range of 60°C–80°C, and the second type works in the range from 110°C to 180°C and is thus termed as the High-Temperature Proton Exchange Membrane Fuel Cell (HT-PEMFCs) (Sharaf and Orhan, 2014). Nafion, a Teflon-based polymer completely fluorinated and developed by DuPont in the 1960s

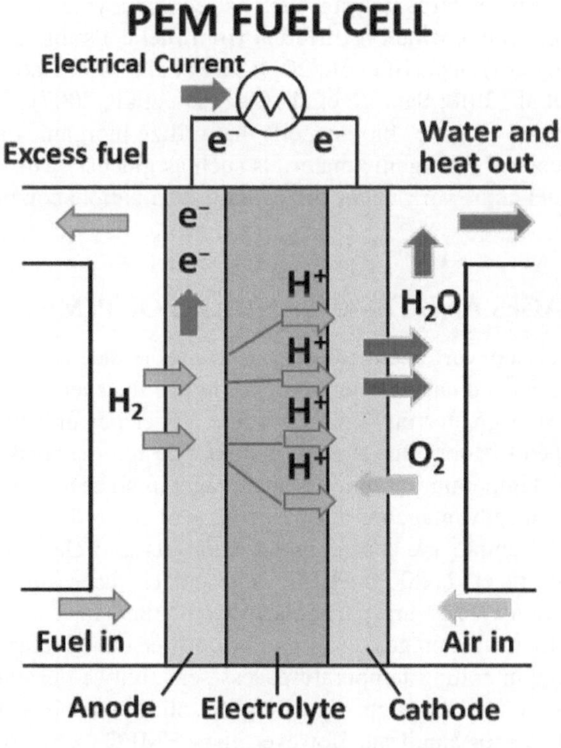

FIGURE 4.8 Schematic diagram of a PEMFC. Adapted from Mekhilef et al. (2012).

for space applications, is the usual electrolyte material employed in low-temperature PEM fuel cells. However, it is also possible to employ Nafion or polybenzimidazole (PBI) doped in phosphoric acid for HT-PEMFCs (Sharaf and Orhan, 2014). The traditional catalyst utilized in LT-PEMFCs at low temperatures is platinum, whereas Platinum–Ruthenium is utilized in catalysts for HT-PEMFCs (Sharaf and Orhan, 2014; Larminie et al., 2003). The electrical efficiency for LT-PEMs fuel cells is approximately 40%–60%, whereas for HT-PEMFCs, it is about 50%–60% (Sharaf and Orhan, 2014).

4.3.7 Biofuel Cells

Biofuel cells produce electricity via electrochemical reactions which are catalyzed either by enzymes or by complete microbial cells (Davis and Higson, 2007; Bullen et al., 2006). Even though most organic substrates produce energy during combustion, oxidizers (for example, molecular oxygen, O_2 or hydrogen peroxide, H_2O_2) may be an alternative way of converting chemical energy into electrical energy by biocatalytic oxidization of organic substances, such as alcohol, organic acids, or sugars. Biocatalytic oxidation occurs explicitly at the anode, generating electrons that flow through the external electric circuit to another electrode (cathode) where the produced electrons serve as the source for the reduction reaction of oxidizer to occur. Organic fuel and oxidizer-aqueous solutions should circulate through the biofuel cell in which the bioelectrochemical processes occur to provide a constant power output. Biofuel cells utilize bioelectrocatalysts, such as redox enzymes or entire biological cells (e.g., microbial cells), which is different from fuel cells that use electrochemical inorganic catalysts (Luckarift et al., 2014; Luz et al., 2014; Hao and Scott, 2010; Das, 2018; Slate et al., 2018; Santoro et al., 2017; Du et al., 2007). A particular biofuel cell type is called "abiotic" biofuel cells that utilize inorganic catalysts to assist the oxidation process of bioorganic materials such as glucose. Unlike ordinary fuel cells, abiotic biofuel cells work under biologically compatible conditions (Katz and Bollella, 2021).

4.4 ADVANTAGES AND DISADVANTAGES OF PEMFCS

PEMFCs effectively convert chemical energy to usable electrical energy via electrochemical reactions and have garnered attention due to several advantages. These power sources have a greater maximum specific power per unit volume or weight than any other type of battery due to their compact design (Bagotskii, 2009). These cells operate at low temperatures, which is advantageous since it enables a fast startup procedure, but also disadvantageous due to the issue of sluggish reaction rates, which are handled via the employment of complex catalysts and electrodes (Sharaf and Orhan, 2014; Larminie et al., 2003). PEMFCs have been successfully used in a wide variety of applications. They are particularly well-suited for transportation applications owing to their advantages, such as high power density, rapid startup time, high efficiency, low operating temperature, less sensitivity to orientation, environmentally friendly, compact design, lightweight, solidity of electrolyte, flexibility in input fuel, and ease of handling. However, the PEMFC technology continues to face many obstacles such as high costs, poor durability, pure hydrogen, and oxygen required, water flooding, platinum catalyst poisoning, and hydrogen storage issues, all of which impede its commercialization.

4.5 APPLICATIONS OF FUEL CELLS

4.5.1 Stationary Power Generation Plants

The PEMFC is the prevalent type of fuel cell used in stationary power applications, though there is increasing use of SOFCs, MCFCs, AFCs, and PAFCs (Carrette et al., 2001). Both low- and high-temperature fuel cells can be employed in stationary applications. While low-temperature fuel cells are typically more efficient during startup, high-temperature systems, like SOFCs and MCFCs, generate heat that can be used directly for other purposes and can also operate directly on fuels without external modification, resulting in higher efficiencies than low-temperature fuel cells (Carrette et al., 2001; Andújar and Segura, 2009). Figure 4.9 displays two main types of power stations: large stationary plants, which have a power output ranging from 300 kW to 20 MW, and small-to-medium stationary power plants, with power output ranging from a few Watts up to 10 kW for small stationary power plants, and from 10 up to 300 kW for medium stationary power plants (Das et al., 2021). Moreover, fuel cells are utilized for a variety of applications in stationary settings, serving as a primary source of power for the grid or remote-area power supply (RAPS) in locations where the grid is inaccessible. Furthermore, fuel cells are employed in hybrid power generation systems, in conjunction with photovoltaic systems, batteries, capacitors, or wind turbines, to provide primary or secondary power (i.e., distributed power or CHP generation), or to act as a backup power source (emergency backup power supply, EPS or UPS), depending on their characteristics (Spiegel, 2007; Sharaf and Orhan, 2014; Lucia, 2014).

By employing fuel cells, it is possible to transition from large-scale centralized power to decentralized distributed generation. Thanks to their static nature, minimal emissions, exceptional load tracking, and high efficiency, fuel cells can be used

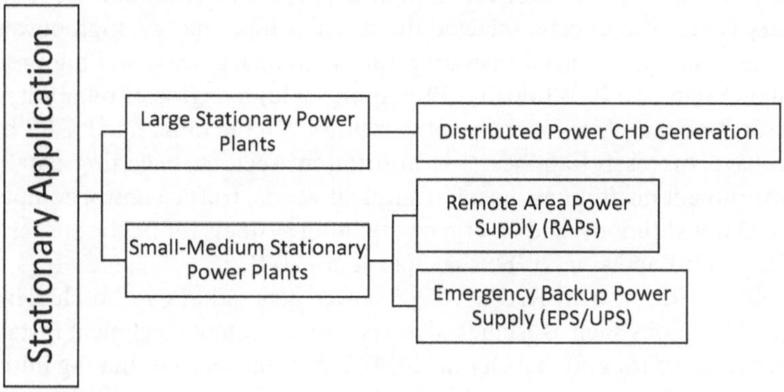

FIGURE 4.9 Applications of fuel cells in stationary power generation plants. Adapted from Alaswad et al. (2016).

to generate residential electricity or CHP power for households or larger residential blocks (Sharaf and Orhan, 2014). The combined heating and power systems boost energy efficiency by using both electrical and thermal energy. For instance, the produced heat can be used for space heating or different uses (Upreti et al., 2012). PEMFCs and PAFCs are used widely for CHP applications, especially for the household generation of CHP, but these fuel cells can only work at lower temperatures, typically less than 100°C and 200°C. Therefore, both types of fuel cells, SOFCs and MCFCs, require an external fuel reformer to use fuels other than pure hydrogen and have a low tolerance for CO, which is a byproduct of natural gas reforming (Zink et al., 2007). Therefore, high-temperature fuel cells, particularly SOFCs, are more appropriate for CHP generation on a larger residential block basis due to their characteristics (Sharaf and Orhan, 2014). At higher temperatures, SOFC systems have more considerable advantages than these technologies for the combined production of heat and electricity (Zink et al., 2007). Combined heat and power (CHP) fuel cell systems can be designed to function independently of the grid or with assistance from the grid. The grid-independent system is more expensive and complicated due to the need to handle dynamic load changes, but this can be resolved by addressing the fuel cell system size and its incorporation with battery banks or ultracapacitors. In contrast, the grid-assisted system exports electricity to the grid during periods of low demand and imports electricity during peak demand periods. In either scenario, an efficient CHP system necessitates a mechanism for storing heat (Sharaf and Orhan, 2014).

Medium-sized power stations are commonly utilized to provide combined electricity and heat to various structures, including cottages, administrative and office buildings, and hospitals. In contrast, small-power fuel cell power plants are constructed to supply electricity to individual residential or administrative areas (Bagotskii, 2009). PEMFCs are the most commonly used fuel cells for low-energy plants, such as the 1 kW power plant established by Ebara Ballard to produce CHP. This unit is designed to function for 10 years and comply with government standards. Similarly, the Japanese company Fuji Electric produced a similar device for 10 years of operation, with a cost of 12,000–16,000 dollars (Bagotskii, 2009). Fuel cells are now becoming a promising alternative to electric power system (EPS) batteries, particularly in the telecommunications market, with PEMFCs and DMFCs being the primary types of fuel cells selected due to their high-energy, high power densities, high modularity, extended operating times, compact sizes, and high reliability (Sharaf and Orhan, 2014). While an EPS requires a high degree of reliability, it does not necessarily require high operational durability. On the other hand, EPS has been proven to have successful applications in different sectors, including data centers, banks, government institutions, hospital surgical wards, traffic control computer systems, financial institutions, and emergency lighting systems for public spaces (Sharaf and Orhan, 2014; Bagotskii, 2009; Cacciola et al., 2001).

Fuel cells are utilized in RAPS where power generation can be challenging in grid-isolated locations such as islands, deserts, forests, remote technical installations, and research centers (Sharaf and Orhan, 2014). For instance, a fascinating initiative on Bechervaise Island in Antarctica, which includes hydrogen production through electrolysis using wind energy and low-power electrical and heat-producing fuel cell units, is funded by a $600,000 grant from the Australian Government (Bagotskii, 2009).

4.5.2 Automotive Industry

Nowadays, the automotive industry is one of the primary areas of clean energy technology development. This is because the transport sector accounts for 17% of the world's greenhouse gas emissions annually (Sharaf and Orhan, 2014). Therefore, significant optimizations are required from this industry to achieve the Kyoto Protocol's objectives (Cacciola et al., 2001). Investing in technology is crucial for addressing the concerns, particularly in reducing harmful emissions and improving energy conversion efficiencies. Fuel cells offer a prime example, as they can enable the automotive industry to achieve nearly zero hazardous emissions while maintaining the efficiency of the vehicle's propulsion system. In addition to that, fuel cells possess efficiencies nearly twice the efficiencies of traditional internal combustion engines. Due to its benefits of smooth operation, fuel flexibility, modularity, and low maintenance needs, the fuel cell is a promising candidate to eventually replace traditional combustion engines in the future. This is primarily because of the absence of moving parts. The fuel cell business currently faces several commercialization issues in the transportation/automotive sector. One crucial difficulty is the cost of the fuel cell, the unit's durability, and its performance. Other than that, hydrogen storage and hydrogen fuel infrastructure must all be considered (Alaswad et al., 2016).

PEMFCs possess a range of characteristics that make them ideal for transportation and automotive applications, including high-power density, high conversion efficiency, compactness, lightweight design, and low operating temperatures. Nevertheless, the use of fuel cells in automobiles is associated with several challenges, including issues with volumetric and gravimetric power density and reliability, as well as costs. Indeed, the high cost of the technology remains the most significant barrier to commercialization. Another promising fuel cell technology for transportation applications is DMFC, but it is not as advanced as PEMFC and has lower power densities due to methanol absorption through the membrane and catalyst toxicity. Apart from that, SOFC is another potential development technology in the field of transport applications. Different from PEMFCs, SOFCs allow the utilization of hydrocarbon fuels, even if they are complex to be used.

On top of that, they have no significant issue with the poisoning of catalysts. Numerous companies have developed SOFCs as auxiliary power units (APUs) for petrol automobiles due to these advantageous properties (Carrette et al., 2001). The application of fuel cells in the automotive industry is summarized in Figure 4.10 (Sharaf and Orhan, 2014).

Fuel cells have significant applications in light traction vehicles (LTVs), such as scooters, personal wheelchairs, motorcycles, airport tugs, golf carts, electric-assisted bicycles, and material handling vehicles (e.g., carts, pallet trucks, and forklifts). One of the most important potential applications of fuel cells in LTVs could be in motorcycles, as they are a significant contributor to pollution in urban areas despite their small size. The successful integration of fuel cells in the transportation industry can be seen in forklifts, with around 1,300 fuel cell-powered forklifts currently in use in the US market (Sharaf and Orhan, 2014). PEMFCs are the most often used fuel cells in research and development for light fuel cell electric vehicles (L-FCEVs) due to their benefits. FCEVs operate on a straightforward principle: The electricity is

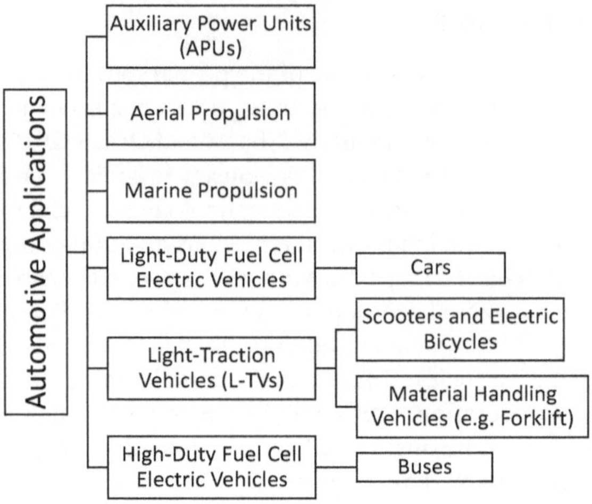

FIGURE 4.10 Automotive applications for fuel cells. Adapted from Sharaf and Orhan (2014).

created from hydrogen using low-temperature fuel cells (most commonly PEM), which are then utilized to power the vehicle or stored in batteries or ultracapacitors (Pollet et al., 2012). Since fuel cells create energy through chemical reactions, they do not emit toxic emissions (only generate water as a byproduct) and generate lesser heat compared to that of internal combustion engines. Despite the benefits of PEMFCs, many automobile manufacturers prefer methanol as a fuel source since it can be used to generate hydrogen on-board via steam reforming. However, the downside of this approach is longer response times and requires considerable gas cleaning before fueling the fuel cell (Alaswad et al., 2016). Due to these problems and the size of the reformer and gas purification unit, DMFCs are more ideal for mobile systems (Carrette et al., 2001). Many car manufacturers are putting much effort into developing L-FCEVs' commercialization (e.g., General Motors, Toyota, Mazda, Daimler AG, Volvo, Volkswagen, Honda, Hyundai, Nissan) (Sharaf and Orhan, 2014). For instance, the car manufacturer Honda introduced the FCX Clarity model at the automotive saloon in Los Angeles in 2007. The world's first fuel cell vehicle platform produced in series has been available in the market since the summer of 2008 (Sharaf and Orhan, 2014). At the 2010 Fuel Cell Expo in Tokyo, Japanese car manufacturers announced a plan to introduce two million FCEVs in Japan by 2025, which would make them completely competitive (Pollet et al., 2012).

The category of heavy-duty electric fuel cell vehicles (H-FCEVs) includes buses, heavy-duty trucks, locomotives, vans, utility trucks, service fleets, and other vehicles that utilize fuel cells as part of their high-capacity electric propulsion systems. In recent times, several fuel cell buses have been operating in different parts of the world (Andújar and Segura, 2009). In 2012, more than 30 buses that utilized fuel cells were deployed in Western Europe and 25 in the United States. The most common applications of fuel cell stacks in the industry are PEMFCs and PAFCs, which exhibit remarkable renewable braking energy recovery and excellent dynamic responsiveness. At present, fuel cell electric buses (FCEBs) are not economically competitive with conventional buses and other alternative technologies due to the advanced nature of fuel cell technology and the lack of mass production and manufacturing capabilities (Sharaf and Orhan, 2014).

Meanwhile, the use of fuel cells in marine transportation has been on the rise in recent years. While fuel cells offer several advantages for ships and ferries, including high efficiency, low emissions, and stationary operation, they face challenges related to reliability, durability, shock resistance, and the corrosive effects of saltwater air (Sharaf and Orhan, 2014). As a result, fuel cell applications in marine transportation have not been explored as extensively as they have in other sectors like automotive or aviation propulsion. On top of that, high cost, harsh marine environment with dynamic stresses and severe corrosion, life cycle, and lack of infrastructure are barriers that prevent the commercialization of fuel cells in this field. In Germany, the first commercial passenger ship with a hybrid PEMFC/battery system was demonstrated in 2003, followed by the world's first commercial passenger ship with the same system in 2008. Additionally, submarines that use oxygen/hydrogen fuel cells for propulsion and auxiliary power have been successfully deployed. The PEMFC technology is the preferred choice for providing air-independent propulsion systems (AIP) for a new generation of conventional submarines in several countries, including Germany, Greece, Italy, Portugal, and Spain (Leo et al., 2010).

The space industry was one of the first areas in which fuel cells were implemented. NASA deployed AFCs and PEMFC technology in the 1960s for the manned space programs. Fuel cells are considered promising for space applications due to their many advantages over other power generation technologies. Of particular interest is the fact that water is a byproduct of the electrochemical reactions that take place within the fuel cell, making it particularly useful for space missions where air, water, and food are crucial. However, due to the high energy density, power density, durability, and reliability requirements, it is currently not practical to commercialize the use of fuel cells in manned military and commercial air vehicles (Sharaf and Orhan, 2014). The summarized application of fuel cells in the transportation industry is shown in Table 4.1 (Alaswad et al., 2016).

4.6 CONCLUSIONS

In conclusion, fuel cell technology stands at the forefront of sustainable energy solutions, offering a versatile and efficient means of electricity generation. With a primary focus on hydrogen fuel cells, which seamlessly convert the chemical potential of hydrogen and oxygen into clean electric power and water, the applications span across diverse sectors including transportation, stationary power generation, and energy storage. Fuel cells offer a potentially non-polluting and renewable way to generate electricity. PEMFCs offer several advantages over the batteries and supercapacitors. Fabrication of a fuel cell with high power density, long durability, and cost-effective properties is essential for the economic application of sustainable energy technologies. The research and development of PEMFCs is still essential to harness the capabilities of hydrogen for energy production to the maximum threshold and operational advantages prior to commercialization. Innumerable challenges and technological breakthroughs are still the obstacles in the R&D of PEMFCs.

TABLE 4.1
The Applications of Fuel Cells in the Transportation Industry

Application	PEMFC	DMFC	SOFC	MCFC	PAFC	AFC	MFC
Stationary	CHP	#		#		#	
	RAPS	#		#	#		
	EPS	#	#	#	#		
Portable	Power generators	#	#				
	Consumer electronics	#	#				
	Battery charges	#	#				
	Army	#	#				
Transportation	APUs	#	#	#	#		
	LTVs	#	#				
	L-FCEVs	#					
	H-FCEVs	#				#	
	Aerial propulsion	#		#			#
	Marine propulsion	#		#	#		

ACKNOWLEDGMENT

This work was partly supported by the Fundamental Research Grant Scheme (FRGS/1/2019/STG07/TARUC/02/1) from the Ministry of Higher Education of Malaysia.

REFERENCES

Alaswad A, Palumbo A, Dassisti M, Olabi AG (2016) Fuel cell technologies, applications, and state of the art. In: *A Reference Guide in Reference Module in Materials Science and Materials Engineering*. Elsevier, Amsterdam.

Andújar JM, Segura F (2009) Fuel cells: History and updating. A walk along two centuries. *Renewable and Sustainable Energy Reviews* 13: 2309–2322.

Bagotskii VS (2009) *Fuel Cells: Problems and Solutions*. John Wiley & Sons, New York.

Borup R, Meyers J, Pivovar B, Kim YS, Mukundan R, Garland N et al. (2007) Scientific aspects of polymer electrolyte fuel cell durability and degradation. *Chemical Reviews* 107: 3904–3951.

Bullen RA., Arnot TC, Lakeman JB, Walsh FC (2006) Biofuel cells and their development. *Biosensors and Bioelectronics* 21: 2015–2045.

Cacciola G, Antonucci V, Freni S (2001) Technology up date and new strategies on fuel cells. *Journal of Power Sources* 100: 67–79.

Carrette L, Friedrich KA, Stimming U (2001) Fuel cells: fundamentals and applications. *Fuel Cells* 1: 5–39.

Das D (2018) Microbial fuel cell: A bioelectrochemical system that converts waste to watts. In: Microbial Fuel Cell: A Bioelectrochemical System that Converts Waste to Watts. Springer International Publishing. pp. 1–506. https://doi.org/10.1007/978-3-319-66793-5

Das HS, Chowdhury MFF, Li S, Tan CW (2021) Fuel cell and hydrogen power plants. In: Kabalci E (ed) *Hybrid Renewable Energy Systems and Microgrids*. Acedemic Press Inc, San Diego, pp. 313–349

Davis F, Higson SPJ. (2007) Biofuel cells-Recent advances and applications. *Biosensors and Bioelectronics* 22: 1224–1235.

Du Z, Li H, Gu T (2007) A state of the art review on microbial fuel cells: A promising technology for wastewater treatment and bioenergy. *Biotechnology Advances* 25: 464–482.

EG and Technical (2016) *Fuel Cell Handbook*, 7th edn. Department of Energy Office of Fossil Energy National Energy Technology Laboratory, United States.

Faghri A, Guo Z (2008) An innovative passive DMFC technology. *Applied Thermal Engineering* 28: 1614–1622.

Giorgi L, Salernitano E, Gagliardi S, Dikonimos T, Giorgi R, Lisi N, De Riccardis F, Martina V. (2011) Electrocatalysts for methanol oxidation based on platinum/carbon nanofibers nanocomposite. *Journal of Nanoscience and Nanotechnology* 11: 8812–7.

Hao YE, Scott K (2010) Enzymatic biofuel cells-fabrication of enzyme electrodes. *Energies* 3: 23–42.

Katz E, Bollella P (2021) Fuel cells and biofuel cells: From past to perspectives. *Israel Journal of Chemistry* 61: 68–84.

Kim DS, Cho HI, Kim DH, Lee BS, Yoon SW, Kim YS, Moon GY, Byun H, Rhim JW (2009) Surface fluorinated poly(vinyl alcohol)/poly(styrene sulfonic acid-co-maleic acid) membrane for polymer electrolyte membrane fuel cells. *Journal of Membrane Science* 342: 138–144.

Kulkarni A, Giddey S (2012) Materials issues and recent developments in molten carbonate fuel cells. *Journal of Solid State Electrochemistry* 16: 3123–3146.

Larminie J, Dicks JLA, Dicks A, Dicks AL, Knovel (2003) *Fuel Cell Systems Explained*. John Wiley & Sons, New York.

Leo T, Durango JA, Navarro E (2010) Exergy analysis of PEM fuel cells for marine applications. *Energy* 35: 1164–1171.

Li X, Faghri A (2013) Review and advances of direct methanol fuel cells (DMFCs) part I: Design, fabrication, and testing with high concentration methanol solutions. *Journal of Power Sources* 226: 223–240.

Liu J, Zhao T, Chen R, Wong C (2005) The effect of methanol concentration on the performance of a passive DMFC. *Electrochemistry Communications* 7: 288–294.

Lucia U (2014) Overview on fuel cells. *Renewable and Sustainable Energy Reviews* 30: 164–169.

Luckarift H, Atanassov P, Johnson GR (2014) *Enzymatic Fuel Cells: From Fundamentals to Applications*. John Wiley & Sons, New York.

Luz RAS, Pereira AR, de Souza JCP, Sales FCPF, Crespilho FN (2014) Enzyme biofuel cells: thermodynamics, kinetics and challenges in applicability. *ChemElectroChem* 1: 1751–1777.

Mehmeti A, McPhail SJ, Pumiglia D, Carlini M (2016a) Life cycle sustainability of solid oxide fuel cells: From methodological aspects to system implications. *Journal of Power Sources* 325: 772–785.

Mehmeti A, Santoni F, Pietra MD, McPhail SJ (2016b) Life cycle assessment of molten carbonate fuel cells: State of the art and strategies for the future. *Journal Power Sources* 308: 97–108.

Mekhilef S, Rahman S, Safari A (2012) Comparative study of different fuel cell technologies. *Renewable & Sustainable Energy Reviews* 16: 981–989.

Munjewar SS, Thombre SB, Mallick RK (2017) Approaches to overcome the barrier issues of passive direct methanol fuel cell: Review. *Renewable and Sustainable Energy Reviews* 67: 1087–1104.

Pollet BG, Staffell I, Shang JL (2012) Current status of hybrid, battery and fuel cell electric vehicles: From electrochemistry to market prospects. *Electrochimica Acta* 84: 235–249.

Santoro C, Arbizzani C, Erable B, Ieropoulos I (2017) Microbial fuel cells: From fundamentals to applications. A review. *Journal of Power Sources* 356: 225–244.

Sharaf O, Orhan M (2014) An overview of fuel cell technology: Fundamentals and applications. *Renewable and Sustainable Energy Reviews* 32: 810–853.

Simya OK, Nair P, Ashok A (2018) Engineered nanomaterials for energy applications: In: Chaudhery MH (ed) *Handbook of Nanomaterials for Industrial Applications*. Elsevier, Amsterdam, pp. 751–767.

Slate A, Whitehead K, Brownson D, Banks C (2018) Microbial fuel cells: An overview of current technology. *Renewable and Sustainable Energy Reviews* 101: 60–81.

Spiegel C (2007) *Designing and Building Fuel Cells*. Citeseer, New York.

Stambouli AB, Traversa E (2002) Solid oxide fuel cells (SOFCs): A review of an environmentally clean and efficient source of energy. *Renewable and Sustainable Energy Reviews* 6: 433–455.

Strickland K, Pavlicek R, Miner Elise, Jia Q, Zoller I, Ghoshal S, Liang W, Mukerjee S (2018) Anion resistant oxygen reduction electrocatalyst in phosphoric acid fuel cell. *ACS Catalysis* 8: 3833–3843.

Upreti G, Greene D, Duleep KG, Sawhney R (2012) Fuel cells for non–automotive uses: Status and prospects. *International Journal of Hydrogen Energy* 37: 6339–6348.

Verma L (2000) Studies on methanol fuel cell. *Journal of Power Sources* 86: 464–468.

Viswanathan B, Scibioh MA (2007) *Fuel Cells: Principles and Applications*. CRC Press, Taylor & Francis Group, UK.

Wang Y, Chen KS, Mishler J, Cho SC, Adroher XC (2011) A review of polymer electrolyte membrane fuel cells: Technology, applications, and needs on fundamental research. *Applied Energy* 88: 981–1007.

Weber AZ, Balasubramanian S, Das PK (2012) Proton exchange membrane fuel cells. In: Sundmacher K (ed) *Advances in Chemical Engineering*. Academic Press, San Diego, pp. 65–144.

Zink F, Lu Y, Schaefer L (2007) A solid oxide fuel cell system for buildings. *Energy Conversion and Management* 48: 809–818.

5 Solid-State Polymer Electrolytes for Fuel Cells Application

Chiam-Wen Liew, Son Qian Liew, and Hieng Kiat Jun

5.1 INTRODUCTION

Several electrochemical devices are emerged as energy storage devices and as power sources in recent years such as fuel cells, solar cells, batteries, and supercapacitors. An electrochemical cell is basically comprised of an electrolyte and a pair of separated electrodes. Two electrodes facilitate the flow of electric current in a cell primarily by allowing ion diffusion to occur in the permeable electrolyte. An electrode serves as an electrical conductor that establishes contact with a non-metallic electrolyte in an electrical circuit. Electrode can be either cathode or anode which involves the reduction–oxidation (redox) reaction. Anode is the electrode where the electron is lost through an oxidation reaction, whereas a cathode is the electrode where the electron is gained through the reduction reaction. On the other hand, electrolyte also plays an important role in the electrochemical devices. Electrolyte is a medium to produce the ionized species or more precisely known as charge carriers. The mobile ions produced from the electrolyte will be transported in the electrolyte from an electrode to the opposite electrode and allowed to perform chemical reaction at the electrodes as long as there is an electric current applied across the cell. As a result, energy is delivered to the external circuit. Electrolytes can be divided into two main classes that are liquid electrolytes and solid-state electrolytes. In this chapter, the transition of electrolytes from liquid type to solid type is examined.

Electrolytes find broad usage in technology, particularly in electrochemical devices, spanning from the manufacture of commercial secondary rechargeable lithium-ion batteries to sophisticated high-energy electrochemical systems like chemical sensors, fuel cells, electrochromic windows (ECWs), solid-state reference electrode systems, memory devices, supercapacitors, and solar cells (Gray, 1991; Rajendran et al., 2004). The diverse applications of electrochemical devices include powering portable electronic devices such as laptops, mobile phones, and music players, as well as serving as the power source for hybrid electric vehicles (EVs) and start-light-ignition (SLI) systems used in traction (Gray, 1997; Ahmad et al., 2005).

DOI: 10.1201/9781003314424-5

5.2 DEVELOPMENT OF ELECTROLYTES

The liquid-based electrolyte faces several problems, especially the leakages of corrosive liquids and harmful gases. Therefore, researchers developed a new type of electrolyte, so-called solid-state electrolyte to overcome those shortcomings.

5.2.1 Liquid Electrolytes

Liquid electrolyte is a substance that conducts electricity through the migration of both cations and anions to the opposite electrodes when an electrical potential is applied. The doping salt is normally dissolved in an aqueous solution to produce an electrolyte solution. The electrolyte is then dissociated into cation and anion, and hence followed by a solvation process. Solvation is a process of association between molecules of solvent and molecules or ions of the salt via intermolecular interactions. In other words, the molecules or ions of the solute will be dissociated and then bounded by the molecules of solvent in the electrolyte solution. Hence, the surrounded ions will be separated and migrate to the opposite electrodes to generate the electric current when an electrical potential is applied to the electrolyte. Sodium chloride (NaCl) is a strong electrolyte because it achieves complete dissociation when it is dissolved in a solution due to the thermodynamic interactions between solvent and solute molecules. In contrast, acetic acid is a weak electrolyte as it cannot be dissociated completely.

Despite their widespread use, traditional liquid electrolytes suffer from several drawbacks including the corrosive nature of the solvent and its tendency of harmful gas leakage, electrolyte degradation, growth of lithium dendrites, unfavorable long-term stability as a result of liquid evaporation, and low safety feature arising from the presence of flammable organic solvents (Ramesh et al., 2011; Yang et al., 2008). These electrolytes also have a limited operating temperature range, are challenging to handle and manufacture due to the liquid phase, and have a lesser shelf life with a likelihood of experiencing internal short circuits (Gray, 1997; Stephan and Nahm, 2006). As a result, researchers have proposed the adoption of solid-state electrolytes as a replacement for traditional liquid electrolytes (Famprikis et al., 2019).

5.2.2 Solid Electrolytes

Solid electrolytes can be defined as non-aqueous materials that exhibit high conductivity due to the rapid diffusion of one ionic species through a lattice formed by immobile counterions. The conduction of solid electrolytes arises from the mobile ions, not electrons as the solid electrolytes have negligible electrical conductivity. Solid electrolytes are also named fast ion conductors. Some literature referred them as superionic conductors and optimized ionic conductors (Takahashi, 1989). Solid electrolytes are typically classified into four categories: crystalline solid electrolytes, glassy electrolytes, molten electrolytes, and polymer electrolytes.

In general, crystalline solid electrolytes are referred to those that utilize multiple divalent and/or trivalent cations based on ionic hopping within the crystal structures. It is noted that conduction pathways are interspersed throughout the immobile ion sub-lattice of the material. The mobile ions move along these channels from site to another site. Common examples of this type of electrolyte include β-alumina, γ-Li_2ZnGeO_4, and $RbAg_4I_5$ (Gray, 1997).

On the other hand, glassy electrolytes are categorized as amorphous solid conductors that are formed when ion-containing liquids are rapidly cooled below their glass transition temperature (T_g). In such event, the crystallization process is avoided. These vitreous materials consist of three main components: network formers, network modifiers, and ionic salts. The conductors in these glassy materials are either pure cation or pure anion (Gray, 1997). Glassy electrolytes exhibit some advantages over crystalline electrolytes because of their unique glassy state. These advantageous features are isotropic conduction, elimination of grain boundaries, high ionic conductivity with extremely low electronic contribution, and ease of preparation (Fusco and Tuller, 1989). These conventional glassy conductors remain in a glassy state at ambient temperature with T_g above room temperature. The mobile ions are transported between different fixed vacant sites in a frozen disordered structure of glassy electrolytes to generate ionic conductivity (Karan et al., 2008). Examples of glass electrolytes are Li_2S-P_2S_5, Li_2S-B_2S_3, Li_2S-SiS_2, and ortho-thiophosphate (Na_3PS_4) (Hayashi et al., 2012; Deshpande 2009).

Molten electrolytes consist of molten single salt or molten eutectic mixtures and exhibit high ionic conductivity, usually above 1 S/cm. Some examples of commonly used molten electrolytes include LiCl-KCl eutectic mixtures and chloroaluminates ($AlCl_3$-MCl, where M is any alkali metal). Despite that, the use of molten electrolytes requires meeting specific technical requirements to prevent cell corrosion and leakage (Gray, 1991). Although crystalline and glassy electrolytes have high ionic conductivity feature, they are brittle. The remedy for this drawback is the application of polymeric materials. With the addition of polymeric materials, the mechanical properties of the electrolyte can be enhanced. Such enhancement is partly due to the ability of the polymer materials to accommodate volume changes. As a result, solid polymer electrolytes (SPEs) have become a viable option for electrochemical applications. SPEs are frequently used along with intercalation materials like electrodes in rechargeable lithium-type batteries. SPEs differ from glassy conductors in terms of T_g. Most of the SPEs have T_g below ambient temperature. A different ion hopping mechanism is observed for SPEs in comparison to glassy electrolytes, as mentioned by Karan and co-workers (Karan et al., 2008).

5.2.3 Solid-State Polymer Electrolytes

After World War II, the industrialized world experienced an economic downturn (Noto et al., 2011). The oil crisis in year 1973 remarked the end of the availability of cheap and abundant energy. Renewable energy sources are the breakthroughs to create new innovative devices for storage and energy conversion including new materials such as polymer electrolytes (Noto et al., 2011). In 1975, Wright, a polymer chemist from Sheffield, first developed conductive solid-state polymer electrolytes to address the limitations of traditional liquid electrolytes (Wright, 1975). SPEs have received an upsurge of interest in the electrochemical devices application since the year 1975 to improve the electrochemical performances of the

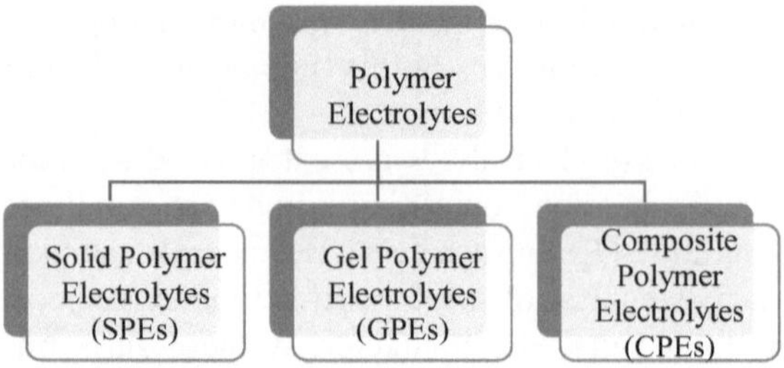

FIGURE 5.1 Sub-classes of polymer electrolytes.

devices. Solid-state polymer electrolytes can be categorized into several classes based on their composition and properties (Figure 5.1).

Solid-state polymer electrolyte plays three important roles in electrochemical devices. Polymer electrolyte serves as an electrode separator by insulating the opposite electrodes from each other. So, the addition of an inert porous spacer between the electrolyte and electrode interface is not required in this case. Polymer electrolyte plays an important role during the charging and discharging processes. It acts as a medium channel to generate ionic conductivity between the electrodes during the processes. Apart from that, it also works as a binder to ensure good interfacial contact with electrodes (Gray, 1991; Kang, 2004).

5.2.3.1 Solid Polymer Electrolytes

SPEs have become increasingly important in various applications, including energy storage and electrochemical displays, due to their flexible ion transportation properties (Wright, 1998). SPEs have been developed to replace conventional liquid electrolytes and offer enhanced safety due to their solvent-free condition. The use of SPEs in lithium-ion batteries has proven to be compatible with lithium metal electrodes, resulting in low self-discharge and closed-circuit discharge. Furthermore, SPEs are highly elastic and easy to handle, allowing for the creation of all solid-state electrochemical cells (Ibrahim et al., 2012). Compared to crystalline or glassy electrolyte-based cells, electrochemical cells based on SPEs exhibit commendable interfacial contact between the electrode and electrolyte, which can be maintained during charging and discharging processes (Gray, 1991, 1997). Additionally, SPEs do not generate internal pressure, which could lead to explosions during charge and discharge processes in electrochemical cells (Liew et al., 2013). According to Gray, the ionic transportation in polymer electrolytes relies on the local relaxation processes that occur within the polymer chains. These relaxation processes can provide similar properties to those of liquid electrolytes (Gray, 1991).

SPEs are unique systems that do not require solvents and instead rely on dissolving low-lattice-energy metal salts in the ion-coordinating molecules with aprotic solvents to create an ionically conducting pathway (Bruce and Vincent, 1993). The solvating group, or donor atom, of the polymer can form a covalent bond with the salt's cations for ion transportation. This coordination occurs through electrostatic interactions between the charges on the cation and the solvating group. Such dissociation of the coordination gives rise to ionic conduction in the polymer electrolytes. Armand's work has demonstrated new possibilities for solid-state ionic conduction, particularly in lithium battery fabrication (Armand, 1986). In his work, electrodes consisting of graphite intercalation compounds were coupled with polymer-salt complexes electrolytes. In addition to high safety performance, SPEs possess numerous advantages ranging from low requirement of vapor pressure to low volatility to high energy density, and other excellent properties (Gray, 1991, 1997; Adebahr et al., 2003; Nicotera et al., 2002; Ramesh and Liew, 2013; Stephan and Nahm, 2006).

SPEs possess numerous superior features, including inherent viscoelasticity, which is a mechanical behavior demonstrating both viscous and elastic properties. Besides having good mechanical properties, SPEs also have the ability to suppress the growth of lithium dendrite. Apart from that, they are lightweight, low in cost, and easy to handle and produce. One of the notable advantages of SPEs is their wide operating temperature range (Rajendran et al., 2004; Ramesh et al., 2011; Baskaran et al., 2007; Imrie and Ingram, 2000). Additionally, SPEs can be configured into various shapes, thanks to their highly flexible polymer membrane structure (Gray, 1991).

The first SPE generation which took in the form of crystalline poly(ethylene oxide) (PEO)–based polymer electrolytes was introduced by Wright's group (Fenton et al., 1973; Quartarone et al., 1998). Initially, they conducted research on the effect of alkali metal salts on PEO (Noto et al., 2011; Wright, 1975; Fenton et al., 1973). Given the good solvating properties exhibited by PEO, its electrolyte form has low ionic conductivity, falling in the range of 10^{-8}–10^{-7} S/cm. Such low value is attributed to its high crystallinity and tendency to recrystallize. Since then, the second generation of SPEs was developed to decrease the crystallinity of polymer electrolytes as the first generation of SPEs reached the same conclusions about the link between the amorphous region and the ionic conductivity. Various techniques have been employed to hinder polymer complexes from recrystallizing as well as to reduce the crystallinity degree. Examples of such techniques include modifying the polymer, blending polymers, and using semi-crystalline or amorphous polymers such as poly(methyl methacrylate) (PMMA). Other techniques include the incorporation of additives such as plasticizers, ionic liquids, and inorganic fillers.

One notable method of minimizing crystallization is structural modification. In this method, cross-linked is induced within the short chains of the ethylene oxide (EO) of the PEO (Quartarone et al., 1998). In one study, poly(dimethylsiloxane) (PDMS) was used to crosslink with PEO using an aliphatic isocyanate, resulting in copolymers with an ionic conductivity of approximately 10^{-5} S/cm after the inclusion of lithium perchlorate ($LiClO_4$) (Bouridah et al., 1985). Another study involved cross-linking PEO with poly(propylene oxide) (PPO), resulting in a LiClO4-doped polymer network. The produced polymer complex exhibited five folds of ionic conductivity compared to that of unlinked PEO polymer electrolytes (Watanabe et al., 1986). Additionally, a polyacrylonitrile-polyethylene oxide (PAN-PEO) copolymer was synthesized by Yuan et al., which showed higher ionic conductivity comparatively. The highest ionic conductivity of polymer electrolytes based on this copolymer and $LiClO_4$ (EO)/(Li) ratio of about 10 was recorded at 6.79×10^{-4} S/cm under ambient temperature (Yuan et al., 2005). Despite their good safety feature and high mechanical strength, SPEs still exhibit low ionic conductivity which hinders their expansion into electrochemical devices (Yang and Wu, 2022).

5.2.3.2 Gel Polymer Electrolytes

The third generation of gel polymer electrolytes (GPEs) was developed in the 1990s to address the low ionic conductivity of SPEs (Noto et al., 2011). GPEs are also known as gel ionic SPEs or plasticized PE. GPEs are formed via dissolution of polymer host with doping salt in organic solvents or plasticizers (Ramesh et al., 2012; Osinska et al., 2009; Rajendran et al., 2008). This immobilizes the liquid electrolyte in the polymer matrix, which gives GPEs its unique property (Han et al., 2002). GPEs possess both the cohesive properties of solids and the diffusive properties of liquids, with local relaxations that provide a liquid-like degree of freedom (Ramesh et al., 2012). Common components of GPEs include carbonate esters such as propylene carbonate (PC) or ethylene carbonate (EC), in high dielectric constant solvents like N,N-dimethylformamide (DMF) and γ-butyrolactone (Ramesh et al., 2011).

GPEs have a variety of inherent characteristics, including low interfacial resistance at the electrode–electrolyte boundary and low electrode materials reactivity. They also have better safety features along with shape flexibility. Additionally, they exhibit good ionic conductivity improvements with just little plasticizers (Ahmad et al., 2008; Pandey and Hashmi, 2009; Aruchamy et al., 2023). Compared to liquid electrolytes, GPEs offer improved electrochemical properties which include extending operating temperature conditions (Ahmad et al., 2005; Stephan et al., 2002). Besides being light in weight, GPEs are leakproof and easy to fabricate into any shape and size (Zhang et al., 2011). To improve GPE performance, plasticizers are commonly used, as seen in PEO-lithium trifluoromethanesulfonate ($LiCF_3SO_3$) electrolytes, where the addition of dibutyl phthalate (DBP) increases ionic conductivity about three orders from 7.13×10^{-7} S/cm to 6.03×10^{-4} S/cm (Sukeshini et al., 1998). Plasticizers can also increase the cationic transport number in the electrolyte system as seen in PEO-sodium metaphosphate ($NaPO_3$) added with PEG (Sukeshini et al., 1998).

However, GPEs also have some drawbacks. These include lower mechanical strength and poor dimensional stability. The slow evaporation of organic solvents during the fabrication process reduces the productivity of the GPE. Besides, they have reduced thermal, electrical, and electrochemical stabilities (Ramesh et al., 2011; Kim et al., 2006; Raghavan et al., 2010). GPEs are less compatible with lithium metal anode in lithium batteries because of their poor interfacial stability (Pandey and Hashmi, 2009). GPEs also exhibit poor electrochemical performances due to their narrow working potential window range and high vapor pressure (Ramesh et al., 2011; Kim et al., 2006; Raghavan et al., 2010). The shortcomings of traditional GPEs have led to the development of the next generation of polymer electrolytes based on composite.

5.2.3.3 Composite Polymer Electrolytes

Recently, researchers have shown a lot of interest in the fourth generation of SPEs, known as composite polymer electrolytes (CPEs). These polymer electrolytes are created by blending small amounts of organic or inorganic fillers with polymer electrolytes, resulting in what are referred to as nanocomposite polymer electrolytes (NCPEs) (Osinska, et al., 2009; Tang et al., 2021). By adding these fillers into GPEs, the physical and mechanical properties of polymer electrolytes are improved. CPEs offer many benefits, including good interfacial contact in the electrode–electrolyte region, high flexibility, enhanced ion transport, high ionic conductivity, and excellent thermodynamic stability toward lithium and other alkali metals (Gray, 1997). Additionally, they exhibit superior interfacial and electrochemical properties when used with a lithium metal anode (Stephan and Nahm, 2006). The capacity of lithium batteries can be increased by incorporating fillers, as shown in a study by Stephan and Nahm in 2006. CPEs are a promising option because of their excellent chemical, mechanical, thermal, and electrochemical stability (Noto et al., 2011). Some commonly used fillers include manganese oxide (MnO_2), titania (TiO_2), and alumina (Al_2O_3).

In the past, SiO_2 has been commonly used as a filler, and it has been found to significantly improve both the ionic conductivity and mechanical properties of PEO-based electrolytes. Fan et al. (2003) compared unmodified SiO_2 with silane-modified SiO_2 and found that the latter resulted in much higher ionic conductivity. Besides that, treatment of conventional fillers can inhibit recrystallization of PEO-based polymer electrolytes. In a study by Xi et al. (2005), solid acid sulfated-zirconia (SZ) was used as a filler in NCPEs based on PEO-$LiClO_4$, resulting in a two-order-of-magnitude increase in ionic conductivity compared to the pristine PEO-$LiClO_4$ polymer electrolytes. Furthermore, the ionic conductivity of polymer electrolytes with SZ was twice as high as untreated NCPEs (Xi et al., 2005). Nanosized organic–inorganic hybrid materials like zinc aluminate ($ZnAl_2O_4$) that have high surface area have also been introduced. When added to PEO-$LiClO_4$, 8 wt.% of $ZnAl_2O_4$ achieved higher ionic conductivity of 2.23×10^{-6} S/cm at ambient temperature (Wang et al., 2009). Such improvement was attributed to the reduction of crystallinity of the polymer membrane and the increase of lithium-ion transference number.

5.2.4 General Requirements of Solid-State Polymer Electrolytes for Electrochemical Devices Applications

Each electrochemical device application requires different basic requirements of solid-state polymer electrolytes. Solid-state polymer electrolytes with strong adhesive and superior chemical stability properties are highly recommended for all electrochemical device applications because these properties can reduce the interfacial contact between electrodes and the electrolyte. For a practical fuel cell application, proton-conducting polymer electrolytes are required. The protons will travel from the anode to the cathode to generate the electrical current in the fuel cell. In contrast, the main requirement of solar cells, particularly dye-sensitized solar cells, is the iodide electrolyte (Sagaidak et al., 2016). The iodide electrolyte will receive an electron from a positive electrode (or known as the cathode). The iodide (I^-)/triiodide (I_3^-) redox couple is regenerated on the cathode and thus the I_3^- is reduced to I^- in the mechanism. GPEs are commonly used as electrolytes in solar cell applications. On the contrary, ECWs application needs a transparent electrolyte film to observe the color changes on the electrolyte when voltage, light, or heat is applied. There are a few requirements that must take note for batteries, especially lithium batteries.

Solid-state polymer electrolytes must be free of water and inert to the reactive electrodes (the lithium electrodes for lithium batteries) for battery application to avoid explosion. The polymer electrolytes must have a wide electrochemical potential window (>3 V) which is the voltage range of a substance before undergoing decomposition. Unlike batteries, fuel cells, or solar cells, supercapacitors do not require any specific requirement. For supercapacitor applications, the charge carriers can be proton, lithium, sodium, magnesium, or potassium.

5.3 DESCRIPTION OF IONIC CONDUCTION

5.3.1 Mechanism of Ionic Conduction

In general, diffusion involves the movement of atoms, molecules, or ions within a solid phase (Raghavan, 2004). This motion allows for the attainment of uniform concentrations of the diffusing species between different locations (Kudo and Fueki, 1990). In an electrolyte, the primary means of diffusion is through the movement of mobile charge carriers when an electric field is present. When there is no electric field, the ions in the electrolyte undergo Brownian motion, but when a voltage is applied, they migrate in the direction of the electric field, a process referred to as ion conduction. However, if the ions are trapped within the crystalline phase of a solid matrix, which is an ordered arrangement of molecules, they must overcome a potential barrier in order to continue the diffusion process (West, 1999). The ion transport is restricted in the crystalline region. Although ions in a lattice structure vibrate constantly, they usually lack sufficient thermal energy to break free from their lattice sites. However, if the ions can surpass the barrier, they can move to neighboring lattice sites, leading to ionic conduction, migration, hopping, or diffusion. In polymer electrolytes, ion diffusion can occur via two primary mechanisms: the vacancy mechanism and the interstitial mechanism. Both mechanisms are graphically represented in Figures 5.2–5.4.

Vacancy mechanism is defined as the hopping mechanism of an ion from its normal position to an adjacent, but empty, site (Souquet et al., 2010). The ions require sufficient energy which arises from the thermal energy of ionic vibrations

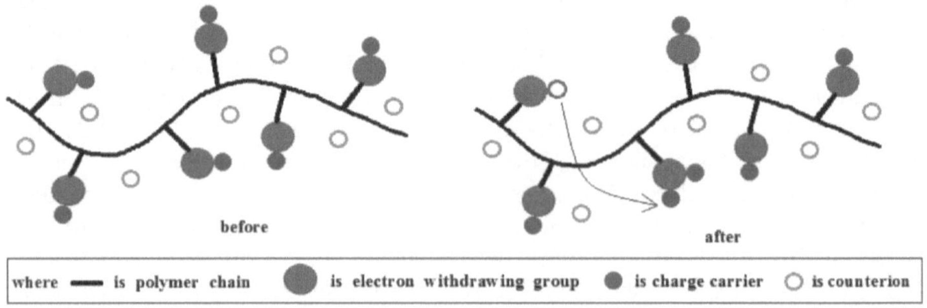

FIGURE 5.2 Illustration of ion diffusion before and after a vacancy mechanism. Adapted from Souquet et al. (2010).

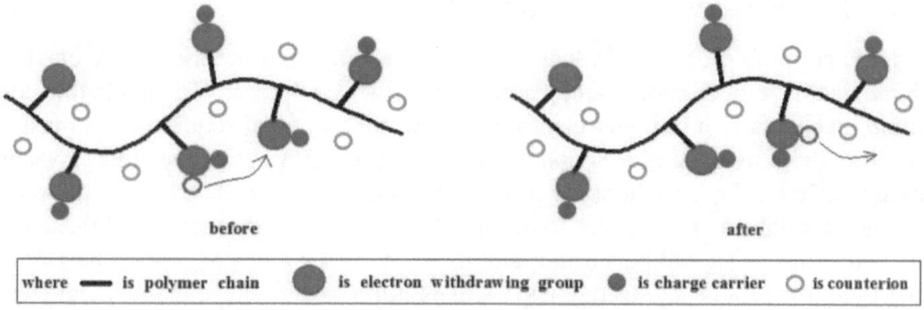

FIGURE 5.3 Graphical representation of ion diffusion of interstitial mechanism (Han et al., 2002).

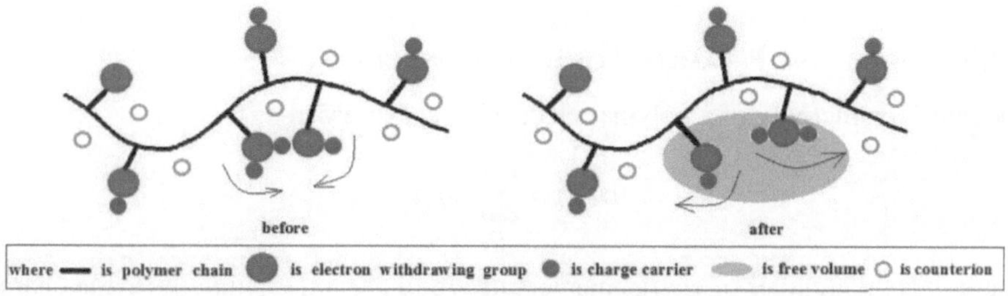

FIGURE 5.4 Graphical representation of ion diffusion of free volume mechanism coupled with the chain movement. Adapted from Souquet et al. (2010).

to break the coordination bonds and jump from a site to an adjacent empty site (Raghavan, 2004). On the other hand, the counterion from the salt can move freely in the electrolyte.

On the contrary, the interstitial mechanism involves the migration of a mobile charge carrier from one interstitial position to another, as represented in Figure 5.3 (Raghavan, 2004; Souquet et al., 2010; Smart and Moore, 2005). Interstitial diffusion rate is faster than vacancy diffusion. This is because the interstitials have weak bonding to the surrounding ions and there is a likelihood of empty adjacent interstitial site for the ions to move through during ion conduction (West, 1999).

Additionally, for polymer electrolytes in the above T_g condition, there is another mechanism of interstitial pair migration called the free volume mechanism, which is a cooperative process involving both interstitial pair migration and polymer segment mobility (Figure 5.4). During the local movement of polymer segments, a cage is created by the polymer's closest neighbors. This cage is subject to random density fluctuations or chain movements, which generate free volume within the polymer matrix. Subsequently, mobile species have the ability to move out of their original cage and into another when the fluctuation in polymer chain density creates an adjacent cage that is spacious enough for ion transport (Souquet et al., 2010).

5.3.2 Requirements of Ionic Conductivity

The ionic hopping process is influenced by the movement of mobile charge carriers, or ions, that have been dissociated from the polymer complexes. To enable this conduction mechanism in polymer electrolytes, five fundamental requirements must be met. First, there must be mobile ions that can migrate. Then, there should be vacant sites available for the ionic hopping process to occur, as ions will be mobile if there is a vacancy availability for the ions to occupy. Next, the potential energies of the empty and occupied sites should be similar, and there should be a low activation barrier for ions to jump from one site to the next. This is important because many vacant sites are of no use if the ions cannot enter due to small vacant sizes. Other than that, the structure should have a certain degree of framework that allows open pathways for mobile ions to diffuse. Finally, the framework should be highly polarizable as well (West, 1999).

5.3.3 Governing Parameters of Ionic Conduction

The ionic conductivity of a polymer electrolyte is expressed as follows:

$$\sigma(T) = \sum_i n_i q_i \mu_i$$

where n_i is the number of charge carriers of type i per unit volume, q_i is the charge of ions of type i, and μ_i is the ionic mobility of type i which is a measure of the drift velocity in a constant electric field (Gray, 1991, 1997; Smart and Moore, 2005). The amount of free charge carriers and their mobility in polymer electrolytes are key factors that strongly influence the ionic conductivity of such materials. To be considered conductive, polymer electrolytes must meet certain criteria, including the ability to detach more mobile charge carriers from coordination bonds and transport these charge carriers with high mobility. There are several factors that can affect these criteria, such as the degree of crystallinity and the dielectric constant of the polymer. In general, a low degree of crystallinity (or more amorphous) with a low T_g value is preferred. Aside from that, high ion mobility with high concentration of mobile ions is also important. In a polymer electrolyte system, crystallinity is referred to as a physical state in which solid materials have ordered and aligned arrangements of atoms, molecules, or ions, while amorphous refers to a state where these arrangements are random and disordered (Steven, 1999).

Pure crystalline polymer electrolytes are challenging for ion transportation because their molecules are tightly packed, making it difficult for ions to break coordinative bonds between cations from salt and electron-withdrawing groups from macromolecules. Therefore, the presence of crystallographic defects in polymer electrolytes can facilitate ionic conduction by creating amorphous regions where charge carriers can diffuse efficiently. When the polymer electrolyte reaches its glass transition temperature, T_g, it transitions from a glassy to a rubbery state in the amorphous region. Despite the short-range vibrations and rotations that occur below T_g, ion mobility remains restricted in the glassy state. Such restriction hinders the

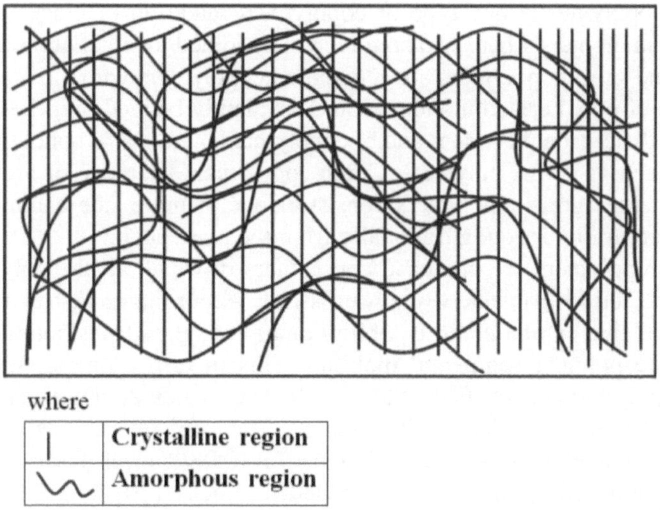

FIGURE 5.5 Illustration of mixed amorphous and crystalline regions in semi-crystalline polymer structure.

ion diffusion in polymer electrolytes. Above T_g, the rubbery state allows for long-range molecular motion, promoting segmental movement among the molecules or ions in the polymer chains. The electrostatic forces between the ions can be reduced by using polymers having high dielectric, which promotes increasing the distance between them and preventing ion aggregation and clustering (Eliasson et al., 2000). In a nutshell, polymer electrolytes with high dielectric constant stimulate ion diffusion and enhance ion hopping. The number density and mobility of charge carriers are largely affected by the flexibility of the polymer chains. This is because solvating ions can readily dissociate from the interactive bonds in polymer chains, thereby increasing ionic conduction in polymer electrolytes.

5.3.4 Methods of Enhancing Ionic Conduction Mechanism

The limited applicability of polymer electrolytes is attributed to the low ionic conductivity. To increase the ionic conductivity to approximately mS/cm, several methods have been employed, including blending different polymers, modifying polymers like irradiating with gamma rays or using mixed salt systems, and adding various additives (Li et al., 2023). Examples of commonly used additives are plasticizers, ionic liquids, and fillers.

5.3.4.1 Modifications of Polymer

The conduction route in electrolytes can be affected by the degree of polymer crystallinity. Because of this reason, researchers have made alterations to the polymer to enhance its amorphousness. To achieve this, they focused on developing various macromolecular architectures, including various forms of copolymers (block and comb copolymers, etc.) and network polymers.

A block copolymer is one type of copolymers made up of blocks of different polymerized monomers where two homopolymers joined together at the ends. Comb copolymer and graft copolymers are segmented copolymers with a linear backbone of one group of polymerized monomers and covalently bonded branches of another group. The difference between comb copolymers and graft copolymers arises from the branching of the polymers. Comb copolymers have two or more different branches, whereas graft copolymers consist of one or more side chains which are different, structurally or configurationally from the main chain.

Network copolymers are branched copolymers in which all the polymer chains are interconnected to form a network by crosslink. According to Fish et al., polymer-salt complexes that exhibit chain flexibility and lower glass transition temperatures could enhance polymer segmental mobility. This in turn promotes ion diffusion through the polymer matrix (Fish et al., 1988). Their study demonstrated that when poly(methylsiloxane) complexes with oligo(oxyethylene) side chains are bound to $LiClO_4$, an ionic conductivity value as high as 7×10^{-5} S/cm was observed. At higher temperature ranges, the conductivity exceeded 10^{-4} S/cm (Fish et al., 1988).

In a separate study, Soo et al. synthesized and electrochemically characterized block copolymer electrolytes based on poly(lauryl methacrylate)-b-poly(oligo (oxyethylene) methacrylate) in the rubbery condition (55). These resultant electrolytes had better ionic conductivity and dimensional stability. Compared to the glassy-type block copolymer systems, potential window could be enhanced up to 5 V. When this polymer was applied in the battery, the device showed high reversible capacity and excellent capacity retention (Soo et al., 1999).

Guilherme et al. developed a block copolymer electrolyte consisting of polyethylene-b-poly(ethylene oxide) (PE-b-PEO) and $LiClO_4$. The highest ionic conductivity of this polymer electrolyte was achieved in the sample with the addition of 15 wt.% $LiClO_4$, reaching 3×10^{-5} S/cm as measured at ambient temperature (Guilherme et al., 2007). At 100°C, the ionic conductivity increased to ~10^{-3} S/cm, but the degree of crystallinity of the polymer electrolytes was reduced due to block copolymerization (Guilherme et al., 2007).

Additionally, Lyons et al. (1996) synthesized polysilane comb polymers, $((CH_3CH_2OCH_2CH_2O(CH_2)_4)Si(CH_3))_n$, with ethoxyethoxybutane incorporated in the polymer's side chain, which served as a host polymer for the preparation of polymer electrolytes. The polymer electrolytes based on this comb polymer host system exhibited ionic conductivity of 1.2×10^{-7} S/cm at (Li)/(O) = 0.25 under ambient temperature. On the other hand, when the comb-branch copolymers were synthesized by copolymerization between poly(ethylene oxide methoxy) acrylate and lithium 1,1,2-trifluorobutane sulfonate acrylate, new fluorinated copolymers would be produced with enhanced ionic conductivity and lower T_g value (Cowie and Spence, 1999).

Polyethers with comb-shaped structures have been synthesized using poly(4-hydroxystyrene) (PHSt) as a multifunctional initiator and graft polymerization of EO or a mixture of EO and propylene oxide (PO). These polyethers, when combined with lithium triflate ($LiCF_3SO_3$), showed an ionic conductivity of approximately 10^{-5} S/cm at room temperature. The grafting process greatly reduced the crystallinity of these polymer electrolytes (Jannasch, 2000). In a separate study, Ding et al. used a combination of stable free radical polymerization (SFRP) and emulsion polymerization to synthesize graft copolymers comprising graft chains of macromonomer poly(sodium styrenesulfonate) (macPSSNa) and polystyrene (PS) backbone (Ding et al., 2002). These graft polymer electrolytes showed better proton conductivity compared to samples synthesized from random copolymers styrenesulfonic acid and styrene (PS–rPSSA), despite lower water uptake (Ding et al., 2002).

Last but not least, a network polymer based on poly(2-(2-methoxyethoxy)ethyl glycidyl ether) (PME2GE) has been used as a host polymer in random copolymer electrolytes containing $LiClO_4$ as the dopant salt. This system achieved a maximum ionic conductivity of approximately 10^{-4} S/cm at 40°C under the optimal composition of EO/ME2GE = 70/30 (Kono et al., 1993).

Polymer membranes with high ionic conductivity were synthesized using network polymers with hyperbranched ether side chains. To achieve this, a monosubstituted-epoxide monomer called 2-(2-methoxyethoxy)ethyl glycidyl ether (MEEGE) was copolymerized with EO. The copolymerization method employed the base-catalyzed anionic ring-opening polymerization with 2-(2-methoxyethoxy) ethanol. The synthesis resulted in the formation of semiterechelic poly(ethylene oxide-co-2-(2-methoxyethoxy)ethyl glycidyl ether) (P(EO/MEEGE)) oligomers, which were esterified with acrylic acid to produce polyether macromonomers. Network polymer electrolytes were prepared via photo cross-linking a mixture of the polyether macromonomer, lithium bis(trifluoromethylsulfonyl)imide (LiTFSI) salt, and a photo initiator. These electrolytes displayed ionic conductivity values of 1×10^{-4} S/cm at 30°C and 1×10^{-3} S/cm at 80°C (Nishimoto et al., 1999). However, this polymer modification technique was not found to be particularly effective in significantly improving the ionic conductivity of polymer electrolytes.

5.3.4.2 Polymer Blending

Blending polymers is a viable method to improve the ionic conductivity of polymer electrolytes. This process entails physically mixing two or more distinct polymers or copolymers without any covalent bonding. Typically, one of the materials in the mixture absorbs the active transporting species, while the other material provides mechanical support to the electrolyte system. In some cases, the inertness feature is contributed. Polymer blending has several benefits, including simplified sample preparation and the ability to regulate the physical properties of the polymer membrane through compositional changes (Rajendran et al., 2002). Additionally, polymer blending is cost-effective and doesn't require a polymerization process, unlike polymer modifications. The properties of polymer blends largely depend on the properties of the participating polymers, both chemical and physical. Not to forget if the resulting blend is homogenous or heterogeneous. If a few distinct polymers are capable of dissolving successfully in a typical solvent, they will create a homogenous blend or intermixing of the dissolved polymers, thanks to the rapid establishment of thermodynamic equilibrium (Braun et al., 2005).

Several dual polymer electrolyte systems have been created and studied, including PMMA-PVC (Stephan et al., 2002; Choi and Park, 2001; Rajendran et al., 2000), PVA-PMMA (Rajendran et al., 2004), PMMA-poly(vinylidene fluoride) (PVdF) (Cui et al., 2008), poly(vinyl acetate) (PVAc)-PMMA (Baskaran et al., 2006b), PVAc-PVdF (Baskaran et al., 2006a), PEO-PVdF (Yang et al., 2008), PVC-poly(ethyl methacrylate)

(PEMA) (Rajendran et al., 2008; Han et al., 2002), and PVA-poly(styrene sulfonic acid) (PSA) (Kumar and Bhat, 2009). In the application of polymer electrolytes, combination of PMMA and poly(vinyl chloride) (PVC) is widely used for the polymer host. PMMA polymer is amorphous in nature with brittle characteristics. To overcome this weakness, PVC is introduced as a mechanical stiffener. PVC's characteristic is attributed to the dipole–dipole interaction between its hydrogen and lone pair electrons of the chlorine atom (Ramesh et al., 2013).

The combination of PMMA and PVC can be observed through the change in the shape of the particles from spherical to rectangular, as observed from SEM images in the work of Ramesh et al. (2010). A polymer electrolyte consists of 70 wt.% PMMA and 30 wt.% PVC polymer blends with 10 wt.% LiTFSI exhibited a room-temperature ionic conductivity of 1.60×10^{-8} S/cm. Beyond this ratio, phase separation occurred in the polymer electrolyte. Consequently, PVC's globular agglomeration as a result of high PVC loadings could obstruct ion transport in the polymer electrolyte system (Ramesh et al., 2010).

Ionic conductivity in polymer electrolytes was found to be higher in other polymer blends compared to PMMA-PVC blend systems. For instance, Sivakumar et al. observed that a PMMA–PVdF blend GPE with a PVdF:PEMA ratio of 90:10 produces 1.50×10^{-4} S/cm of ionic conductivity value (Sivakumar et al., 2007). However, the ionic conductivity value decreased at higher PEMA content due to increased crystallinity in PEMA domains. Nevertheless, this GPE displayed favorable transport properties and better interfacial stability with Li electrodes.

Wu et al. (2009) developed a promising LiTFSI-based polymer electrolyte by blending hyperbranched polyether with poly(3-{2-(2-(2-hydroxyethoxy) ethoxy) ethoxy}methyl-3′-methyloxetane) (PHEMO), and poly(vinylidene fluoride-hexafluoropropylene) (PVDF-HFP) as a host polymer. This novel polymer electrolyte exhibited a maximum ionic conductivity of 1.64×10^{-4} S/cm at 30°C and has a wide electrochemical potential window up to 4.2 V vs. Li+/Li and high decomposition temperature above 400°C, making it a suitable candidate for use as an electrolyte in lithium-ion batteries (Rahaman et al., 2014). Unfortunately, the ionic conductivity is still less than mS/cm, prompting research in exploring other methods that can significantly increase the ionic conductivity of polymer electrolytes, in place of the polymer blending method.

5.3.4.3 Gamma Irradiation

Utilizing gamma-ray radiation as a means of enhancing the ionic conductivity of polymer electrolytes is a viable approach. When polymeric materials are exposed to ionizing radiation, it can generate reactive intermediate products like excited states, ions, and free radicals (Rahaman et al., 2014). These free radicals, which are produced due to gamma irradiation, have the ability to impact the structure of the polymer chains by means of intermolecular cross-linking or chain scission (Nanda et al., 2010; Sinha et al., 2008). Additionally, gamma irradiation can modify various properties of the polymer complexes (Rahaman et al., 2014). By absorbing high energy, gamma irradiation can restrain the crystalline region, alter the molecular weight distribution, boost the ionic conductivity, or even enhance the mechanical strength of the polymer electrolytes (Nanda et al., 2010; Damle et al., 2008; Ghosal et al., 2013).

In Song et al. work, PEO was cross-linked with $LiClO_4$ using gamma radiation, and subsequently, PVdF was blended with cross-linked PEO to produce polymer blend electrolytes (Song et al., 1997). The blend was then exposed to gamma radiation to create a simultaneous interpenetrating network (SIN). It is noted that these SIN polymer electrolytes induced by gamma radiation not only showed a high mechanical modulus of 10^7 Pa but also a room-temperature ionic conductivity exceeding 10^{-4} S/cm.

Tarafdar et al. also conducted gamma radiation research, developing polymer electrolytes based on gamma-irradiated PEO, and observed that ionic conductivity was substantially improved at 35 kGy of gamma radiation dose, which is attributed to the decreased crystallinity of the polymer (Tarafdar et al., 2010). An SPE consisting of γ-irradiated PVdF and lithium bis(oxalato)borate (LiBOB) achieved a maximum ionic conductivity of 3.05×10^{-4} S/cm, which is 15% higher than the polymer electrolyte prepared without gamma radiation (Rahaman et al., 2014). While gamma radiation is a promising method for enhancing ionic conductivity, high doses of it may degrade the polymer electrolytes (Akiyama et al., 2010).

5.3.4.4 Mix Salt System

The ionic conductivity of polymer electrolytes can be improved by mixing dual salts, which can prevent the formation of aggregates and clusters, thereby increasing the mobility of ion carriers (Gray, 1997). Additionally, this mixed dual-salt system can deliver more mobile charge carriers for ion transport, resulting in higher ionic conductivity compared to a single-salt system. Arof and Ramesh developed polymer electrolytes based on a dual-salt system consisting of PVC and $LiCF_3SO_3$, with lithium tetrafluoroborate ($LiBF_4$) as doping salts. Although the ionic conductivity only increased slightly compared to that single-salt system, the enhancement is attributed to the improved mobility of charge carriers by avoiding the aggregation effect (Arof and Ramesh, 2000).

Yang et al. also prepared a mixed salt system of GPEs consisting of PVdF/poly((ethylene glycol) diacrylate) (PEDGA)/PMMA and a mixture of salts comprising of lithium hexafluorophosphate ($LiPF_6$)/$LiCF_3SO_3$ (Yang et al., 2006). The ionic conductivity of the mixed salt system polymer electrolytes was five times higher than that of the $LiCF_3SO_3$-based polymer electrolyte system. With polymer electrolytes having 10 wt.% of $LiPF_6$ and 1 wt.% of $LiCF_3SO_3$, ionic conductivity as high as 1.5 mS/cm was observed. This polymer electrolyte also had a stable electrochemical potential range. However, the limitation of this technique is the solubility of dual salts in the same solvent, which requires researchers to ensure that both salts can be solubilized in the same solvent.

5.3.4.5 Additives

To enhance the ionic conductivity, different kinds of additives can be employed, including plasticizers, fillers, and ionic liquids.

5.3.4.5.1 Plasticizers

Plasticization is commonly acknowledged as an efficient technique to significantly enhance the ionic conductivity of polymer electrolytes by reducing their degree of crystallinity (Suthanthiraraj et al., 2009). A plasticizer, which is a low molecular weight and non-volatile aprotic organic solvent with a glass transition temperature close to −50°C (Ramesh et al., 2012), is used in plasticized GPE. Plasticizers such as PC and EC as shown in Figure 5.6 are widely used (Suthanthiraraj et al., 2009; Ning et al., 2009; Pradhan et al., 2005).

The impact of plasticizers on polymer electrolytes is determined by their properties such as its viscosity, dielectric constant, interaction with the polymer, and coordinative bonding with the ions (Rajendran and Sivakumar, 2008).

In general, the presence of plasticizers in polymer electrolytes improves the solvating degree of the salt, boosts ion mobility, and improves the contact between the electrolyte and the electrode interfaces (Rajendran et al., 2004; Ramesh and Arof, 2001). Moreover, plasticizers are advantageous additives due to their high miscibility with polymers and low viscosity which enhance processability (Ramesh and Chao, 2011). Besides that, they usually have a high dielectric constant. Adding plasticizers to the polymer electrolyte system is a proven technique for improving ionic conductivity without affecting other properties of the polymer electrolytes (Ganesan et al., 2008). Understanding the roles of plasticizers in enhancing the ionic conductivity of polymer electrolytes is crucial. The addition of plasticizers is expected to increase the ionic conductivity of polymer electrolytes by modifying the local structure, increasing the amorphous segment, and altering the local electric field distribution of the polymer matrix.

The primary role of plasticizers is to lower the T_g of polymer electrolytes. The effect of this is the reduction of polymer's modulus at the defined temperature. An ideal solid-state polymer electrolyte must have high ionic conductivity and excellent

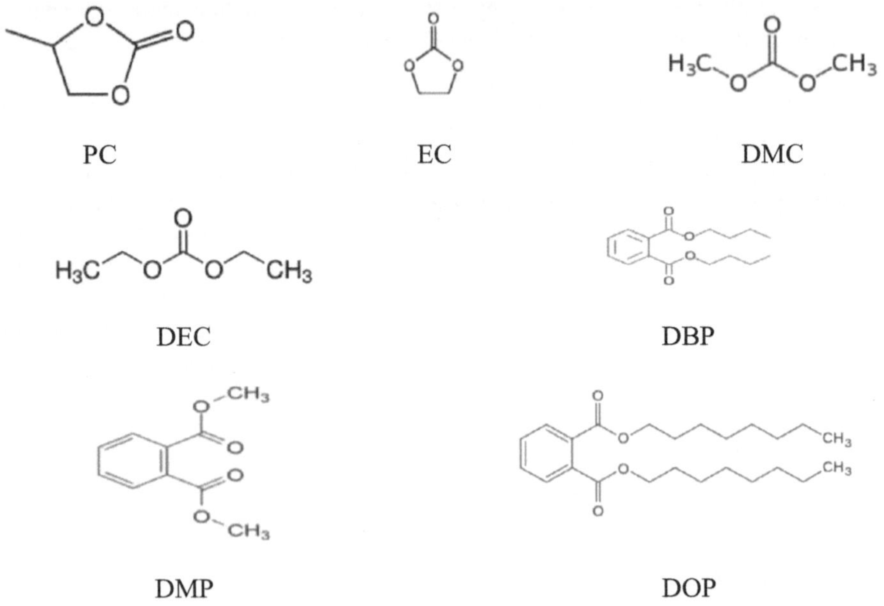

FIGURE 5.6 The chemical structure of some common plasticizers.

mechanical properties. However, the mechanical behavior of polymer electrolytes is reduced with mass loadings of plasticizers. Plasticizer undergoes a transition from a glassy state to a rubbery state at lower temperatures. In addition, plasticizers decrease the viscosity of the polymer system, facilitating the transport of ions within the polymer complexes. Besides, plasticizers can reduce the coordinative interactions within the polymer chains, improving their flexibility within the polymer matrix which favors ionic migration. This encourages the development of free volume within the polymer matrix, thereby enhancing the long-range segmental motion of the polymer system when it is swollen with a plasticizer (Ganesan et al., 2008).

To examine the effect of plasticizers on the ionic conductivity of the PEO-$LiClO_4$ polymer complex, three types of ester class plasticizers – dioctyl phthalate (DOP), DBP, and dimethyl phthalate (DMP) – were reviewed. Among them, DOP has better thermal stability, as demonstrated in differential thermal analysis. Michael's findings showed that the weight loss reduced as the plasticizer concentration increased (Michael et al., 1997).

In a study by Ali et al. (2007), plasticized-polymer electrolytes were synthesized from PMMA with PC or EC as the plasticizer, and LiTf or $LiN(CF_3SO_2)_2$ as the dopant salt. Their result showed that ionic conductivity increased with the concentration of the plasticizer. They also observed that PC plasticized-polymer electrolytes exhibited higher ionic conductivity compared to that of EC plasticized-polymer electrolytes.

Incorporating plasticizers in polymer electrolytes composed of PVA/PMMA-$LiBF_4$ was investigated by Rajendran et al. (2004). The highest ionic conductivity was observed at 1.29 mS/cm for the EC complex due to its high dielectric constant ($\varepsilon = 85.1$). Moreover, a maximum electrical conductivity of 2.60×10^{-4} S/cm at 300 K was achieved for electrolyte with 30 wt.% PEG as the plasticizer. Compared to the pure PEO-NaClO4 system, the ionic conductivity of the plasticized-polymer electrolyte increased by two orders of magnitude from 1.05×10^{-6} S/cm. This indicates that incorporation of plasticizer enhances the amorphous phase and, at the same time, reduces the energy barrier for ion transport, ultimately leading to higher mobility of lithium ions (Kuila et al., 2007). Despite its ability to extensively improve ionic conductivity, plasticization has some limitations. Among them are low safety performances and poor mechanical and thermal stabilities. Besides that, plasticization has a slow evaporation effect and produces a poor interfacial stability with lithium electrodes. In some cases, it also results in a narrow electrochemical window and low flash point (Ramesh et al., 2011; Pandey and Hashmi, 2009). Below is the summary of findings of some selected plasticized-polymer electrolyte systems (Table 5.1).

5.3.4.5.2 Ionic Liquids

Efforts have been made to address issues surrounding inadequate safety measures, subpar interfacial stability, and deficient electrochemical properties. To replace plasticizers, researchers have synthesized and created room-temperature ionic liquids (RTILs) in the past few years (Yim et al., 2015). Ionic liquids (ILs) are defined as non-volatile molten salts which usually have melting temperature below 100°C (Pandey and Hashmi, 2009). They are also regarded as molten salts that remain liquid at room temperature (Quartarone and Mustarelli, 2011). ILs consist of a bulky and asymmetric

TABLE 5.1
The Ionic Conductivity of Selected Plasticized-Polymer Electrolyte System

Polymer Electrolyte System	Ionic Conductivity (S/cm)	References
PEO/PPO-$LiCF_3SO_3$-EC	3.00×10^{-4} S/cm at 303 K	Morales and Acosta (1997)
PEO/PPO-$LiCF_3SO_3$-PC	3.80×10^{-4} S/cm at 303 K	Morales and Acosta (1997)
PEO/PPO-$LiCF_3SO_3$-(γ-butyrolactone (γ-BLA))	1.53×10^{-6} S/cm at 303 K	Morales and Acosta (1997)
PEO/PPO-$LiCF_3SO_3$-ethyleneglycol dimethylether (EGDME)	3.40×10^{-6} S/cm at 303 K	Morales and Acosta (1997)
PEO/polyphosphazene (PPz)-$LiCF_3SO_3$-EC	5.94×10^{-4} S/cm at 303 K	Morales and Acosta (1997)
PEO/polyphosphazene (PPz)-$LiCF_3SO_3$-PC	5.30×10^{-4} S/cm at 303 K	Morales and Acosta (1997)
PEO/polyphosphazene (PPz)-$LiCF_3SO_3$-γ-BLA)	1.23×10^{-6} S/cm at 303 K	Morales and Acosta (1997)
PEO/polyphosphazene (PPz)-$LiCF_3SO_3$-EGDME	1.06×10^{-6} S/cm at 303 K	Morales and Acosta (1997)
PVC/PMMA-$LiAsF_6$-DBP	3.97×10^{-5} S/cm at 304 K	Rajendran et al. (2000)
P(VdF-co-HFP)/PVAc-$LiClO_4$-(EC+PC)	2.30×10^{-3} S/cm at 298 K	Choi et al. (2001)
PEO/P(VdF-HFP)-$LiClO_4$-(EC+PC)	1.30×10^{-4} S/cm at 303 K	Fan et al. (2002)
P(VdF-co-HFP)-$LiClO_4$-(EC+PC)	1.25×10^{-3} S/cm at 303 K	Fan et al. (2002)
Chitosan-lithium acetate (LiOAc)-oleic acid (OA)	1.05×10^{-5} S/cm at room temperature	Yahya and Arof (2003)
$(PEO)_8$-$LiAsF_6$-dibutyl sebacate (DBS)	~10^{-6} S/cm at room temperature	Ragavendran et al. (2004)
Poly(epichlorohydrin-co-ethylene oxide), P(EPI-EO)-NaI-I_2-poly(ethylene glycol) methyl ether (P(EGME))	1.70×10^{-4} S/cm at room temperature	Nogueira et al. (2006)
PAN-$LiCF_3SO_3$-EC	1.32×10^{-3} S/cm at room temperature	Ahmad et al. (2011)
PAN-$LiCF_3SO_3$-PC	8.64×10^{-4} S/cm at room temperature	Ahmad et al. (2011)
PMMA-$NaClO_4$-DMF	1.94×10^{-6} S/cm at room temperature	Sekhar et al. (2012)
PMMA-$LiClO_4$-(PC+DEC)	0.40×10^{-5} S/cm at room temperature	Sharma et al. (2012)
PMMA-$LiClO_4$-DBP	5.64×10^{-3} S/cm at 353 K	Kuo et al. (2013)
Poly (viny pyrrolidone) (PVP)-sulfamic acid (SA)-PEG	4.26×10^{-5} S/cm at 303 K	Bella et al. (2014)
Carboxy methylcellulose (CMC)-OA-glycerol	1.64×10^{-4} S/cm at 303 K	Chai and Isa (2014)

organic cation and an anion. A range of ionic liquids exist, with examples of organic cations include 1,3-dialkylimidazolium, N-methyl-N-alkylpyrrolidinium, tetraalkyl-ammonium, and guanidinium, to name a few (Jain et al., 2005; Ye et al., 2013). In contrast, common inorganic anions such as acetate (CH_3COO^-), nitrate (NO^{3-}), triflate (Tf^-), tetrafluoroborate (BF^{4-}), bis(trifluoromethylsulfonyl imide) ($TFSI^-$), bis(perfluoroethyl sulfonyl) imide ($N(C_2F_5SO_2)^{2-}$), hexafluorophosphate (PF^{6-}), and halides (Cl^-, Br^-, and I^-) have been widely adopted in ILs.

ILs exhibit unique properties that are influenced by the combination of cations and anions (Jain et al., 2005; Vioux et al., 2010). These properties make them a promising choice for use in electrolyte systems. Among their notable features are a wide electrochemical potential window of up to 6 V, a broad temperature range for decomposition, and their eco-friendliness as they are non-toxic and non-volatile (Ramesh et al., 2011; Cheng et al., 2007; Patel et al., 2011; Pandey and Hashmi, 2013). Furthermore, RTILs possess outstanding characteristics such as excellent chemical, thermal, and electrochemical stabilities, high ionic conductivity due to a high concentration of ions, good oxidative stability, superior ion mobility, and a high cohesive energy density. These properties make them ideal for use in polymer electrolytes (Bella et al., 2014). ILs also have exceptional solubility for a wide range of organic, inorganic, and organometallic compounds, as well as superior safety performance (Vioux et al., 2010; Reiter et al., 2006).

The strong plasticizing effect of ILs makes the polymer chains in the electrolyte system to be flexible via softening of the polymer backbone. This promotes the movement of mobile charge carriers in the polymer matrix. As a result, ionic conductivity is increased. On top of that, the low-viscosity nature of the ILs is responsible for the disruption of the orderly arrangement of the backbone. This gives rise to reduced crystalline regions of the polymer matrix (Singh et al., 2009). This creates more voids and free spaces for ion migration, enhancing the ionic mobility in the polymer system. And likewise for the ionic conductivity value.

When the bulky cations are paired with anions in ILs, packing efficiency tends to be low. This facilitates ion detachment from the ionic compound, resulting in better ionic conductivity of the polymer electrolytes. The incorporation of ionic liquids also produces sticky GPE. Such GPEs are beneficial for the design and application of electrochemical devices where better contact between the electrolyte and electrode interfaces is made possible (Reiter et al., 2006). The immobilization of ILs within the polymer matrices enables the application in the solid state, where drawbacks related to shaping and the risk of leakage are minimized. The incorporation of ionic liquid in the polymer electrolytes can produce a highly conducting ionic liquids-integrated polymer framework as shown in Figure 5.7 (Park et al., 2013). The mobile charge carriers are thus dissociated from the temporary coordination bond with the electron-withdrawing group of polymers. These charge carriers are eventually transported to the adjacent empty site.

In the past few years, many researchers have extensively studied the impact of incorporating ionic liquids into polymer electrolytes. For example, Sirisopanaporn et al. (2009) developed flexible, transparent, and freestanding GPEs by embedding N-n-butyl-N-ethylpyrrolidinium N,N-bis(trifluoromethane)sulfonimide-lithium N,N-bis (trifluoromethane) sulfonamide (Py_{24}TFSI-LiTFSI) ionic liquid solutions in PVdF-co-HFP copolymer matrices. The produced membranes exhibited good ionic conductivity ranging from 0.34 to 0.94 mS/cm at room temperature and could operate up to 110°C without physical degradation or leakage of IL within 4 months of storage time. In another study, a proton-conducting PVdF-co-HFP copolymer membrane containing

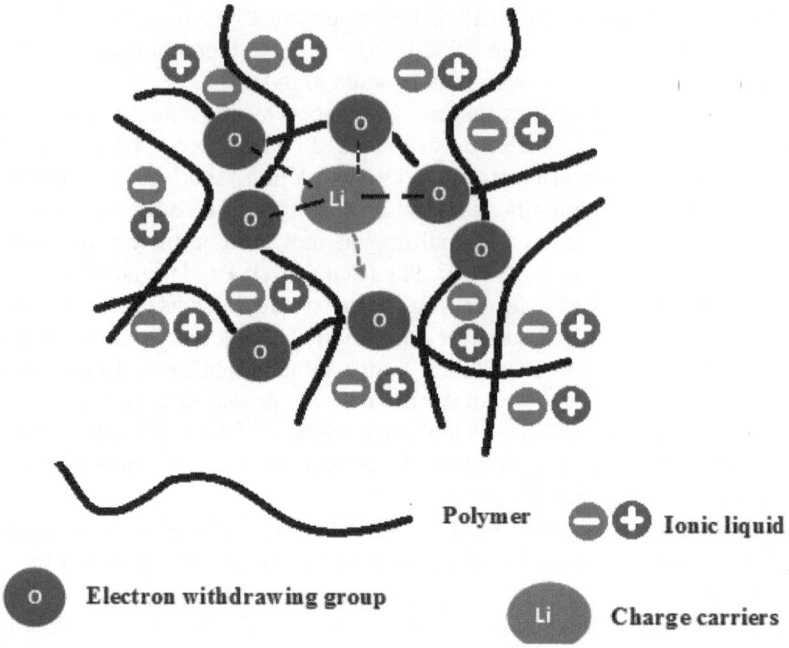

FIGURE 5.7 Schematic diagram of ion transport along the ionic liquid–integrated polymer framework. Adapted from Park et al. (2013).

2,3-dimethyl-1-octylimidazolium trifluromethanesulfonylimide (DMOImTFSI) was prepared. It showed good mechanical stability and achieved an ionic conductivity value of 2.74 mS/cm at 130°C (Sekhon et al., 2006).

IL is also incorporated into biodegradable polymers to form biopolymer electrolytes. For instance, IL plasticized-corn starch films were synthesized by Ning et al (Akiyama et al., 2010). The addition of 30 wt.% of 1-ally-3-methylimidazolium chloride (AmImCl) resulted in a maximum conductance of $10^{-1.6}$ S/cm. In our previous works, we prepared biopolymer electrolytes containing corn starch, $LiPF_6$, and ILs, namely 1-butyl-3-methylimidazolium hexafluorophosphate ($BmImPF_6$) or 1-butyl-3-methylimidazolium trifluoromethanesulfonate (BmImTf), using the solution casting technique (Ramesh et al., 2011; Liew and Ramesh, 2013, 2014). In both systems, the addition of ILs increased the ionic conductivity by three orders of magnitude, with the highest value observed at 1.47×10^{-4} S/cm for the sample with 50 wt.% of BmImTf (Singh et al., 2009). The Tf-based system exhibited even higher ionic conductivity, with a maximum of 3.21×10^{-4} S/cm (Liew and Ramesh, 2013, 2014). Doping ILs is an effective way to significantly enhance the ionic conductivity of polymer electrolytes without compromising their integrity.

5.3.4.5.3 Fillers and Nanofillers

Filler is a substance used to enhance the physical and mechanical characteristics of materials. When fillers are introduced into polymer electrolytes, they can create CPEs. NCPEs, on the other hand, are produced by dispersing nanometer-sized fillers into the polymer matrix. Nanotechnology, a novel field concerned with manipulating materials at nanoscale with dimension of smaller than 100 nm, has recently gained attention. Nanoparticles have a high surface area to volume ratio, making them an important driver in various fields due to their unique properties at nanoscale. Similarly, fillers also exhibit high activity and chemical stability (Yang et al., 2010). Krawiec and colleagues discovered that the size of filler particles is a crucial factor that determines the conductivity of polymer electrolytes (Krawiec et al., 1995). They reported that adding nanosized Al_2O_3 to polymer electrolytes increased the conductivity by about an order of magnitude compared to that of micrometer-sized Al_2O_3. This is attributed to the small size of the fillers that enhance the uniformity and electrochemical properties of the sample. The higher conductivity of samples with the addition of nanoscale fillers is also due to the rapid development of the space charge region between the grains in the system (Mei et al., 2008).

Fillers can be broadly categorized as either inorganic or organic. Some examples of inorganic fillers are calcium carbonate, fly ash, mica, clay, manganese oxide (MnO_2), TiO_2, ZrO_2, SiO_2, and alumina (Al_2O_3), while organic fillers include graphite fiber, aromatic polyamide, and cellulosic rigid rods (whiskers) (Samir et al., 2005). The growing interest in the growth of fillers has led to the creation of a new type of filler, the organic–inorganic hybrid, which combines both organic and inorganic phases. Examples of these hybrids include poly(cyclotri-phosphazene-co-4,40-sulfonyldiphenol) (PZS) microspheres (Zhang et al., 2010). Inorganic fillers can be further classified as either active or passive. Active fillers are materials that participate in the ionic conduction process, such as lithium-nitrogen (Li_2N), lithium aluminate ($LiAlO_2$), and ceramics-containing lithium, titanium nitrides (TiN_x), titanium carbides (TiC_x), and titanium carbonitrides (TiC_xN_y) (Ishkov and Sagalakov; 2005). In contrast, passive fillers do not have a role in the transport mechanism of the charge carrier (Adebahr et al., 2003; Giffin et al., 2012).

Various types of inorganic ceramic fillers are utilized in the preparation of polymer electrolytes. These include inert metal oxides such as titanium dioxide (TiO_2), zirconium dioxide (ZrO_2), alumina (Al_2O_3), and manganese dioxide (MnO_2), as well as other substances like treated silica (SiO_2), molecular sieves and zeolites, rare earth oxides, ferroelectric materials, solid superacids, nanoclay such as montmorillonite (MMT), carbon nanotubes (CNTs), and heteropolyacids such as silicotungstic acid (SiWA) and phosphotungstic acid (Jung et al., 2009; Noto et al., 2012; Zapata et al., 2012). Fillers can enhance the mechanical strength of polymer electrolytes, promote ionic conduction, and improve electrochemical properties while providing enhanced membrane stability (Zhang et al., 2010; Lue et al., 2008; Polu and Kumar, 2013). In addition, fillers offer several other benefits in polymer electrolyte development such as improving interfacial stability between electrode and electrolyte, reducing the glass transition temperature (T_g) of polymer membrane, increasing cationic diffusivity, improving the physical and morphological properties of polymer matrix, lowering interfacial resistance, reducing water retention, enhancing thermal stability, and improving the long-term electrochemical stability of polymer electrolytes (Krawiec et al., 1995; Samir et al., 2005; Kim et al., 2002; Saikia et al., 2009; Hammami et al., 2013; Jian-Hua et al., 2008).

Adding nanofillers to amorphous polymers allows them to maintain their liquid-like characteristics, resulting in faster ionic mobility at the microscopic level. This behavior not only improves the performance of energy storage devices by reducing capacity fading but also increases the electrochemical stability and long-term cycle life, as reported in several studies using CPEs or NCPEs (Raghava et al., 2008; Yang et al., 2012). For example, PEG-magnesium acetate ($Mg(CH_3COO)_2$) containing 10 wt.% of alumina achieved a maximum ionic conductivity of 3.45×10^{-6} S/cm, leading to a fabricated battery with a current density of 13.91 μA/cm^2, discharge capacity of 1.721 mA h, power density of 13.14 mW/kg, and energy density of 1.84 W h/kg with an open circuit voltage (OCV) of 1.85 V (Jung et al., 2009). In another study, nanoscale TiO_2 was distributed into a porous membrane made up of two copolymers, PVdF-co-HFP and poly (ethylene oxide-co-ethylene carbonate) (P(EO-co-EC)). Ionic conductivity as high as 5.1×10^{-5} S/cm was achieved at room temperature, slightly higher than the polymer electrolyte without the addition of TiO_2 nanoparticles. Jeon et al. reported these findings in 2006 (Jeon et al., 2006).

Ketabi and Lian examined the impact of amorphous SiO_2 nanofillers on the ionic conductivity and crystallinity of PEO-based electrolytes (Ketabi and Lian, 2013). They found that the ionic conductivity of the polymer-ionic liquid electrolyte system, which consisted of 1-ethyl-3-methylimidazolium hydrogensulfate ($EMIHSO_4$) and nanosized SiO_2, was able to achieve 2.15 mS/cm. This represents a more than two-fold increase over the electrolyte that did not contain fillers (Nordström et al., 2012). SiO_2 is a hydrophilic compound and prefers to form silanol (Si–OH) bonding on the surface of the silicon atom of silica as illustrated in Figure 5.8 (Ahmad et al., 2005).

Formation of hydrogen bonding between SiO_2 grains is favorable with the presence of Si–OH bond when a sufficient amount of SiO_2 is embedded into the polymer matrix. This will lead to the formation of a three-dimensional network whereby silica is immobilized in the liquid electrolyte (Ahmad et al., 2005). The charge carriers are thus accumulated on the silica surface as shown in Figure 5.9a. Therefore, the charge carriers are expelled from the surface easily when the hydrogen bonding between SiO_2 grains is formed as illustrated in Figure 5.9b (Nordström et al., 2012). As a result, the charge carriers can transport freely in the electrolyte.

Johan and Ting incorporated a novel nano-filler, specifically nanosized MnO_2, into a PEO-$LiCF_3SO_3$-DBP polymer system. The resulting NCPE achieved an optimal

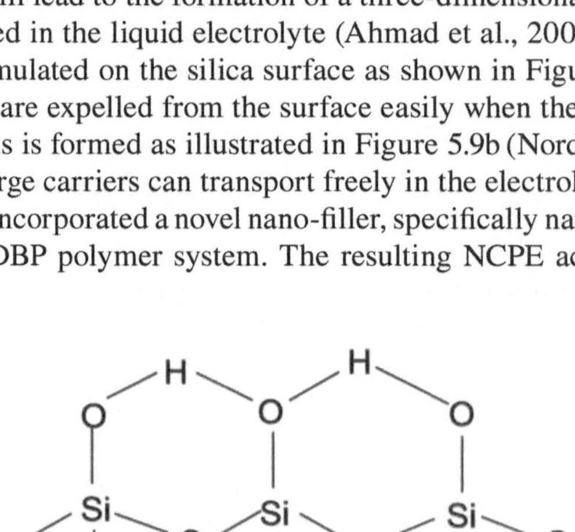

FIGURE 5.8 The formation of three-dimensional network with the silanol bonding.

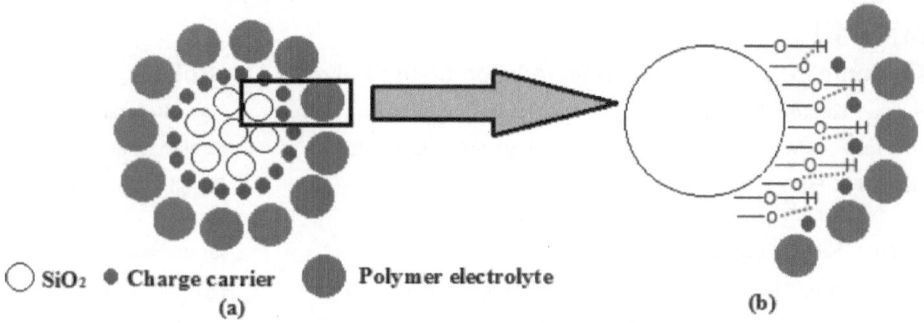

FIGURE 5.9 (a) The immobilization of silica in the electrolyte and (b) the formation of hydrogen bonding between silica grains. Adapted from Nordström et al. (2012).

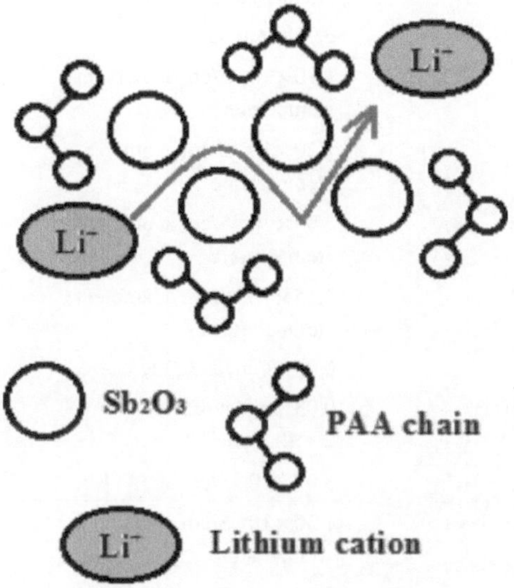

FIGURE 5.10 Formation of a conducting pathway between Sb_2O_3 particles. Adapted from Kam et al. (2014).

conductivity of 4.2×10^{-4} S/cm at room temperature when the MnO_2 was dispersed at a concentration of 12 wt.% (Johan and Ting, 2011). A new material namely antimony trioxide (Sb_2O_3) is also investigated by our peer (Kam et al., 2014). We found out that the ionic conductivity of poly(acrylic acid) (PAA)–based polymer electrolytes is increased rapidly at high concentration of fillers as reported in our publication. The filler grains could be distributed evenly throughout the film when more fillers are added to the polymer system. This presents an extra conducting path (also known as tunnel) which is accessible for ionic transportation around the surface of particles. As a result, the mobile charge carriers could diffuse within the vicinity of grains through this conducting trail as illustrated in Figure 5.10. The ionic conductivity of some selected CPEs is listed in Table 5.2.

TABLE 5.2
The Ionic Conductivity of Selected Composite Polymer Electrolytes

Polymer Electrolyte System	Ionic Conductivity (S/cm)	References
PEO-$LiBF_4$-MMT	1.60×10^{-6} S/cm at 303 K	Krawiec et al. (1995)
PEO-$LiN(CF_3SO_3)_2$-SiO_2	1.40×10^{-4} S/cm at ambient temperature	Capiglia et al. (1999)
$(PEO)_9$ $LiN(CF_3SO_3)_2$-Al_2O_3	2.20×10^{-4} S/cm at 298 K	Jayathilaka et al. (2002)
$(PEO)_9Cu(CF_3SO_3)_2$-Al_2O_3	9.07×10^{-7} S/cm at 301 K	Dissanayake et al. (2005)
PVdF-co-HFP-lithium fluoroalkyl phosphate (LiFAP)-(EC+DEC)-SiO_2	1.13×10^{-3} S/cm at ambient temperature	Aravindan and Vickraman (2007)
$(PEO)_9LiCF_3SO_3$-EC-Al_2O_3	1.50×10^{-4} S/cm at ambient temperature	Pitawala et al. (2007)
PVA-KOH-TiO_2	0.048 S/cm at 303 K	Yang et al. (2008)
PEO-$LiClO_4$-hydrotalcite	1.10×10^{-5} S/cm at ambient temperature	Borgohain et al. (2010)
PEO-$LiClO_4$-PZS	1.20×10^{-5} S/cm at ambient temperature	Zhang et al. (2010)
PMMA-$LiCF_3SO_3$-EC-Al_2O_3	2.05×10^{-4} S/cm at ambient temperature	Chew and Tan (2011)
PMMA-$LiCF_3SO_3$-EC-SiO_2	2.15×10^{-5} S/cm at ambient temperature	Chew and Tan (2011)
PVA-KOH-SiO_2	0.035 S/cm at 303 K	Yang et al. (2011)
Poly(vinylphosphonic acid) (PVPA)-sulfated TiO_2	0.03 S/cm at 423 K	Aslan and Bozkurt (2012)
Nafion-sulfated ZrO_2	3.00×10^{-3} S/cm at 393 K	Griffin et al. (2012)
PEO-$LiPF_6$-EC-amorphous CNTs	1.30×10^{-3} S/cm at ambient temperature	Ibrahim and Johan (2012)
Poly(ethyleneimine) (PEI)-$LiN(CF_3SO_3)_2$-SiO_2	3.80×10^{-5} S/cm at ambient temperature	Pehlivan et al. (2012)
Polysaccharide (agarose)-LiI-I_2-TiO_2	5.12×10^{-4} S/cm at ambient temperature	Raghava et al. (2008)
PEO-$LiClO_4$-Al_2O_3	6.45×10^{-4} S/cm at 293 K	Masoud et al. (2013)
PVdF-$LiPF_6$-(EC+DEC)-MMT	3.08×10^{-3} S/cm at ambient temperature	Prasanth et al. (2013)
Poly(ethylene oxide-co-2-(2-methoxyethoxy)ethyl glycidylether) P(EO/EM_2)-$LiClO_4$-TiO_2	1.40×10^{-5} S/cm at 303 K	Tominaga and Endo (2013)
PEO-tetraethylammonium tetrafluoroborate ($TEABF_4$) salt-organically modified nanoclay (MNclay)	1.0×10^{-4} S/cm at 303 K	Sivaraman et al. (2015)
PVA-PMMA-NH_4SCN-nanosized multiferrioc ($BiFeO_3$) (BFO)	2.9×10^{-2} S/cm at 293 K	Nileshrai et al. (2015)

Although the addition of fillers is a valuable approach for enhancing ionic conduction, the degree of improvement is currently limited. Based on the findings, the combined effect of plasticizer and filler can have a good impact on the ionic conductivity of polymer electrolytes.

5.4 FUTURE ASPECTS

5.4.1 Liquid Crystal Polymer Electrolytes (LCPEs)

Liquid crystal polymer electrolytes (LCPEs) represent a novel type of polymer electrolytes that replace common polymers with liquid crystal polymers (LCPs) as host polymers (Álvarez Moisés et al., 2023). In conventional polymer electrolytes, the transport of ions is achieved through segmental motions of the host polymers, leading to a significant increase in ionic conductivity above the glass transition temperature (T_g) only (Imrie et al., 2004). However, lowering the T_g to sub-ambient temperatures greatly reduces the mechanical stability of these electrolytes (Imrie and Ingram, 2000). To address this issue, LCPs have been developed, which exhibit high ionic conductivity in both glassy and liquid crystal phases (Imrie et al., 2004). Unlike amorphous polymer complexes, where ionic motion is coupled to structural relaxations, LCPs' liquid crystalline behavior can decouple the ion mobility from the structural relaxations (McHattie et al., 1998). Furthermore, LCPs offer additional benefits due to a combination of anisotropic and exceptional bulk properties, along with new possibilities for polymer processing (Park et al., 2010).

LCPs are polymers that contain mesogen, which are the fundamental functional groups of liquid crystals that promote structural order in crystals. These LCPs can be classified into two main categories based on whether the mesogens are located in the backbone or side chain of the polymer. These categories include main chain-liquid crystal polymers (MCLCPs) and side chain-liquid crystal polymers (SCLCPs) (refer Figure 5.11).

LCPs that contain mesogenic units in the backbone are referred to as MCLCPs. The mesomorphic properties of MCLCPs depend on the polymer's chain flexibility and structure. By incorporating mesogenic units into an ionically conducting polymer backbone, MCLCP electrolytes can be created (McHattie et al., 1998).

In contrast, SCLCPs are LCPs that have mesogenic groups in the side chain. The preparation of SCLCPs requires three main components: the polymer backbone, mesogenic units, and a space connecting the mesogenic units (Felipe, 2009).

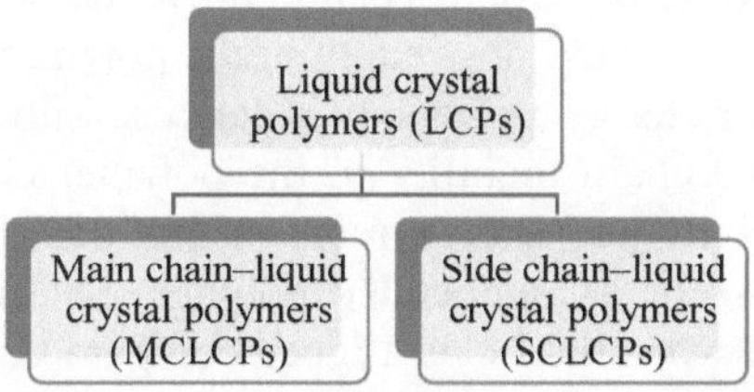

FIGURE 5.11 Two main types of liquid crystal polymers.

When an ion-coordinating cyclic macromolecule is inserted into the mesogenic units that are attached to the polymer backbone through flexible spacers, SCLCP electrolytes can be produced (McHattie et al., 1998). The mesomorphic properties of SCLCPs are influenced by various structural factors such as the flexibility of the polymer backbone, the repeating unit length, and the spacer length between the side chains of the polymer. In SCLCPs, the molecular mobility of the polymeric backbone and mesogenic groups are linked through the flexible spacer, which requires the decoupling of ionic motion from structural relaxation for effective ionic transport in polymer electrolytes. Increasing the mentioned structural factors can enhance the degree of decoupling, ultimately leading to a decrease in the T_g of polymer electrolytes and promoting the ionic hopping mechanism within the polymer matrix.

The development of ionically conducting LCPEs has been significantly advanced by Imrie and her research group (Imrie and Ingram, 2000; Imrie et al., 2004; McHattie et al., 1998). McHattie et al. created a new mesogenic liquid crystalline side-chain polymer with predominantly PEO backbone and mesogenic groups attached as pendants through flexible alkyl spacers (McHattie et al., 1998). PEO-$LiClO_4$-based liquid crystalline systems exhibit higher ionic conductivity and different conductivity behavior compared to pristine PEO-$LiClO_4$ polymer complexes. Tong et al. synthesized novel star-branched amphiphilic liquid crystal copolymers based on PEO with cyanobiphenyl mesogenic pendants (MA_xLC) using atom transfer radical polymerization (ATRP) (Tong et al., 2012). The insertion of mesogenic groups significantly increased the ionic conductivity of polymer electrolytes by providing an efficient ion-conducting pathway and suppressing polymer crystallization to promote movement (Tong et al., 2012). LCPEs have also been employed in electrochemical devices, with Park et al. fabricating dye-sensitized solar cells (DSSCs) using a new series of SCLCPs that achieved a maximum power conversion efficiency (PCE) of 4.11% and better photovoltaic performance (Park et al., 2010). Extensive research has been conducted to understand the impact of the liquid crystalline region on the ionic conductivity of polymer electrolytes for use in electrochemical devices (Chai et al., 2021).

5.4.2 Liquid Crystals-Embedded Polymer Electrolytes

Polymer electrolytes containing liquid crystals (LCs) are a new type of SPE that offers promising characteristics for use in electrochemical devices. LCs, which can form a mesophase with desirable properties such as high charge carrier mobility and conductivity, are a potential replacement for volatile liquids in electrolytes. The alignment of LCs in polymer electrolytes can improve the photovoltaic performance of DSSCs by increasing the ordering strength of the polymeric material, which provides a better pathway for charge carriers. LCs can also initiate the iodide exchange reaction and increase the chain mobility in the polymer electrolytes, resulting in improved ionic conductivity and short-circuit current density (J_{sc}). Several studies have reported on the use of LC-embedded polymer electrolytes in DSSCs, with increased PCE observed when the LC is aligned compared to when it is not aligned or in conventional DSSCs. Further research is needed to fully understand the effects of the LC region on the ionic conductivity of these polymer electrolytes.

There are literature works discussing polymer electrolyte system that incorporates LC and its application in DSSCs (Ahn et al., 2012; Kim et al., 2010; Vijayakumar et al., 2009). The study shows that the addition of E7 LC to PVdF-co-HFP-based polymer electrolytes significantly enhances the ionic conductivity of the GPE, with a value of 2.9 mS/cm at ambient temperature, 37% higher than the polymer gel electrolyte without E7. The assembled DSSCs using this LC-embedded polymer electrolyte show comparable performance to those using liquid electrolytes, with a PCE of 6.82% at 1 sun intensity. Karim et al. also fabricated DSSCs using LC-embedded polymer electrolytes, achieving a PCE of 4.70% at 1 sun intensity, which is much higher than the PAN-based polymer electrolyte (Karim et al., 2010). Although the effect of LCs on the ionic conductivity of polymer electrolytes is not fully understood, the incorporation of LC in polymer electrolytes is a promising approach to improve their ionic conductivity in the future.

5.5 CONCLUSIONS

Polymer electrolytes are a viable alternative to hazardous liquid electrolytes. The requirements of an ideal polymer electrolyte are high ionic conductivity (> mS/cm), excellent thermal stability, wide electrochemical potential window, and superior electrochemical performances in cell fabrication. Development and improvement are still needed to overcome the current shortcomings of SPEs. Perhaps, SPE can be a good replacement of the current liquid electrolytes. Fuel cells offer a potentially non-polluting and renewable way to generate electricity. PEMFCs offer several advantages over the batteries and supercapacitors. Fabrication of a fuel cell with high power density, long durability, and cost-effective properties is essential for the economic application of sustainable energy technologies. The research and development (R&D) of PEMFCs is still essential to harness the capabilities of hydrogen for energy production to the maximum threshold and operational advantages prior to commercialization. Innumerable challenges and technological breakthroughs are still the obstacles in the R&D of PEMFCs.

ACKNOWLEDGMENT

This work was partly supported by the Fundamental Research Grant Scheme (FRGS/1/2019/STG07/TARUC/02/1) from the Ministry of Higher Education of Malaysia.

REFERENCES

Adebahr J, Byrne N, Forsyth M, MacFarlane DR, Jacobsson P (2003) Enhancement of ion dynamics in PMMA-based gels with addition of TiO_2 nano-particles. *Electrochim Acta* 48: 2099–2103.

Ahmad A, Isa KB, Osman Z (2011) Conductivity and structural studies of plasticized polyacrylonitrile (PAN)-lithium trifate polymer electrolyte films. *Sains Malaysiana* 40: 691–694.

Ahmad S, Ahmad S, Agnihotry SA (2005) Nanocomposite electrolytes with fumed silica in poly(methyl methacrylate): Thermal, rheological and conductivity studies. *J Power Sources* 140: 151–156.

Ahmad S, Deepa M, Agnihotry SA (2008) Effect of salts on the fumed silica-based composite polymer electrolytes. *Solar Energy Mater Solar Cells* 92: 184–189.

Ahn SK, Ban T, Sakthivel P, Lee JW, Gal Y-S, Lee J-K, Kim M-R, Jin S-H (2012) Development of dye-sensitized solar cells composed of liquid crystal embedded, electrospun poly(vinylidene fluoride-co-hexafluoropropylene) nanofibers as polymer gel electrolytes. *ACS Appl Mater Interfaces* 4: 2096–2100.

Akiyama Y, Sodaye H, Shibahara Y, Honda Y, Tagawa S, Nishijima S (2010) Study on gamma-ray-induced degradation of polymer electrolyte by pH titration and solution analysis. *Polym Degrad Stab* 95: 1–5.

Ali AMM, Yahya MZA, Bahron H, Subban RHY, Harun MK, Atan I (2007) Impedance studies on plasticized PMMA-LiX (X: CF_3SO_3-, $N(CF_3SO_2)^{2-}$) polymer electrolytes. *Mater Lett* 61: 2026–2029.

Álvarez Moisés I, Innocenti A, Somville M, Notredame B, Passerini S, Gohy, JF (2023). Liquid crystals as additives in solid polymer electrolytes for lithium metal batteries. *MRS Adv* 8: 797.

Aravindan V, Vickraman P (2007) Polyvinylidenefluoride-hexafluoropropylene based nanocomposite polymer electrolytes (NCPE) complexed with $LiPF_3(CF_3CF_2)_3$. *Eur Polym J* 43: 5121–5127.

Armand MB (1986) Polymer electrolytes. *Ann Rev Mater Res* 16: 245–261.

Arof AK, Ramesh S (2000) Electrical conductivity studies of poly(vinylchloride) based electrolytes with double salt system. *Solid State Ionic*s 136–137: 1197–1200.

Aruchamy K, Ramasundaram S, Divya S, Chandran M, Yun K, OhTH (2023). Gel polymer electrolytes: Advancing solid-state batteries for high-performance applications. *Gels* 9: 585.

Aslan A, Bozkurt A (2012) Nanocomposite polymer electrolyte membranes based on poly(vinylphosphonic acid)/sulfated nano-titania. *J Power Sources* 217: 158–163.

Baskaran R, Selvasekarapandian S, Kuwata N, Kawamura J, Hattori T (2006a) Conductivity and thermal studies of blend polymer electrolytes based on PVAc-PMMA. *Solid State Ionics* 177: 2679–2682.

Baskaran R, Selvasekarapandian S, Kuwata N, Kawamura J, Hattori T (2006b) Ac impedance, DSC and FT-IR investigations on (x)PVAc-(1- x)PVdF blends with $LiClO_4$. *Mater Chem Phys* 98: 55–61.

Baskaran R, Selvasekarapandian S, Kuwata N, Kawamura J, Hattori T (2007) Structure, thermal and transport properties of PVAc-$LiClO_4$ solid polymer electrolytes. *J Physics Chem Solids* 68: 407–412.

Bella RSD, Hirankumar G, Devaraj P (2014) Characterization of plasticized proton conducting polymer electrolyte and its application in primary proton battery. *Int J ChemTech Res* 6: 5372–5377.

Borgohain MM, Joykumar T, Bhat SV (2010) Studies on a nanocomposite solid polymer electrolyte with hydrotalcite as a filler. *Solid State Ionics* 181: 964–970.

Bouridah A, Dalard F, Deroo D, Cheradame H, LeNest JF (1985) Poly(dimethylsiloxane)-poly(ethylene oxide) based polyurethane networks used as electrolytes in lithium electrochemical solid state batteries. *Solid State Ionics* 15: 233–240.

Braun D, Cherdron H, Rehahn M, Ritter H, Voit B (2005) Methods and techniques for synthesis, characterization, processing, and modification of polymers. In: *Polymer Synthesis: Theory and Practice*, Springer, Berlin, Germany, Vol. 1, pp 33–147.

Bruce PG, Vincent CA (1993) Polymer electrolytes. *J Chem Soc Faraday Trans* 89: 3187–3203.

Capiglia C, Mustarelli P, Quartarone E, Tomasi C, Magistris A (1999) Effects of nanoscale SiO_2 on the thermal and transport properties of solvent-free, poly(ethylene oxide) (PEO)-based polymer electrolytes. *Solid State Ionics* 118: 73–79.

Chai MN, Isa MIN (2014) Electrical study of plasticized carboxy methylcellulose based solid polymer electrolyte. *Int J Phys Sci* 9: 397–401.

Chai S, Xu F, Zhang R, Wang X, Zhai L, Li X, Qian HJ, Wu L, Li H (2021). Hybrid liquid-crystalline electrolytes with high-temperature-stable channels for anhydrous proton conduction. *J Am Chem Soc* 143, 21433–21442.

Cheng H, Zhu C, Huang B, Lu M, Yang Y (2007) Synthesis and electrochemical characterization of PEO-based polymer electrolytes with room temperature ionic liquids. *Electrochim Acta* 52: 5789–5794.

Chew KW, Tan KW (2011) The effects of ceramic fillers on PMMA-based polymer electrolyte salted with lithium triflate, $LiCF_3SO_3$. *Int J Electrochem Sci* 6: 5792–5801.

Choi NS, Lee YG, Park JK, Ko JM (2001) Preparation and Electrochemical characteristics of the plasticized polymer electrolyte based on P(VdF-co-HFP)/PVAc blend. *Electrochim Acta* 46: 1581–1586.

Choi NS, Park JK (2001) New polymer electrolytes based on PVC/PMMA blend for plastic lithium-ion batteries. *Electrochim Acta* 46 1453–1459.

Cowie JMG, Spence GH (1999) Novel single ion, comb-branched polymer electrolytes. *Solid State Ionics* 123: 233–242.

Cui ZY, Xu YY, Zhu LP, Wei XZ, Zhang CF, Zhu BK (2008) Preparation of PVDF/PMMA blend microporous membranes for lithium ion batteries via thermally induced phase separation process. *Mater Lett* 62: 3809–3811.

Damle R, Kulkarni PN, Bhat SV (2008) Study of effect of composition, irradiation and quenching on ionic conductivity in $(PEG)_x$:NH_4NO_3 solid polymer electrolyte. *Bull Mater Sci* 31: 869–876.

Deshpande VK (2009) Science and technology of glassy solid electrolytes. *IOP Conf Ser Mater Sci Eng* 2: 12011–12015.

Ding J, Chuy C, Holdcroft S (2002) Enhanced conductivity in morphologically controlled proton exchange membranes: Synthesis of macromonomers by SFRP and their incorporation into graft polymers. *Macromolecules* 35: 1348–1355.

Dissanayake MAKL, Jayathilaka PARD, Bokalawela RSP (2005) Ionic conductivity of PEO_9:$Cu(CF_3SO_3)_2$:Al_2O_3 nano-composite solid polymer electrolyte. *Electrochim Acta* 50: 5602–5605.

Eliasson H, Albinsson I, Mellander BE (2000) Conductivity and dielectric properties of $AgCF_3SO_3$-PPG. *Mater Res Bull* 35: 1053–1065.

Famprikis T, Canepa P, Dawson JA, Islam MS and Masquelier C (2019). Fundamentals of inorganic solid-state electrolytes for batteries. *Nat Mater* 18, 1278–1291.

Fan L, Dang Z, Nan C-W (2002) Thermal, electrical and mechanical properties of plasticized polymer electrolytes based on PEO/P(VDF-HFP) blends. *Electrochim Acta* 48: 205–209.

Fan L, Nan C-W, Zhao S (2003) Effect of modified SiO_2 on the properties of PEO based polymer electrolytes. *Solid State Ionics* 164: 81–86.

Felipe AM (2009) Preparation and characterisation of new materials for electrolytes used in direct methanol fuel cells. Doctoral Dissertation, University of Aberdeen and Univesitat Politècnica de València, Valencia, Spain. https://riunet.upv.es/bitstream/handle/10251/8327/tesisUPV3183.pdf?

Fenton DE, Parker JM, Wright PV (1973) Complexes of alkali metal ions with poly(ethylene oxide). *Polym* 14: 589.

Fish D, Khan IM, Smid J (1988) Conductivity of solid complexes of lithium perchlorate with poly{(ω-methoxyhexa(oxyethylene)ethoxy)methylsiloxane). *Makromolekul Chem Rapid Comm* 7: 115–120.

Fusco FA, Tuller HL (1989) Fast ion transport in glasses In Laskar AL, Chandra S (eds) *Superionic Solids and Solid Electrolytes Recent Trends*. Academic Press, San Diego, pp. 43–110.

Ganesan S, Muthuraaman B, Mathew V, Madhavan J, Maruthamuthu P, Suthanthiraraj SA (2008) Performance of a new polymer electrolyte incorporated with diphenylamine in nanocrystalline dye-sensitized solar cell. *Sol Energy Mater Sol C* 92: 1718–1722.

Ghosal S, Ray R, Ballabh TK, Tarafdar S (2013) Study of diffusion and conduction in gamma irradiated solid polymer electrolytes by fractal model structure. *Indian J Pure Appl Phys* 51: 324–327.

Giffin GA, Piga M, Lavina S, Navarra MA, D'Epifanio A, Scrosati B, Noto VD (2012) Characterization of sulfated-zirconia/Nafion composite membranes for proton exchange membrane fuel cells. *J Power Sources* 198: 66–75.

Gray FM (1991) *Solid Polymer Electrolytes: Fundamentals of Technological Applications*. Wiley-VCH, New York.

Gray FM (1997) *Polymer Electrolytes*. The Royal Society of Chemistry, London.

Guilherme LA, Borges RS, Moraes EMS, Silva GG, Pimenta MA, Marletta A, Silva RA (2007) Ionic conductivity in polyethylene-b-poly(ethylene oxide)/lithium perchlorate solid polymer electrolytes. *Electrochim Acta* 53: 1503–1511.

Hammami R, Ahamed Z, Charradi K, Beji Z, Assaker IB, Naceur JB, Auvity B, Squadrito G, Chtourou R (2013) Elaboration and characterization of hybrid polymer electrolytes Nafion-TiO_2 for PEMFCs. *Int J Hydrogen Energy* 38: 11583–11590.

Han HS, Kang HR, Kim SW, Kim HT (2002) Phase separated polymer electrolyte based on poly(vinyl chloride)/poly(ethyl methacrylate) blend. *J Power Sources* 112: 461–468.

Hayashi A, Noi K, Sakuda A, Tatsumisago M (2012) Superionic glass-ceramic electrolytes for room-temperature rechargeable sodium batteries. *Nat Comm* 3: 856–860.

Ibrahim S, Johan MR (2012) Thermolysis and conductivity studies of poly(Ethylene Oxide) (PEO) based polymer electrolytes doped with carbon nanotube. *Int J Electrochem Sci* 7: 2596–2615.

Ibrahim S, Yasin SMM, Ahmad R, Johan MR (2012) Conductivity, thermal and morphology studies of PEO based salted polymer electrolytes. *Solid State Sci* 14: 1111–1116.

Imrie CT, Ingram M (2000) Bridging the gap between polymer electrolytes and inorganic glasses: Side group liquid crystal polymer electrolytes. *Mol Cryst Liq Cryst* 347: 199–210.

Imrie CT, Inkster RT, Lu Z, Ingram M (2004) Discotic side group liquid crystal polymer electrolytes. *Mol Cryst Liq Cryst* 408: 33–43.

Ishkov AV, Sagalakov AM (2005) Effect of active fillers on properties of heat-resistant composites. *Russ J Appl Chem* 78: 1512–1516.

Jain N, Kumar A, Chauhan S, Chauhan SMS (2005) Chemical and biochemical transformations in ionic liquids. *Tetrahedron* 61: 1015–1060.

Jannasch P (2000) Synthesis of novel aggregating comb-shaped polyethers for use as polymer electrolytes. *Macromolecules* 33: 8604–8610.

Jayathilaka PARD, Dissanayake MAKL, Albinsson I, Mellander B-E (2002) Effect of nanoporous Al_2O_3 on thermal, dielectric and transport properties of the $(PEO)_9$LiTFSI polymer electrolyte system. *Electrochim Acta* 47: 3257–3268.

Jeon J-D, Kim M-J, Kwak S-Y (2006) Effect of addition of TiO_2 nanoparticles on mechanical properties and ionic conductivity of solvent-free polymer electrolytes based on porous P(VdF-HFP)/P(EO-EC) membranes. *J Power Sources* 162: 1304–1311.

Jian-Hua T, Gao P-f, Zhang Z-y, Luo W-h, Shan Z-q (2008) Preparation and performance evaluation of a Nafion-TiO_2 composite membrane for PEMFCs. *Int J Hydrogen Energy* 33: 5686–5690.

Johan MR, Ting LM (2011) Structural, thermal and electrical properties of nano manganese-composite polymer electrolytes. *Int J Electrochem Sci* 6: 4737–4748.

Jung S, Kim DW, Lee SD, Cheong M, Nguyen DQ, Cho BW, Kim HS (2009) Fillers for solid-state polymer electrolytes: Highlight. *Bull Korean Chem Soc* 30: 2355–2361.

Kam W, Liew C-W, Lim JY, Ramesh S (2014) Electrical, structural, and thermal studies of antimony trioxide-doped poly(acrylic acid)-based composite polymer electrolytes. *Ionics* 20: 665–674.

Kang X (2004) Nonaqueous liquid electrolytes for lithium-based rechargeable batteries. *Chem Rev* 104: 4303–4417.
Karan NK, Pradhan DK, Thomas R, Natesan B, Katiyar RS (2008) Solid polymer electrolytes based on polyethylene oxide and lithium trifluoromethanesulfonate (PEO-$LiCF_3SO_3$): Ionic conductivity and dielectric relaxation. *Solid State Ionics* 179: 689–696.
Karim MA, Song M, Park JS, Kim YH, Lee MJ, Lee JW, Lee CW, Cho Y-R, Gal Y-S, Lee JH, Jin S-H (2010) Development of liquid crystal embedded in polymer electrolytes composed of click polymers for dye-sensitized solar cell applications. *Dyes Pigm* 86: 259–265.
Ketabi S, Lian K (2013) Effect of SiO_2 on conductivity and structural properties of PEO-$EMIHSO_4$ polymer electrolyte and enabled solid electrochemical capacitors. *Electrochim Acta* 103: 174–178.
Kim H-K, Hur S-Y, Kang W-S, Lee G-D (2010) Homeotropic aligned liquid crystal molecules in dye-sensitized solar cells for high efficiency. *J Korean Phys Soc* 56: 1519–1522.
Kim KM, Park N-G, Ryu KS, Chang KS (2002) Characterization of poly(vinylidenefluoride-co-hexafluoropropylene)-based polymer electrolytes filled with TiO_2 nanoparticles. *Polym* 43: 3951–3957.
Kim KS, Park SY, Choi S, Lee H (2006) Ionic liquid-polymer gel electrolytes based on morpholinium salt and PVdF(HFP) copolymer. *J Power Sources* 155: 385–390.
Kitaura H, Hayashi A, Ohtomo T, Hama S, Tatsumisago M. (2011) Fabrication of electrode-electrolyte interfaces in all-solid-state rechargeable lithium batteries by using a supercooled liquid state of the glassy electrolytes. *J Mater Chem* 21: 118–124.
Kono M, Furuta K, Mori, S, Watanabe M, Ogata N (1993) Synthesis of polymer electrolytes based on poly(2-(2-methoxyethoxy)ethyl glycidyl ether) and their high ionic conductivity. *Polym Advan Technol* 4: 85–91.
Krawiec W, Jr Scanlon LG, Fellner JP, Vaia RA, Vasudevan S, Giannelis EP (1995) Polymer nanocomposites: A new strategy for synthesizing solid electrolytes for rechargeable lithium batteries. *J Power Sources* 54: 310–315.
Kudo T, Fueki K (1990) *Solid State Ionics*. VCH and Kodansha Ltd, Tokyo, Vol. 1, pp. 29–42.
Kuila T, Acharya H, Srivastava SK, Samantaray BK, Kureti S (2007) Enhancing the ionic conductivity of PEO based plasticized composite polymer electrolyte by $LaMnO_3$ nanofiller. *Mater Sci Eng B* 137: 217–224.
Kumar MS, Bhat DK (2009) Polyvinyl alcohol-polystyrene sulphonic acid blend electrolyte for supercapacitor application. *Phys B* 404: 1143–1147.
Kuo C-W, Li W-B, Chen P-R, Liao J-W, Tseng C-G, Wu T-Y (2013) Effect of plasticizer and lithium salt concentration in PMMA based composite polymer electrolytes. *Int J Electrochem Sci* 8: 5007–5021.
Li Z, Fu J, Zhou X, Gui S, Wei L, Yang H, Li H, Guo X (2023). Ionic conduction in polymer-based solid electrolytes. *Adv Sci* 10, 18p..
Liew C-W, Ong YS, Lim JY, Lim CS, Teoh KH, Ramesh S (2013) Effect of ionic liquid on semi-crystalline poly(vinylidene fluoride-co-hexafluoropropylene) solid copolymer electrolytes. *Int J Electrochem Sci* 8: 7779–7794.
Liew C-W, Ramesh S (2013) Studies on ionic liquid-based corn starch biopolymer electrolytes coupling with high ionic transport number. *Cellulose* 20: 3227–3237.
Liew C-W, Ramesh S (2014) Comparing triflate and hexafluorophosphate anions of ionic liquids in polymer electrolytes for supercapacitor applications. *Mater* 7: 4019–4033.
Lue SJ, Lee D-T, Chen J-Y, Chiu C-H, Hu C-C, Jean YC, Lai J-Y (2008) Diffusivity enhancement of water vapor in poly(vinyl alcohol)-fumed silica nano-composite membranes: Correlation with polymer crystallinity and free-volume properties. *J Membrane Sci* 325: 831–839.
Lyons LJ, Southworth BA, Stam D, Yuan C-H, West R (1996) Polymer electrolytes based on polysilane comb polymer. *Solid State Ionics* 91: 169–173.

Masoud EM, Bellihi A-A El, Bayoumy WA, Mousa MA (2013) Organic-inorganic composite polymer electrolyte based on PEO-$LiClO_4$ and nano-Al_2O_3 filler for lithium polymer batteries: Dielectric and transport properties. *J Alloy Compd* 575: 223–228.
McHattie GS, Imrie CT, Ingram MD (1998) Ionically conducting side chain liquid crystal polymer electrolytes. *Electrochim Acta* 43: 1151–1154.
Mei A, Wang X-L, Feng Y-C, Zhao S-J, Li G-J, Geng H-X, Lin Y-H, Nan C-W (2008) Enhanced ionic transport in lithium lanthanum titanium oxide solid state electrolyte by introducing silica. *Solid State Ionics* 179:, 2255–2259.
Michael MS, Jacob MME, Prabaharan SRS, Radhakrishana S (1997) Enhanced lithium ion transport in PEO-based solid polymer electrolytes employing a novel class of plasticizers. *Solid State Ionics* 98: 167–174.
Morales E, Acosta JL (1997) Thermal and electrical characterization of plasticized polymer electrolytes based on polyethers and polyphosphazene blends. *Solid State Ionics* 96: 99–106.
Nanda P, De SK, Manna S, De U, Tarafdar S (2010) Effect of gamma irradiation on a polymer electrolyte: Variation in crystallinity, viscosity and ion-conductivity with dose. *Nucl Instrum Methods Phys Res Sec B* 268: 73–78.
Nicotera I, Ranieri GA, Terenzi M, Chadwick AV, Webster MI (2002) A study of stability of plasticized PEO electrolytes. *Solid State Ionics* 146: 143–150.
Nileshrai, Markandaysingh, Abas M, Agarwal SL (2015) Experimental studies on blend based polymer nanocomposite electrolytes: PVA:PMMA:NH_4SCN:BFO System. *Int J Inno Appl Res* 3: 23–38.
Ning W, Xingxiang Z, Haihui L, Benqiao H (2009) 1-Allyl-3-methylimidazolium chloride plasticized-corn starch as solid biopolymer electrolytes. *Carbohydr Polym* 76: 482–484.
Nishimoto A, Agehara K, Furuya N, Watanabe T, Watanabe M (1999) High ionic conductivity of polyether-based network polymer electrolytes with hyperbranched side chains. *Macromolecules* 32: 1541–1548.
Nogueira VC, Longo C, Nogueira AF, Soto-Oviedo MA, Paoli M-AD (2006) Solid-state dye-sensitized solar cell: Improved performance and stability using a plasticized polymer electrolyte. *J Photochem Photobio A Chem* 181: 226–232.
Nordström J, Aguilera L, Matic A (2012) Effect of lithium salt on the stability of dispersions of fumed dilica in the ionic liquid $BMImBF_4$. *Langmuir* 28: 4080–4085.
Noto VD, Bettiol M, Bassetto F, Boaretto N, Negro E, Lavina S, Bertasi F (2012) Hybrid inorganic-organic nanocomposite polymer electrolytes based on Nafion and fluorinated TiO_2 for PEMFCs. *Int J Hydrogen Energy* 37: 6169–6181.
Noto VD, Lavina S, Giffin GA, Negro E, Scrosati B (2011) Polymer electrolytes: Present, past and future. *Electrochim Acta* 57: 4–13.
Osinska M, Walkowiak M, Zalewska A, Jesionowski T (2009) Study of the role of ceramic filler in composite gel electrolytes based on microporous polymer membranes. *J Membr Sci* 326: 582–588.
Pandey GP, Hashmi SA (2009) Experimental investigations of an ionic-liquid-based, magnesium ion conducting, polymer gel electrolyte. *J Power Sources* 187: 627–634.
Pandey GP, Hashmi SA (2013) Ionic liquid 1-ethyl-3-methylimidazolium tetracyanoborate-based gel polymer electrolyte for electrochemical capacitors. *J Mater Chem A* 1: 3372–3378.
Park JS, Kim YH, Song M, Kim C-H, Karim MA, Lee JW, Gal Y-S, Kumar P, Kang S-W, Jin S-H (2010) Synthesis and photovoltaic properties of side-chain liquid-crystal click polymers for dye-sensitized solar-cells application. *Macromol Chem Phys* 211: 2464–2473.
Park MJ, Choi I, Hong J, Kim O (2013) Polymer electrolytes integrated with ionic liquids for future electrochemical devices. *J Appl Polym Sci* 129: 2363–2376.
Patel M, Gnanavel M, Bhattacharyya AJ (2011) Utilizing an ionic liquid for synthesizing a soft matter polymer "gel" electrolyte for high rate capability lithium-ion batteries. *J Mater Chem* 21: 17419–17424.

Pehlivan İB, Granqvist CG, Marsal R, Georén P, Niklasson GA (2012) (PEI-SiO_2):(LiTFSI) nanocomposite polymer electrolytes:Ion conduction and optical properties. *Sol Energy Mater Sol C* 98: 465–471.

Pitawala HMJC, Dissanayake MAKL, Seneviratne VA (2007) Combined effect of Al_2O_3 nanofillers and EC plasticizer on ionic conductivity enhancement in the solid polymer electrolyte $(PEO)_9$LiTf. *Solid State Ionics* 178: 885–888.

Polu AR, Kumar R (2013) Effect of Al_2O_3 ceramic filler on PEG-based composite polymer electrolytes for magnesium batteries. *Advan Mater Lett* 4: 543–547.

Pradhan DK, Samantaray BK, Choudhary RNP, Thakur AK (2005) Effect of plasticizer on structure-property relationship in composite polymer electrolytes. *J Power Sources* 139: 384–393.

Prasanth R, Shubha N, Hng HH, Srinivasan M (2013) Effect of nano-clay on ionic conductivity and electrochemical properties of poly(vinylidene fluoride) based nanocomposite porous polymer membranes and their application as polymer electrolyte in lithium ion batteries. *Eur Polym J* 49: 307–318.

Quartarone E, Mustarelli P (2011) Electrolytes for solid-state lithium rechargeable batteries: Recent advances and perspectives. *Chem Soc Rev* 40: 2525–2540.

Quartarone E, Mustarelli P, Magistris A (1998) PEO-based composite polymer electrolytes. *Solid State Ionics* 110: 1–14.

Ragavendran K, Kalyani P, Veluchamy A, Banumathi S, Thirunakaran R, Benedict TJ (2004) Characterization of plasticized PEO based solid polymer electrolyte by XRD and AC impedance methods. *Portugaliae Electrochim Acta* 22: 149–159.

Raghava P, Zhao X, Kim J-K, Manuel J, Chauhan GS, Ahn J-H, Nah C (2008) Ionic conductivity and electrochemical properties of nanocomposite polymer electrolytes based on electrospun poly(vinylidenefluoride-co-hexafluoropropylene) with nano-sized ceramic fillers. *Electrochim Acta* 54: 228–234.

Raghavan P, Zhao X, Manuel J, Chauhan GS, Ahn JH, Ryub HS, Ahn HJ, Kim KW, Nah C (2010) Electrochemical performance of electrospun poly(vinylidene fluoride-co-hexafluoropropylene)-based nanocomposite polymer electrolytes incorporating ceramic fillers and room temperature ionic liquid. *Electrochim Acta* 55: 1347–1354.

Raghavan V (2004) *Materials Science and Engineering: A First Course*. Prentice Hall of India Private Limited, New Delhi, Vol. 5, pp 332–354.

Rahaman MHA, Khandaker MU, Khan ZR, Kufian MZ, Noor ISM, Arof AK (2014) Effect of gamma irradiation on poly(vinyledene difluoride)-lithium bis(oxalato)borate electrolyte. *Phys Chem Chem Phys* 16: 11527–11537.

Rajendran S, Mahendran O, Kannan R (2002) Characterisation of ((1) PMMA PVdF) poly mer blend electrolyte with Li^+ ion. *Fuel* 81: 1077–1081.

Rajendran S, Prabhu MR, Rani MU (2008) Ionic conduction in poly(vinyl chloride)/poly(ethyl methacrylate)-based polymer blend electrolytes complexed with different lithium salts. *J Power Sources* 180: 880–883.

Rajendran S, Sivakumar M, Subadevi R (2004) Investigations on the effect of the various plasticizers in PVA-PMMA solid polymer blend electrolytes. *Mater Lett* 58: 641–649.

Rajendran S, Sivakumar P (2008) An investigation of PVdF/PVC-based blend electrolytes with EC/PC as plasticizers in lithium battery applications. *Phys B* 403: 509–516.

Rajendran S, Uma T, Mahalingam T (2000) Conductivity studies on PVC±PMMA±LiAsF6±DBP polymer blend electrolyte. *Eur Polym J* 36: 2617–2620.

Ramesh S, Arof AK (2001) Ionic conductivity studies of plasticized poly(vinyl chloride) polymer electrolytes. *Mater Sci Eng B* 85: 11–15.

Ramesh S, Chao LZ (2011) Investigation of dibutyl phthalate as plasticizer on poly(methyl methacrylate)-lithium tetraborate based polymer electrolytes. *Ionics* 17: 29–34.

Ramesh S, Liew C-W (2013) Development and investigation on PMMA-PVC blend-based solid polymer electrolytes with LiTFSI as dopant salt. *Polym Bull* 70: 1277–1288.

Ramesh S, Liew C-W, Arof AK (2011) Ion conducting corn starch biopolymer electrolytes doped with ionic liquid 1-butyl-3-methylimidazolium hexafluorophosphate. *J Non-Cryst Solids* 357: 3654–3660.

Ramesh S, Liew C-W, Lau PY, Ezra M (2012) Characterization of high molecular weight poly (vinyl chloride)-lithium tetraborate electrolyte plasticized by propylene carbonate. In: Luqman M (ed) *Recent Advances in Plasticizers*. In Tech, Crotia, Rijeka, pp. 165–190.

Ramesh, S, Liew C-W, Morris E, Durairaj R (2010) Effect of PVC on ionic conductivity, crystallographic structural, morphological and thermal characterizations in PMMA-PVC blend-based polymer electrolytes. *Thermochim Acta* 511: 140–146.

Ramesh S, Liew C-W, Ramesh K (2011) Evaluation and investigation on the effect of ionic liquid onto PMMA-PVC gel polymer blend electrolytes. *J Non-Cryst Solids* 357: 2132–2138.

Ramesh S, Liew C-W, Ramesh K (2013) Ionic conductivity: Dielectric behavior and HATR-FTIR analysis onto PMMA-PVC binary solid polymer blend electrolytes. *J Appl Polym Sci* 127: 2380–2388.

Reiter J, Vondrak J, Michalek J, Micka Z (2006) Ternary polymer electrolytes with 1-methylimidazole based ionic liquids and aprotic solvents. *Electrochim Acta* 52: 1398–1408.

Sagaidak I, Huertas G, Van NhienAN, Sauvage F (2016). Towards renewable iodide sources for electrolytes in dye-sensitized solar cells. *Energies* 9, 241.

Saikia D, Chen-Yang YW, Chen YT, Li YK, Lin SI (2009) ^{7}Li NMR spectroscopy and ion conduction mechanism of composite gel polymer electrolyte: A comparative study with variation of salt and plasticizer with filler. *Electrochim Acta* 54: 1218–1227.

Samir MASA, Alloin F, Sanchez J-Y, Dufresne A (2005) Nanocomposite polymer electrolytes based on poly(oxyethylene) and cellulose whiskers. *Polímeros: Ciência e Tecnologia* 15: 109–113.

Sekhar PC, Kumar PN, Sharma AK (2012) Effect of plasticizer on conductivity and cell parameters of (PMMA+$NaClO_4$) polymer electrolyte system. *IOSR J Appl Phys* 2: 1–6.

Sekhon SS, Krishnan P, Singh B, Yamada K, Kim CS (2006) Proton conducting membrane containing room temperature ionic liquid. *Electrochim Acta* 52: 1639–1644.

Sharma R, Sil A, Ray S (2012) Characterization of Plasticized PMMA-$LiClO_4$ Solid Polymer Electrolytes. *Advan Mater Res* 585: 185–189.

Singh PK, Kim K-W, Rhee H-W (2009) Development and characterization of ionic liquid doped solid polymer electrolyte membranes for better efficiency. *Synth Met* 159: 1538–1541.

Sinha M, Goswami MM, Mal D, Middya TR, Tarafdar S, De U, Chaudhuri SK, Das D (2008) Effect of gamma irradiation on the polymer electrolyte PEO-NH_4ClO_4. *Ionics* 14: 323–327.

Sirisopanaporn C, Fernicola A, Scrosati B (2009) New ionic liquid-based membranes for lithium battery application. *J Power Sources* 186: 490–495.

Sivakumar M, Subadevi R, Rajendran S, Wu H-C, Wu N-L (2007) Compositional effect of PVdF-PEMA blend gel polymer electrolytes for lithium polymer batteries. *Eur Polym J* 43: 4466–4473.

Sivaraman P, Shashidhara K, Thakur AP, Samui AB, Bhattacharyya AR (2015) Nanocomposite solid polymer electrolytes based on polyethylene oxide, modified nanoclay, and tetraethylammonium tetrafluoroborate for application in solid-state supercapacitor. *Polym Eng Sci* 55: 1536–1545.

Smart L, Moore EA (2005) Defects and nonstoichiometry. In: *Solid State Chemistry: An Introduction*. Taylor & Francis, New York, Vol. 4, pp 201–270.

Song Y, Wu S, Jing X, Sun J, Chen D (1997) Thermal, mechanical and ionic conductive behaviour of gamma-radiation induced PEO/PVDF(SIN)-$LiClO_4$ polymer electrolyte system. *Radiat Phys Chem* 49: 541–546.

Soo PP, Huang B, Jang Y, Chiang Y-M, Sadoway DR, Mayes AM (1999) Rubbery block copolymer electrolytes for solid-state rechargeable lithium batteries. *J Electrochem Soc*, 146: 32–37.

Souquet J-L, Nascimento MLF, Rodrigues ACM (2010) Charge carrier concentration and mobility in alkali silicates. *J Chem Phys* 132, 034704-1–034704-7.

Stephan AM, Nahm KS (2006) Review on composite polymer electrolytes for lithium batteries. *Polym* 47: 5952–5964.

Stephan AM, Saito Y, Muniyandi N, Renganathan NG, Kalyanasundaram S, Elizabeth RN (2002) Preparation and characterization of PVC/PMMA blend polymer electrolytes complexed with $LiN(CF_3SO_2)_2$. *Solid State Ionics* 148: 467–473.

Steven MP (1999) *Polymer Chemistry: An Introduction*. Oxford University Press, New York.

Sukeshini AM, Kulkarni AR, Sharma A (1998) PEO based solid polymer electrolyte plasticized by dibutyl phthalate. *Solid State Ionics* 113–115: 179–186.

Suthanthiraraj SA, Sheeba DJ, Paul BJ (2009) Impact of ethylene carbonate on ion transport characteristics of PVdF-$AgCF_3SO_3$ polymer electrolyte system. *Mater Res Bull* 44: 1534–1539.

Takahashi T (1989) Recent trends in high conductivity solid electrolytes and their applications: An overview. In: Laskar AL, Chandra S (eds) *Superionic Solids and Solid Electrolytes Recent Trends*. Academic Press, San Diego, pp. 1–36.

Tang S, Guo W, Fu Y (2021). Advances in composite polymer electrolytes for lithium batteries and beyond. *Adv Energy Mater* 11, 2000802.

Tarafdar S, De SK, Manna S, De U, Nanda P (2010) Variation in viscosity and ion conductivity of a polymer-salt complex exposed to gamma irradiation. *Pramana-J Phys* 74: 271–279.

Tominaga Y, Endo M (2013) Ion-conductive properties of polyether-based composite electrolytes filled with mesoporous silica, alumina and titania. *Electrochim Acta* 113: 361–365.

Tong Y, Chen L, Chen Y, He X (2012) Enhanced conductivity of novel star branched liquid crystalline copolymer based on poly(ethylene oxide) for solid polymer electrolytes. *Appl Surf Sci* 258: 10095–10103.

Vijayakumar G, Lee MJ, Song M, Jin S-H (2009) New liquid crystal-embedded PVdF-co-HFP-based polymer electrolytes for dye-sensitized solar cell applications. *Macromol Res* 17: 963–968.

Vioux A, Viau L, Volland S, Bideau JL (2010) Use of ionic liquid in sol-gel; ionogels and applications. *Comptes Rendus Chim* 13: 242–255.

Wang L, Yang W, Wang J, Evans DG (2009) New nanocomposite polymer electrolyte comprising nanosized $ZnAl_2O_4$ with a mesopore network and PEO-$LiClO_4$. *Solid State Ionics* 180: 392–397.

Watanabe M, Nagano S, Sanui K, Ogata N (1986) Ion conduction mechanism in network polymers from poly(ethylene oxide) and poly(propylene oxide) containing lithium perchlorate. *Solid State Ionics* 18&19: 338–342.

West AR (1999) Electrical properties. In: *Basic Solid State Chemistry*. John Wiley & Sons West Sussex, Vol. 2, pp. 293–372.

Wright VP (1975) Electrical conductivity in ionic complexes of poly(ethylene oxide). *Brit Polym J* 7: 319–327.

Wright VP (1998) Polymer electrolytes-the early days. *Electrochim Acta* 43: 137–1143.

Wu F, Feng T, Bai Y, Wu C, Ye L, Feng Z (2009) Preparation and characterization of solid polymer electrolytes based on PHEMO and PVDF-HFP. *Solid State Ionics* 180: 677–680.

Xi J, Qiu X, Zheng S, Tang, X. (2005) Nanocomposite polymer electrolyte comprising PEO/$LiClO_4$ and solid super acid: Effect of sulphated-zirconia on the crystallization kinetics of PEO. *Polym* 46: 5702–5706.

Yahya MZA, Arof AK (2003) Effect of oleic acid plasticizer on chitosan-lithium acetate solid polymer electrolytes. *Eur Polym J* 39: 897–902.

Yang C-C, Chien W-C, Li YJ (2010) Direct methanol fuel cell based on poly(vinyl alcohol)/titanium oxide nanotubes/poly(styrene sulfonic acid) (PVA/nt-TiO_2/PSSA) composite polymer membrane. *J Power Sources* 195: 3407–3415.

Yang C-C, Chiu S-J, Lee K-T, Chien W-C, Lin C-T, Huang C-A (2008) Study of poly(vinyl alcohol)/titanium oxide composite polymer membranes and their application on alkaline direct alcohol fuel cell. *J Power Sources* 184: 44–51.

Yang C-C, Li YJ, Liou T-H (2011) Preparation of novel poly(vinyl alcohol)/SiO_2 nanocomposite membranes by a sol-gel process and their application on alkaline DMFCs. *Desalination* 276: 366–372.

Yang C-C, Lin C-T, Chiu S-J (2008) Preparation of the PVA/HAP composite polymer membrane for alkaline DMFC application. *Desalination* 233: 137–146.

Yang H, Wu N (2022). Ionic conductivity and ion transport mechanisms of solid-state lithium-ion battery electrolytes: A review. *Energy Science and Engineering* 10, 1643–1671.

Yang Y, Guo X-Y, Zhao X-Z (2012) A novel composite polysaccharide/inorganic oxide electrolyte for high efficiency quasi-solid-state dye-sensitized solar cell. *Procedia Eng* 36: 13–18.

Yang Y, Zhou CH, Xu S, Hu H, Chen BL, Zhang J, Wu SJ, Liu W, Zhao XZ (2008) Improved stability of quasi-solid-state dye sensitized solar cell based on poly(ethylene oxide)-poly(vinylidene fluoride) polymer-blend electrolytes. *J Power Sources* 185: 1492–1498.

Yang, C.-M.; Kim, H.-S.; Na, B.-K.; Kum, K.-S.; Cho, B.W. Gel-type polymer electrolytes with different types of ceramic fillers and lithium salts for lithium-ion polymer batteries. *J Power Sources* 2006, 156, 74–580.

Ye Y-S, Rick J, Hwang B-J (2013) Ionic liquid polymer electrolytes. *J Mater Chem A* 1: 2719–2743.

Yim T, Kwon MS, Mun J, Lee KT (2015). Room temperature ionic liquid-based electrolytes as an alternative to carbonate-based electrolytes. *Israel J Chem* 55, 586–598.

Yuan F, Chen H-Z, Yang H-Y, Li H-Y, Wang M. (2005) PAN-PEO solid polymer electrolytes with high ionic conductivity. *Mater Chem Phys* 89: 390–394.

Zapata P, Lee J-H, Meredith JC (2012) Composite proton exchange membranes from zirconium-based solid acids and PVDF/acrylic polyelectrolyte blends. *J Appl Polym Sci* 124: E241–E250.

Zhang J, Huang X, Wei H, Fu J, Huang Y, Tang X (2010) Novel PEO-based solid composite polymer electrolytes with inorganic-organic hybrid polyphosphazene microspheres as fillers. *J Appl Electrochem* 40: 1475–1481.

Zhang P, Yang LC, Li LL, Ding ML, Wua YP, Holze R (2011) Enhanced electrochemical and mechanical properties of P(VDF-HFP)-based composite polymer electrolytes with SiO_2 nanowires. *J Membr Sci* 379 80–85.

6 Next-Generation Supercapacitors with Sustainably Processed Carbon Quantum Dots

Grishika Arora, Tejas Sharma, Foo Wah Low, Chai Yan Ng, Pramod K. Singh, Chiam-Wen Liew, and Hieng Kiat Jun

6.1 INTRODUCTION

The advanced version of batteries has become possible in recent years with the introduction of supercapacitors (SCs), which possess improved power density (PD) up to 10 KW/kg, energy density (ED), and specific capacitance value (Gupta et al., 2018; Hashmi et al., 2007). Based on the charge system, SCs can be divided into electric double-layer capacitors (EDLCs), pseudocapacitors, and hybrid capacitors.

EDLCs utilize carbon electrodes or their derivatives, which exhibit significantly higher capacitance due to the electrostatic double-layer formation rather than electrochemical pseudocapacitance. This phenomenon occurs at the interface between the conductive electrode surface and the electrolyte, leading to the separation of charges in the Helmholtz double layer (Shi et al., 2011). As such, the specific capacity of the device can be increased by increasing the area of the electrode surface, thus enhancing the conductivity and the specific capacitance of the electrode.

Metallic oxides like V_2O_5, Fe_3O_4, and RnO_2 are commonly used as active layer materials in pseudocapacitors. In comparison to EDLCs, pseudocapacitors exhibit higher energy storage capabilities and a specific level of capacitance. These are attributed to the occurrence of Faradaic charge transfer processes between the electrode and the electrolyte ions, which contribute to the overall capacitance of the pseudocapacitor system. However, due to the limitations of this Faradic reaction, pseudocapacitor electrodes show poor electrical conductivity, attributed to their inferior solidity (Xiao et al., 2009; Yang et al., 2013).

Carbon quantum dots (CQDs) have been gaining research interest in recent years. CQDs offer several advantages such as low toxicity, excellent solubility, strong and long-lasting photoluminescence (PL), exhibit high specific surface area, making them more commercially promising than their semiconductor counterparts (Lim et al., 2015; Hu et al., 2016). These advantages have received significant attention for CQDs in the field of material science. They also exhibit remarkable liquid dispersibility,

DOI: 10.1201/9781003314424-6

ease of functionalization, visual characteristics, and can be produced on a large scale at low cost. Due to these properties, CQDs have been employed in diverse sectors, encompassing cell molecular biology, drug delivery systems, carbon capture technologies, gas preservation applications, and cancer cell imaging (Ezugbe and Rathilal, 2020; Abbas et al., 2018; Ren et al., 2018).

As reported, there are several advantages to utilizing CQDs in biomedical applications compared to conventional semiconductor quantum dots and organic dyes (Wang et al., 2019). The extremely small size, remarkable biocompatibility, high biochemical inertness, and adjustable hydrophilic nature of CQDs make them widely used tools for scientific research such as biomedical and cancer studies (Ezugbe and Rathilal, 2020; Abbas et al., 2018; Li et al., 2017; Tuerhong et al., 2017). The small size of CQDs, particularly those with a particle size of less than 10 nm offers benefits such as good biocompatibility, low luminescence, enhanced molecule-membrane stability, biological inertness, favorable water mobility, and flexibility. These properties make CQDs highly sought after in research endeavors. CQDs outperform in organic dye-based systems that are dyes made from organic compounds and semiconductor quantum dots with metallic cores in biological systems, establishing their dominance in biosensor and bioimaging applications (Ren et al., 2018). Owing to their diminutive dimensions, CQDs demonstrate an expanded surface area and harness the quantum confinement effect (QCE), which manifests as discrete energy levels. This phenomenon yields heightened energy storage capabilities, enabling the EDLC to accommodate a greater charge. CQDs also can function as conductive additives, augmenting the conductivity of the electrode material. Furthermore, their environmentally benign nature contributes to the enhanced performance of the EDLC SC.

6.2 GENERAL CHARACTERISTICS OF SUPERCAPACITORS AND EDLCS

The efficacy of an SC hinges on factors such as the composition of its electrode and electrolyte components, and the mechanisms involved in charge storage (Iro, 2016; Braga et al., 2023). In response to challenges arising from the overreliance on non-renewable fossil fuels and the pursuit of cost-effective and environmentally friendly energy alternatives, the need for new and sustainable energy sources has become paramount. Despite this, existing energy storage solutions fall short of accommodating extensive storage demands. As a result, the demand for high-power and high-energy-density storage technologies is growing progressively (Iro, 2016). In general, SC technologies can be categorized into EDLC, hybrid capacitor, and pseudocapacitor (refer Figure 6.1).

SCs have emerged as energy storage devices that solve most of the problems associated with traditional batteries. Due to the quick electrostatic charge storage mechanism at the electrode–electrolyte interface, SCs like EDLCs are capable of being charged and discharged quickly, frequently in a matter of seconds. This makes them perfect for uses that call for brief bursts of electricity, such as regenerative braking in automobiles. Due to their high PD, EDLCs can deliver a considerable amount of energy in a brief length of time. The performance of EDLCs can withstand

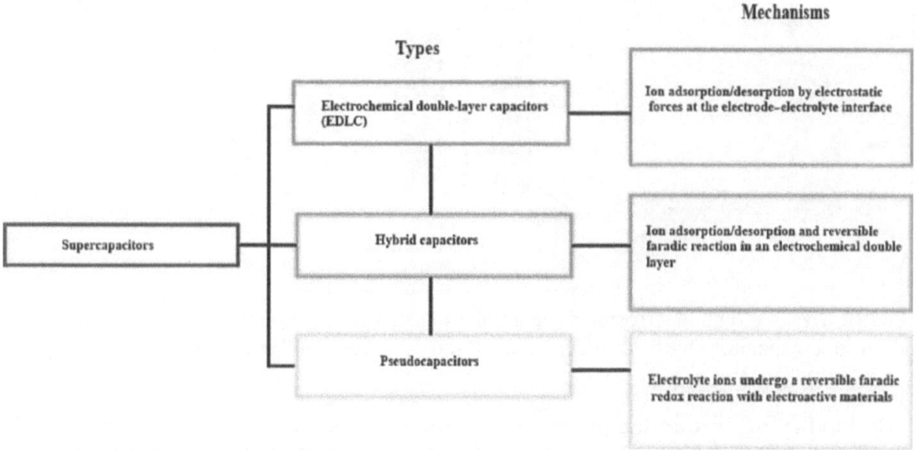

FIGURE 6.1 Varieties of supercapacitors and their energy storage mechanisms.

numerous charge and discharge cycles without noticeably degrading and EDLCs are safer, light, and environmentally friendly. SCs are found in a variety of applications, including electric buses and renewable energy systems. SCs are rated in Farads, with a nominal number of 500 Farad available (Braga et al., 2023).

EDLCs offer higher ED than ordinary capacitors due to their wide cross-sectional area and shorter lengths of the charge separation. As a result, they can achieve high capacitance values. When referring to EDLC devices, there are a lot of benefits and drawbacks. For instance, because of their non-Faradaic electrical mechanism, filling and discharging cycles are reversible. Additionally, this results in a high rate of charging and discharging, highly steady cycling capacity of at least 10^6 cycles or more, and minimal degradation. A fundamental disadvantage of EDLCs is the limited availability of electrode materials, where highly conductive electrodes are required in EDLC devices. Ionic conductive electrolytes, recently developed, can address this weakness. Because of this, the study of ionic electrolytes, especially sturdy ones, is currently popular research (De Klerk et al., 2018; Najib and Erdem, 2019).

6.3 WORKING PRINCIPLE AND CLASSIFICATION OF SUPERCAPACITORS

Electrostatic capacitors are the foundation of the SC operating system (refer Figure 6.2a). Equation 6.1 is the basic formula for all types of capacitors:

$$C=(\varepsilon_0 \times \varepsilon_r \times A)/d \tag{6.1}$$

Electrode surface area, the interval within two electrodes, the dielectric material's relative permittivity, and the permittivity of empty space with opposing biases are all given in this equation by the letters $\varepsilon_0 \times \varepsilon_r$, A, and d, respectively.

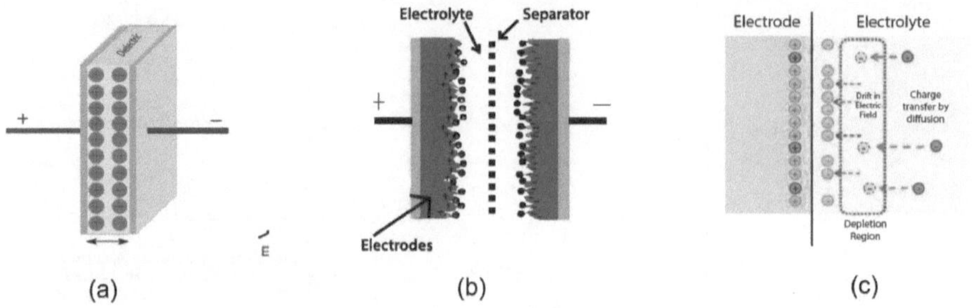

FIGURE 6.2 (a) The electrostatic capacitor's construction, (b) supercapacitor's structure, and (c) supercapacitor charge transfer between the diffusion and Helmholtz layers.

By modifying certain factors of a capacitor, such as the dielectric constant of the material used, the surface area of the electrodes, and the interplanar thickness, it is possible to increase the capacitance value according to Equation 6.1 (Najib and Erdem, 2019; Sahin et al., 2020). However, in the case of an SC, the structure differs from a traditional capacitor.

An SC's basic structure consists of current collectors, typically made of aluminum, as the electrodes. These electrodes are separated by a separator, which is usually made of a material that can be dissolved in an organic or aqueous electrolyte, as depicted in Figure 6.2b. The functioning of an SC relies on energy storage through the diffusion of ions from the electrolyte into the outermost layers of its two electrodes (Wang et al., 2017). The electrical double layer (EDL) forms at the two interfaces, leading to the generation of an EDLC.

When the electrodes and electrolytes come into contact, a thin layer is formed, known as the Helmholtz capacitance layer. At the electrode/electrolyte interface, the Helmholtz layer is assumed to move within an electric field in close proximity to the contact area. This movement of charge within the Helmholtz layer is what gives rise to the capacitance of an SC. The charge from the bulk electrolyte diffuses into the depletion zone created by this drift of charge in the Helmholtz layer (Wasterlain et al., 2006; Kuparowitz, 2010).

The Helmholtz theory utilizes two superficial charge distributions to illustrate the physical processes occurring at the interface between an ionic electrode and an electronic electrode. This layer plays a crucial role in the distribution of charge over time during the charging process as SCs excel in applications demanding swift energy transfer and elevated power performance due to their ability to effectively store and discharge electrical energy, as shown in Figure 6.2c.

SCs differ from ceramic or electrolyte capacitors in that they do not rely on a dielectric material. Instead, SCs utilize the formation of an EDL at the interface between a solid electrode, a carbon-based powder, and a liquid electrolyte. In order to prevent direct contact between the electrodes, a separator is used to keep them apart.

The electrodes in SCs are coated with activated carbon (AC) powder and placed on current collectors. During the charging process, the positive electrode accumulates ions and electrons, while the negative electrode accumulates vacancies.

This alignment of charge occurs at the contact interface between the electrodes. This arrangement forms the structure of an EDLC, as illustrated in Figure 6.3a. The discharge process, shown in Figure 6.3b, involves the dispersion of ions throughout the electrolyte, as they are not strongly attracted to the current collectors. Consequently, the charge states of both current collectors decrease (Thangaraj et al., 2019; Kumari et al., 2023; Najib and Erdem, 2019).

When considering the advantages and disadvantages of SCs, it is important to view them as complementary energy storage options alongside other technologies (Glavin and Hurley, 2012). SCs are voltage-regulated devices that require responsible usage to ensure that the voltage remains within specified limits. Capacitors with organic solvents typically have a nominal voltage range of 2.5–2.7 V, while standard SCs with aqueous electrolytes have a maximum voltage range of 2.1–2.3 V (Salanne, 2017). To achieve higher voltages, SC cells must be connected in series. The maximum capacitance of an SC can vary from 1 to 1,000 F, and for larger applications, parallel connections of cells are required to achieve the desired capacitance (Sahin et al., 2020).

In Figure 6.4, SCs could either act as batteries or capacitors as they have inter-crossed power/energy density values (Kim et al., 2015). Conventional capacitors have a higher power efficiency than the SC, despite the SC having a higher energy efficiency. Compared to SCs, batteries have a higher ED, but the PD of the SCs is higher. Galvanostatic charge-discharge (GCD) and cyclic voltammetry (CV) electrochemical techniques are two prominent ways for the purpose of characterizing material capacitance (Kim et al., 2015; Najib and Erdem, 2019). Equation 6.2 represents the specific ED:

$$ED = \frac{1}{2}Cs(\Delta V)^2 = \frac{1C}{2m}(\Delta V)^2 \tag{6.2}$$

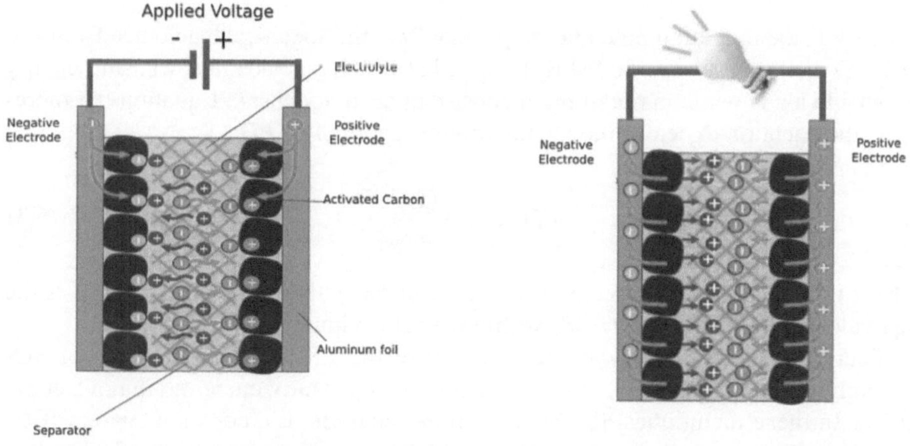

FIGURE 6.3 (a) Charging phase of supercapacitor and (b) discharging phase of supercapacitor.

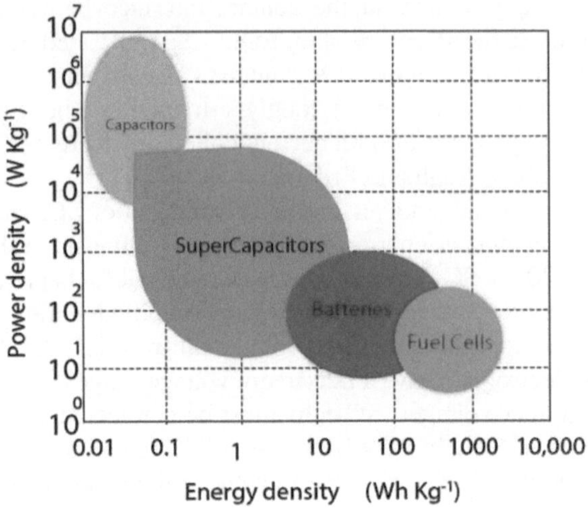

FIGURE 6.4 Power and energy densities of storage systems are compared.

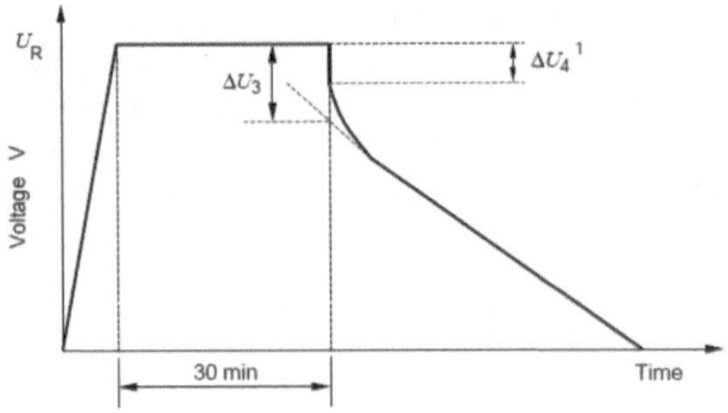

FIGURE 6.5 Current charging and discharging.

where V is the operating potential range and Cs is the specific capacitance from CV or GCD methods. A specific PD is the speed at which a device may transmit energy to outside loads while maintaining a consistent current density. Equation 6.3 represents the equation to determine the maximum specific PD, PD:

$$PD_{\max} = \frac{(V)^2}{4mR_{ESR}} \tag{6.3}$$

where V is the potential range, m is the mass of the active elements, and R_{ESR} is the equivalent series resistance (ESR) within the cell (Kim et al., 2015).

There are techniques available to measure the ESR and capacitance of SCs through direct current (DC) characterization using steady current (Gualous et al., 2016). In these techniques, the SC is initially charged at a constant current (CC) until it reaches a very low voltage threshold. Then, it is connected to a DC source with a predefined voltage. During the discharge cycle, the SC is discharged at a CC, as shown in Figure 6.5 (Lawu et al., 2017; Kraev et al., 2016). For DC

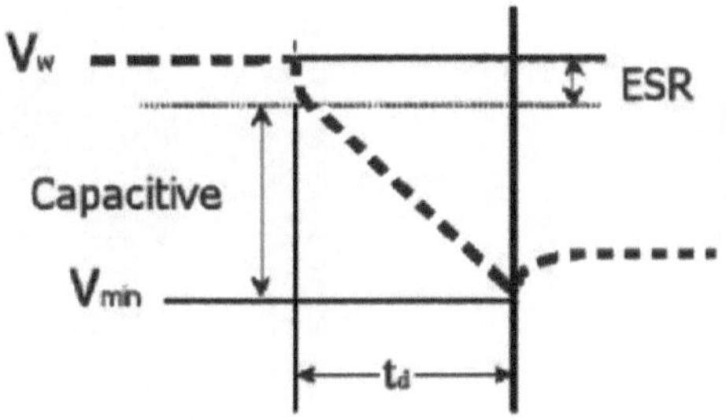

FIGURE 6.6 Supercapacitor discharging.

characterization, ESR and capacitance measurement techniques have been developed based on the charging and discharging of SCs at a CC. In this approach, the SC is initially charged using a CC until it reaches a specific voltage threshold. Subsequently, it is connected to a fixed-voltage DC source. The discharge process involves discharging the SC at a CC until an initial voltage drop is observed. By adjusting certain parameters, such as charging time, the duration of the charging process can be modified. The discharge profile, illustrated in Figure 6.6, provides insights into the discharging behavior (Glavin and Hurley, 2012; Prasad et al., 2019).

The voltage across the SC comprises both capacitive and resistive elements. The capacitive component accounts for voltage changes during the charging and discharging phases. On the other hand, the resistive component contributes to voltage variations due to internal resistance, which is quantified by the ESR.

The reactive component in the measurement represents the voltage fluctuations caused by the charge and discharge processes. Both the ESR and the resistive component provide insights into the voltage shift caused by the internal resistance of the SC. Unlike self-discharge, the redistribution of charge in SC leads to a relatively rapid voltage drop or increase within a short period of time (Prasad et al., 2019).

6.3.1 Electric Double-Layer Capacitors (EDLCs)

When it comes to energy storage, EDLCs, a specific subset of SCs, primarily utilize electric double-layer capacitance. In contrast, other types of SCs may employ additional mechanisms such as pseudocapacitance. The diverse designs of SCs result in a range of energy storage capabilities and characteristics, and these distinctions can be attributed to the variations in underlying mechanisms. EDLCs consist of an electrolyte, two electrodes made of carbon-based materials, and a separator. EDLC can keep the charges either electrostatically or by a non-Faradic mechanism that eliminates the requirement for charge transfers between the electrode and the electrolyte (Kiamahalleh et al., 2012). EDLCs utilize the idea of electrochemical double layers for energy storage. When the opposing charge attraction resulting in potential difference, the voltage is induced and there is no build-up of charges on the electrode surface. Instead, electrolyte ions are spread to the charged-opposing electrode pores and over the separator as shown in Figure 6.7a. An additional charge layer is formed

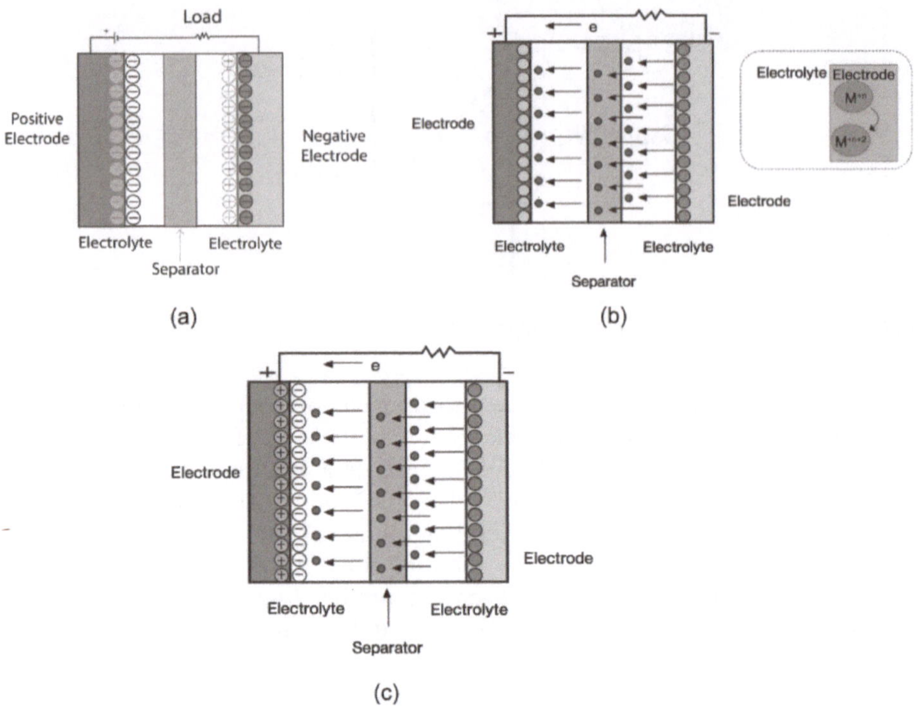

FIGURE 6.7 Illustration of several SC types for energy storage: (a) EDLCs, (b) pseudocapacitors, and (c) hybrid capacitors. Electrochemical capacitors or EDLCs, commonly referred to as supercapacitors, Boost Caps, or Gold Caps, have high capacitance and low internal resistance. Therefore, perfect for a wide range of applications.

to prevent ion recombination within the electrodes. The presence of a double layer in EDLCs facilitates a high ED, enhances the specific surface area, and reduces the electrode separation (Jalal et al., 2021; Choi and Yoon, 2015).

Furthermore, EDLCs' storage method enables rapid absorbance of energy, distribution, and efficient power generation. The non-Faradic phase, characterized by the absence of chemical changes, allows EDLCs to endure millions of cycles, whereas batteries are typically limited to just a few thousand cycles at best. This distinction could explain some of the disparities between batteries and EDLCs. Additionally, when high potential cathodes or graphite anodes are used in Li-ion batteries, the solid electrolyte interphase is formed. However, there is no need for an electrolyte solvent during charging (Jalal et al., 2021; Iro, 2016; Simon and Gogotsi, 2008).

6.3.2 Pseudocapacitors

This type of SC is characterized by employing high electrochemical pseudocapacitance materials and the electrodes are made from metal oxides or conducting polymers. The charge storage mechanism involves Faradaic processes, such as oxidation–reduction reactions, which include charge transfer between the electrode and the electrolyte as shown in Figure 6.7b (Jalal et al., 2021).

Applying an electric field to a pseudocapacitor triggers reduction and oxidation on the electrode material due to charge passage through two layers, resulting in the Faradic current across an SC cell. Additionally, compared to EDLCs, pseudocapacitors can attain high specific capacitance and energy densities, thanks to the Faradic mechanism (Jalal et al., 2021; Chen et al., 2014). In the early days, efforts have been made to develop novel reaction processes for efficient and ultrafast charging/discharging by utilizing rechargeable batteries and SCs in the creation of electrode materials such as conducting polymers and metal oxides. In fact, it must be extremely energetic SCs and also have high power and long cycle life. Through the physical manipulation of electrode materials, researchers have discovered an intercalation pseudocapacitance contribution, akin to the pseudocapacitance found in certain electrode materials for metal-ion batteries (Wang et al., 2017).

6.3.3 Hybrid Capacitors

In a particular type of SC, the electrodes are asymmetric, with one primarily showcasing electrostatic capacitance, while the other demonstrates the transmission of chemical capacitance signals. Capacitors that are in hybrid mode have combined performance characteristics that were previously impossible. Additionally, they are fusing the most advantageous aspects of pseudocapacitors and EDLCs into an individual SC (Jalal et al., 2021). Although hybrid capacitors have not garnered as much attention as pseudocapacitors or EDLCs, ongoing efforts are dedicated to enhancing the hybrid capacitor technology and establishing more precise quantitative models for their performance.

Hybrid capacitors have surpassed EDLCs as the primary SCs class due to their enormous simplicity in tuning their performance and design, as well as the increased focus on developing high energy and high cycle life SCs (Sharma and Kumar, 2020). The mechanism for storing charge is shown in Figure 6.7c. The charge storage mechanism in hybrid capacitors is a synergy of electrostatic double-layer capacitance and pseudocapacitance. The EDLC mechanism operates at the electrode–electrolyte interface, where electrical charges accumulate due to ion separation within the electrolyte. This leads to the formation of a dual charge layer and, consequently, elevated capacitance. In contrast, the pseudocapacitance mechanism involves redox reactions within the electrode material itself, resulting in reversible Faradaic processes that contribute to additional charge storage. By combining these two mechanisms, hybrid capacitors exhibit higher energy density than standard EDLC capacitors, making them suitable for diverse energy storage applications. Asymmetric, composite, and battery-type hybrid capacitors are categorized into these three groups, which could be distinguished by their electrode configuration.

By incorporating the polymerized carbon-based materials conduction or metal oxides in a single electrode via composite electrodes, electrodes will have both chemical and physical processes for storing energy. Binary and ternary composites are the two primary types. Binary composites are electrodes made of only two components; ternary composites are electrodes built of three different materials

(Jalal et al., 2021). Through the integration of an EDLC with asymmetric hybrid capacitors, a combination of Faradic and non-Faradic processes is harnessed to form a pseudocapacitor electrode. In this approach, the carbon material functions as the negative electrode while the metal oxide or conducting polymer acts as the positive electrode (Jalal et al., 2021).

On the other hand, batteries are a unique fusion of an SC electrode with a battery electrode. By fusing SC recharge times and battery qualities to obtain both battery and SC properties in a single cell, this approach highlights the need for greater power batteries and high energy capacitors. Battery-powered hybrids haven't been the subject of many investigations though (Jalal et al., 2021; Costa et al., 2022).

6.4 CARBON QUANTUM DOTS (CQDS)

When combining sp^2 and sp^3 carbons, CQDs often have a quasi-spherical crystalline core with a surface coating of chemical functional groups as shown in Figure 6.8. Most of the carboxyl groups on the surface are both biocompatible and highly hydrophilic (Liu et al., 2021).

CQD synthesis has been accomplished using a variety of demonstrated techniques. The primary goal of a lot of research is to identify the most effective, environmentally friendly, and high-yielding synthetic approach (Magesh et al., 2022). Top-down and bottom-up techniques are the two broad categories used to categorize the routes for CQD synthesis (Gonzalez et al., 2016). Using top-down method techniques like electrochemical oxidation, chemical oxidation, and ultrasonic technologies, bigger carbonaceous materials and structures are broken down or divided into smaller pieces (Lim et al., 2015; Wang et al., 2019). In contrast, the bottom-up approach involves the formation of small CQDs by structuring carbon materials to achieve the desired size. Typical bottom-up methods for CQD creation include solvothermal, hydrothermal, ultrasonic, thermal decomposition, pyrolysis, and carbonization processes (Figure 6.9) (Desmond et al., 2021; Costa et al, 2022).

For the effective creation of CQDs, several top-down techniques have been introduced, such as etching with bigger carbon materials, laser treatment, thermal

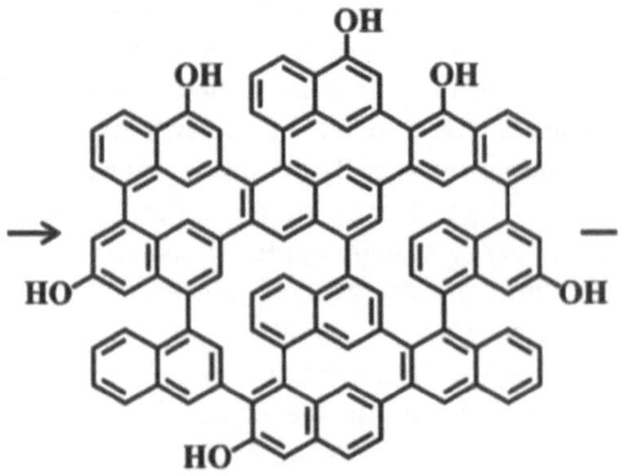

FIGURE 6.8 The general chemical structure of CQDs. Adapted from Desmond et al. (2021).

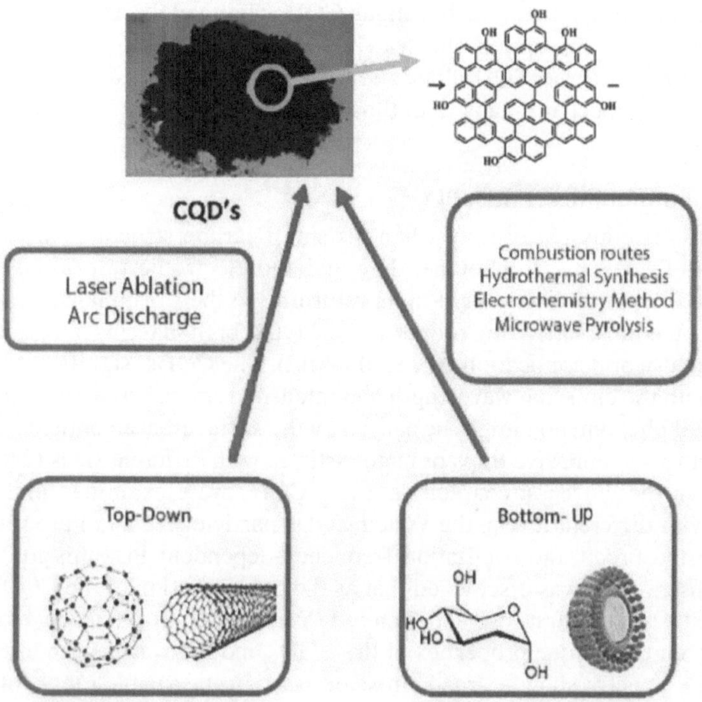

FIGURE 6.9 A general diagram demonstrating the creation of CQDs from a number of efficient top-down and bottom-up techniques. Adapted from Wang et al. (2019). The common beginning supplies for the techniques presented in the bundles are displayed in the top-down and bottom-up approaches. The completed CQD product and its general chemical composition are displayed at the top of the illustration.

oxidation, electrolytic oxidation, hot injection, distillation, and microwave irra diation. Bottom-up procedures typically produce consistent-sized particles that are preferred for utilization in biological applications whereas top-down approaches typically produce tiny flakes or particles that come in a variety of sizes (Desmond et al., 2021; Tuerhong et al., 2017).

6.4.1 Properties of CQDs

6.4.1.1 Absorbance

The typical characteristic of the absorbance of CQDs is a broad and strong absorption peak in the UV-visible spectrum. Overall, it is anticipated that the reflected light peaks of CQDs in the UV-visible range are expected to be π-π^* transition of sp^2 conjugated carbon and n-π^* transition and heteroatoms hybridization such as N, S, P, and so on. Passivation of the surface or other types of modification can change the absorption characteristic (Sadat et al., 2019). Jiang et al. used three isomers of phenylenediamines to establish a simple hydrothermal method for making red, green, and blue luminous CQDs (Jiang et al., 2015). The obtained CQDs' UV-visible absorption spectra showed an analogous pattern. It is interesting to note that three CQD absorption transitions were red-shifted, revealing lower electronic spectrum gaps than those of their equivalent antecedents (Wang et al., 2019).

6.4.1.2 Photoluminescence (PL)

PL refers to the emission of light when the small carbon structures are stimulated by light. As CQDs absorb photons, they undergo electronic transitions, releasing energy in the form of visible light upon returning to their original state. PL represents one of the most intriguing properties of CQDs, both in terms of scientific study as well as real-world applications (Li et al., 2012). The CQDs' significant PL dependence on both the emission wavelength and intensity remains consistent throughout. This unusual phenomenon might be caused by the surface-based optical selection of CQDs with varied emissive traps or nanoparticles with different sizes (Zhang et al., 2017). When the emission characteristics of CQDs were examined under 470 nm radiation with different levels, the varied CQDs particle size and PL emission may be seen in the broad and oscillation frequency-dependent PL emission spectrum that was observed. It was discovered that as the concentration of the CQDs solution grew, the PL strength first rose and then fell (Wang et al., 2019; Zhang et al., 2017). Depending on the unique properties of the CQDs and the wavelength of excitation, the usual PL spectra show a broad emission peak in the visible range, often spanning from blue to green.

6.4.1.3 Electroluminescence

Electroluminescence (ECL) pertains to the phenomenon where the carbon nanostructures emit light when subjected to an electric field. This emission occurs due to electron energy level transitions triggered by the passage of an electric current through CQD-based materials. Since semiconductor nanocrystals are known to exhibit ECL, it should come as no surprise that CQDs have sparked a variety of research interests in ECL that can be useful in electrochemical domains as disclosed in a light-emitting diode device based on CQDs, where the driving current may be used to regulate the emission color (Sk et al., 2014). Under various working voltages, the same CQDs showed color-switchable ECL that ranged from blue to white.

Two concepts were developed based on the emission from the band gap of the conjugated "p"" domain and the reduction effect arising from surface issues. These concepts were developed by the researchers to explain the luminescence process of CQDs (Yuan et al., 2015). The QCE of p-conjugated electrons in the sp^2 atomic framework is what gives CQDs' conjugated p domain fluorescence emission its PL features. These characteristics can be changed by varying the size, edge configuration, and shape of the CQDs. Among other surface faults in CQDs, sp^2 and sp^3 hybridized carbon produce fluorescence emission in these materials. This defect also depends on peak position and fluorescence intensity (Wang et al., 2019; Chen et al., 2021).

6.4.2 Synthesis of CQDs

6.4.2.1 Bottom-Up Approaches

6.4.2.1.1 Hydrothermal/Solvothermal Method

The process of hydrothermal carbonization (HTC), which is a way to create new carbon-based compounds from a variety of precursors, is cost-effective, safe, and nontoxic. Generally, a sealed organic precursor solution is heated to a high temperature in a hydrothermal reactor (Wang and Hu, 2014). Small organic molecules are suspended with the reaction precursor in water or an organic solvent. It is then put into the Teflon-lined stainless-steel autoclave, which is frequently utilized in laboratories for the hydrothermal process (He et al., 2018). Organic molecules are then combined at a reasonably high temperature to create carbon-seeded cores that grow into 10 nm or smaller CQDs (Azam et al., 2021).

Given its uncomplicated synthesis procedure and adaptable hetero-atom doping, this technique offers a promising avenue for producing and synthesizing novel catalysts tailored for energy applications, offering the flexibility to customize doping ratios and electronic structures. According to Atchudan et al., the resulting CQDs have a near-spherical morphology with an average diameter of 5 nm and an adequate level of graphitization (Magesh et al., 2022; Atchudan et al., 2021). Because of their functionality, the water solubility of CQDs is very good and exhibits bright fluorescence without further surface treatment, with an acceptable quantum yield (QY) of 20%. CQDs have high biocompatibility and fluorescence stability, making them ideal candidates in bioimaging applications. To enhance water solubility and facilitate binding to specific cell receptors, Mahani et al. employed a modified approach involving transferrin (TF) and HTC of citric acid (Magesh et al., 2022; Mahani et al., 2021). For the purpose of delivering the drug (doxorubicin) to breast cancer cells, the drug was applied onto a carrier composed of TF-CQDs (MCF-7). The results also indicate the potential utility of Dox-loaded TF-CQDs for precise drug delivery.

When evaluating the benefits of the strategies employed in CQD production, the hydrothermal method stands out as a straightforward approach. It yields CQDs with high QY and controlled particle size, thereby reducing the need for expensive chemicals. Additionally, CQDs produced through hydrothermal methods find extensive applications in the fabrication of sensors, SCs, solar cells, and drug delivery systems (Magesh et al., 2022).

6.4.2.1.2 Acidic Oxidation

A commonly used method involves exfoliating and deconstructing hydrophilic functional groups, like hydroxyl or carboxyl groups, which are incorporated into the structure to convert bulk carbon into nanoparticles (Shen and Xia, 2014). It was reported that acidic oxidation and hydrothermal reduction were used to create hetero-atom-doped CQDs on a massive scale. To begin, carbon nanoparticles made using Chinese ink were oxidized using a mixture of H_2SO_4, HNO_3, and $NaClO_3$ (Magesh et al., 2022). Then, nitrogen, sulfur, and silicon sources were hydrothermally reacted with the oxidized CQDs. Di-methyl formamide (DMF), sodium hydrosulfide (NaHS), and sodium selenide (NaHSe) have all been used as precursors. The produced N-CQDs, S-CQDs, and Se-CQDs had higher QY, greater PL performance, and longer adjustable fluorescence lifetimes when compared to pure CQDs. The results showed that the heavily doped heteroatoms can modify the PL properties that are advantageously related to the electro-negativities of N, S, and Se. The active heteroatoms on the surface would modify the electronic structure of the associated CQDs when used as electrocatalysts, resulting in good electrocatalytic activity (Magesh et al., 2022; Feng and Zhang, 2019). Using this technique, Feng et al. produced CQDs from coke, which revealed 9.2% of QY. After a one-step chemical oxidation utilizing coke as the starting material and hydrogen peroxide as the oxidizing agent, the product had a blue-green-red spectral makeup of up to 48%. This study demonstrated the intriguing uses of coke-based CQDs as possible phosphors in light-emitting devices (Monte-Filho et al., 2019).

The acidic oxidation process is a quick and efficient way to make CQDs on a wide scale. However, the use of some potent oxidizers in this procedure could result in burns or explosions, and the post-processing is also exceedingly challenging (Magesh et al., 2022).

6.4.2.1.3 *Microwave Pyrolysis*

Because of its flexibility of use and commercial viability, microwave pyrolysis is a well-known bottom-up approach (refer Figure 6.10) (Feng and Zhang, 2019). Microwave irradiation induces atomic-level heating through the interaction of polar solvent molecules with electric and magnetic fields, resulting in rapid and energy-efficient reactions without physical contact or intermediaries. Compared to existing methods, this approach offers efficient CQD synthesis, rapid processing, eco-friendliness, energy/time efficiency, low impurity levels, controlled size/temperature, consistency, and precise reaction parameter control (Magesh et al., 2022; Tungare et al., 2020).

To create extremely luminous CQDs, Liu et al. discussed about microwave-mediated pyrolysis of citric acid with different amine compounds. The simultaneous roles that the amine molecules, particularly primary amine molecules, play in the N-doping process and surface passivation of the CQDs improve the PL performance. For CQDs made from citric acid and 1,2-ethylenediamine, the QY values significantly increased as the N content increased, exhibiting a QY up to 30.2%.

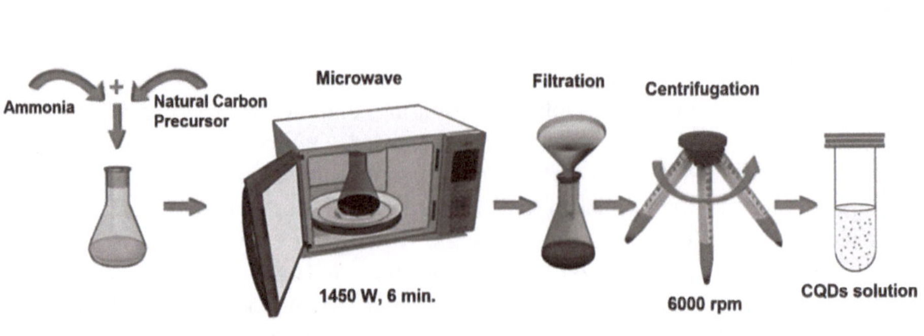

FIGURE 6.10 Synthesis method for CQDs using microwave-assisted pyrolysis. Adapted from Feng and Zhang (2019).

The resulting CQDs have significant potential for biomedical applications and are extremely biocompatible (Zhai et al., 2012).

With this technique, the synthesized CQDs have good physiochemical and optical characteristics, such as tunable emission, wide spectrum absorption, high charge carrier extraction, and high extraction of charge carriers. Although it is an easy one-stop method and is good for the environment, there is no way to regulate the size of the CQDs (Magesh et al., 2022).

6.4.2.2 Top-Down Approaches

6.4.2.2.1 Laser Ablation Method

In this method, the surface of the sample is illuminated by a powerful laser pulse. It then enters a thermodynamic condition with a high temperature (>900°C) and pressure (75 kPa), which rapidly warms up to a liquid phase where vapor crystallizes to produce carbon nanomaterials (Magesh et al., 2022; Hu et al., 2011). Radiating a precursor to carbon distributed in several typical organic solvents with a laser is an easy way to create CQDs (Figure 6.11). The synthesized carbon nanoparticles are aggregated with different-sized particles and lack any clear physical PL. After treating the sample with a weak HNO_3 solution, it is refluxed for 12 hours before polyethylene glycol is added to react with it (Zuo et al., 2015).

It was reported that the CQDs have adjustable and visible PL (Li et al., 2011). Modifying the PL properties of freshly synthesized CQDs and altered CQDs with strong PL and a diameter of 5 nm can alter the surface characteristics of CQDs. Laser ablation is a potent method for producing CQDs characterized by a narrow size distribution, excellent water solubility, and strong fluorescence (Magesh et al., 2022; Hu et al., 2009). When stimulated at 400 nm, CQDs produced fluorescence QYs that varied from 4% to 10% (Atchudan et al., 2021).

Typically, according to the solvent and precursor, CQDs produce more PL; hence, this method results in increased PL emissions (Cui et al., 2020). The shortcomings of this strategy are the huge number of raw materials needed and the inability to control particle sizes, which led to low QY (Magesh et al., 2022). It is

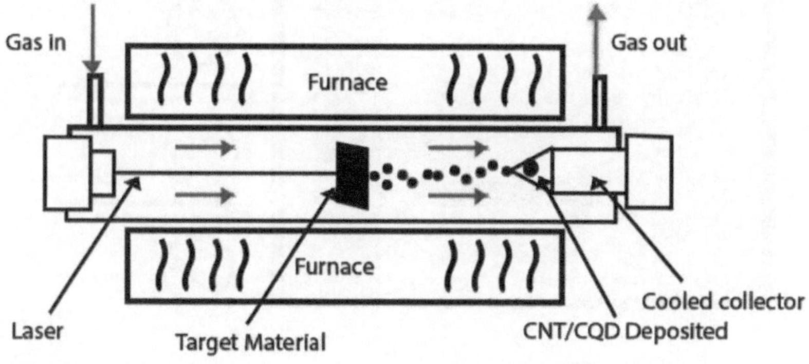

FIGURE 6.11 Diagram showing the development of CNT/CQD creation through laser ablation.

important to emphasize the speed and effectiveness of this technology as well as the customizable surface state of CQDs.

6.4.2.2.2 Arc-Discharge Method

Using gas plasma created in a sealed reactor inside of an anodic electrode, this technique breaks down the bulk carbon sources while reordering the carbon molecules within them (Figure 6.12) (Arora and Sharma, 2014). Under the influence of the electric current, the reactor's temperature can reach 4,000 K, producing plasma with enormous energy. CQDs are created in the cathode by the accumulation of carbon vapor. Different forms of carbon nanoparticles with various molecular weights and luminescence can be when synthesizing single-walled carbon nanotubes using the arc-discharge method (Dey et al., 2014). Under 365 nm excitation, the initially synthesized CQDs exhibit emissions in shades of orange, blue-green, and yellow. Through the introduction of the carboxyl group (hydrophilic) onto the CQDs' surface using HNO_3, enhanced water solubility is achieved. However, the resultant CQDs often possess a larger particle size of approximately 18 nm. This outcome can be attributed to the diverse scales and low yield of carbon particles generated during the arc-discharge method. Moreover, nanoparticles produced via the arc-discharge route may encompass intricate constituents that pose challenges to filtration and elimination processes (Zuo et al., 2015).

With this method, it is possible to create CQDs for bioimaging applications. However, the synthesized CQDs exhibit a limited QY and lack of particle size scalability. Nevertheless, they show a strong PL property (Magesh et al., 2022). Among all the approaches, the laser ablation approach is suggested as one of the top-down techniques for making CQDs since it allows for exact control over particle size, a reduction in contaminants, and the possibility to create CQDs with specific qualities. By employing laser irradiation to disintegrate a carbon target, this technique produces CQDs with the appropriate properties. Table 6.1 summarizes the unique features of the top-down and bottom-up approaches in synthesizing CQDs.

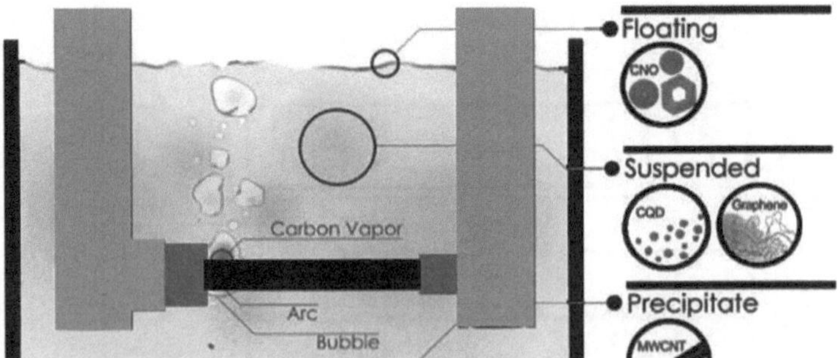

FIGURE 6.12 The graphic depicts the immersed arc-discharge in water method for creating CQDs.

TABLE 6.1
According to Reports, CQDs Have Been Synthesized Using Various Top-Down and Bottom-Up Approaches from a Diverse Range of Carbonaceous Starting Materials

Method (Top-Down/Bottom-Up)	Course of Action	Pros	Cons
Hydrothermal/solvothermal method	To create CQD, in water or an organic solvent, tiny organic molecules or polymers are dissolved and then transferred to an autoclave made of stainless steel lined with Teflon.	The CQDs' QY can reach 80%, and there are numerous carbon sources that can be used in the process.	Minimal QY. Gas is a byproduct of the reactor that might cause losses. inadequate control of size.
Acidic oxidation	After being oxidized by a solution of H_2SO_4, $NaClO_3$, and HNO_3 carbon nanoparticles are DMF, NaHS, and NaHSe hydrothermally interacted with each other.	Created CQDs outperform other CQDs in terms of prolonged fluorescence lifetime, high QY, and tunable PL performance.	Strongly doped heteroatoms may be caused by the electronegativity of S, N, and Se to have an impact on the PL characteristics.
Microwave pyrolysis	Forming a transparent solution by putting a carbon source and a sugar solution in water and heating it in a microwave in an inert environment.	A straightforward, ecologically harmless, commercially viable, and quick way to synthesis.	The highest QY reported is 34%, which is less than that for other methods in the table.
Laser ablation method	Laser irradiation is used to create the desired effect, toluene serves as the carbon source. Laser weapons are used to regulate CQD size.	Changing parameters like the irradiation period will allow you to easily modify the size and PL characteristics of CQDs.	This approach produces a modest yield of CQDs (4%–10%), making it non-green. Low QY, inadequate size control, adjustment required.
Arc-discharge method	An arc of current and voltage is created between two graphite rods that are employed as the anode and cathode electrodes, respectively, in an open vessel.	This procedure generates MWCNTs, which, in theory, might be further broken down to generate CQDs.	The soot from nanotubes contains a number of contaminants. Compared to nanotubes, the graphitic sheets in soot have stronger oxidative stability.

6.4.3 Synthesis of CQDs from Biomass Wastes

Despite the widespread availability at little cost, waste biomass generated from food and animal waste, forest byproducts, agricultural residues, and other sources remains largely unnoticed and underutilized as a potential raw resource. The entire synthesis process is specifically engineered to transform waste into an affordable and renewable final product, which holds promise for large-scale commercial production. Notably, no chemicals are employed during the preparation of CQDs from biomass (Thangaraj et al., 2019).

Among the biomass wastes, spent coffee grounds (CGs) are a prevalent organic waste byproduct known for their elevated carbon content. Studies have confirmed that the carbon obtained from spent CG exhibits exceptional characteristics, such as a high surface area, a well-developed graphitic structure, and a hierarchical porous arrangement, all of which are essential for promoting efficient capacitive behavior (Kumari et al., 2023). In comparison to heat drying, the hydrothermal approach utilized to produce CQD has been proven to be a more cost-effective method. The following sections discuss on practical methods of fabricating CQDs from biomass waste of spent CG.

6.4.3.1 Hydrothermal Method

In order to create fluorescent CQDs from the spent coffee beans as a carbon source, HTC is chosen. According to the work by Costa et al., the coffee beans were first cleaned in deionized water, air dried, and made in the powdered form (Costa et al., 2022). A stainless-steel autoclave with a 100mL inner volume and Teflon coating was filled with some powdered coffee beans and heated at a continuous temperature of 200°C for 4hours. It was then given time to cool down to room temperature. To get the clear, brownish solution, a 0.45m membrane filter was used to filter out any undesirable big particles and unreacted organic molecules. The purified fluorescent CQD was then collected, and it was kept at 5°C for additional characterization and testing (Costa et al., 2022).

Carbon dots (CDs) were synthesized through a process involving the dissolution of CGs (approximately 5.0g) in 100mL of water (Chen et al., 2021). An amine additive was introduced at specific temperatures and durations within a sealed container. Following cooling to room temperature, the mixture was subjected to filtration using a cellulose membrane. The resulting brown filtrate was passed through syringe filters. Subsequently, measurements were conducted to determine insoluble residue concentrations, organic substance concentrations, and the quantity of CDs in aqueous dispersions. These measurements were executed in accordance with the previously described methods, wherein proportional volumes of organic solvents were employed for extraction.

In a separate study, Chen et al. concluded the CQDs produced from 3.125g/L of CGs have the best fluorescence intensity, as determined by contrasting the fluorescence intensities of the three CQD concentrations under the identical circumstances, as shown in Figure 6.13a, and while investigating the reaction time, CQDs' fluorescence intensity in 6hours was comparable to that in 12hours, but it was much higher in 10hours. Consequently, 10hours is the ideal reaction time as shown in

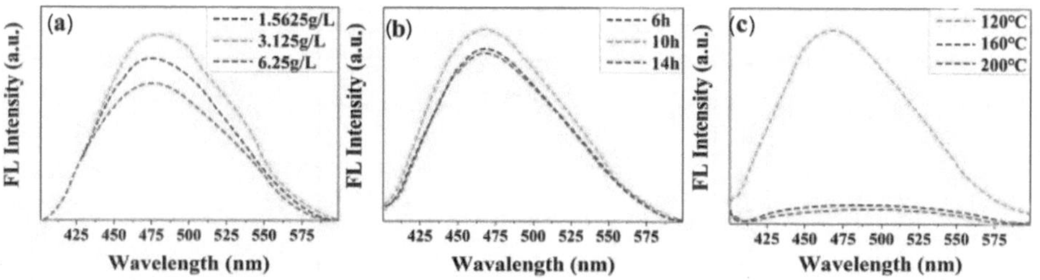

FIGURE 6.13 The ideal reaction conditions' emission spectra: (a) concentration, (b) time, and (c) temperature of the reaction. Adapted from Chen et al. (2021).

Figure 6.13b. While investigating the reaction temperature at 120°C, the intensity of this system peaked and remained constant. Additionally, at reaction temperature of 120°C, CQD fluorescence intensity is significantly higher than that at 160°C and 200°C as shown in Figure 6.13c (Chen et al., 2021).

6.4.3.2 Microwave-Assisted Hydrothermal Carbonization (Mw-HTC) Method

Coffee beans, a plentiful and environmentally unfavorable heavy trash, were found to be an appropriate source of vegetable biomass for the quick and efficient HTC technique that produces fluorescent CQDs. According to Shen et al., in order to make highly luminescent CQDs in the best chemical yields possible, an expanded criteria for the HTC and Mw-HTC's reactions were explored (Shen and Xia, 2014). Good reaction yields of CQDs with blue luminescence (up to 18%) were produced under carefully designed conditions. In conclusion, this research revealed CQDs from used CGs show outstanding toxic nitroaniline detecting capabilities for the initial period.

The CQDs made from CGs had outstanding fluorescence characteristics, with emission maxima that varied depending on an excitation wavelength (340–420 nm). Maximum fluorescence emission that shifts with decreasing excitation energies are typically linked to the presence of various particle sizes and/or surface states, which results in a distribution of visual gaps on the array (Figure 6.14). After 380 nm excitation, the maximum emission (HTC) is produced at 461 nm (Shen and Xia, 2014).

In the work by Arumugham et al., they used the Catharanthus roseus (a white blooming plant) leaves that were hydrothermally carbonized to create an eco-friendly fluorescence CQD probe, without the need of chemicals or laborious post-processing (Arumugham et al., 2020). This method was successfully proven. At the 330 nm excitation wavelength, they demonstrated a maximum QY of 28.2% providing stable photostability and an intriguing pH-dependent behavior of the produced CQDs. Studies on bioimaging has shown its capacity to diffuse across MCF-7 cells without any fluorescence emission degradation. These intriguing findings lead to the conclusion that Catharanthus roseus leaf-derived CQD may be a novel type of fluorescence probe with considerable possibilities in the study of cancer treatment and the reduction of environmental pollution (Arumugham et al., 2020).

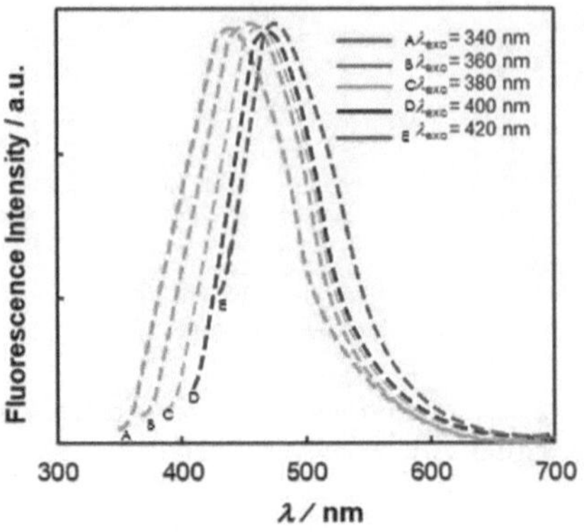

FIGURE 6.14 C-dot (HTC) aqueous dispersions (0.1 mg/mL) with standardized emission spectra derived from coffee grounds (excitation in the 340–420 nm region). Adapted from Shen and Xia (2014).

Utilizing electrode materials based on CQDs is widespread due to their economic viability and availability. Additionally, the flexibility to alter and control factors like pore configuration, surface traits, and surface area offers an added benefit when incorporating carbon materials into energy storage apparatus. In SCs, the storage mechanism hinges on the electrostatic allure of electrolyte ions to the surface of the electrode material. Thus, the presence of appropriately sized pores that align with the ions' dimensions becomes essential. Here are a few recent research endeavors aimed at enhancing SC performance.

6.5 APPLICATION OF "GREEN" CQDS IN SUPERCAPACITORS

There are number of works reported on CQDs applied in SC in recent years. A few notable works are discussed here. In the work by Inayat et al., they employed a hydrothermal approach to synthesize CQDs using tea leaves as a precursor (Inayat et al., 2023). These CQDs were intended for use as electrode materials in EDLC SCs. CV and GCD measurements were performed over a voltage range of −0.2 to +0.8 V compared to the Standard Calomel Electrode (SCE) reference electrode. These measurements were carried out using various sweep rates in a 1M H_2SO_4 electrolyte solution. Additionally, electrochemical impedance spectroscopy (EIS) experiments were conducted in a frequency range spanning from 0.01 to 10^5 Hz. The CQDs were prepared with a crystalline arrangement, exhibiting an average diameter of approximately 2–5 nm. Evaluations of the electrochemical performance of the CQDs revealed exceptional stability when employed as an SC electrode.

The CQD-based SC demonstrated remarkable characteristics, including a high specific capacitance of 302.0 F/g at a current of 0.5 A/g, impressive cyclability with a capacitance of 144.4 F/g at 20 A/g after undergoing 5,000 cycles, and a notable rate performance of 186.4 F/g at 20 A/g. Moreover, the specific-capacity retention values after 5,000 cycles stood at 93.8%, 89.8%, and 77.4% at 10, 15, and 20 A/g, respectively, as shown in Figure 6.15.

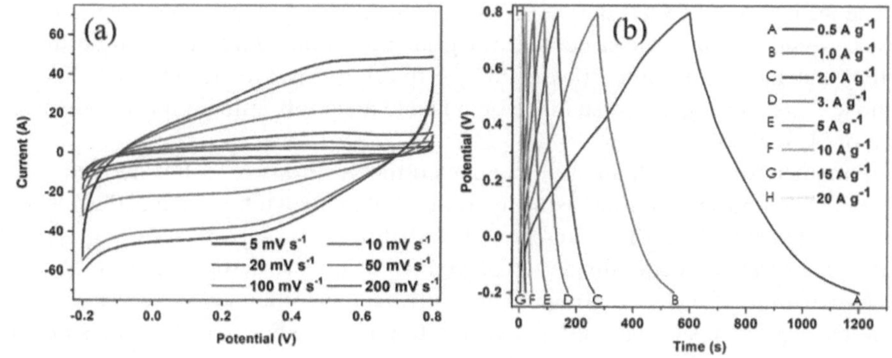

FIGURE 6.15 The electrochemical behavior of CQDs in supercapacitors: (a) through cyclic voltammetry (CV) curves illustrate the response of the CQD electrode across a range of scan rates (5–200 mV/s), covering a potential range of −0.2 to 0.8 V. These experiments were conducted in a 1 M H_2SO_4 electrolyte solution. (b) Galvanostatic charge-discharge (GCD) curves showcase the behavior of the CQDs electrode across various current densities ranging from 0.5 to 20 A/g. The experiments were conducted in a 1 M H_2SO_4 electrolyte solution. Adapted from Inayat et al. (2023).

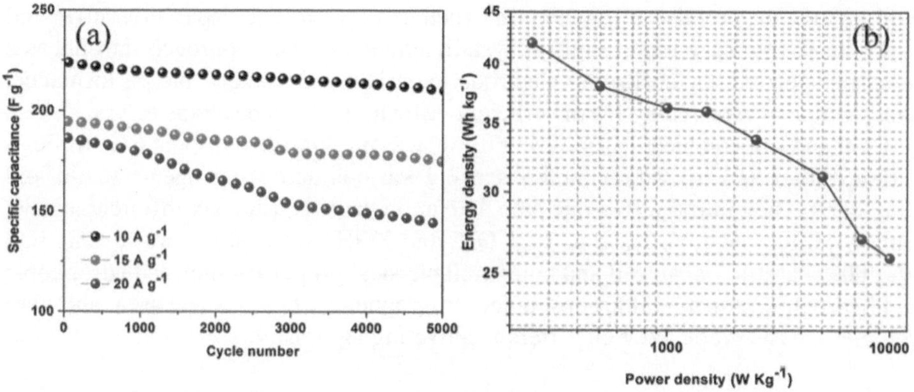

FIGURE 6.16 (a) The durability of the CQD electrode was evaluated through cyclic stability tests conducted over 5,000 cycles at current density levels of 10, 15, and 20 A/g. (b) The Ragone plot illustrates the performance of the CQD electrode across different current densities. Adapted from Inayat et al. (2023).

The notable preservation of high capacity can be attributed to the exceptional electronic and ionic conductivity of the CQDs, along with their structural adaptability, effectively preventing mechanical degradation during continuous insertion and extraction cycles. Additionally, the PD and ED reached impressive values of 25.8 and 10,000 Wh/kg, respectively, at a current of 20 A/g as shown in Figure 6.16. The introduction of nano-sized CQDs significantly improved the active sites of the particles, as well as surface roughness and electrical conductivity. This improvement further facilitated enhanced ion transfer and improved accessibility to electrolyte ions (Inayat et al., 2023).

The utilization of CQDs derived from spent tea leaves enables the integration of these CQDs into composite structures alongside transition-metal oxides and sulfides, thereby enhancing the properties of SCs. Exploring this avenue further holds the promise of generating a diverse range of valuable materials tailored for the advancement of SC technology.

In another study, H. Quan et al. conducted the synthesis of CQDs derived from Camellia Oleifera shell powder using hydrothermal method (Quan et al., 2022). Their findings indicated that CQDs integrated with hierarchical porous carbon (HPC) exhibited a notable specific surface area, well-suited pore structure, and favorable N- and O-containing functional groups. These attributes were found to be advantageous in enhancing active sites, as well as ameliorating the kinetics of charge transfer and the rate of ion transfer.

The resultant CQDs/HPC-2 material demonstrated remarkable electrochemical performance, with a substantial capacitance of 259 F/g at 1 A/g and exceptional rate capability in a 1 M H_2SO_4 electrolyte. Furthermore, when used in a symmetric SC configuration, CQDs/HPC-2 exhibited an impressive energy density of 8.61 Wh/kg at a power density of 477 W/kg in a 1 M H_2SO_4 solution, along with a remarkable energy density of 17.86 Wh/kg at a power density of 954 W/kg in a 1 M Na_2SO_4 solution as shown in Figure 6.17 (Quan et al., 2022).

The outcome of the research suggests that porous carbons derived from biomass hold significant promise in augmenting their energy storage capability, attributed to the uniform dispersion of highly crystallized GQDs. This approach introduces a novel perspective for fabricating electrode materials with remarkable performance, along with a novel strategy for the efficient utilization of biomass waste.

Chiu and Lin conducted a study where AC was produced from spent CGs through carbonization and activation methods using various activating agents (Chiu and Lin, 2019). They introduced a one-step activation process using six different agents, namely H_3PO_4, HCl, $FeCl_3$, $ZnCl_2$, NaOH, and KOH, to transform waste CGs into AC. They carefully analyzed and compared physical properties such as hydrophobic and hydrophilic functional groups, defect-to-graphene ratios, surface area, and pore volumes for AC produced with different activating agents.

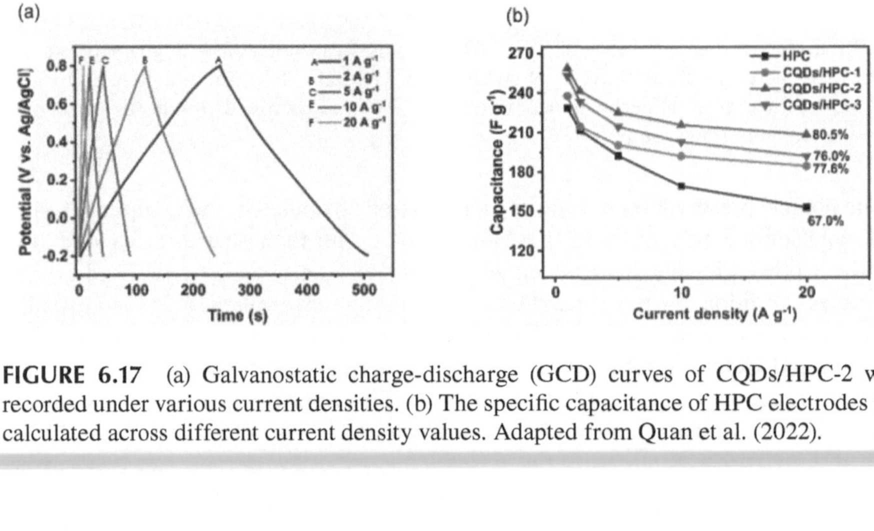

FIGURE 6.17 (a) Galvanostatic charge-discharge (GCD) curves of CQDs/HPC-2 were recorded under various current densities. (b) The specific capacitance of HPC electrodes was calculated across different current density values. Adapted from Quan et al. (2022).

The research found that the CV curve integrating area of the CG electrode was significantly smaller compared to the CV curves of AC electrodes. This indicated that the energy storage capability of the non-activated CG was notably inferior. The CV curves displayed increased oxidation and reduction currents at high and low potential regions, respectively, attributed to water-splitting reactions. Furthermore, GCD curves were presented for AC electrodes synthesized using various activating agents. These curves exhibited symmetric charge and discharge patterns with no plateaus, suggesting dominant EDLC behavior for energy storage in these AC electrodes. The AC electrodes prepared with different activating agents resulted in C_F of 51.0, 0.6, 41.3, 72.9, 69.5, and 105.3 F/g for H_3PO_4, HCl, $FeCl_3$, $ZnCl_2$, NaOH, and KOH, respectively. Notably, AC synthesized using KOH displayed the best performance with a C_F value of 105.3 F/g, surpassing the commercial carbon electrode with a C_F value of 21.8 F/g shown in Figure 6.18.

Zhang (2019) conducted an experiment in which they successfully synthesized mesoporous carbon materials, denoted as "Original-SE," through a high-temperature carbonization process using biomass materials derived from silkworms. These materials were developed for the purpose of serving as high-performance electrodes with a wide potential window in SC applications. They determined that under visible light exposure, they achieved a specific capacitance of 312 F/g at 0.1 A/g, surpassing more than twice the capacitance achieved in the absence of light (Zhang, 2019).

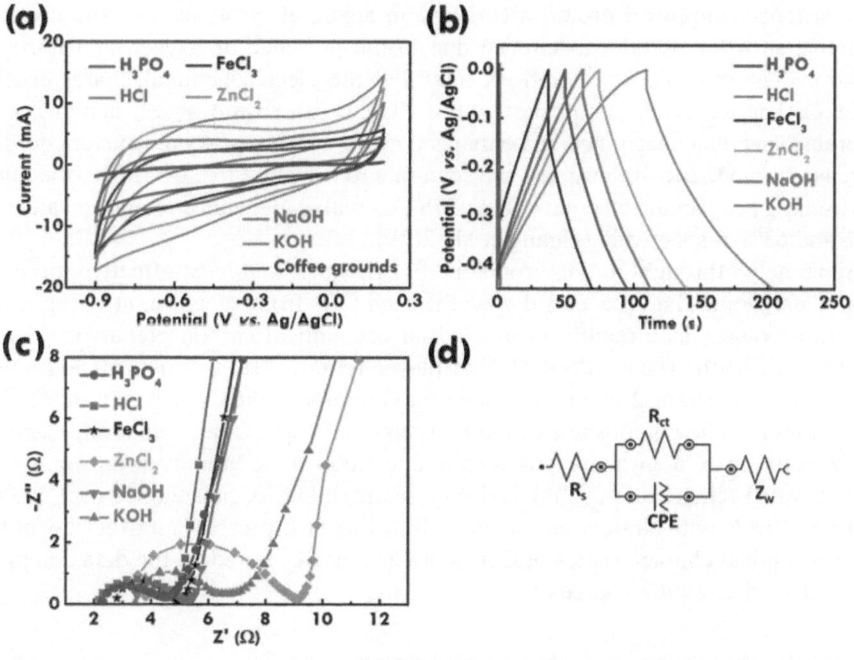

FIGURE 6.18 (a) The CV curves were obtained at a scan rate of 10 mV/s. (b) The GCD curves were recorded at a current density of 0.5 A/g. (c) The Nyquist plots were generated, and (d) the corresponding equivalent circuit was established for the AC electrodes synthesized using H_3PO_4, HCl, $FeCl_3$, $ZnCl_2$, NaOH, and KOH. Adapted from Chiu and Lin (2019).

TABLE 6.2
Electrochemical Behavior of CQDs or CDs in Supercapacitor

Electrode	Specific Capacitance	Energy Density	Power Density	References
Tea leaves	302 F/g	41.9 Wh/g	250 W/g	Inayat et al. (2023)
Spent coffee grounds	105.3 F/g	6.94 Wh/kg	350 W/Kg	Chiu and Lin (2019)
Silkworm	312 F/g	-	-	Zhang (2019)
Cauliflower	278 F/g	-	-	Hoang et al. (2019)
Zucchini	374 F/g	-	-	Hoang and Gomes (2019)

Meanwhile, Hoang et al. had outlined an eco-friendly, one-step synthesis method for producing CDs derived from biomass, using waste cauliflower leaves as a naturally occurring carbon source (Hoang et al., 2019). These synthesized CDs were then used to create composites with reduced graphene oxide (RGO) via a straightforward hydrothermal process. In this process, the CDs served as effective spacers, preventing the restacking of graphene nanosheets, which resulted in increased surface area and pore volume. The introduction of inherently nitrogen-doped CDs had a significant impact on various electrochemical properties. It led to reduced charge transfer resistance, improved proton diffusivity in acidic electrolytes, and the creation of active sites for pseudocapacitance due to the presence of oxygen and nitrogen functional groups. These alterations modified the electrochemical characteristics of RGO. The RGO/CD composite, with a GO:CD mass ratio of 2:1, demonstrated favorable specific capacitance and enhanced cycling stability. As the current density increased from 0.2 to 100 A/g and the scan rate rose from 2 to 1,000 mV/s, the highest discharge capacitance achieved was 278 F/g, with capacitance retention values of 60% and 63%, respectively (Hoang et al., 2019).

Separately, through an environmentally friendly and cost-effective process, Hoang and Gomes had successfully synthesized CDs from zucchini, utilizing waste zucchini biomass as a readily available and economical carbon precursor (Hoang and Gomes, 2019). The resulting CDs displayed remarkable electrochemical properties. At a scan rate of 2 m/Vs, a notable maximum specific capacitance of 374 F/g was achieved. The composite exhibited exceptional capacitance retention, reaching 71.7% as the scan rate was increased from 2 to 1,000 mV/s. Moreover, the composite demonstrated remarkable cycling stability, maintaining its capacitance over 10,000 cycles at 10 A/g with a retention rate of 93.8%. Table 6.2 presents an overview of the electrochemical characteristics of CQDs or CDs for SC, based on the data compiled from selected literature sources.

6.6 CONCLUSIONS AND PERSPECTIVES

The SCs and EDLCs, as well as a rundown of current developments, were presented in this article. According to the methodology they store charge, these SC designs can be split into three classes, including EDLC, pseudocapacitors, and hybrid capacitors. SCs have a lot of potential for applications due to their long storage life, strong cycling

TABLE 6.3
Comparing Energy Storing Technologies (Roy and Rengarajan, 2015; Arumugham et al., 2020; Sharma and Kumar, 2020)

Property	Battery	Supercapacitor
Energy density	2–100	1–10
Power density	50–200	7,000–18,000
Efficiency	85%–95%	>95%
Cycle life	10^5 times	>105 times

reliability, and short load time (Kiamahalleh et al., 2012). SCs must be made cheap while still keeping their extended cycle life and high concentration of energy. SCs feature up to 90% recycling stability and symmetric charge and discharge throughout their entire nominal voltage range, which is a significant benefit. As the SC has very low energy density compared to batteries (comparison shown in Table 6.3), one of the most important strategies for addressing the low energy density issue is the adoption of new, better materials for electrochemical SC electrodes (Conway, 1999).

We learned from the above discussion that numerous researchers have developed and synthesized CQDs using a variety of methodologies. Excellent characteristics of the CQDs include good biocompatibility, lower toxicity, fluorescent nature, straightforward synthesis, excellent electrical conductivity, and affordable price. CQDs have also been utilized in numerous significant applications, such as involving sensing, bio-medicine, optoelectronics, and the energy sector (Magesh et al., 2022). Based on the outcomes of this research, formulating innovative electrode materials using CQDs could serve as a viable and environmentally friendly approach for energy storage (Inayat et al., 2023).

In this review, we have outlined the applications of nanomaterials based on CQDs in SCs, with a focus on their fabrication, structure, and energy storage potential. Below is a synopsis of the key roles that CQDs fulfill in energy storage and conversion systems, leveraging their unique physical attributes in tandem with the current state of research.

1. Enhanced SC properties can be achieved through augmented specific surface area and nitrogen doping in conventional carbon materials, leading to improved rates and capacitance. This effect is attributed to the facilitation of electrolyte ion storage within deep pores.
2. Substantially reducing volume change stress and enhancing electrode dynamics can be accomplished by downsizing electrode materials to the scale of QDs.

Extensive efforts have been dedicated by researchers to formulate diverse synthesis methods for QDs. Furthermore, various design and assembly strategies have been established to enhance the quality of QD-based electrode materials. However, prior to transitioning to practical application, there remain numerous scientific and technical challenges within this field that require resolution.

Research on CQD composites in the field of electrochemical energy storage is still in its early stages. The lack of comprehensive studies on the electrochemical behavior and energy storage mechanisms of CQDs and other components hinders a complete understanding of how different CQD attributes impact their performance in energy storage and conversion systems. Given this scenario, conducting thorough investigations into aspects such as the interactions between CQDs and other materials, and their influence on composite material properties, is crucial. These investigations should encompass both theoretical and experimental analyses.

This research concludes by highlighting the significance and immense potential of quantum dot composites in the development of high-performance energy storage and catalytic systems. It can be deduced that CQDs are progressively emerging as essential multifunctional materials for energy storage and conversion technologies. With the advancement of state-of-the-art technologies and characterization methods, novel physicochemical attributes of CQDs will likely be uncovered, further broadening their utility across diverse research fields. We anticipate that the exploration of enhanced electrode materials based on CQDs and their potential utilization in energy-related industries will witness a rapid and extensive expansion in the near future.

ACKNOWLEDGMENT

The research was supported by the Ministry of Higher Education (MoHE), through the Fundamental Research Grant Scheme (FRGS/1/2022/TK08/UTAR/02/7).

REFERENCES

Abbas, A., Mariana, L. T., & Phan, A. N. (2018). Biomass-waste derived graphene quantum dots and their applications. *Carbon*, 140, 77–99.

Arora, N., & Sharma, N. N. (2014). Arc discharge synthesis of carbon nanotubes: Comprehensive review. *Diamond and Related Materials*, 50, 135–150.

Arumugham, T., Manikandan, A., Amimodu, R. G., Munusamy, S., & Iyer, S. K. (2020). A sustainable synthesis of green carbon quantum dot (CQD) from Catharanthus roseus (white flowering plant) leaves and investigation of its dual fluorescence responsive behavior in multi-ion detection and biological applications. *Sustainable Materials and Technologies*, 23, e00138.

Atchudan, R., Edison, T. N. J. I., Manivannan, S., Perumal, S., Somanathan, T., & Lee, Y. R. (2021). Sustainable synthesis of carbon quantum dots from banana peel waste using hydrothermal process for in vivo bioimaging. *Physical E-Low-Dimensional Systems & Nanostructures*, 126, 114417.

Azam, N., Ali, M. N., & Khan, T. J. (2021). Carbon quantum dots for biomedical applications: Review and analysis. *Frontiers in Materials*, 8: 700403.

Braga, R. F., Fernandes, D. M., Adán-Más, A., Silva, T. M., & Montemor, M. (2023). The role of the precursor on the electrochemical performance of N,S Co-Doped graphene electrodes in aqueous electrolytes. *Batteries*, 9(3), 168.

Chen, J., Du, H., Xu, Y., Ma, B., Zheng, Z., Li, P., & Jiang, Y. (2021). A turn-on fluorescent sensor based on coffee-ground carbon dots for the detection of sodium cyclamate. *Journal of Materials Science: Materials in Electronics*, 32(10), 13581–13587.

Chen, S., Ramachandran, R., Mani, V., & Saraswathi, R. (2014). Recent advancements in electrode materials for the highperformance electrochemical supercapacitors: A review. *International Journal of Electrochemical Science*, 9(8), 4072–4085.

Chiu, Y., & Lin, L. (2019). Effect of activating agents for producing activated carbon using a facile one-step synthesis with waste coffee grounds for symmetric supercapacitors. *Journal of the Taiwan Institute of Chemical Engineers*, 101, 177–185.

Choi, H., & Yoon, H. (2015). Nanostructured electrode materials for electrochemical capacitor applications. *Nanomaterials*, 5(2), 906–936.

Conway, B. E. (1999). *Electrochemical Supercapacitors*. Springer, New York.

Costa, A. I., Barata, P. D., De Moraes, B. G. S., & Prata, J. V. (2022). Carbon dots from coffee grounds: Synthesis, characterization and detection of noxious nitroanilines. *Chemosensors*, 10(3), 113.

Cui, L., Ren, X., Wang, J., & Sun, M. (2020). Synthesis of homogeneous carbon quantum dots by ultrafast dual-beam pulsed laser ablation for bioimaging. *Materials Today Nano*, 12, 100091.

De Klerk, N. J. J., Van Der Maas, E., & Wagemaker, M. (2018). Analysis of diffusion in solid-state electrolytes through md simulations, improvement of the li-ion conductivity in β-Li3PS4 as an example. *ACS Applied Energy Materials*, 1(7), 3230–3242.

Desmond, L. J., Phan, A. N., & Gentile, P. (2021). Critical overview on the green synthesis of carbon quantum dots and their application for cancer therapy. *Environmental Science. NANO*, 8(4), 848–862.

Dey, S., Govindaraj, A., Biswas, K., & Rao, C. N. R. (2014). Luminescence properties of boron and nitrogen doped graphene quantum dots prepared from arc-discharge generated doped graphene samples. *Chemical Physics Letters*, 595-596, 203–208.

Ezugbe, E. O., & Rathilal, S. (2020). Membrane technologies in wastewater treatment: A review. *Membranes*, 10(5), 89.

Feng, X., & Zhang, Y. (2019). A simple and green synthesis of carbon quantum dots from coke for white light-emitting devices. *RSC Advances*, 9(58), 33789–33793.

Glavin, M., & Hurley, W. (2012). Optimisation of a photovoltaic battery ultracapacitor hybrid energy storage system. *Solar Energy*, 86(10), 3009–3020.

Gonzalez, A., Goikolea, E., Barrena, J. A., & Mysyk, R. (2016). Review on supercapacitors: Technologies and materials. *Renewable & Sustainable Energy Reviews*, 58, 1189–1206.

Gualous, H., Chaoui, H., & Gallay, R. (2016). Supercapacitor calendar aging for telecommunication applications. In: *INTELEC, International Telecommunications Energy Conference (Proceedings)*. Institute of Electrical and Electronics Engineers Inc. Austin, TX, USA.

Gupta, S., Singh, P. K., & Bhattacharya, B. (2018). Low-viscosity ionic liquid-doped solid polymer electrolytes. *High Performance Polymers*, 30(8), 986–992.

Hashmi, S., Kumar, A., & Tripathi, S. K. (2007). Experimental studies on poly methyl methacrylate based gel polymer electrolytes for application in electrical double layer capacitors. *Journal of Physics D*, 40(21), 6527–6534.

He, M., Zhang, J., Wang, H., Kong, Y., Xiao, Y., & Xu, W. (2018). Material and optical properties of fluorescent carbon quantum dots fabricated from lemon juice via hydrothermal reaction. *Nanoscale Research Letters*, 13(1).

Hoang, V. C., & Gomes, V. G. (2019). High performance hybrid supercapacitor based on doped zucchini-derived carbon dots and graphene. *Materials Today Energy*, 12, 198- 207.

Hoang, V. C., Nguyen, L. H., & Gomes, V. G. (2019). High efficiency supercapacitor derived from biomass based carbon dots and reduced graphene oxide composite. *Journal of Electroanalytical Chemistry*, 832, 87–96.

Hu, L., Zhang, C., Zeng, G., Chen, G., Wan, J., Guo, Z., Wu, H., Yu, Z., Zhou, Y., & Liu, J. (2016). Metal-based quantum dots: Synthesis, surface modification, transport and fate in aquatic environments and toxicity to microorganisms. *RSC Advances* 6(82), 78595–78610.

Hu, S., Li, J., Yang, J., Wang, Y., & Cao, S. (2011). Laser synthesis and size tailor of carbon quantum dots. *Journal of Nanoparticle Research*, 13(12), 7247–7252.

Hu, S., Niu, K., Sun, J., Yang, J., Zhao, N., & Du, X. (2009). One-step synthesis of fluorescent carbon nanoparticles by laser irradiation. *Journal of Materials Chemistry*, 19(4), 484–488.

Inayat, A., Albalawi, K., Rehman, A. U., Adnan, A., Saad, A. Y., Saleh, E. a. M., Alamri, M. A., El-Zahhar, A. A., Haider, A., & Abbas, S. M. (2023). Tunable synthesis of carbon quantum dots from the biomass of spent tea leaves as supercapacitor electrode. *Materials Today Communications*, 34, 105479.

Iro, Z. S. (2016). A brief review on electrode materials for Supercapacitor. *International Journal of Electrochemical Science*, 10628–10643.

Jalal, N. I., Ibrahim, R. I., & Oudah, M. K. (2021). A review on supercapacitors: Types and components. *Journal of Physics*, 1973(1), 012015.

Jiang, K., Sun, S., Zhang, L., Lu, Y., Wu, A., Cai, C., & Lin, H. (2015). Red, green, and blue luminescence by carbon dots: Full-Color emission tuning and multicolor cellular imaging. *Angewandte Chemie*, 54(18), 5360–5363.

Kiamahalleh, M. V., Zein, S. H. S., Najafpour, G., Sata, S. A., & Buniran, S. (2012). Multiwalled carbon nanotubes based nanocomposites for supercapacitors: A review of electrode materials. *Nano*, 7(2), 1230002.

Kim, B. K., Sy, S., Yu, A., & Zhang, J. (2015). Electrochemical supercapacitors for energy storage and conversion. In: *Handbook of Clean Energy Systems*, Yan J. (ed.), 1–25.

Kraev, G., Hinov, N., Arnaudov, D., Rangelov, N. R., & Gilev, B. (2016). Serial ZVS DC-DC converter for supercapacitor charging. In: *2016 19th International Symposium on Electrical Apparatus and Technologies, SIELA 2016.* Institute of Electrical and Electronics Engineers Inc.

Kumari, R., Singh, V., & Ravikant, C. (2023). Enhanced performance of activated carbon-based supercapacitor derived from waste soybean oil with coffee ground additives. *Materials Chemistry and Physics*, 305, 127882.

Kuparowitz, T. Charge transport and storage in a supercapacitor structure. Brno University of Technology, The Faculty of Electrical Engineering and Communication. Ph.D. Thesis (2010)

Lawu, B. L., Fuada, S., Ramadhan, S., Sabana, A. F., & Sasongko, A. (2017). Charging supercapacitor mechanism based-on bidirectional DC-DC converter for electric ATV motor application. In: *2017 International Symposium on Electronics and Smart Devices, ISESD 2017.* Institute of Electrical and Electronics Engineers Inc, pp. 129–132.

Li, H., Kang, Z., Liu, Y., & Lee, S. (2012). Carbon nanodots: Synthesis, properties and applications. *Journal of Materials Chemistry*, 22(46), 24230.

Li, H., Tao, L., Huang, F., Sun, Q., Zhao, X., Han, J., Shen, Y., & Wang, M. (2017). Enhancing efficiency of perovskite solar cells via surface passivation with graphene oxide interlayer. *ACS Applied Materials & Interfaces*, 9(44), 38967–38976.

Li, X., Wang, H., Shimizu, Y., Pyatenko, A., Kawaguchi, K., & Koshizaki, N. (2011). Preparation of carbon quantum dots with tunable photoluminescence by rapid laser passivation in ordinary organic solvents. *Chemical Communications*, 47(3), 932–934.

Lim, S. Y., Shen, W., & Gao, Z. (2015). Carbon quantum dots and their applications. *Chemical Society Reviews*, 44(1), 362–381.

Liu, Y., Roy, S., Sarkar, S., Xu, J., Zhao, Y., & Zhang, J. (2021). A review of carbon dots and their composite materials for electrochemical energy technologies. *Carbon Energy*, 3(5), 795–826.

Magesh, V., Sundramoorthy, A. K., & Dhanraj, G. (2022). Recent advances on synthesis and potential applications of carbon quantum dots. *Frontiers in Materials*, 9: 906838.

Mahani, M., Pourrahmani-Sarbanani, M., Yoosefian, M., Divsar, F., Mousavi, S. M., & Nomani, A. (2021). Doxorubicin delivery to breast cancer cells with transferrin-targeted carbon quantum dots: An in vitro and in silico study. *Journal of Drug Delivery Science and Technology*, 62, 102342.

Monte-Filho, S. S., Andrade, S. I. E., Lima, M. B., & De Araújo, M. C. U. (2019). Synthesis of highly fluorescent carbon dots from lemon and onion juices for determination of riboflavin in multivitamin/mineral supplements. *Journal of Pharmaceutical Analysis*, 9(3), 209–216.

Najib, S., & Erdem, E. (2019). Current progress achieved in novel materials for supercapacitor electrodes: Mini review. *Nanoscale Advances*, 1(8), 2817–2827.

Prasad, G. G., Shetty, N., Thakur, S., Rakshitha, & Bommegowda, K. B. (2019). Supercapacitor technology and its applications: A review. *IOP Conference Series*, 561(1), 012105.

Quan, H., Tao, W., Wang, Y., & Chen, D. (2022). Enhanced supercapacitor performance of Camellia oleifera shell derived hierarchical porous carbon by carbon quantum dots. *Journal of Energy Storage*, 55, 105573.

Ren, W., Chen, S., Li, S., Zhang, Y., Liu, J., Guan, M., Yang, H., Li, N., Han, C., Li, T., Zhao, Z., & Ge, J. (2018). Photoluminescence enhancement of carbon dots by surfactants at room temperature. *Chemistry: A European Journal*, 24(59), 15806–15811.

Roy, B. P., & Rengarajan, N. (2015). Application of Supercapacitors for Short term Energy Storage Requirements. *Asian Journal of Applied Sciences*, 8(2), 158–164.

Sadat, A., Ding, H., Xu, M., Hu, X., Li, Z., Wang, J., Liu, L., Jiang, L., Wang, D., Dong, C., Yan, M., Wang, Q., & Hong, B. (2019). Recent advances in synthesis, optical properties, and biomedical applications of carbon dots. *ACS Applied Bio Materials*, 2(6), 2317–2338.

Şahin, M. E., & Blaabjerg, F. (2020). A hybrid PV-battery/supercapacitor system and a basic active power control proposal in MATLAB/Simulink. *Electronics*, 9(1), 129.

Sahin, M.E., Blaabjerg, F., & Sangwongwanich, A. (2020). A review on supercapacitor materials and developments. *Turkish Journal of Materials* 5, 10–24.

Salanne, M. (2017). Ionic liquids for supercapacitor applications. *Topics in Current Chemistry*, 375(3).

Sharma, P. R., & Kumar, V. (2020). Current technology of supercapacitors: A review. *Journal of Electronic Materials*, 49(6), 3520–3532.

Shen, P., & Xia, Y. (2014). Synthesis-modification integration: One-step fabrication of boronic acid functionalized carbon dots for fluorescent blood sugar sensing. *Analytical Chemistry*, 86(11), 5323–5329.

Shi, W., Zhu, J., Sim, D., Tay, Y. Y., Lu, Z., Zhang, X., Sharma, Y., Srinivasan, M., Zhang, H., Hng, H. H., & Yan, Q. (2011). Achieving high specific charge capacitances in Fe3O4/reduced graphene oxide nanocomposites. *Journal of Materials Chemistry*, 21(10), 3422.

Simon, P., & Gogotsi, Y. (2008). Materials for electrochemical capacitors. *Nature Materials*, 7(11), 845–854.

Sk, M. A., Ananthanarayanan, A., Huang, L., Lim, K. H., & Chen, P. (2014). Revealing the tunable photoluminescence properties of graphene quantum dots. *Journal of Materials Chemistry C*, 2(34), 6954–6960.

Thangaraj, B., Solomon, P. R., & Ranganathan, S. (2019). Synthesis of carbon quantum dots with special reference to biomass as a source: A Review. *Current Pharmaceutical Design*, 25(13), 1455–1476.

Tuerhong, M., Xu, Y., & Yin, X. (2017). Review on carbon dots and their applications. *Chinese Journal of Analytical Chemistry*, 45(1), 139–150.

Tungare, K., Bhori, M., Racherla, K. S., & Sawant, S. (2020). Synthesis, characterization and biocompatibility studies of carbon quantum dots from Phoenix dactylifera. *3 Biotech*, 10(12).

Wang, F., Wu, X., Yuan, X., Liu, Z., Zhang, Y., Fu, L., Zhu, Y., Zhou, Q., Wu, Y., & Huang, W. (2017). Latest advances in supercapacitors: From new electrode materials to novel device designs. *Chemical Society Reviews*, 46(22), 6816–6854.

Wang, X., Feng, Y., Dong, P., & Huang, J. (2019). A mini review on carbon quantum dots: Preparation, properties, and electrocatalytic application. *Frontiers in Chemistry*, 7: 671.

Wang, Y., & Hu, A. (2014). Carbon quantum dots: Synthesis, properties and applications. *Journal of Materials Chemistry C*, 2(34), 6921.

Wasterlain, S., Gualous, H., & Fauvarque, J.F. (2006). Hybrid power source with batteries and supercapacitor for vehicle applications. *Esscap2006*.

Xiao, W., Xia, H., Fuh, J., & Li, L. (2009). Growth of single-crystal α-MnO2 nanotubes prepared by a hydrothermal route and their electrochemical properties. *Journal of Power Sources*, 193(2), 935–938.

Yang, Z., Li, Z., Xu, M., Ma, Y., Zhang, J., Su, Y., Gao, F., Wei, H., & Zhang, L. (2013). Controllable synthesis of fluorescent carbon dots and their detection application as nanoprobes. *Nano-Micro Letters* 5, 247–259.

Yuan, F., Ding, L., Li, Y., Li, X., Fan, L., Zhou, S., Fang, D., & Yang, S. (2015). Multicolor fluorescent graphene quantum dots colorimetrically responsive to all-pH and a wide temperature range. *Nanoscale*, 7(27), 11727–11733.

Zhai, X., Zhang, P., Liu, C., Bai, T., Li, W., Dai, L., & Liu, W. (2012). Highly luminescent carbon nanodots by microwave-assisted pyrolysis. *Chemical Communications*, 48(64), 7955.

Zhang, C., Zhu, F., Xu, H., Liu, W., Yang, L., Ma, J., Kang, Z., & Liu, Y. (2017). Significant improvement of near-UV electroluminescence from ZnO quantum dot LEDs via coupling with carbon nanodot surface plasmons. *Nanoscale*, 9(38), 14592–14601.

Zhang, Q., Sun, X., Hong, R., Yin, K., & Li, H. (2017). Production of yellow-emitting carbon quantum dots from fullerene carbon soot. *Science China. Materials*, 60(2), 141–150.

Zhang, X. (2019). Mesoporous biochar derived from silkworm excrement for high performance supercapacitors. *International Journal of Electrochemical Science*, 14(9), 8793–8804.

Zuo, P., Xiuhua, L., Sun, Z., Guo, Y., & He, H. (2015). A review on syntheses, properties, characterization and bioanalytical applications of fluorescent carbon dots. *Mikrochimica Acta*, 183(2), 519–542.

7 Tin-Based Anodes for Next-Generation Lithium-Ion Batteries

L. P. Teo, M. H. Buraidah, and A. K. Arof

7.1 INTRODUCTION

Batteries are inseparable from modern daily lives. They are found in small items such as watches, kitchen scales, clocks, remote controls, toy cars, calculators, mobile phones, and laptops, and in larger objects including cars, electric buses, and satellites, as well as in space applications.

Batteries, in general, are distributed into two broad classifications: primary and secondary. The former is non-rechargeable and must be disposed of after a single usage, whereas the latter are rechargeable and can be used multiple times. A very good representative of secondary batteries is lithium-ion batteries (LIBs). Apart from the above-mentioned items, LIB can also be found in iPads, digital cameras, video camcorders, hybrid and electric vehicles, and so on.

Basically, LIBs are electrochemical cells that produce electrical energy from their stored chemical energy. The main constituents of LIBs are the anode and cathode, along with the electrolyte and separator separating the two electrodes. Here, we focus solely on the anode. The anode of a LIB is the component that donates electrons to the exterior circuit and is oxidized when the electrochemical reaction occurs. An ideal anode in an LIB should have features such as the ability to intercalate/deintercalate lithium ions at low potential with great capacity, diffuse Li^+ ions quickly and conduct them fast, undergo minimal structural and redox potential variation during intercalation/deintercalation of Li^+ ions (Tao et al., 2011). Anodes should also form a good solid-electrolyte interface (SEI) with the electrolyte during the initial cycle and remain compatible in subsequent cycles.

Essentially, anode materials can be sectionalized into two categories: carbon and non-carbonaceous materials. Graphite can be acknowledged as the most commonly used carbon-based anode owing to its distinctive features, which include (1) abundance in nature, (2) environmental benignity, (3) easy intercalation/deintercalation of Li^+ ions into its structure, (4) flat and low working potential versus lithium, (5) good cycle life, and (6) relatively cheap (Zhang, 2011; Tao et al., 2011; Manthiram, 2011; Chen, 2013). The theoretical specific capacity of an anode can be obtained from the following formula:

$$\text{Theoretical capacity} = \frac{\text{Faraday constant (Ah/mol)}}{3{,}600\,\text{s}} \div \left(\text{Molar mass} \times \text{Lithiated number}\right)$$

DOI: 10.1201/9781003314424-7

As an example, let's consider graphite. Graphite has a molar mass of 12 g/mol and is a type of carbon material consisting of 6C atoms, which can lithiate with 1 Li. Based on this, it is known that its theoretical capacity is 372 mAh/g. Therefore, it is crucial to identify how many atoms of the active material will be lithiated with 1 Li. To ease readers' understanding, the following Table 7.1 provides the values of the lithiated phase of some anode materials along with their properties.

Graphite also has other disadvantages, such as (1) low energy density, (2) a low Li^+ ion insertion potential (0.05 V versus lithium), resulting in lithium adhering to the surface of graphite and potentially leading to safety issues, (3) a low Li^+ ion chemical diffusion coefficient (10^{-7} to 10^{-9} cm^2/s), and (4) the possibility of organic solvent co-intercalation in graphite (Wakihara, 2001; Zhang, 2011; Tao et al., 2011; Manthiram, 2011; Zhang et al., 2013; Goriparti et al., 2014). These disadvantages have motivated researchers to explore alternative anode materials, such as silicon (Si), aluminum (Al), tin (Sn), magnesium (Mg), lithium titanate ($Li_4Ti_5O_{12}$), and various transition metal oxides like NiO, Co_3O_4, Fe_2O_3, etc. In this chapter, special attention is given to tin-based compounds, which show promising potential as anodes for the next generation of LIB. Herein, some achievements with tin-based materials used as anodes in LIB are presented.

From the periodic table, tin is positioned in Group IV along with other elements: carbon (C), silicon, germanium (Ge), and lead (Pb). Although silicon possesses a higher theoretical capacity than tin (Table 7.1), the latter is less brittle and has higher electrical conductivity compared to the former (Zhao et al., 2015). In addition, tin-based materials are cheaper, abundant in nature, environmentally acceptable, and easier to process compared to silicon-based materials (Kamali and Fray, 2011; Chen, 2013; Wan et al., 2014).

On the other hand, germanium is expensive, whereas lead is toxic, making it less attractive. Moreover, tin is non-toxic, and its working potential of approximately 0.6 V versus Li makes it safe and attractive without compromising the operating voltage of the cell. In 1997, Idota and et al. (1997) pronounced that amorphous tin composite oxide containing metallic elements, i.e., boron (B), phosphorus (P), and aluminum (Al), attained greater capacity (1,030 mAh/g) than graphite. This amorphous tin-based oxide delivered ~100% coulombic efficiency with a reversible capacity of 650 mAh/g upon the initial charge in the potential window 0–1.2 V versus Li/Li^+. It has also been reported that the amorphous nanostructured Sn–Co–C anode

TABLE 7.1
Suitable Anode Materials in Lithium-Ion Batteries (Wu and Cui, 2012)

Materials	Li	C	Si	Sn	Sb	Al	Mg
Density (g cm^{-3})	0.53	2.25	2.23	7.29	6.7	2.70	1.30
Lithiated phase	Li	LiC_6	$Li_{4.4}Si$	$Li_{4.4}Sn$	Li_3Sb	LiAl	Li_3Mg
Theoretical capacity (mAh/g)	3,862	372	4,200	994	660	993	3,350
Volume change (%)	100	12	320	260	200	96	100
Potential vs. Li (V)	0	0.05	0.4	0.6	0.9	0.3	0.1

exhibited a 30% increment in volumetric capacity compared to a typical carbon anode (Whittingham, 2008). This makes tin-based materials more attractive for study. The Sn–Co–C anode has been marketed by Sony.

One crucial benefit of tin over graphite and carbon-based materials is that there is no co-intercalation of solvents in tin-containing substances, making it safe (Hu et al., 2012). Besides this, tin also has a higher theoretical capacity and relatively high lithium insertion potential compared to graphite (Wang et al., 2011; Yoon and Manthiram, 2011).

Tin is known to belong to a tetragonal crystal structure with lattice constants, a = 0.5831 nm and c = 0.3182 nm (Hu et al., 2012). As mentioned earlier, the theoretical capacity of tin without any impurity is 994 mAh/g, which is about three times greater than graphite (372 mAh/g), based on the final lithiated phase $Li_{4.4}Sn$. Tin can interact with lithium to form numerous lithium alloys, namely LiSn, Li_2Sn_5, Li_3Sn_2, Li_5Sn_2, Li_7Sn_2, Li_7Sn_3, and $Li_{22}Sn_5$ following the reaction (Winter and Besenhard, 1999; Zhang et al., 2019):

$$Li_xSn \leftrightarrow xLi^+ + xe^- + Sn$$

Crystallographic studies have suggested that $Li_{17}Sn_4$ (4.25 Li per Sn) is the realistic form of the end lithiated phase and not the previously assumed $Li_{22}Sn_5$ phase (Chen, 2013). Hence, its maximum gravimetric capacity is 959.5 mAh/g, which is still much higher than that of graphite. It has been reported that pure tin foil (bulk) delivered a reversible capacity of 600 mAh/g for only 10–15 cycles (Yang et al., 2003). After 15 cycles, the expansion and contraction of the electrode crystalline lattice occurred, resulting in an increase in cell impedance and a decrement in capacity. An electroplated Sn thin film on copper foil exhibited an initial capacity of 860 mAh/g and a reversible capacity of 200 mAh/g after 20 cycles (Morimoto and Tobishima, 2005). The highest capacity of 940 mAh/g has been recorded for electroplated tin, which is close to the theoretical capacity (Morimoto and Tobishima, 2005).

However, a pure tin anode undergoes a large volume change (260%) due to lithium-ion insertion/extraction (Kamali and Fray, 2011). This causes breakup/cracking of the tin anode, limiting its cycle life and resulting in an increase in cell impedance. To overcome this problem, tin in nanostructured form, tin-based alloys, and nanocomposites containing tin have been extensively studied, as a smaller particle size can help prevent the aggregation of tin into large clusters. Generally, when the size of tin particles is at the nanometer level, the lithium-ion diffusion length can be significantly reduced, thus reducing the volume expansion/contraction during repeated cycling while maintaining structural integrity.

7.2 TIN-BASED ALLOYS

By dispersing tin particles in an inactive matrix, the volume change problem can be alleviated. The electrochemically conducting and mechanically ductile inactive matrix provides a structurally stable form to counter the volume change of tin and decrease irreversible capacity loss. Many intermetallic alloys, such as Cu_6Sn_5, $FeSn_2$, FeSn, and others, have shown improved performance compared to pure tin.

$CoSn_3$, $FeSn_2$, Ni_3Sn_4, and Cu_6Sn_5 have theoretical capacities of 852, 804, 725, and 605 mAh/g, respectively, to name a few. Kepler et al. (1999) reported that Cu_6Sn_4 exhibited better performance compared to Cu_6Sn_5 and Cu_6Sn_6 for all 20 cycles under a constant current of 0.1 mA in the potential window 0–1 V. On the first cycle, the Cu_6Sn_4 cell delivered a capacity of ~450 mAh/g and ~150 mAh/g at the 20th cycle, while the Cu_6Sn_6 cell exhibited capacities of ~350 mAh/g and ~80 mAh/g on the first and 20th cycles, respectively. The Cu_6Sn_5 cell delivered a capacity of ~345 mAh/g at the end of the first cycle and ~70 mAh/g at the end of the 20th cycle. Zou and Wang (2013) prepared tin sulfide (SnS_2) via a microwave-assisted solvothermal technique, which delivered capacities of 1,755 and 370 mAh/g at the first and 50th cycles, respectively, at a current density of 100 mA/g in the voltage range from 0.005 to 1.1 V. A recent review in the literature is dedicated to tin sulfides (Mou et al., 2020). Zhang et al. (2008) reported that both $FeSn_2$ prepared by solvothermal and chemical reduction methods exhibited initial discharge capacities of ~640 mAh/g and ~750 mAh/g at a current density of 80 mA/g. However, solvothermal-prepared $FeSn_2$ maintained a reversible discharge capacity of ~435 mAh/g, whereas $FeSn_2$ prepared via the chemical reduction technique yielded a capacity of ~400 mAh/g at the 20th cycle. In another report, an LIB using $FeSn_2$ nanospheres attained a capacity of ~500 mAh/g for almost 15 cycles continuously (Wang et al., 2012a). The commercialization of the Nexelion battery, which consists of amorphous nano-SnCo inserted in carbon, delivered an initial capacity of 300 mAh/g but then decreased to ~200 mAh/g after 10 cycles (Fan et al., 2007). Sn/Fe/C composite made from mechanical milling utilizing Al, Ti, and Mg as reducing agents with various grinding media exhibited a specific capacity of 600 mAh/g (Zhang and Whittingham, 2010). The Sn-Fe carbon cell also showed good capacity retention upon cycling for 200 cycles, reaching a specific capacity of 250 mAh/g in the potential window between 0.01 and 1.5 V at a current density of 5 mA/cm^2. It is interesting to note that Dang and co-workers (2015) prepared micro-$Sn_{0.9}Se_{0.1}$ alloy via the jet-milling method instead of nanosize, maintaining a capacity of ~500 mAh/g after 100 cycles at a 0.5 C rate.

7.2.1 Tin-Based Oxides

Tin-based oxide materials have garnered considerable attention due to their stability, ease of synthesis with varying morphologies, and controllable particle sizes. These oxides are known for exhibiting reduced capacity fading compared to pure tin. In this context, we will explore various tin oxide materials used as anodes in LIBs.

7.2.2 Tin (II) Oxide (SnO)

SnO crystallizes in a tetragonal structure with lattice parameters, $a=b=3.803$ Å and $c=4.849$ Å (Macías et al., 2011). From the literature, it can be understood that the reaction mechanism between SnO and Li^+ ion can be summarized as follows (Uchiyama et al., 2008):

- $SnO + 2Li^+ + 2e^- \rightarrow Sn + Li_2O$
- $Sn + xLi^+ + xe^- \leftrightarrow Li_xSn\ (0 \leq x \leq 4.4)$

As seen in the first equation, the electrochemical reaction of SnO with Li^+ ions leads to the formation of metallic tin and Li_2O, which is an irreversible process. After that, the metallic tin reacts reversibly with Li^+ ions to form Li_xSn alloy, as shown in the second equation. SnO is said to possess a high theoretical specific capacity of 875 mAh/g (Uchiyama et al., 2008). However, the synthesis of SnO is challenging since Sn^{2+} can easily oxidize to Sn^{4+}, making it less studied compared to its counterpart, SnO_2.

Zhang and co-workers succeeded in obtaining ultrathin SnO nanosheets using a facile one-step hydrothermal method, utilizing tin dichloride dihydrate ($SnCl_2{\cdot}2H_2O$) and hexamethylenetetramine as precursors (Zhang et al., 2014). The SnO nanosheets delivered high specific discharge capacities of 1,496, 882, 559, and 305.4 mAh/g at the first, fifth, 20th, and 40th cycles, respectively, under a constant specific current of 100 mA/g in the potential window between 0.005 and 3 V. Even at higher current densities of 200 and 500 mA/g, the SnO nanosheets achieved capacities of 785.4 and 606.3 mA/g, respectively, indicating their good rate capability (Zhang et al., 2014). The SnO nanosheets have a short diffusion length for Li^+ ion insertion and a large surface area to facilitate fast Li^+ ion transport. In addition, the space between the nanosheets, built up in a disordered manner, can act as a buffer against volume expansion/contraction during Li+ ion insertion/de-insertion (Zhang et al., 2014).

SnO prepared in the form of nanoplatelets was reported for the first time using a reducing atmosphere flame spray pyrolysis technique without a catalyst and surfactant (Hu et al., 2014). The SnO nanoplatelets, which are in a rounded square shape and have an average diameter of ~250 nm with a thickness of ~30 nm, exhibited initial discharge and charge capacities of ~1,900 mAh/g and ~950 mAh/g, respectively, under a specific current drain of 100 mA/g between 0.005 and 2 V. The discharge capacity was maintained at 330 mAh/g after 40 cycles. Similarly, hydrothermal-derived SnO nanoflowers exhibited an initial discharge capacity of ~2,000 mAh/g and charge capacity of 950 mAh/g in the potential window between 0.01 and 2 V, with a reversible capacity of ~200 mAh/g at the end of the 25th cycle (Iqbal et al., 2012). Therefore, it can be inferred that the morphology of SnO will affect the electrochemical performance. It has been reported that SnO morphology can be altered by using different solvents in the wet chemistry route (Jaśkaniec et al., 2021). SnO prepared using solvents, viz., ethanol, deionized water, 1-hexanol, and a mixture of water/methanol (30:70 ratio), displayed flower-like, square-shaped, cracked squares, and platelets-like morphologies (Jaśkaniec et al., 2021). Their capacity performance decreases in the order of SnO (methanol) > SnO (ethanol) > SnO (hexanol) > SnO (deionized water) at a 0.1C rate.

Although there is relatively little research involving SnO compared to research on SnO_2, SnO shows good electrochemical performance in some cases. Thomas and Rao (2014) reported that SnO nanoplates embedded in graphene nanosheets (GNSs) exhibited better performance than SnO_2 nanoparticles surrounded by GNS. The two tin oxide-based materials were deposited by the evaporation of tin granules in oxygen. The phase and morphology of the tin oxide-based materials were controlled by the evaporation rate of tin granules. The initial discharge capacity of 4,004 and 2,461 mAh/g were obtained by SnO-GNS and SnO_2-GNS, respectively. SnO-GNS also showed a higher reversible capacity of 1,022 mAh/g than that of SnO_2-GNS

(715 mAh/g) after 40 cycles at a current density of 23 μA/cm^2. The unique characteristics of GNS, such as high flexibility, good conductivity, a large surface, and chemical stability, make it a good matrix to embed the tin oxide materials to reduce strain during lithiation/de-lithiation processes and thus minimize volume change.

On the other hand, Hong and Kang (2015) prepared core-shell-structured tin oxide-carbon composite powders via a one-pot spray pyrolysis using a solution containing tin (II) oxalate (C_2O_4Sn) and polyvinylpyrrolidone (PVP). The tin oxide-carbon (SnO_x-C) composite with a mole ratio SnO:SnO_2 of 24:76 (6:19) exhibited an initial discharge capacity of 1,667 mAh/g, whereas bare tin dioxide (SnO_2) only delivered 1,473 mAh/g at a current density of 1 A/g. At a higher current density of 2 A/g, the tin oxide-carbon composite showed a stable discharge capacity of 1,033 mAh/g at the end of the 500th cycle, but the discharge capacity of bare SnO_2 decreased continuously to 78 mAh/g at the 500th cycle. SnO_x-C also showed better rate capability than bare SnO_2 when cycled at different current densities. The excellent performance of the tin oxide-carbon composite can be attributed to the low porosity property of carbon, which enables the formation of a stable SEI layer and thus prevents the electrolyte from penetrating into the composite (Hong and Kang,, 2015).

7.2.3 Tin (IV) Oxide (SnO_2)

Similar to SnO, SnO_2 also has a tetragonal crystal structure and lattice parameters of $a=b=4.738$ Å and $c=3.188$ Å (Chandra Bose et al., 2002; Liu et al., 2017). SnO_2 forms an alloy with lithium in a Li/SnO_2 half-cell in which the reactions can be written as follows (Sivashanmugam et al., 2005; Kim et al., 2014):

- $SnO_2 + 4Li^+ + 4e^- \rightarrow Sn + 2Li_2O$
- $Sn + xLi^+ + xe^- \leftrightarrow Li_xSn\ (0 < x < 4.4)$

The first equation illustrates the irreversible reduction of tin dioxide to tin, which is dispersed in a lithium oxide (Li_2O) matrix during the first discharge process. The Li_2O matrix is said to help alleviate volume change but not for prolonged cycling (Wang et al., 2011). On the other hand, the second equation indicates the reversible alloying/dealloying processes, where the metallic tin reacts with Li-ion to produce $Li_{4.4}Sn$ alloy. This shows that one Sn atom is capable of storing 4.4 Li, resulting in a theoretical specific capacity of 783 mAh/g. Experimentally, an initial capacity of ~1,689 mAh/g was obtained by SnO_2 nanoparticles prepared via the sol-gel technique at a constant current of 0.1 C in the 0.01–2.00 V range (Liu et al., 2017). The capacity declined to ~571 mAh/g after the 20th cycle. Ultrafine SnO_2 nanoparticles recorded a discharge capacity of ~1,197 mAh/g and ~273 mAh/g at the first and 50th cycles, respectively, under a 100 mA/g current density (Yin et al., 2016). Using the pulse laser deposition technique, SnO_2 thin film anode first attained the specific capacity of ca. 642 mAh/g and then decreased progressively from 400 mAh/g (100th cycle) to 200 mAh/g (250th cycle) before maintaining to this value for the next 6,250 cycles under a 2 C current density in the potential window 0.01–3 V (Dai et al., 2021). Nonetheless, SnO_2 is still affected by volume change. Therefore, the performance of SnO_2 can be improved by coating with carbon (Wang et al., 2011) and polypyrrole

(PPy) as well as doping with copper (Cu), nitrogen (N), indium (In), to name a few (Wan et al., 2014;).

Deposition of a small quantity of cobalt (Co) element into SnO_2 can hinder particle aggregation and thus enhances lithium-storage capability (Mei et al., 2012). Co-SnO_2 delivered high capacities of 1,956, 1,346, and 810 mAh/g at the first, second, and 50th cycles, which are higher than that of pure SnO_2 (first cycle—1,584 mAh/g, second cycle—983 mAh/g, 50th cycle—499 mAh/g). Nanocrystalline $Sn_{0.9}In_{0.1}O_2$ showed better performance with a reversible capacity of ~500 mAh/g at the 10th cycle than the undoped nano-SnO_2 (~300 mAh/g) (Subramanian et al., 2004). Wan et al. (2014) revealed that co-doping of Cu and N into SnO_2 via the hydrothermal method exhibited enhanced capacity retention when compared to Cu-doped SnO_2 and pure SnO_2. The charge capacities of SnO_2, Cu/SnO_2, and Cu-N/SnO_2 were recorded at 1,624, 1,846, and 1,939 mAh/g, respectively, at the first cycle and 200, 462, and 664 mAh/g after the 50th cycle. It is worth mentioning that antimony-doped SnO_2 nanopowders can maintain a reversible capacity of 637 mAh/g at the end of the 100th cycle with a high initial discharge capacity of 2,400 mAh/g (Wang et al., 2009a). Fluorine-doped SnO_2 nanoparticles attached to reduced graphene oxide (RGO) sheets have been reported with good electrochemical properties (100 mA/g: 1,037 mAh/g at the 150th cycle; 500 mA/g: 733 mAh/g at the 250th cycle; 1 A/g: 860 mAh/g; 2 A/g: 770 mAh/g) (Cui et al., 2017). In another report, SnO_2 nanoparticles have been doped with molybdenum (Mo) and then implanted in graphite nanosheets by Feng et al. (2020). This design produced astonishing results (1,317 mAh/g capacity at the 200th cycle under a 0.2 A/g current density; 759 mAh/g capacity at the 950th cycle under a 1 A/g current density).

Carbon coated on SnO_2 acts as a shield, minimizing large volume alteration and improving electronic conductivity, resulting in better performance (Wang et al., 2011). Carbon-coated SnO_2 hollow microspheres, produced via the hydrothermal method, showed improved cycling performance than uncoated SnO_2 (Guo et al., 2013). At a current density of 100 mA/g and a potential window between 0.005 and 3 V, the initial discharge and charge capacities were ~1,650 mAh/g and ~900 mAh/g for SnO_2 hollow microspheres. Under similar conditions, SnO_2 coated with carbon exhibited an irreversible discharge capacity of ~1,580 mAh/g and a charge capacity of ~1,100 mAh/g in the first cycle. After cycling 20 times, the carbon-coated SnO_2 delivered a discharge capacity of ~680 mAh/g, which was ~180 mAh/g more than uncoated SnO_2 at the same cycle. The hollow structure of SnO_2 has provided prereserved space to accommodate the volume expansion (Guo et al., 2013). A three-dimensional (3D) nanostructured SnO_2-carbon composite encapsulated in copper foam attained a slow capacity decline from a reversible capacity of ~1,099 mAh/g (second cycle) to 846 mAh/g (100th cycle) at a 0.2 A/g current density in the voltage window 0.01–1.50 V (Chen et al., 2016). Wang et al. (2021) confined nanoparticles of SnO_2 and Sn in a porous carbon network and recorded high-capacity retention of 1,105 mAh/g at the 290th cycle under a current rate of 200 mA/g. The electrode also showed decent performance at different current rates (500 mA/g: 453 mAh/g; 5,000 mA/g: 171 mAh/g; 1 A/g: 940 mAh/g; 10 A/g: 107 mAh/g) (Wang et al., 2021). In another paper, a ternary design of Sn with SnO_2 in the carbon matrix presented an initial reversible capacity of 2,248 mAh/g (100 mA/g) and a 475 mAh/g capacity at the 100th cycle (500 mA/g) (Saddique et al., 2022).

On the other hand, conducting polymers, namely PPy and polyaniline (PANI), can also help alleviate the volume expansion by providing a soft matrix that can accommodate the stress induced during lithium insertion and extraction (Cui et al., 2011). However, conducting polymers have only moderate conductivity, whereas the conductivity of SnO_2 is low (Huang et al., 2014). Therefore, Huang et al. (2014) prepared hierarchical SnO_2 with double carbon coating, which showed higher capacity and better cycling performance compared to uncoated SnO_2 and single coating of SnO_2 with PPy-derived carbon and RGO. The SnO_2 was first coated with carbon derived from PPy and then with RGO. Under the same specific current drain of 250 mA/g and potential window between 0.005 and 3 V, uncoated SnO_2 only exhibited a capacity of 37.4 mAh/g, whereas the capacities of 161.6, 291, and 550 mAh/g were obtained for PPy-derived carbon/SnO_2, RGO/SnO_2, and RGO/PPy-derived carbon/SnO_2, respectively, at the end of the 50th cycle (Huang et al., 2014). Recently, Shin and co-workers (2022) prepared an anode from porous nano-SnO_2 coated with PANI on a Cu substrate that can deliver a capacity of 1,644, 1,002, and 440 mAh/g at the first, second, and 50th cycles under a current density of 0.1 A/g and voltage region 0.01–3 V. No binder was used in such an electrode by the authors. The nano-SnO_2 anode without PANI can only achieve ~200 mAh/g at the 50th cycle, which is attributed to its higher charge-transfer resistance than the PANI-coated SnO_2 electrode (Shin et al., 2022).

In addition to coating and doping, researchers have also explored different morphologies of SnO_2, such as nanorods, nanowires, nanotubes, nanosheets, or hollow or porous nanostructures as ways to relieve the severe volume change problem. Nanostructured SnO_2 in the forms of wire, rod, tube, and sheet can withstand the large volume variation during Li+ ion insertion/extraction processes. Nano-SnO_2 in a one-dimensional (1D) structure (nanowires, nanorods, nanotubes) shows a large surface area to volume ratio, good electrical and optical characteristics, better mechanical stability, efficient electron transport, and excellent capacity.

7.2.4 SnO_2 Nanowires

Park et al. (2008) examined the morphology effect on the electrochemical performance of SnO_2 nanowires, SnO_2 nanotubes, and SnO_2 nanoparticles. Under similar conditions, i.e., a voltage window from 0.05 to 1.50 V and a specific current of 100 mA/g, both SnO_2 nanowires and nanotubes exhibited a higher reversible specific capacity of 300 and 100 mAh/g, respectively, compared to SnO_2 nanopowders with an average particle size of 100 nm at the end of the 15th cycle. SnO_2 nanowires, prepared by the self-catalysis growth technique with diameters between 200 and 500 nm, demonstrated improved performance compared to SnO_2 in powder form, attributed to the high surface area that provided more reaction sites in the nanowires (Park et al., 2007). Following the report by Meduri et al. (2009), SnO_2 nanowires achieved an initial capacity of 2,400 mAh/g with a reversible capacity of 166 mAh/g at the end of the 40th cycle. In comparison, a capacity of 2,800 and 490 mAh/g at the first cycle and 40th cycle, respectively, was observed for metal tin nanoclusters embedded in SnO_2 nanowire networks. Using the atomic layer deposition technique, Guan and co-workers (2014) prepared nanowired SnO_2 surrounded by TiO_2 nanotubes, with carbon cloth as the substrate. The TiO_2 nanotubes aimed to stabilize the SEI layer and endure the volume expansion. This formation achieved better electrochemical performance in terms of rate capability and cyclability compared to the electrode with SnO_2 nanowires only. SnO_2 nanowires doped with antimony (Sb) and coated with a nanocarbon shield exhibited excellent capacity retention, with a 585 mAh/g capacity at the 500th cycle under 100 mA/g, compared to anodes containing Sb-doped SnO_2 (152 mAh/g) and SnO_2 nanowires only (52 mAh/g) (Mousavi et al., 2021).

7.2.5 SnO_2 Nanorods

Similar to nanowires, the nanorod morphology is advantageous for the good cycling performance of SnO_2 due to its 1D structure, which can provide high inter-porosity and sufficient space to sustain the volume change. It is worth mentioning that Wen and co-workers (Wen et al., 2013) employed the hydrothermal technique to produce flower-like SnO_2 nanorod packets formed by adding oleic acid to a mixture of sodium hydroxide (NaOH) and tin tetrachloride pentahydrate ($SnCl_4{\cdot}5H_2O$). The flower-like SnO_2 nanorod bundles exhibited a first discharge capacity of 1,673 mAh/g and a charge capacity of 815 mAh/g in the potential window from 0 to 2.5 V at a 0.1C rate (78 mA/g). Upon continuous cycling for 40 cycles, its capacity could still maintain 694 mAh/g, which is ~2 times higher than that of graphite (Wen et al., 2013). In another report, the initial discharge capacity of SnO_2 nanorods by the hydrothermal method almost reached 1,600 mAh/g at a 0.2 C rate in the potential window 0.01–3.00 V (Zhang et al., 2019). Nonetheless, a capacity of ~400 mAh/g was obtained after the 40th cycle. There is a controversial report that revealed SnO_2 nanorods performed poorly (641 mAh/g) than SnO_2 nanoparticles (526 mAh/g) upon continuous cycling for 50 times at a 200 mA/g current density (Hameed et al., 2017). The former also suffered from a bigger irreversible capacity loss (first discharge capacity 1,529 mAh/g; first charge capacity 872 mAh/g) than the latter (first discharge capacity 1,529 mAh/g; first charge capacity 872 mAh/g), although it showed a higher initial discharge capacity. This is attributed to the lower value of the lithium diffusion coefficient ($4.36 \times 10^{-6} cm^2/s$) obtained by SnO_2 nanorods as compared to SnO_2 nanoparticles ($9.14 \times 10^{-6} cm^2/s$) (Hameed et al., 2017).

Nanorods of SnO_2 have been adhered to carbon to avoid direct contact with the electrolyte to prevent SEI cracks and showed better electrochemical performance (initial capacity 1,881 mAh/g, 884 mAh/g at 10th cycle) compared to SnO_2 nanorods without carbon at 100 mA/g current density and a voltage range of 0.005–1.5 V (Palaniyandy et al., 2019). Yu and co-workers (2013) have reported that hydrothermal-derived SnO_2 nanorods delivered discharge and charge capacities of 1,963 and 862 mAh/g at a 100 mA/g current density with a voltage range of 0.005–3 V. The charge and discharge capacities of 167 and 626 mAh/g were lower for carbon-coated SnO_2 nanorods under similar cycling conditions. After 10 times cycling at a specific current of 100 mA/g, the uncoated and carbon-coated SnO_2 exhibited reversible capacities of 690 and 653 mAh/g, respectively. Thereafter, the uncoated and carbon-coated SnO_2 were cycled at a specific current of 500 mA/g for another 10 cycles. It was found that the capacity of uncoated SnO_2 decreased from 474 mAh/g at the 11th cycle to 374 mAh/g at the 20th cycle, whereas carbon-coated SnO_2 showed stable cycling performance, maintaining the capacity between 190 and 200 mAh/g throughout the cycling. It is interesting to note that carbon coating in this case does not show improved performance as expected but instead shows capacity decline and poor rate capability. The authors have indicated that the lower conductivity of carbon-coated SnO_2 nanorods (4.2×10^{-6} S/m) than that of uncoated SnO_2 (9.9×10^{-4} S/m) may be the reason behind this (Yu et al., 2013). Another style of SnO_2 in the form of nanoribbons has been synthesized on a stainless-steel foil and used directly as an anode without a binder (Faramarzi et al., 2018). The initial irreversible capacity of ~1,818 mAh/g was obtained, followed by a sharp decline to ~929 mAh/g, thereafter gradually decreased to ~676 mAh/g after 20 times cycling at a 300 mA/g current density in the potential range 0.05–2.00 V. The authors have claimed that nano-SnO_2 in a ribbon arrangement showed less capacity fading than SnO_2 in nanowire form due to smaller thickness, shorter Li-ion diffusion length, and a large surface area (Faramarzi et al., 2018).

7.2.6 SnO_2 Nanotubes

The nanotube morphology has a beneficial effect on the electrochemical performance of SnO_2, where its 1D tubular structure restricts the movement of SnO_2, and thus prevents the SnO_2 particles from agglomerating. Also, the porous surface and cavities of the nanotubes provide empty space that can accommodate the structural strain and withstand the volume change. Ye et al. (2010) investigated the size effect of SnO_2 nanotubes on electrochemical properties. They found that smaller SnO_2 nanotube sizes prepared via the hydrothermal method using mesostructured silica nanorods as a template give high reversible capacity and good cycling performance (468 mAh/g after 30 cycles), which may be due to its more robust structure. In contrast, long SnO_2 nanotubes can only retain a capacity of ~395 mAh/g at the end of the 30th cycle. Zhu et al. (2015) designed an approach for preparing carbon-coated SnO_2 nanotubes by a hydrothermal route using zinc oxide (ZnO) nanorods as a template. The process involves forming ZnO-SnO_2 nanorods from the template ZnO, followed by hydrothermal carbonization with glucose as the carbon precursor to obtain carbon-coated SnO_2 nanotubes. The carbon-coated SnO_2 nanotubes have a capacity of ~1,711 mAh/g, ~834 mAh/g, ~454 mAh/g, and ~406 mAh/g at the first, second, 50th, and 100th cycles, respectively.

7.2.7 SnO_2 Nanosheets

SnO_2 nanosheets have a two-dimensional (2D) structure that offers structural flexibility to accommodate the large volume change upon lithium insertion/extraction. However, it is quite complex to prepare well-defined SnO_2 nanosheets due to the difficulty in controlling the reaction rate of the oxidation process of Sn^{2+} to Sn^{4+}, anisotropic crystal growth, and the hydrolysis of tin salt (Chen and Lou, 2012). Liu et al. (2013) prepared SnO_2 nanosheets by a hydrothermal route without utilizing both surface-active agents and organic solvents. The material showed good electrochemical performance with a large reversible capacity and excellent cycling stability. Cycling between 0 and 3 V, the specific capacity of SnO_2 nanosheets was maintained at 604 mAh/g over 50 cycles at a specific current of 100 mA/g, with initial discharge and charge capacities of 1,508 and 1,053 mAh/g, respectively. Flower-like spheres assembled from SnO_2 nanosheets have been reported by Wu et al. (2011) to possess good rate capability and a high reversible capacity of 400 mAh/g under a high current density of 800 mA/g. The authors believed that good electrochemical performance was attributed to a shorter lithium diffusion path. Other SnO_2 nanosheets also displayed promising performance at different current rates (731 mAh/g at 1 C; 678 mAh/g at 2 C; 424 mAh/g at 4 C) (Wang et al., 2021). It is worth mentioning that the SnO_2 nanosheets were very thin (ca. 20 nm) and utilized straightaway as an electrode without mixing with carbon black super P and polyvinylidene fluoride (PVDF) binder. Also, it is noteworthy that SnO_2 nanopillars prepared via pulsed laser deposition showed a 3D irregular cylindrical structure with a pyramid shape at the upper part, and accomplished admirable long-term cycling performance (Dai et al., 2021). SnO_2 nanopillars deposited on copper foil directly without PVDF binder exhibited 1,082, 832, 524, and 313 mAh/g at first, 300th, 1,100th, and 6,500th cycle, respectively, under 2 C in the 0.01–3 V range. Furthermore, good rate cyclability is also shown by this material (Dai et al., 2021). In short, such a design allows more space for expansion.

Wang et al. (2019) reported that SnO_2 nanospheres-implanted RGO forming a 2D architecture was capable of delivering the initial capacities of 1,212 and 1,063 mAh/g under 0.1 and 1 A/g specific currents, respectively, in the 0.01–3 V window. This anode material achieved a reversible capacity of 1,335 mAh/g at 1 A/g specific current after 500 times cycling, thus demonstrating superior long-term cycling at high current density. Another anode material showing good long-cycle life with high-rate capability was made from SnO_2 hollow nanospheres coated with PPy (Yuan et al., 2017). At various current rates of 100, 200, 500, and 1,000 mA/g, the mean specific capacities were 368, 307, 229, and 169 mAh/g, respectively. After cycling at different current densities, the electrode managed to obtain a capacity of 408 mAh/g when subjected again to a specific current of 100 mA/g. Under the same rate, ~900 mAh/g capacity was attained after 600 times cycling (Yuan et al., 2017). As expected, the nano-SnO_2 electrode without PPy coating performed poorly. The conducting polymer offers mechanical integrity and good conductivity. A more detailed review on SnO_2 (Mou et al., 2020) and specifically on SnO_2 nanosheets (Chen and Lou, 2012) is available elsewhere.

7.2.8 SnO_2/Carbon Composites

Whether it is coating or incorporating carbon into SnO_2, carbon provides good electronic conductivity to the SnO_2 and also buffers against volume expansion during Li^+ insertion. Hence, this will improve the capacity and cyclability. There are many ways to prepare SnO_2/carbon composite. Liu et al. (2012) synthesized SnO_2 nanorods grown on graphite by a hydrothermal method, which showed better electrochemical performance. The SnO_2/graphite nanocomposites have an initial discharge capacity of 1,197 mAh/g and a charge capacity of 872 mAh/g at a specific current of 50 mA/g and a voltage window between 0 and 3 V. However, under similar conditions, the lithium insertion capacity of SnO_2 nanorods is higher (~1,650 mAh/g) than that of both SnO_2/graphite and graphite (~600 mAh/g). Nevertheless, after 10 cycles, SnO_2/graphite delivers the highest reversible capacity of 627 mAh/g compared to graphite and SnO_2, with specific capacities of ~390 and ~300 mAh/g, respectively. The SnO_2/graphite composite also shows a high specific capacity at higher rates with a capacity of ~600 mAh/g at 100 mA/g and ~410 mAh/g at 500 mA/g, much higher than graphite and SnO_2. The graphite in the composite can obstruct SnO_2 nanorods from agglomeration and thus enhances the lithium-storage properties.

Chen and Yano (2013) designed and prepared a nanocomposite containing 80 wt.% SnO_2 nanoparticles and 20 wt.% carbon spheres via a simple infiltration procedure. An irreversible discharge capacity and charge capacity of 2,038 and 1,468 mAh/g were obtained at the first cycle, and the capacity was retained at 1,200 mAh/g after 30 cycles at a specific current of 80 mA/g and a potential window between 0.005 and 3 V. The high surface area mesoporous carbon spheres have open pore channels that enable the electrolyte to penetrate into SnO_2 and give fast Li^+ ion diffusion. The empty space at the interior of carbon can tolerate the volume expansion. Using the microwave-hydrothermal technique, Liu et al. (2015) prepared SnO_2/graphene composite which exhibits first discharge and charge capacities of 2,213 and 1,402 mAh/g at a specific current of 100 mA/g with a cutoff voltage window between 0.01 and 3 V. At the end of the 100th cycle, capacity remained at 1,359 mAh/g. At a large specific current of 1,000 mA/g, the SnO_2/graphene composite can still deliver a high discharge capacity of 1,502 and 677 mAh/g at the first and 1,000th cycles, respectively.

Ternary nanocomposite containing SnO_2-titanium dioxide (TiO_2) in GNSs has been prepared by a simple reflux method. It exhibits better performance with an initial discharge capacity of ~1,595 mAh/g and a reversible capacity of ~1,500 mAh/g at 100 mA/g between 0.01 and 3 V (Chen and Qin, 2014). A specific capacity of ~1,177 mAh/g can still be delivered even after 100 cycles. On the contrary, the discharge capacity of the binary SnO_2-graphene nanocomposite is ~1,384 mAh/g and ~208 mAh/g at the first and 100th cycles, respectively. The incorporation of TiO_2 hinders the cracking and prevents the loss of interparticle contact and indirectly improves the stability of SnO_2 during volume change, whereas the function of GNSs is as mentioned above. A much more comprehensive review of SnO_2-based nanomaterials can be found in references (Goriparti et al., 2014; Zhao et al., 2015; Zoller et al., 2019).

7.3 TERNARY TIN OXIDES

To the best of our knowledge, there is no single article that summarizes the literature review on almost all ternary tin oxides and their electrochemical properties, unlike SnO_2. Nonetheless, one recent article focuses only on three tin-based materials: manganese orthostannate (Mn_2SnO_4), cobalt orthostannate (Co_2SnO_4), and zinc orthostannate (Zn_2SnO_4), along with their applications in LIB and sodium-ion batteries (Duan et al., 2023). Another article provides an overview specifically for Zn_2SnO_4

and zinc metastannate ($ZnSnO_3$) (Sun and Liang, 2017). The latter emphasizes the synthesis–morphology relationship with a brief description of LIB. Herein, we provide a brief account of ternary tin-based oxide materials, namely lithium tin oxide (Li_2SnO_3), calcium metastannate ($CaSnO_3$), barium metastannate ($BaSnO_3$), cobalt metastannate ($CoSnO_3$), strontium metastannate ($SrSnO_3$), cadmium metastannate ($CdSnO_3$), calcium orthostannate (Ca_2SnO_4), magnesium orthostannate (Mg_2SnO_4), Mn_2SnO_4, Co_2SnO_4, and Zn_2SnO_4, as anodes in Li-ion batteries.

7.3.1 Lithium Tin Oxide or Lithium Stannate (Li_2SnO_3)

According to the literature (Lang, 1966; Tarakina et al., 2009), lithium stannate, or lithium tin oxide (Li_2SnO_3), exists in two forms: α-Li_2SnO_3, stable up to 800°C, and β-Li_2SnO_3, produced at 1,000°C. β-Li_2SnO_3 crystallizes in a monoclinic structure (space group of C2/c) with lattice parameters $a = 5.289$ Å, $b = 9.187$ Å, $c = 10.026$ Å, $\beta = 100.35°$ (Hodeau et al., 1982; Fu et al., 2016). Table 7.2 summarizes the electrochemical performance of Li_2SnO_3 reported by many researchers. It is evident from the table that the synthesis method affects the electrochemical performance. Li_2SnO_3 prepared by the sol-gel technique exhibited better performance than that prepared via solid-state reaction. This difference may be attributed to the morphology of the Li_2SnO_3 product. Li_2SnO_3 derived from the sol-gel method has uniform nanocrystallites with a narrow particle size distribution (200–300 nm), whereas an agglomerate of particles with different sizes is observed for Li_2SnO_3 prepared by the solid-state reaction method. In the sol-gel method, a homogeneous mixing of the starting materials can occur at the atomic or molecular level, resulting in a decrease in the product's particle size (Lin et al., 2013). On the other hand, in the solid-state reaction method, the reactants are simply ground together in an agate mortar and undergo a heating process. As seen in Table 7.2, it is noteworthy that the hydrothermal-derived Li_2SnO_3 exhibited the highest specific capacity at the first cycle and still attained a capacity of 613 mAh/g after 30 cycles (Wang et al., 2013), compared to those prepared via sol-gel and solid-state reaction techniques.

To further improve the electrochemical properties of Li_2SnO_3, researchers have explored coating and doping Li_2SnO_3 with materials such as PPy (Huang et al., 2012), carbon (Wang et al., 2012b, 2012c), graphene (Zhao et al., 2013a), and PANI (Wang et al., 2012d). A composite of bamboo charcoal/Li_2SnO_3, prepared via the sol-gel technique, exhibited an improvement in capacity (1,689 mAh/g at the first cycle and ~674 mAh/g at the 30th cycle) compared to a Li//Li_2SnO_3 cell with a capacity of ~1,196 and 202 mAh/g at the first and 30th cycles, respectively, under a specific current of 60 mA/g in the potential window from 0.01 to 2 V (Zhao et al., 2013b). The porous structure and large surface area of bamboo charcoal can reduce volume expansion/contraction during Li^+ intercalation/deintercalation reactions. Wang et al. (2012b) first synthesized Li_2SnO_3 by the sol-gel method before doping it with carbon via a carbothermic reduction process. The carbon-doped Li_2SnO_3 exhibited an initial discharge capacity of 1,671 mAh/g and maintained a reversible capacity of 576.9 mAh/g at the end of the 50th cycle. It is worth highlighting that carbon-doped Li_2SnO_3 has been employed by Wang et al. (2012b) as a cathode material instead of an anode material for LIBs.

TABLE 7.2
Literature Review on Li_2SnO_3

Method	Starting Materials	Capacity (mAh/g)	Potential Window (V)	Current or Current Density (mA or mA/g)	References
Solid-state reaction	SnO_2 and Li_2CO_3	1,200-first cycle; 550-second cycle	0–2.5	9.3 mA/g	Courtney and Dahn (1997)
Solid-state reaction	SnO_2 and Li_2CO_3 Calcined at 1,000°C	1,177-first cycle	0.01–2	0.05 mA	Belliard and Irvine (2001)
Solid-state reaction and ball-milling	SnO_2 and Li_2CO_3 Calcined at 650°C	1,167-first cycle	0.01–2	0.05 mA	Belliard and Irvine (2001)
Solid-state reaction	SnO_2 and Li_2CO_3 – ball-milled 15 h – first calcination at 800°C for 6 hours – second calcination at 1,000°C for 12 h	1,074-first cycle; ~363-30th cycle	0–1	20 mA/g-first cycle; 60 mA/g subsequent cycle	Zhang et al. (2006)
Sol-gel	$SnCl_4{\cdot}5H_2O$ and Li_2CO_3 – citric acid as complexing agent	1,157-first cycle; ~400-30th cycle	0–1	20 mA/g-first cycle; 60 mA/g subsequent cycle	Zhang et al. (2006)
Solid-state reaction	SnC_2O_4 and Li_2CO_3 – heating 900°C 12 h	1,250-first cycle; ~400-10th cycle	0–1.5	0.1 mA	Vaughey et al. (2007)
Hydrothermal	$SnCl_4{\cdot}5H_2O$ and LiOH	2,105-first cycle; ~613-30th cycle	0–2	60 mA/g	Wang et al. (2013)
Hydrothermal	$SnCl_4{\cdot}5H_2O$ and $LiOH.H_2O$ – calcined at 800°C for 6 hours	850-first cycle; <50-200th cycle	0.01–3	150 mA/g	Wang et al. (2019)
Hydrothermal	SnO_2 and $LiOH.H_2O$ – vacuum drying at 60°C overnight	2,582-first cycle; ~179-50th cycle	0.1–2	0.1C	Zakuan et al. (2019)

As reported by Zhao et al. (2013a), Li_2SnO_3 powders distributed on GNSs demonstrated good cycling performance with a reversible capacity of 582 mAh/g at the 50th cycle at 60 mA/g, which is higher than the capacity of GNSs (408 mAh/g). It is said that the GNSs are flexible and can accommodate the volume change, thereby improving the structural stability of the composite. On the other hand, a slight increment in Li-ion storage capability with a reversible capacity of 560 mAh/g after 50 cycles has been observed in PPy-coated Li_2SnO_3 compared to that of uncoated Li_2SnO_3 (510 mAh/g) (Huang et al., 2012). A similar observation in Li_2SnO_3/PANI nanocomposites (569 mAh/g at the 50th cycle) and pure Li_2SnO_3 (510 mAh/g at the 50th cycle) has been reported (Wang et al., 2012a). Both PPy and PANI, which are conducting polymers, can serve as a backbone for Li_2SnO_3, while their soft structure acts as a shield to Li_2SnO_3 in relieving the volume change during discharge/charge processes. According to literature (Yang et al., 2019), nanocomposites comprised of Li_2SnO_3, SnO_2, and Li_2SiO_3 produced improved cycling and rate performance than Li_2SnO_3 only due to the former having a better framework to alleviate volume expansion and improve electrode–electrolyte contact. Another work has reported an anode made of Li_2SnO_3, Li_2SiO_3, and graphene which gave ~1,017 mAh/g capacity at a current density of 150 mA/g in the voltage window 0.01–3 V (Wang et al., 2019). Lower capacity values were obtained for Li_2SnO_3-Li_2SiO_3 (935 mAh/g) and Li_2SnO_3 (860 mAh/g) electrodes. Li_2SnO_3-Li_2SiO_3-graphene electrode also showed slow capacity loss after 200 times cycling compared to the other two. It should be noted that in the work of Wang et al. (2019), both Li_2SnO_3 and Li_2SiO_3 were the active materials and contributed to the electrochemical process. A full cell has also been fabricated using the Li_2SnO_3-Li_2SiO_3-graphene anode and $LiMn_2O_4$ cathode. At a current rate of 40 mA/g, the recorded capacities were 263 and ~92 mAh/g at the first and 50th cycles, respectively (Wang et al., 2019).

In order to further enhance the electrochemical performance, a ternary structure incorporating carbon-based materials and conducting polymers in addition to Li_2SnO_3 has been proposed. Li-ion half cells employing ternary Li_2SnO_3 composites, i.e., carbon-doped Li_2SnO_3/graphene and graphene/Li_2SnO_3/PPy, have been reported to exhibit better electrochemical properties compared to binary composites and pure Li_2SnO_3 material (Zhao et al. 2013c, 2013d). Reversible capacities of ~825 and 699 mAh/g have been obtained for carbon-doped Li_2SnO_3/graphene and graphene/ Li_2SnO_3/PPy at the end of the 30th cycle under a current density of 60 mA/g in the potential window 0–2 V (Zhao et al., 2013c, 2013d). The improved performance of ternary Li_2SnO_3 composite is attributed to the presence of carbon/graphene and PPy, which can serve as a double protection for Li_2SnO_3 when volume change takes place during Li^+ ion insertion/extraction. It is interesting to note that nano-Li_2SnO_3 has been coated on $LiNi_{0.5}Co_{0.2}Mn_{0.3}O_2$ cathode and displayed better performance than the non-coated $LiNi_{0.5}Co_{0.2}Mn_{0.3}O_2$ cathode (Hu et al., 2016). According to the authors, Li_2SnO_3 worked as a shield to safeguard $LiNi_{0.5}Co_{0.2}Mn_{0.3}O_2$ particles from electrolyte decomposition. Other than that, it has been reported that Li_2SnO_3 prepared via vapor-solid technique showed luminescent properties and thus has potential in light-emitting devices (García-Tecedor et al., 2019).

7.3.2 Calcium Metastannate ($CaSnO_3$)

The crystal structure of $CaSnO_3$ is orthorhombic, with a unit cell having lattice parameters $a = 11.09$ Å, $b = 7.91$ Å, and $c = 11.34$ Å (Sharma et al., 2002). Table 7.3 lists the research conducted by other scientists on LIBs using $CaSnO_3$. It can be observed from Table 7.3 that a higher capacity was obtained in $CaSnO_3$ prepared by the sol-gel technique compared to that prepared by the solid-state reaction method. This is due to the uniform particle size of sol-gel-derived $CaSnO_3$, whereas solid-state-prepared $CaSnO_3$ has large clusters of particles with different sizes. The uniformity of the particle size results in a smaller diffusion length, thus leading to a higher capacity. On the other hand, $CaSnO_3$ prepared via the hydrothermal technique, which showed a porous flower-like morphology (Zhao et al., 2010), exhibited a higher capacity at the 30th cycle under a specific current drain of 60 mA/g compared to sol-gel-derived and solid-state-prepared $CaSnO_3$ at the same cycle (Sharma et al., 2002). It is interesting to note that by using different concentrations of PVP as a surfactant, different morphologies of $CaSnO_3$ can be produced, although the starting materials used in the hydrothermal method are the same (Kim et al., 2014). Higher capacity is found in cubic-shaped $CaSnO_3$, which was prepared using 5 mL PVP, than the star-shaped $CaSnO_3$ produced with 1 mL PVP. Based on these observations, it can be concluded that the morphology of the materials influences the electrochemical properties. Carbon-coated $CaSnO_3$ nanotubes have been prepared by the solvothermal method to achieve better electrochemical properties (Hu et al., 2012). The carbon-coated $CaSnO_3$ nanotubes exhibited an initial capacity of 1,296 mAh/g at a specific current of 60 mA/g, whereas the uncoated $CaSnO_3$ nanotubes gave only a capacity of ~1,020 mAh/g. At the 30th cycle, the uncoated $CaSnO_3$ nanotubes delivered a capacity of 60 mAh/g, whereas ~190 mAh/g was achieved by the carbon-coated $CaSnO_3$ in the potential range of 0.005–1.3 V. It is said that coating with amorphous carbon acts as a barrier to relieve the volume expansion-contraction and also helps to maintain the dimensional stability of the electrode during the alloying/dealloying reaction (Hu et al., 2012).

7.3.3 Strontium Metastannate ($SrSnO_3$)

Table 7.4 summarizes the methods of preparation for $SrSnO_3$ and the electrochemical properties in LIB applications available in the literature. It is worth mentioning that $SrSnO_3$ in powder form, prepared by both sol-gel and solid-state reaction methods, crystallizes in an orthorhombic crystal structure with lattice parameters $a = 5.71$ Å, $b = 5.69$ Å, and $c = 8.05$ Å (Sharma et al., 2005). However, Li et al. (2011) synthesized $SrSnO_3$ nanorods via cetyltrimethylammonium bromide (CTAB)-assisted hydrothermal method, which has a cubic structure with a lattice constant, $a = 8.04$ Å. The $SrSnO_3$ nanorods that formed by subjecting the precursor, i.e., $SrSn(OH)_6$ nanowires, to hydrothermal treatment in ethanol at 180°C for 8 hours and calcined in air at 700°C for 3 hours display the orthorhombic crystal structure (Hu et al., 2010). The $SrSn(OH)_6$ nanowires were previously prepared via the sonochemical method at room temperature without a surfactant. The $SrSnO_3$ nanorods prepared by this method exhibit an impurity peak of strontium carbonate ($SrCO_3$), but it does not

TABLE 7.3
Preparation Methods of $CaSnO_3$ and Their Electrochemical Performance in Li-Ion Batteries

Method	Starting Materials	Morphology	Capacity (mAh/g)	Potential Window (V)	Current or Current Density (mA or mA/g)	References
Solid-state reaction	SnO_2 and $Ca(NO_3)_2{\cdot}4H_2O$	Large clusters of different sizes particles	~1,054-first cycle; ~265–30th cycle	0.005–1	10 mA/g-first cycle; 60 mA/g subsequent cycle	Sharma et al. (2002)
Sol-gel	$CaCO_3$ and $SnCl_2$	Uniform nanocrystallites (200–300 nm)	~1,083-first cycle; ~340–30th cycle	0.005–1	10 mA/g-first cycle; 60 mA/g subsequent cycle	Sharma et al. (2002)
Hydrothermal	$Ca(NO_3)_2{\cdot}4H_2O$ and $Na_2SnO_3{\cdot}3H_2O$	Porous flower-like	1,725-first cycle; 544–30th cycle	0.005–1	60 mA/g	Zhao et al. (2010)
Hydrothermal	$CaCl_2$ and Na_2SnO_3 – 1 mL PVP as surfactant	Star-shaped particles with size of ~ µm	~872-first cycle; ~228–30th cycle	0–1	50 mA/g	Kim et al. (2014)
Hydrothermal	$CaCl_2$ and Na_2SnO_3 – 5 mL PVP as surfactant	Cubic-shaped with smaller particles size (~500 nm)	~1,238-first cycle; ~386–30th cycle	0–1	50 mA/g	Kim et al. (2014)
Electrospinning	$SnCl_2{\cdot}2H_2O$ and $Ca(NO_3)_2{\cdot}4H_2O$ Metal: PAN = 1.30:1	Eggroll-like nanotubes Average diameter: 560 nm	~1,153-first cycle; ~648–50th cycle	0.005- 3	60 mA/g	Li et al. (2013)
Electrospinning	$SnCl_2{\cdot}2H_2O$ and $Ca(NO_3)_2{\cdot}4H_2O$ Metal: PAN = 0.65:1	Nanorods average diameter: 380 nm	~1,177-first cycle; ~410–50th cycle	0.005–3	60 mA/g	Li et al. (2013)

$Ca(NO_3)_2{\cdot}4H_2O$, calcium nitrate tetrahydrate; $CaCO_3$, calcium carbonate; $SnCl_2$, tin (II) chloride; $Na_2SnO_3{\cdot}3H_2O$, sodium stannate trihydrate; $CaCl_2$, calcium chloride; Na_2SnO_3, sodium stannate; PVP, polyvinylpyrrolidone; $SnCl_2{\cdot}2H_2O$, tin (II) chloride dihydrate; PAN, polyacrylonitrile.

TABLE 7.4
Preparation Methods of $SrSnO_3$ and Their Electrochemical Performance in Li-Ion Batteries

Method	Starting Materials	Morphology	Capacity (mAh/g)	Potential Window (V)	Current Density (mA/g)	References
Solid-state reaction	SnO_2 and $SrCO_3$	Well-separated smooth particles with average size of 0.5–1 μm	738-first cycle; 144-20th cycle; 82-40th cycle	0.005–1	10 mA/g-first cycle; 30 mA/g-20th cycle; 60 mA/g-40th cycle	Sharma et al. (2005)
Sol-gel	$SrCO_3$ and $SnCl_2$	Aggregates of particles with size 200–300 nm	870-first cycle; 222-20th cycle; 184-40th cycle	0.005–1	10 mA/g-first cycle; 30 mA/g-20th cycle; 60 mA/g-40th cycle	Sharma et al. (2005)
Hydrothermal	$SnCl_4{\cdot}2H_2O$ and $Sr(NO_3)_2$ – CTAB as surfactant	Nanorods with diameter 100–150 nm	1,400-first cycle; ~150-10th cycle	0–2	80 mA/g	Li et al. (2011)
Hydrothermal	$SrSn(OH)_6$ nanowires	Nanorods with uniform diameter of ~100 nm	714-first cycle; ~200-50th cycle	0.005–2	50 mA/g	Hu et al. (2010)

$SrCO_3$, strontium carbonate; $SnCl_2{\cdot}2H_2O$, tin (IV) chloride dihydrate; $Sr(NO_3)_2$, strontium nitrate; CTAB, cethyltrimethyl ammonium bromide; $SrSn(OH)_6$, strontium tin hydroxide.

involve in the alloying/de-alloying reactions. It is also interesting to point out that in ref (Li et al., 2011), $SrSnO_3$ in the nanorods form can be produced at hydrothermal temperatures of 160°C and above with CTAB as a surfactant. At a temperature of 140°C, $SrSnO_3$ exists in the micelles form (Li et al., 2011). Therefore, in this case, it can be inferred that the hydrothermal reaction temperature and CTAB plays an important role in influencing the crystal structure formation of $SrSnO_3$. CTAB is known to act as a template for the 1D directional growth of nanostructured material (Li et al., 2011). On the other hand, Hu et al. (2010) have proven that $SrSnO_3$ nanorods achieve better electrochemical performance than $SrSnO_3$ in the bulk form since the nanorods can tolerate the stress that occurs during volume changes during charging/discharging processes.

7.3.4 Barium Metastannate ($BaSnO_3$)

Barium metastannate ($BaSnO_3$) exhibits a cubic structure with a lattice constant *a* of 4.11 Å. It has found applications in gas sensors, ceramic capacitors, dielectric ceramics, and photocatalytic processes, in addition to its use in LIB (Nithyadharseni et al., 2016). Table 7.5 lists the synthesis processes of $BaSnO_3$, the starting materials used, and their electrochemical properties as reported in the literature. It is noteworthy that $BaSnO_3$, when prepared by the solid-state reaction method, exhibited a higher capacity than that synthesized via the sol-gel technique, regardless of the current density. This result contrasts with the $CaSnO_3$, $SrSnO_3$, and Li_2SnO_3 systems, where sol-gel-derived materials demonstrated better electrochemical performance than those prepared by the solid-state reaction method, as observed in Tables 7.2–7.5. It is also observed in Table 7.5 that $BaSnO_3$ exhibits fast capacity fading when cycling at 0.005–3.0 V compared to $BaSnO_3$ in the potential window of 0.005–1.0 V. However, a higher capacity of $BaSnO_3$ can be seen at the end of the 50th cycle in the potential range of 0.005–3.0 V (Nithyadharseni et al., 2016). As is customary for performance improvement, $BaSnO_3$ was mixed with RGO (Veerappan et al., 2016). The nanocomposites of $BaSnO_3$ and RGO showed the best performance in terms of the highest capacity, rate capability, and cycling stability when compared to pure $BaSnO_3$ and RGO.

7.3.5 Cobalt Metastannate ($CoSnO_3$)

Huang et al. (2003) synthesized $CoSnO_3$ using a pyrolysis reaction with starting materials, i.e., cobalt (II), sulfate ($CoSO_4$), and sodium stannate (Na_2SnO_3). They found that $CoSnO_3$ is amorphous in nature and has a square flake shape with an average particle size of 70 nm. Li//$CoSnO_3$ cells delivered an initial irreversible capacity of ~1,840 mAh/g and a reversible capacity of ~1,050 mAh/g at the 10th cycle under a specific current of 100 mA/g in a 0–3 V potential window (Huang et al., 2003). Additionally, the authors observed that $CoSnO_3$ prepared by the solid-state route suffered from a fast capacity degradation compared to pyrolysis-made $CoSnO_3$ (Huang et al., 2003). In another report, $CoSnO_3$ synthesized from a solvothermal method gave an initial capacity of ~1,408 mAh/g at a 50 mA/g current rate in a 0.005–3 V voltage window (Cao et al., 2014). When the current density varied to

TABLE 7.5
Preparation Methods of $BaSnO_3$ and Their Electrochemical Performance in Li-Ion Batteries

Method	Starting Materials	Morphology	Capacity (mAh/g)	Potential Window (V)	Current Density (mA/g)	Reference
Solid-state reaction	SnO_2 and $BaCO_3$	–	721-first cycle; 190-20th cycle; 142-40th cycle	0.005–1	10 mA/g-first cycle; 30 mA/g-20th cycle; 60 mA/g-40th cycle	Sharma et al. (2005)
Sol-gel	$BaCO_3$ and $SnCl_2$	–	715-first cycle; 156-20th cycle; 122-40th cycle	0.005–1	10 mA/g-first cycle; 30 mA/g-20th cycle; 60 mA/g-40th cycle	Sharma et al. (2005)
Molten salt	$SnCl_2 \cdot 2H_2O$ and $Ba(OH)_2$ – KOH as alkali flux	Non-uniform aggregates of spherical, rod-shaped with particles size 100–300 nm	1,293-first cycle; 252-50th cycle 1,077-first cycle; 289-50th cycle	0.005–1 0.005–3	60 mA/g	Nithyadharseni et al. (2016)
Molten salt	$SnCl_2 \cdot 2H_2O$ and $BaCl_2$ – KOH as alkali flux	Aggregated oval-shaped particles size <100 nm in layer-like form	846-first cycle; 183-50th cycle 914-first cycle; 255-50th cycle	0.005–1 0.005–3	60 mA/g	Nithyadharseni et al. (2016)
Hydrothermal	Na_2SnO_3 and $Ba(NO_3)_2$	Columnar shape 20–30 µm diameter, 50–100 µm length	~870-first cycle; ~150-50th cycle; ~75-100th cycle	0.005–2.5	60 mA/g	Cha and Kim (2020)
Hydrothermal	$Na_2SnO_3 \cdot 3H_2O$ and $BaCl_2 \cdot 2H_2O$	Hollow rods 200 nm diameter, 1.5–2 µm length	~700-first cycle; ~125-50th cycle; ~100-100th cycle	0.005–2.5	60 mA/g	Cha and Kim (2020)
Purchased commercially and used	–	Spherical shape 2–3 µm diameter	~675-first cycle; ~125-50th cycle; ~90-100th cycle	0.005–2.5	60 mA/g	Cha and Kim (2020)

$BaCO_3$, barium carbonate; $Ba(OH)_2$, barium hydroxide; KOH, potassium hydroxide; $BaCl_2$, barium chloride.

100 mA/g, a capacity of ~920 mAh/g was obtained at the 10th cycle (Cao et al., 2014). Once again, for enhanced LIB performance, nanocomposites of $CoSnO_3$, carbon nanotubes, and graphene gave better results than nano-$CoSnO_3$ (Zhao et al., 2017). Meanwhile, $CoSnO_3$ nanoboxes (~230 nm particle size) surrounded by RGO pieces also showed high cycling and rate performance (Huang et al., 2017). Stable cycling for 400 cycles has been achieved by coating $CoSnO_3$ nanoboxes with nanocarbon (Wang et al., 2013).

It should be clarified here that the Sn component of $CoSnO_3$, $CaSnO_3$, $SrSnO_3$, and $BaSnO_3$ plays an active role in the electrochemical process, while Co, Ca, Sr, and Ba are inactive and serve as safeguards in alleviating volume expansion (Sharma et al., 2005). Sharma et al. (2005) investigated the electrochemical performance among sol-gel-derived $CaSnO_3$, $SrSnO_3$, and $BaSnO_3$ at different current rates (10 and 60 mA/g). At both rates, the capacity values decreased in the order of $CaSnO_3 > SrSnO_3 > BaSnO_3$. The authors attributed the superior performance of calcium stannate to the lighter weight and smaller radius of calcium (Ca) than the other two. According to the literature (Sharma et al., 2002, 2005; Zhao et al., 2010; Li et al., 2011; Hu et al., 2012), the electrochemical reactions that occur during discharge are as follows:

- $ASnO_3 + 4Li^+ + 4e^- \rightarrow AO + 2Li_2O + Sn$ (A = Co, Ca, Sr, Ba)
- $Sn + xLi^+ + xe^- \leftrightarrow Li_xSn$ $(0 < x < 4.4)$

The first reaction, which is irreversible, indicates the crystal structure's demolition and results in the creation of lithium oxide (Li_2O) and tin metal implanted in AO. The second process is reversible, showing the formation of a lithium-tin alloy.

7.3.6 Cadmium Metastannate ($CdSnO_3$)

Unlike amorphous $CoSnO_3$, $CdSnO_3$ is crystalline, with a hexagonal unit cell (lattice constants $a = 5.43$ Å, $c = 14.95$ Å) (Sharma et al., 2009). In contrast to the above-mentioned stannates, both tin and cadmium of $CdSnO_3$ are electrochemically active. This indicates that a lithium-cadmium (Li_3Cd) alloy is also formed, in addition to the LiSn alloy, involving the participation of 3 mol of Li^+ ions and resulting in a higher capacity (Sharma et al., 2009; Wang et al., 2014a). It has been reported that $CdSnO_3$ delivered the first and 40th capacities at 1,330 mAh/g and ~475 mAh/g under a current density of 60 mA/g in the 0.005–1.0 V potential range (Sharma et al., 2009). Under a constant current rate of 70 mA/g in the 0.005–1.0 V voltage window, $CdSnO_3$ nanoparticles exhibited specific capacities of 1,546 and 515 mAh/g at the first and 40th cycles, respectively (Wang et al., 2014a). Its capacity was 506 mAh/g when the current density was 150 mA/g (Wang et al., 2014a). The authors believe that the hydrothermal-derived $CdSnO_3$ structure was porous, enabling it to accommodate volume changes while retaining structural stability upon cycling, thus inhibiting capacity degradation (Wang et al., 2014a). It is noticeable that in the past five years, there have not been many works done on $CdSnO_3$, $SrSnO_3$, $CoSnO_3$, and $BaSnO_3$ as anodes in LIB.

7.3.7 Zinc Metastannate ($ZnSnO_3$)

Crystalline $ZnSnO_3$, forming nanocubes and nanosheets, was examined for its electrochemical performance (Chen et al., 2012). $ZnSnO_3$ sheets displayed less capacity fading than cubic $ZnSnO_3$, even though both had almost similar first specific capacity. In contrast, the performance of amorphous $ZnSnO_3$, having hollow and solid microcubes, was compared by Xie et al. (2014). The former exhibited higher initial charge and discharge capacities, as well as a reversible capacity over the 50th cycle at a current density of 100 mA/g. Researchers also coated carbon onto the hollow $ZnSnO_3$ microcubes, resulting in larger capacity values over 50 continuous cycles (Xie et al., 2014). The carbon-coated hollow $ZnSnO_3$ also showed superior rate capability compared to the other two. $ZnSnO_3$ nanofibers, with and without carbon, were successfully prepared via the electrospinning technique (Wei et al., 2021). $ZnSnO_3$-carbon nanofibers were obtained by calcination at 873 K for 4 hours in an argon gas condition, whereas nanofibrous pure $ZnSnO_3$ was sintered at 723 K for one day in an air atmosphere. The former displayed a much higher capacity of around 586 mAh/g after 100 cycles than the latter at a current density of 100 mA/g (Wei et al., 2021). Very fine nanoparticles of $ZnSnO_3$ were inserted into RGO sheets and provided a steady capacity over 100 cycles at a current density of 100 mA/g (Ma et al., 2018). Recently, it has been reported that a 3D framework comprised of nanofibers of $ZnSnO_3$ and carbon recorded a good capacity of 695 mAh/g after 200 cycles at a constant current of 100 mA/g (Wang et al., 2023). Another 3D structure of $ZnSnO_3$ and TiO_2 nanocomposites documented the first capacity of 1,590 mAh/g, and the capacity was maintained at 780 mAh/g after 200 cycles at a current density of 100 mA/g (Qin et al., 2015). An anode made from $ZnSnO_3$ only performed below par compared to the $ZnSnO_3$-TiO_2 nanomaterial (Qin et al., 2015).

7.3.8 Zinc Stannate (Zn_2SnO_4)

There are several reports on crystalline Zn_2SnO_4 in the literature. In this chapter, we have selected a few for discussion. Similar to $CdSnO_3$, zinc in Zn_2SnO_4 and $ZnSnO_3$, along with tin, are active elements that produce respective alloys with lithium. For solid-state-derived Zn_2SnO_4, the irreversible and reversible capacities were recorded at 1,150 and 550 mAh/g in the voltage window 0.02–1.5 V under a steady current of 0.12 mA (Belliard et al., 2000). Another solid-state-prepared Zn_2SnO_4 achieved a first specific capacity of 2,054 mAh/g and ~690 mAh/g capacity upon 50 cycles (current density: 50 mA/g, potential range: 0.05–3 V) (Hou et al., 2010). On the other hand, hydrothermally procured Zn_2SnO_4 exhibited capacities of 1,384, 988, and 580 mAh/g at the first, second, and 50th cycles, respectively, under a 100 mA/g constant rate in the 0.05–3.0 V voltage (Rong et al., 2006). The shape of Zn_2SnO_4 was cubic. Another cube-shaped Zn_2SnO_4 demonstrated a first cycle capacity of ca. 1,700 mAh/g and a 50th cycle capacity of ~366 mAh/g at 300 mA/g current density in the 0.01–2.0 V voltage range (Wang et al., 2014b). However, in the same paper, flower-shaped Zn_2SnO_4 gave better performance (capacities of 1,750 and 501 mAh/g at the first and 50th cycles, respectively). To improve electrochemical performance, the same authors surrounded Zn_2SnO_4 nanoflowers with a graphene matrix that could produce a 50th cycle capacity of 540 mAh/g with an initial capacity of 1,967 mAh/g under the same conditions (Wang et al., 2014c). Another group of researchers prepared composites of Zn_2SnO_4 with carbon, displaying ~564 mAh/g capacity at the 40th cycle under a current density of 60 mA/g in the 0.05–3.0 V voltage range, with an initial capacity of ~1,437 mAh/g (Yuan et al., 2010). A lower capacity (~405 mAh/g) was obtained for pure Zn_2SnO_4 after the 40th cycle, even though it initially had a higher capacity of 1,634 mAh/g (Yuan et al., 2010).

In another study, a comparison was made between cubes and hollow boxes of nano-Zn_2SnO_4, finding that the latter exhibited higher rate capability and better capacity retention (Zhao et al., 2013a). In an unrelated report, fast capacity decline was detected in nano-Zn_2SnO_4 in plate form, while nanowires of Zn_2SnO_4 revealed a rather stable capacity during initial cycling. However, the outcome was reversed after cycling for 50 times. Zhu and co-workers (2009) reported that hydrothermal-derived Zn_2SnO_4 displayed a capacity of ~645 mAh/g upon 20 cycles, with initial capacities of ~1,046 and ~1,904 mAh/g during charging and discharging. Song et al. (2013), on the other hand, implanted Zn_2SnO_4 nanocrystals into GNSs to enhance performance. Another approach to improvement is coating Zn_2SnO_4 cubes with PANI, which verified that the capacity increased from 366 for pure Zn_2SnO_4 cubes to 491 mAh/g with PANI at the 50th cycle under a 600 mA/g constant current rate in the 0.01–2.0 V voltage (Zhao et al., 2013a). In addition, a higher initial specific capacity was obtained for that with PANI. The same effect was observed when coating hollow Zn_2SnO_4 boxes with PPy and N-doped carbon, as reported by Wang et al. (2014d) and Zhao et al. (2014).

7.4 OTHER STANNATES

Connor and Irvine (2001) investigated the potential use of magnesium tin oxide (Mg_2SnO_4), cobalt tin oxide (Co_2SnO_4), and manganese tin oxide (Mn_2SnO_4) as anode materials for Li-ion batteries. Initial findings from galvanostatic cycling for 1.5 cycles in the 0.02–1.5 V voltage window revealed that the best reversibility decreased in the order of $Mn_2SnO_4 > Co_2SnO_4 > Mg_2SnO_4$. Mn_2SnO_4 nanoparticles delivered an initial capacity of 1,320 mAh/g in the potential window of 0.01–2.0 V at a current density of 0.067 mA/cm^2 (Lei et al., 2009). Despite the high capacity achieved in the first cycle, Mn_2SnO_4 suffered from significant capacity degradation, producing a capacity of only ~503 mAh/g at the second cycle. The capacity further declined to ~150 mAh/g after the 100th cycle (Lei et al., 2009). Zhu and co-workers (2010) prepared an amorphous tin–manganese oxide film via electrostatic spray deposition, which exhibited a 3D reticular structure. The tin–manganese oxide film exhibited an initial discharge capacity of 1188.3 mAh/g and showed good capacity retention of 656.2 mAh/g up to the 30th cycle. On the other hand, Xiao et al. (2009) prepared SnO_2/Mg_2SnO_4 from hydrothermal-derived $MgSn(OH)_6$ by calcination at 850°C for 5 hours. This SnO_2/Mg_2SnO_4 anode had a high initial capacity of 1,245 mAh/g but poor cyclability, with a reversible capacity of only 350 mAh/g at the end of the 20th cycle. In our opinion, although magnesium and manganese are eco-friendly and low cost, there is not much attention on Mg_2SnO_4 and Mn_2SnO_4 due to their poor performance.

Co_2SnO_4 prepared by the hydrothermal route displayed high capacity and good capacity retention, with an initial capacity of 1533.1 and 555.9 mAh/g at the 50th cycle under a 30 mA/g current density in the 0.005–3.0 V voltage window (Wang et al., 2009b). On the other hand, solid-state-derived Co_2SnO_4 only achieved capacities of 1400.9 mAh/g (first cycle) and 112.8 mAh/g (50th cycle) under the same conditions (Wang et al., 2009b). Unlike its counterpart $CoSnO_3$, it is thought that both cobalt and tin are active electrochemically (Connor and Irvine, 2001; Wang et al., 2009b). To enhance performance, Co_2SnO_4 was coated with carbon (Yuvaraj et al., 2014). The first capacity of the uncoated and carbon-coated Co_2SnO_4 was 1,854 and 1,868 mAh/g, respectively. After the 20th cycle, their capacities declined to 777 mAh/g for the former and 780 mAh/g for the latter at a current density of 40 mA/g in the voltage window of 0–3.0 V. Although not much improvement was observed, the carbon-coated Co_2SnO_4 showed lower charge-transfer resistance and a higher lithium-ion diffusion coefficient based on impedance studies than the uncoated one.

Kishore and co-workers (2004) prepared lithium iron tin oxide, lithium stannoferrite ($LiFeSnO_4$), and lithium indium tin oxide ($LiInSnO_4$) composites using the solid-state reaction method. The starting materials included Li_2CO_3, SnO_2, iron (III) oxide (Fe_2O_3), and indium (III) oxide (In_2O_3). Both compounds are indexed to an orthorhombic unit cell, with lattice parameters for $LiFeSnO_4$ as $a = 3.06$ Å, $b = 5.06$ Å, and $c = 9.85$ Å, and for $LiInSnO_4$ as $a = 3.18$ Å, $b = 5.13$ Å, and $c = 10.17$ Å. An initial capacity of 1,175 mAh/g was achieved by the Li//$LiFeSnO_4$ cell, which gradually decreased to ~105 mAh/g at the 25th cycle in the voltage range from 0.05 to 2.0 V. In contrast, the Li//$LiInSnO_4$ cell delivered a capacity of 1,125 mAh/g in the first cycle, dropping drastically to ~0 mAh/g at the end of the 10th cycle in the voltage window between 0.05 and 2.0 V. Although $LiInSnO_4$ exhibited poor electrochemical performance, it is noteworthy that it was synthesized for the very first time, contributing to the discovery of novel materials in the fields of materials science and crystallography. The authors also attempted to synthesize $LiMSnO_4$, where M = Mn, Al, Ga, Cr, and Ni, but were unsuccessful, as their X-ray patterns exhibited main peaks of SnO_2 instead.

Lithium manganese tin oxide (Li_2MnSnO_4) was synthesized via an oxalyl dihydrazide-assisted combustion method by Mani et al. (2014). Li_2MnSnO_4 was also coated with carbon. Interestingly, the X-ray diffractogram of the carbon-coated Li_2MnSnO_4 calcined at 700°C under an argon atmosphere showed peaks ascribed to hybrid $MnSnO_4$ and Li_2SnO_3 phases instead of Li_2MnSnO_4. A Li-ion half-cell with the uncoated and carbon-coated Li_2MnSnO_4 delivered an initial capacity of ~1,300 mAh/g and ~800 mAh/g, respectively, at a 100 mA/g current density in the 0.05–3 V voltage range. After the 30th cycle, a capacity of ~150 mAh/g was obtained by the Li_2MnSnO_4 anode, while carbon-coated Li_2MnSnO_4 exhibited a capacity of ~600 mAh/g (Mani et al., 2014). As far as we observe, no other literature on Li_2MnSnO_4 compound can be found, similar to $LiMnSnO_4$ and $LiFeSnO_4$. Another anode material tested is tin antimony oxide, $Sn_{0.918}Sb_{0.109}O_2$ (Xiao et al., 2022), prepared via the hydrothermal route, with graphene attached before employing it in LIB. At a 0.1 mA/g current density, the capacities were ~1,176 and 648 mAh/g for the first and 365th cycles, respectively (Xiao et al., 2022). These initial reports indicate that researchers are constantly exploring alternative electrode materials in LIB.

7.5 SUMMARY

In a summary, tin-containing compounds have demonstrated encouraging results and show potential as anodes in LIB. Despite facing a major obstacle—namely, suffering from significant volume expansion—we remain optimistic that this hurdle can be overcome, ultimately improving their performance. We also believe that the potential of tin-based materials is yet to be fully realized to its maximum extent. There is plenty of room for improvement, and much work can still be done to achieve this. Hence, more developments on these materials are foreseen, and we believe they have a bright future in being employed in the next generation of LIBs.

ACKNOWLEDGMENTS

This work was supported by the Malaysian Ministry of Science, Technology and Innovation (MOSTI) under the Nanofund Grant (No. 53-02-03-1089).

REFERENCES

Belliard, F., Connor, P.A., Irvine, J.T.S. (2000). Novel tin oxide-based anodes for Li-ion batteries. *Solid State Ionics* 135, 163–167

Belliard, F., Irvine, J.T.S. (2001). Electrochemical comparison between SnO_2 and Li_2SnO_3 synthesized at high and low temperatures. *Ionics* 7, 16–21

Cao, Y., Zhang, L., Tao, D., Huo, D., Su, K. (2014) Facile synthesis of $CoSnO_3$/graphene nanohybrid with superior lithium storage capability. *Electrochimica Acta* 132, 483-489

Cha, Y.L., Kim, S.H. (2020). Morphological effect on electrochemical properties of $BaSnO_3$ as an anode material for lithium ion batteries. *Journal of Nanoscience and Nanotechnology* 20, 5498–5501

Chandra Bose, A., Kalpana, D., Thangadurai, P., Ramasamy, S. (2002). Synthesis and characterization of nanocrystalline SnO_2 and fabrication of lithium cell using nano-SnO_2. *Journal of Power Sources* 107, 138–141

Chen, J. (2013). Recent progress in advanced materials for lithium ion batteries. *Materials* 6, 156–183

Chen, J., Yano, K. (2013). Highly monodispersed tin oxide/mesoporous starbust carbon composite as high performance li ion battery anode. *ACS Applied Materials and Interfaces* 5, 7682–7687

Chen, J.S., Lou, X.W. (2012). SnO_2 and TiO_2 nanosheets for lithium-ion batteries. *Materials Today* 15, 246–254

Chen, J.S., Lou, X.W. (2013). SnO_2-based nanomaterials: Synthesis and application in lithium-ion batteries. *Small* 9, 1877–1893

Chen, S.S., Qin, X. (2014). Tin oxide-titanium oxide/graphene composited as anode materials for lithium-ion batteries. *Journal of Solid State Electrochemistry* 18, 2893–2902

Chen, W., Maloney, S., Wang, W. (2016). Three-dimensional SnO_2/carbon on Cu foam for high-performance lithium ion battery anodes. *Nanotechnology* 27, Article 415401

Chen, Y., Qu, B., Mei, L., Lei, D., Chen, L., Li, Q., Wang, T. (2012). Synthesis of $ZnSnO_3$ mesocrystals from regular cube-like to sheet-like structures and their comparative electrochemical properties in Li-ion batteries. *Journal of Materials Chemistry* 22, 25373–25379

Cherian, C.T., Zheng, M., Reddy, M.V., Chowdari, B.V.R., Sow, C.H. (2013). Zn_2SnO_4 nanowires versus nanoplates: Electrochemical performance and morphological evolution during Li-cycling. *ACS Applied Materials and Interfaces* 5, 6054–6060

Connor, P.A. and Irvine, J.T.S. (2001) Novel tin oxide spinel-based anodes for Li-ion batteries. *Journal of Power Sources* 97-98, 223-225

Courtney, I.A., Dahn, J.R. (1997). Electrochemical and in situ X-ray diffraction studies of the reaction of lithium with tin oxide composites. *Journal of The Electrochemical Society* 144, 2045–2052

Cui, D., Zheng, Z., Peng, X., Li, T., Sun, T., Yuan, L. (2017). Fluorine-doped SnO_2 nanoparticles anchored on reduced graphene oxide as a high-performance lithium ion battery anode. *Journal of Power Sources* 362, 20–26

Cui, L., Shen, J., Cheng, F., Tao, Z., Chen, J. (2011). SnO_2 nanoparticles@polypyrrole nanowires composite as anode materials for rechargeable lithium-ion batteries. *Journal of Power Sources* 196, 2195–2201

Dai, L., Zhong, X., Zou, J., Fu, B., Su, Y., Ren, C., Wang, J., Zhong, G. (2021) Highly ordered SnO_2 nanopillar array as binder-free anodes for long-life and high-rate Li-ion batteries. *Nanomaterials*, 11, 1307

Dang, H.X., Klavetter, K.C., Meyerson, M.L., Heller, A., Mullins, C.B. (2015). Tin microparticles for a lithium ion battery anode with enhanced cycling stability and efficiency derived from Se-doping. *Journal of Materials Chemistry A* 3, 13500–13506

Duan, Y.-K.; Li, Z.-W.; Zhang, S.-C.; Su, T.; Zhang, Z.-H.; Jiao, A.-J.; Fu, Z.-H. (2023) Stannate-based materials as anodes in lithium-ion and sodium-ion batteries: A review. *Molecules*, 28, 5037.

Fan, Q., Chupas, P.J., Whittingham, M.S. (2007) Characterization of amorphous and crystalline tin-cobalt anodes. *Electrochemistry Solid-State Letters*, 10, A274–A278

Faramarzi, M.S., Abnavi, A., Ghasemi, S., Sanaee, Z. (2018) Nanoribbons of SnO_2 as high performance Li-ion anode material. *Materials Research Express*, 5, 065040

Feng, Y., Wu, K., Sun, Y., Guo, Z., Ke, J., Huang, X., Bai, C., Dong, H., Xiong, D., He, M. (2020) Mo-doped SnO_2 nanoparticles embedded in ultrathin graphite nanosheets as a high-reversible capacity, superior-rate and long-cycle-life anode material for lithium-ion batteries. *Langmuir*, 36, 9276–9283

Fu, Z., Liu, P., Ma, J., Guo, B., Chen, X., Zhang, H. (2016) Microwave dielectric properties of low-fired Li_2SnO_3 ceramics co-doped with MgO-LiF. *Materials Research Bulletin*, 77, 78–83

García-Tecedor, M., Bartolomé, J., Maestre, D., Trampert, A., Cremades, A. (2019). Li_2SnO_3 branched nano- and microstructures with intense and broadband white-light emission. *Nano Research* 12, 441–448

Goriparti, S., Miele, E., Angelis, F.D., Fabrizio, E.D., Zaccaria, R.P., Capiglia, C. (2014) Review on recent progress of nanostructured anode materials for Li-ion batteries. *Journal of Power Sources* 257, 421–443

Guan, C., Wang, X., Zhang, Q., Fan, Z., Zhang, H., Fan, H.J. (2014). Highly stable and reversible lithium storage in SnO_2 nanowires surface coated with a uniform hollow shell by atomic layer deposition. *Nano Letters* 14, 4852–4858

Guo, X.W., Fang, X.P., Sun, Y., Shen, L.Y., Wang, Z.X., Chen, L.Q. (2013) Lithium storage in carbon-coated SnO_2 by conversion reaction. *Journal of Power Sources*, 226, 75–81

Hameed, M.U., Dar, S.U., Ali, S., Liu, S., Akram, R., Wu, Z., Butler, I.S. (2017) Facile synthesis of low-dimensional SnO_2 nanostructures: An investigation of their performance and mechanism as anode materials for lithium-ion batteries. *Physica E*, 91, 119–127

Hodeau, J.L., Marezio, M., Santoro, A., Roth, R.S. (1982) Neutron profile refinement of the structures of Li_2SnO_3 and Li_2ZrO_3. *Journal of Solid State Chemistry* 45, 170–179

Hong, Y.J. and Kang, Y.C. (2015) One-pot synthesis of core-shell-structured tin oxide-carbon composite powders by spray pyrolysis for use as anode materials in Li-ion batteries. *Carbon*, 88, 262–269

Hou, X., Cheng, Q., Bai, Y., Zhang, W.F. (2010) Preparation and electrochemical characterization of Zn_2SnO_4 as anode materials for lithium ion batteries. *Solid State Ionics*, 181, 631–634

Hu, G., Zhang, M., Wu, L., Peng, Z., Du, K., Cao, Y. (2016) Enhanced electrochemical performance of $LiNi_{0.5}Co_{0.2}Mn_{0.3}O_2$ cathodes produced via nanoscale coating of Li+-conductive Li_2SnO_3. *Electrochimica Acta*, 213, 547–556

Hu, R.Z., Liu, H., Zeng, M.Q., Liu, J.W., Zhu, M. (2012) Progress on Sn-based thin-film anode materials for lithium-ion batteries. *Chinese Science Bulletin* 57, 4119–4130

Hu, X., Tang, Y., Xiao, T., Jiang, J., Jia, Z., Li, D., Li, B., Luo, L. (2010). Rapid synthesis of single-crystalline SrSn(OH)6 nanowires and the performance of $SrSnO_3$ nanorods used as anode materials for li-ion battery. *Journal of Physical Chemistry C* 114, 947–952

Hu, Y., Xu, K., Kong, L., Jiang, H., Zhang, L., Li, C. (2014) Flame synthesis of single crystalline SnO nanoplatelets for lithium-ion batteries. *Chemical Engineering Journal*, 242, 220–225

Huang, B., Yang, J., Zhou, X. (2014). Hierarchical SnO_2 with double carbon coating composites as anode materials for lithium ion batteries. *Journal of Solid State Electrochemistry* 18, 2443–2449

Huang, F., Yuan, Z., Zhan, H., Zhou, Y., Sun, J. (2003) A novel tin-based nocomposite oxide as negative-electrode materials for Li-ion batteries. *Materials Letters* 57, 3341-3345

Huang, J., Ma, Y., Xie, Q., Zheng, H., Yang, J., Wang, L., & Peng, D.-L. (2017). 3D graphene encapsulated hollow $CoSnO_3$ nanoboxes as a high initial coulombic efficiency and lithium storage capacity anode. *Small*, 14(10), 1703513.

Huang, Y., Wang, Q., Wang, Y. (2012) Preparation and electrochemical characterisation of polypyrrole-Li_2SnO_3 anode materials for lithium-ion batteries. *Micro & Nano Letters* 7, 1278–1281

Idota, Y., Kubota, T., Matsufuji, A., Maekawa, Y., Miyasaka, T. (1997) Tin-based amorphous oxide: A high-capacity lithium-ion-storage material. *Science* 276, 1395–1419

Iqbal, M.Z., Wang, F., Zhao, H., Rafique, M.Y., Wang, J., Li, Q. (2012) Structural and electrochemical properties of SnO nanoflowers as an anode material for lithium ion batteries. *Scripta Materialia*, 67, 665–668

Jaśkaniec, S., Kavanagh, S.R., Coelho, J., Ryan, S., Hobbs, C., Walsh, A., Scanlon, D.O., Nicolosi, V. (2021). Solvent engineered synthesis of layered SnO for high-performance anodes. *Npj 2D Materials and Applications,* 5. 27 (9 pp)

Kamali, A.R., Fray, D.J. (2011). Tin-based materials as advanced anode materials for lithium ion batteries: A review. *Reviews on Advanced Materials Science* 27, 14–24

Kepler, K.D., Vaughey, J.T., Thackeray, M.M. (1999) Copper-tin anodes for rechargeable lithium batteries: an example of the matrix effect in an intermetallic system. *Journal of Power Sources* 81–82, 383–387

Kim, H.S., Park, S.-S., Kang, S.H. Kang, Sung, Y.-E. (2014) Effect of particle shape on the electrochemical properties of $CaSnO_3$ as an anode material for lithium-ion batteries. *Journal of Applied Electrochemistry* 44, 789–796

Kim, I.-S., Blomgren, G.E., Kumta, P.N. (2004) Sn/C composite anodes for Li-ion batteries. *Electrochemistry Solid-State Letters*, 7, A44–A48.

Kishore, M. V. V. M.s., Varadaraju, U. V., Raveau, B. (2004) Electrochemical performance of LiMSnO4 (M = Fe, In) phases with ramsdellite structure as anodes for lithium batteries. *Journal of Solid State Chemistry*, 177(11), 3981–3986

Lang, G. (1966) Strukturvergleiche an ternären und quarternären Oxiden. *Zeitschrift für anorganische und allgemeine Chemie* 348, 246–256

Lei, S., Tang, K., Chen, C., Jin, Y., Zhou, L. (2009) Preparation of Mn_2SnO_4 nanoparticles as the anode material for lithium secondary battery. *Materials Research Bulletin* 44, 393–397

Li, C., Zhu, Y., Fang, S., Wang, H., Gui, Y., Bi, L., Chen, R. (2011) Preparation and characterization of $SrSnO_3$ nanorods. *Journal of Physics and Chemistry of Solids* 72, 869-874

Li, L., Peng, S., Cheah, Y.L., Wang, J., Teh, P., Ko, Y., Wong C., Srinivasan, M. (2013) Electrospun eggroll-like $CaSnO_3$ nanotubes with high lithium storage performance. *Nanoscale* 5, 134–138

Lin, Y., Wu, J., Chen, W. (2013) Enhanced electrochemical performance of $LiFePO_4$/C prepared by sol-gel synthesis with dry ball-milling. *Ionics* 19, 227–234

Liu, H., Huang, J., Li, X., Liu, J., Zhang, Y. (2012) SnO_2 nanorods grown on graphite as high-capacity anode material for lithium ion batteries. *Ceramics International*, 38, 5145–5149

Liu, H., Huang, J., Li, X., Liu, J., Zhang, Y. (2013) Green synthesis of SnO_2 nanosheets and their electrochemical properties. *Ceramics International* 39, 3413–3415

Liu, L., An, M., Yang, P., Zhang, J. (2015). Superior cycle performance and high reversible capacity of SnO_2/graphene composite as an anode material for lithium-ion batteries. *Scientific Reports* 5

Liu, X.-Y., Han, Y.-L., Li, Q., Pan, D. (2017) SnO_2 nanoparticles for lithium-ion batteries prepared by sol-gel method. *Key Engineering Materials*, 727, 719–725

Ma, Y., Jiang, R., Li, D., Dong, Y., Liu, Y., Zhang, J. (2018) Embedding ultrafine $ZnSnO_3$ nanoparticles into reduced graphene oxide composites as high-performance electrodes for lithium ion batteries. *Nanotechnology* 29, 195401

Macías, M.A., Martínez, J.A.H., Gauthier, G.H., Rodriguez, J.E., Avila, H., Pinto, J., Pinilla, J. (2011) Structural and microstructural characterization of tin(II) oxide useful as anode material in lithium rechargeable batteries obtained from a different synthesis route at room temperature. *Materials Research*, 14, 172–177

Mani, V., Babu, G., Kalaiselvi, N. (2014) $Li2MnSnO_4$/C anode for high capacity and high rate lithium battery applications. *Electrochimica Acta* 133, 347–353

Manthiram. A. (2011) Materials challenges and opportunities of lithium ion batteries. *Journal of Physical Chemistry Letters* 2, 176–184

Meduri, P., Pendyala, C., Kumar, V., Sumanasekera, G.U., Sunkara, M.K. (2009) Hybrid tin oxide nanowires as stable and high capacity anodes for Li-ion batteries. *Nano Letters* 9, 612–616

Mei, L., Li, C., Qu, B., Zhang, M., Xu, C., Lei, D., Chen, Y., Xu, Z., Chen, L., Li, Q., Wang, T. (2012) Small quantities of cobalt deposited on tin oxide as anode material to improve performance of lithium-ion batteries. *Nanoscale*, 4, 5731–5737

Morimoto, H. and Tobishima, S.I. (2005) Anode behavior of electroplated rough surface Sn thin films for lithium-ion batteries. *Journal of Power Sources*, 146, 469–472

Mou, H., Xiao, W., Miao, C., Li, R., Yu, L. (2020). Tin and tin compound materials as anodes in lithium-ion and sodium-ion batteries: A review. *Frontiers in Chemistry*, 8, Article 141

Mousavi, M.R., Abolhassani, R., Hosseini, M., Akbarnejad, E., Mojallal, M.H., Ghasemi, S., Mohajerzadeh, S., Sanaee, Z. (2021). Antimony doped SnO_2 nanowire@C core-shell structure as a high-performance anode material for lithium-ion battery. *Nanotechnology* 32

Nithyadharseni, P., Reddy, M.V., Ozoemena, K.I., Ezema, F.I., Balakrishna, R.G., R. Chowdari, B.V. (2016) Electrochemical performance of $BaSnO_3$ anode material for lithium-ion battery prepared by molten salt method. *Journal of The Electrochemical Society*, 163 (3) A540–A545

Palaniyandy, N., Kebede, M.A., Ozoemena, K.I., Mathe, M.K. (2019) Rapidly microwave-synthesized SnO2 nanorods anchored on onion-like carbons (OLCs) as anode material for lithium-ion batteries. *Electrocatalysis* 10, 314–322

Park, M.-S., Kang, Y.-M., Wang, G.-X., Dou, S.-X., Liu, H.-K. (2008) The effect of morphological modification on the electrochemical properties of SnO2 nanomaterials. *Advanced Functional Materials* 18, 455–461

Park, M.S., Wang, G.X., Kang, Y.M., Wexler, D., Dou, S.X., Liu, H.K. (2007). Preparation and electrochemical properties of SnO_2 nanowires for application in lithium-ion batteries. *Angewandte Chemie-International Edition* 46, 750–753

Qin, Y.L., Zhang, F.F., Du, X.C., Huang, G., Liu, Y.C., Wang, L.M. (2015). Controllable synthesis of cube-like $ZnSnO_3$@TiO_2 nanostructures as lithium ion battery anodes. *Journal of Materials Chemistry A* 3, 2985–2990

Rong, A., Gao, X.P., Li, G. R., Yan, T. Y., Zhu, H. Y., Qu, J. Q., Song, D. Y. (2006) Hydrothermal synthesis of Zn_2SnO_4 as anode materials for Li-ion battery. *Journal of Physical Chemistry B* 110, 14754–14760

Saddique, J., Shen, H., Ge, J., Huo, X., Rahman, N., Mushtaq, M., Althubeiti, K., Al-Shehri, H. (2022) Synthesis and characterization of $Sn/SnO_2/C$ nano-composite structure: high-performance negative electrode for lithium-ion batteries. *Materials* 15, 2475.

Sharma, N., Shaju, K.M., Subba Rao, G.V., Chowdari, B.V.R. (2002) Sol-gel derived nano-crystalline $CaSnO_3$ as high capacity anode material for Li-ion batteries. *Electrochemistry Communications* 4, 947-952

Sharma, N., Shaju, K.M., Subba Rao, G.V., Chowdari, B.V.R. (2005) Anodic behaviour and X-ray photonelectron spectroscopy of ternary tin oxides. *Journal of Power Sources* 139, 250-260

Sharma, Y., Sharma, N., Subba Rao, G.V., Chowdari, B.V.R. (2009) Lithium-storage and cycleability of nano-$CdSnO_3$ as an anode material for lithium-ion batteries. *Journal of Power Sources 192, 627-635*

Shin, N., Kim, M., Ha, J., Kim, Y.-T., Choi, J. (2022) Flexible anodic SnO_2 nanoporous structures uniformly coated with polyaniline as a binder-free anode for lithium ion batteries. *Journal of Electroanalytical Chemistry* 914, 116296

Sivashanmugam, A., Kumar, T.P., Renganathan, N.G., Gopukumara, S., Wohlfahrt-Mehrens, M., Garche, J. (2005) Electrochemical behavior of Sn/SnO_2 mixtures for use as anode in lithium rechargeable batteries. *Journal of Power Sources* 144, 197–203

Song, W., Xie, J., Hu, W., Liu, S., Cao, G., Zhu, T., Zhao, X. (2013) Facile synthesis of layered Zn_2SnO_4/graphene nanohybrid by a one-pot route and its application as high-performance anode for Li-ion batteries. *Journal of Power Sources* 229, 6–11

Sun, S. and Liang, S. (2017). Morphological zinc stannate: synthesis, fundamental properties and applications. *Journal of Materials Chemistry A* 5(39), 20534–20560

Tao, H., Feng, Z., Liu, H., Kan, X., Chen, P. (2011) Reality and future of rechargeable lithium batteries. *The Open Materials Science Journal* 5 (Suppl 1: M2), 204–214

Tarakina, N.V., Denisova, T.A., Maksimova, L.G., Baklanova, Y.V., Tyutyunnik, A.P., Berger, I.F., Zubkov, V.G., van Tendeloo, G. (2009) Investigation of stacking disorder in Li_2SnO_3. *Z Kristallogr* Suppl. 30, 375–380

Thomas, R. and Rao, G.M. (2014) Phase and dimensionality of tin oxide at graphene nanosheet array and its electrochemical performance as anode for lithium ion battery. *Electrochimica Acta* 125, 380–385

Uchiyama, H., Hosono, E., Honma, I., Zhou, H., Imai, H. (2008) A nanoscale meshed electrode of single-crystalline SnO for lithium-ion rechargeable batteries. *Electrochemistry Communications* 10, 52–55

Vaughey, J.T., Geyer, A.M., Fackler, N., Jonhson, C.S., Edstrom, K., Bryngelsson, H., Benedek, R., Thackeray, M.M. (2007) Studies of layered lithium metal oxide anodes in lithium cells. *Journal of Power Sources* 174, 1052–1056

Veerappana, G., Yoo, S., Zhang, K., Ma, M., Kang, B., Park, J.H. (2016) High-reversible capacity of perovskite $BaSnO_3$/rGO composite for lithium-ion battery anodes. *Electrochimica Acta* 214, 31–37

Xiao, T., Tang, Y., Jia, Z., Feng, S. (2009) Synthesis of SnO_2/Mg_2SnO_4 nanoparticles and their electrochemical performance for use in Li-ion battery electrodes. *Electrochimica Acta* 54, 2396-2401

Wakihara, M. (2001) Recent developments in lithium ion batteries. *Materials Science and Engineering* R33, 109–134

Wan, N., Yu, P., Sun, S., Wu, Q., Li, T., Bai, Y. (2014) Copper and nitrogen co-doped SnO_2 hierarchical microspheres as a novel anode material for lithium ion batteries. *Materials Letters* 133, 168–170

Wang, G., Gao, X.P., Shen, P.W. (2009b) Hydrothermal synthesis of Co_2SnO_4 nanocrystals as anode materials for Li-ion batteries. *Journal of Power Sources* 192, 719–723

Wang, J., Zhao, H., Liu, X., Wang, J., Wang, C. (2011) Electrochemical properties of SnO_2/carbon composite materials as anode material for lithium-ion batteries. *Electrochimica Acta* 56, 6441–6447

Wang, Q., Huang, Y., Miao, J., Zhao, Y., Wang, Y. (2012a) Synthesis and properties of carbon-doped Li_2SnO_3 nanocomposite as cathode materials for lithium-ion batteries. *Materials Letters* 71, 66–69

Wang, Q., Huang, Y., Miao, J., Wang, Y., Zhao, Y. (2012b) Hydrothermal derived Li_2SnO_3/C composite as negative electrode materials for lithium-ion batteries. *Applied Surface Science* 258, 6923–6929

Wang, Q., Huang, Y., Zhao, Y., Wang, Y. (2012c) Synthesis and properties of Li_2SnO_3/polyaniline nanocomposites as negative electrode material for lithium-ion batteries. *Applied Surface Science* 258, 9896–9901

Wang, Q., Huang, Y., Zhao, Y., Zhang, W., Wang, Y. (2013) Preparation of Li_2SnO_3 and its application in lithium-ion batteries. *Surface and Interface Analysis* 45, 1297–1303

Wang, L., Zhang, W., Wang, C., Wang, D., Liu, Z., Hao, Q., Wang, Y., Tang, K., Qian, Y. (2014a) A facile synthesis of highly porous $CdSnO_3$ nanoparticles and their enhanced performance in lithium-ion batteries. *Journal of Materials Chemistry A* 2, 4970–4974

Wang, K., Huang, Y., Huang, H., Y. Zhao, Qin, X., Sun, X., Wang, Y. (2014b) Hydrothermal synthesis of flower-like Zn_2SnO_4 composites and their performance as anode materials for lithium-ion batteries. *Ceramics International* 40, 8021–8025

Wang, K., Huang, Y., Shen, Y., Xue, L., Huang, H., Wu, H., Wang, Y. (2014c) Graphene supported Zn_2SnO_4 nanoflowers with superior electrochemical performance as lithium-ion battery anode. *Ceramics International* 40, 15183–15190

Wang, K., Huang, Y., Han, T., Zhao, Y., Huang, H., Xue, L. (2014d) Facile synthesis and performance of polypyrrole-coated hollow Zn_2SnO_4 boxes as anode materials for lithium-ion batteries. *Ceramics International* 40, 2359–2364

Wang, Q., Yang, S., Miao, J., Zhang, Y., Zhang, D., Chen, Y., Li, Z. (2019) Synthesis of graphene supported Li2SiO3/Li_2SnO_3 anode material for rechargeable lithium ion batteries. *Applied Surface Science* 469, 253–261

Wang, X., Zhao, X., Wang, Y. (2021) A Nanosheet-assembled SnO2-integrated anode. *Molecules* 26, 6108

Wang, X., Zheng, T., Cheng, Y.-J., Yin, S., Xia, Y., Ji, Q., Xu, Z, Liang, S., Ma, L., Zuo, X., Meng, J.-Q., Zhu, J., Müller-Buschbaum, P. (2021) SnO_2/Sn/Carbon nanohybrid lithium-ion battery anode with high reversible capacity and excellent cyclic stability. *Nano Select* 2, 642–653

Wang, X.L., Han, W.Q., Chen, J., Graetz, J. (2010). Single-crystal intermetallic M-Sn (M = Fe, Cu, Co, Ni) nanospheres as negative electrodes for lithium-ion batteries. *ACS Applied Materials and Interfaces* 2, 1548–1551

Wang, Y., Djerdj, I., Smarsly, B., Antonietti, M. (2009a) Antimony-doped SnO_2 nanopowders with high crystallinity for lithium-ion battery electrode. *Chemistry of Materials* 21, 3202–3209

Wang, Z., Chen, L., Feng, J., Liu, S., Wang, Y., Fan, Q., Zhao, Y. (2019) In-situ grown SnO_2 nanospheres on reduced GO nanosheets as advanced anodes for lithium-ion batteries. *ChemistryOpen* 8, 712–718

Wang, Z., Wang, Z., Liu, W., Xiao, W., Lou, X.W. (2013). Amorphous $CoSnO_3$@C nanoboxes with superior lithium storage capability. *Energy and Environmental Science* 6, 87–91

Wang, K., Zhang, S., Hou, Z., Wang, L., An, P., Jia, J., Li, Y., Zhang, P. (2023) Nanofibrous $ZnSnO_3$/C composite derived from natural cellulose substance as an enhanced lithium-ion battery anode. *Materials Letters* 331, 133435

Wei, J.-L., Jin, X.-Y., Yu, M.-C., Wang, L., Guo, Y.-H., Dong, S.-T., Zhang, Y.-M. (2021). Electrospun $ZnSnO_3$/C nanofibers as an anode material for lithium ion batteries. *Journal of Electronic Materials* 50(8), 4945–4953

Wen, Z., Zheng, F., Yu, H., Jiang, Z., Liu, K. (2013) Hyrothermal synthesis of flower-like SnO_2 nanorods bundles and their application for lithium ion battery. *Materials Characterization* 76, 1–5

Whittingham, M.S. (2008) Materials challenges facing electrical energy storage. *MRS Bulletin* 33, 411–419

Winter, M. and Besenhard, J.O. (1999) Electrochemical lithiation of tin and tin-based intermetallics and composites. *Electrochimica Acta* 45, 31–50

Wu, H. and Cui, Y. (2012) Designing nanostructured Si anodes for high energy lithium ion batteries. *Nano Today* 7(5), 414–429

Wu, H.B., Chen, J.S., Lou, X.W., Hng, H.H. (2011). Synthesis of SnO_2 hierarchical structures assembled from nanosheets and their lithium storage properties. *Journal of Physical Chemistry C* 115, 24605–24610

Xiao, B., Wu, G., Wang, T., Wei, Z., Sui, Y., Shen, B., Qi, J., Wei, F., Meng, Q., Ren. Y., Xue, X., Zheng, J., Mao, J., Dai, K., Yan, Q. (2022) Tin antimony oxide @graphene as a novel anode material for lithium ion batteries. *Ceramics International* 48(2), 2118–2123

Xie, Q., Ma, Y., Zhang, X., Guo, H., Lu, A., Wang, L., G. Yue, Peng, D.-L. (2014) Synthesis of amorphous $ZnSnO_3$-C hollow microcubes as advanced anode materials for lithium ion batteries. *Electrochimica Acta* 141, 374–383

Yang, S., Zavalij, P.Y., Whittingham, M.S. (2003). Anodes for lithium batteries: Tin revisited. *Electrochemistry Communications* 5, 587–590

Yang, S., Zhang, J., Wang, Q., Miao, J., Zhang, C., Zhao, L., Zhang, Y. (2019) Li_2SiO_3@Li_2SnO_3/SnO_2 as a high performance lithium-ion battery material. *Materials Letters* 234, 375–378

Ye, J., Zhang, H., Yang, R., Li, X., Qi, L. (2010) Morphology-controlled synthesis of SnO_2 nanotubes by using 1D silica mesostructures as sacrificial templates and their applications in lithium-ion batteries. *Small* 6, 296–306

Yin, L., Chai, S., Wang, F., Huang, J., Li, J., Liu, C., Kong, X. (2016) Ultrafine SnO_2 nanoparticles as a high performance anode material for lithium ion battery. *Ceramics International* 42, 9433–9437

Yoon, S. and Manthiram, A. (2011) Nanostructured Sn-Ti-C composite anodes for lithium ion batteries. *Electrochimica Acta* 56, 3029–3035

Yu, L., Cai, D., Wang, H., Titirici, M.-M. (2013) Hydrothermal synthesis of SnO_2 and SnO_2@C nanorods and their application as anode materials in lithium-ion batteries. *RSC Advances* 3, 17281–17286

Yuan, J., Chen, C., Hao, Y., Zhang, X., Zou, B., Agrawal, R., Wang, C., Yu, H., Zhu, X., Yu, Y., Xiong, Z., Luo, Y., Li, H., Xie, Y. (2017) SnO_2/polypyrrole hollow spheres with improved cycle stability as lithium-ion battery anodes. *Journal of Alloys and Compounds* 691. 34–39

Yuan, W.S., Tian, Y.W., Liu, G.Q. (2010) Synthesis and electrochemical properties of pure phase Zn2SnO4 and composite Zn_2SnO_4/C. *Journal of Alloys and Compounds* 506, 683–687

Yuvaraj, S., Amaresh, S., Lee, Y. S., Kalai Selvan, R. (2014) Effect of carbon coating on the electrochemical properties of Co_2SnO_4 for negative electrodes in Li-ion batteries. *Royal Society of Chemistry Advances* 4, 6407–6416

Zakuan, N.S., Woo, H.J., Teo, L.P., Kufian, M.Z., Osman, Z. (2019). Li2 SnO_3 anode synthesized via simplified hydrothermal route using eco-compatible chemicals for lithium-ion battery. *Journal of The Electrochemical Society* 166, A2341–A2348

Zhang, C.Q., Tu, J.P., Huang, X.H., Yuan, Y.F., Wang, S.F., Mao F. (2008) Preparation and electrochemical performances of nanoscale $FeSn_2$ as anode material for lithium ion batteries. *Journal of Alloys and Compounds* 457, 81–85

Zhang, D.W., Zhang, S.Q., Jin, Y., Yi, T.H., Xie, S., Chen, C.H. (2006) Li_2SnO_3 derived secondary Li-Sn alloy electrode for lithium-ion batteries. *Journal* of *Alloys and Compounds* 415, 229–233

Zhang, H., He, Q., Wei, F., Tan, Y., Jiang, Y., Zheng, G., Ding, G., Jiao, Z. (2014) Ultrathin SnO nanosheets as anode materials for rechargeable lithium-ion batteries. *Materials Letters* 120, 200–203

Zhang, H.-J. Shu, J., Wei, X., Wang, K.-X., Chen, J.-S. (2013) Cerium vanadate nanoparticles as a new anode material for lithium ion batteries. *RSC Advances* 3, 7403–7407

Zhang, R., Whittingham, M.S. (2010) Electrochemical behavior of the amorphous tin-cobalt anode. *Electrochemistry Solid-State Letters* 13, A184–A187

Zhang, W., Feng, L., Chen, H., Zhang, Y. (2019) Hydrothermal synthesis of SnO_2 nanorod as anode materials for lithium-ion battery. *Nano: Brief Reports and Reviews* 14, 1950109 (9 pages)

Zhang, W.-J. (2011) A review of the electrochemical performance of alloy anodes for lithium-ion batteries. *Journal of Power Sources* 196, 13–24

Zhao, S., Bai, Y., Zhang, W.-F. (2010) Electrochemical performance of flowerlike $CaSnO_3$ as high capacity anode material for lithium-ion batteries. *Electrochimica Acta* 55, 3891-3896

Zhao, Q., Ma, L., Zhang, Q., Wang, C., Xu, X. (2015) SnO2-based nanomaterials: synthesis and application in lithium-ion batteries and supercapacitors. *Journal of Nanomaterials* 2015, 850147

Zhao, X., Wang, G., Zhou, Y., Wang, H. (2017) Flexible free-standing ternary $CoSnO_3$/graphene/carbon nanotubes composite papers as anodes for enhanced performance of lithium-ion batteries, *Energy* 118, 172–180

Zhao, Y., Li, X., Yan, B., Li, D., Lawes, S., Sun, X. (2015). Significant impact of 2D graphene nanosheets on large volume change tin-based anodes in lithium-ion batteries: A review. *Journal of Power Sources* 274, 869–884

Zhao, Y., Huang, Y., Wang, Q., Wang, K., Zong, M., Wang, L., Sun, X. (2014) Hollow Zn_2SnO_4 boxes coated with N-doped carbon for advanced lithium-ion batteries, *Ceramics International* 40, 2275–2280

Zhao, Y., Huang, Y., Wang, Q., Wang, K., Zong, M., Wang, L., Zhang, W., Sun, X. (2013a). Preparation of hollow Zn_2SnO_4 boxes for advanced lithium-ion batteries. *RSC Advances* 3, 14480–14485.

Zhao, Y., Huang, Y., Wang, Q., Wang, X., Zong, M., Wu, H., Zhang, W. (2013b) Graphene supported Li_2SnO_3 as anode material for lithium-ion batteries. *Electronic Materials Letters* 9, 683–686

Zhao, Y., Huang, Y., Wang, Q., Zhang, W., Wang, K., Zong, M. (2013c) The study on Li-storage performances of bamboo charcoal (BC) and BC/Li_2SnO_3 composites. *Journal of Applied Electrochemistry* 43, 1243–1248

Zhao, Y., Huang, Y., Xue, L., Sun, X., Wang, Q., Zhang, W., Wang, K., Zong, M. (2013d) Polyaniline (PANI) coated Zn_2SnO_4 cube as anode materials for lithium batteries, *Polymer Testing* 32, 1582–1587

Zhu, X., Zhu, J.,Yao, Y., Zhou, Y., Tang, Y., Wu, P. (2015) Facile template-directed synthesis of carbon-coated SnO_2 nanotubes with enhanced Li-storage capabilities, *Materials Chemistry and Physics* 163, 581–586

Zhu, X.J., Geng, L.M., Zhang, F.Q., Liu, Y.X., Cheng, L.B. (2009) Synthesis and performance of Zn_2SnO_4 as anode materials for lithium ion batteries by hydrothermal method. *Journal of Power Sources* 189, 828–831

Zhu, X.J., Guo, Z.P., Zhang, P., Du, G.D., Poh, C.K., Che, Z.X., Li, S., Liu, H.K. (2010) Three-dimensional reticular tin-manganese oxide composite anode materials for lithium ion batteries. *Electrochimica Acta* 55, 4982–4986

Zoller, F., Böhm, D., Bein, T., Fattakhova-Rohlfing, D. (2019). Tin oxide based nanomaterials and their application as anodes in lithium-ion batteries and beyond. *ChemSusChem* 12, 4140–4159

Zou, Y., Wang, Y. (2013) Microwave solvothermal synthesis of flower-like SnS_2 and SnO_2 nanostructures as high-rate anodes for lithium ion batteries. *Chemical Engineering Journal* 229, 183–189

8 Sustainable Materials for Energy Conversion System

Grishika Arora and Hieng Kiat Jun

A society grows great when old men plant trees whose shade they know they shall never sit in.

– Greek Proverb

8.1 INTRODUCTION

The saying "Sustainable development is development that meets the needs of the present without compromising the ability of future generations to meet their own needs" shows that the idea of sustainable development is not new (from the 1987 publication of the Brundtland Commission Report) (Park, 2018). The so-called "cradle-to-grave" strategy, according to Narayan, mandates that produced goods be conceived and engineered from "conception to reincarnation." It is important to figure out how biomass is used in the context of the entire carbon cycle. The lithosphere (for example, limestone), the biosphere (for example, plants and animals), the hydrosphere (for example, bicarbonate dissolved in the oceans), and the atmosphere (CO_2) are the four main reservoirs of the carbon on the planet known as the carbon cycle. Recent human activities (such as the burning of fossil fuels and extensive deforestation) cause a significant imbalance in the carbon cycle, resulting in a large-scale release of CO_2 into the atmosphere that cannot be entirely offset by photosynthesis or oceanic dissolution. It causes a significant buildup of CO_2 in the atmosphere, which aids in the process of global warming. People are increasingly aware that measures must be taken to lessen CO_2 generation to rebalance the carbon cycle (Song et al., 2009).

Since only a certain number of energy sources (solar, wind, hydro, tidal, and geothermal) are renewable and none of our natural resources are limitless, the implications of sustainable development affect every aspect of human effort. Sustainability, handling resources, sustainable production, technology for renewable and clean energy, and management of water and air are all included in the vast subject of sustainable development. It is the procedure by which scientific analysis ("the triple bottom line") is used to inform strategies for ongoing improvements in the economy, environment, and society (Green et al., 2012).

DOI: 10.1201/9781003314424-8

Sustainable development issues are inherently extremely interconnected as the three parts of the triple bottom line overlap as illustrated in Figure 8.1.

We have found it helpful to use the 2×3 taxonomy of the objectives outlined in the extensive literature that defines or discusses sustainable development in order to clarify the definitional ambiguities connected with it as shown in Table 8.1 (Parris and Kates, 2003; Ness, 2020). There are three main categories listed in the first column under the heading "what is to be sustained": community, health care, and environment. Most of the literature focuses on sustaining life support systems/health care/medical care, in which the ecosystem or nature provides resources for the real-world existence of humanity, such as products and services (Lawton, 1998; Costanza et al., 1997).

However, a substantial portion of literature gives more weight to the beauty and richness of the natural world than to its practical value. Finally, there are assertions that unique and threatened communities should be preserved, including cultural diversity, livelihoods, groups, and locations (Parris and Kates, 2003; Muehlebach, 2001).

Three distinct categories—people, business, and society—should also be created. Most of the early writing was focused on the economy, with its producing sectors providing both the needed labour and the desired consumption and wealth. In this literature, economic growth is presented as offering incentives and methods for investment as well as financial resources for upkeep and restoration of the ecology (Solow, 1993). People are the focus now, with a focus on human development, improved life expectancy, education, equity investment, and prosperity. Furthermore, there are also calls for the development of societies that place a strong emphasis on the stability and well-being of national states, regions, and organizations as well as the social capital of ties within communities (Parris and Kates, 2003; Adeney, 2002).

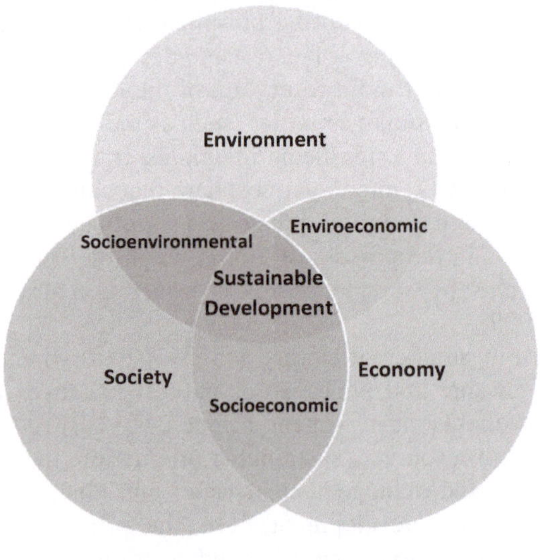

FIGURE 8.1 Sustainable growth focuses on the "triple bottom line" of social, economic, and environmental factors. Sustainable growth happens when all three of them exist side by side. Adapted from Green et al. (2012).

On the other hand, although the connections between sustainable development and materials are clear-cut and fundamental, consumers of cutting-edge technologies frequently fail to notice them. Materials often play a supporting role in technology and are not seen as being crucial by end consumers. For instance, when operating a modern car, the owner appreciates the increased fuel efficiency while frequently being unaware of the material advances in technology (composite and alloys that are lightweight, digital control systems, higher-density rechargeable batteries, etc.) which made gain feasible (Green et al., 2012). Figure 8.2 shows some examples of materials for sustainable development. In order for technologists, policymakers, and business leaders, this chapter intends to emphasize the significant role that materials contribute to environmentally friendly growth and to define and plan the future responsibilities of advanced materials in the sustainable development area. The sustainable development is focused on materials involved in energy conversion systems, especially solar energy devices and energy storage systems.

Perhaps the biggest obstacle to sustainable growth is changing people's behaviour. Regional and global strategies will need to be developed by governing bodies, politicians, and scientists; Apelian covered this in the first article of the bulletin. Technology alone won't be the solution. Additionally, he contrasts the open loop of human civilization with the closed cycle of the ecological system, in which there is no "waste" (LeStar et al., 2012) As the world's population increases, the most crucial issue may not be how many people the planet can sustain, but rather how many people it can support sustainably while maintaining an adequate standard of living (Green et al., 2012). Let's briefly first understand how do some energy conversion and storage systems function.

TABLE 8.1
Sustainable Development Objectives' Taxonomy

What Is to Be Sustained	What Is to Be Developed
Nature	People
Earth	Child survival
Biodiversity	Life expectancy
Ecosystems	Education
	Equity
	Equal opportunity
Life support	Economy
Ecosystem services	Wealth
Resources	Productive sectors
Environment	Consumption
Community	Society
Cultures	Institutions
Groups	States
Places	Regions
	Social capital

Adapted from Parris and Kates (2003) and Ness (2020).

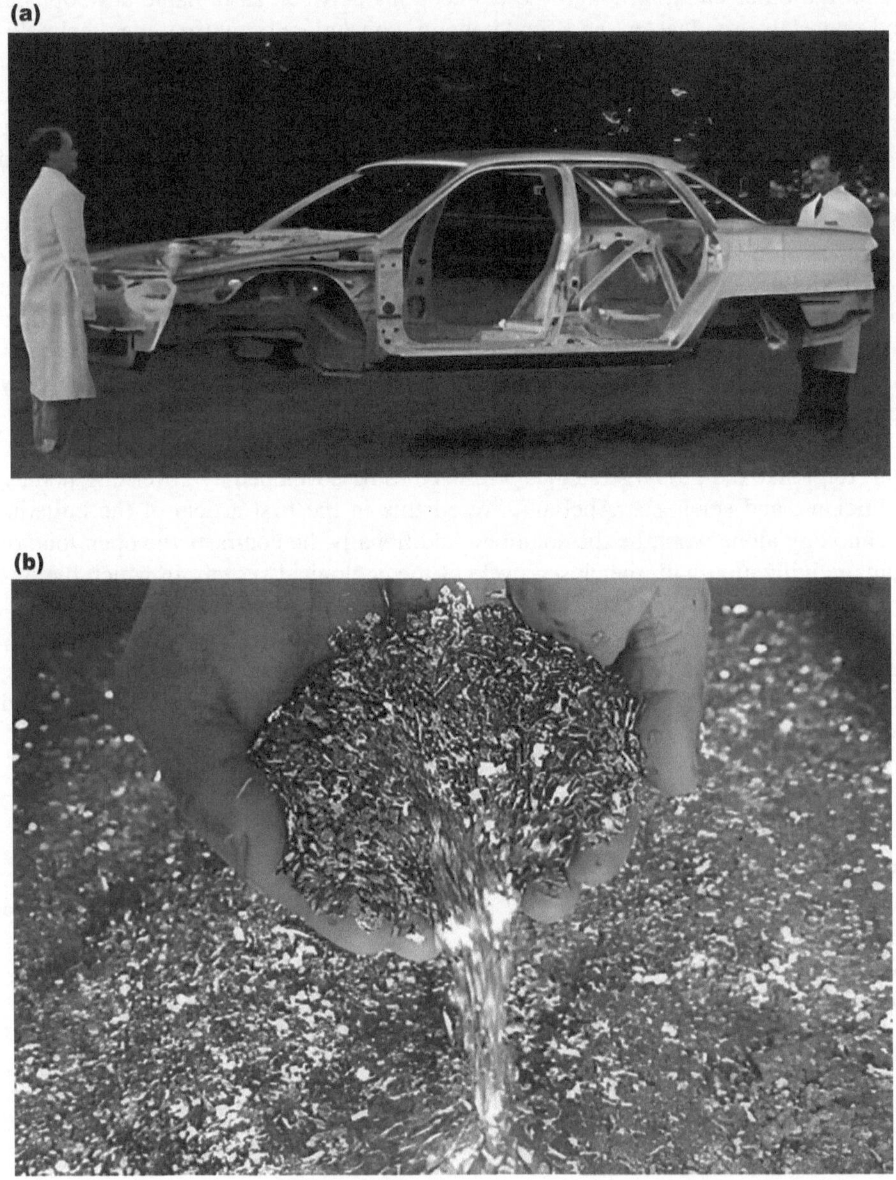

FIGURE 8.2 Sustainable development in materials: (a) the apparent solution for all modes of transportation is to make them lighter; (b) "urban mining" includes the recycling of electronic waste, which can contain more important materials than a mine itself. Adapted from Green et al. (2012).

8.2 ENERGY CONVERSION AND STORAGE SYSTEMS

8.2.1 Solar Energy

Based on the various energy conversion and storage, the integrated energy conversion and storage systems (IECSSs) can be classified into two classes as indicated in Figure 8.3, which are photovoltaic (PV) charging based and photocatalytic charging based. During the photo-charging and photo-discharging operations, the energy conversion and storage sections are typically electrochemical separation. The performance activities of these sections are independent (Figure 8.3a). The PV-charging procedure does not involve any photo-induced redox processes. The solar energy conversion sections of the IECSSs are usually from semiconductor solar cell types, which include but not limited to silicon solar cells, organic solar cells, and perovskite solar cells (PSCs). PV solar cells may fully or partially provide the charging voltage for the energy storage component. In contrast, the method used in a photocatalytic charged arrangement for energy storage (photo-charging) will use photo-induced redox processes as shown in Figure 8.3b. The photoelectrodes hold a significant substance that will be stimulated during the photo-charging procedure to produce the electron hole pairs. The resultant electrons will then move to the energy storage. The electrons originating from the counter electrode will subsequently balance the redox reactions in which both the electrode and the holes participate. The electrolyte that exists between each of these electrodes' cations or anions will serve as a counterweight when the charges from the energy storage electrode return to the counter electrode during the discharge process (Luo and Wang, 2017).

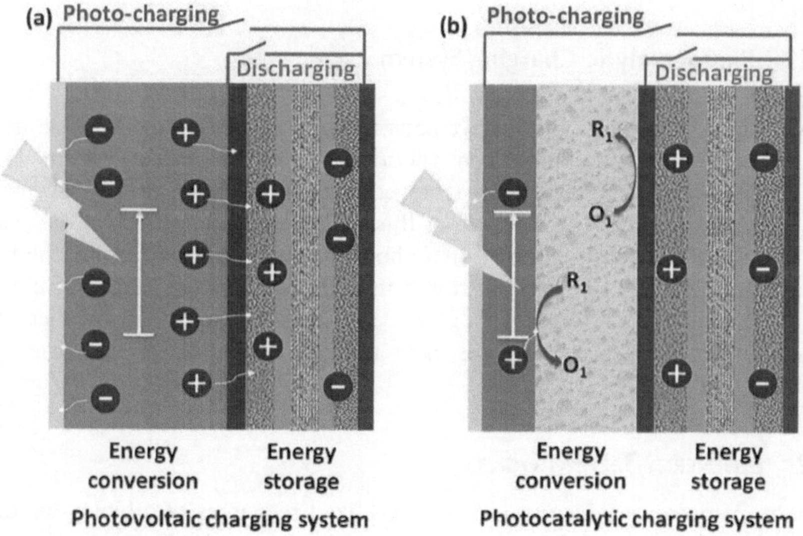

FIGURE 8.3 Two categories of PV-based IECSSs are shown schematically: (a) photovoltaic charging technologies and (b) photocatalytic charging technologies.

8.2.1.1 Photovoltaic Charging System

To achieve practical solar energy storage, recent developments have seen the development of a wide range of integrated systems that combine different PV cell technologies (such as silicon-based solar cells, organic solar cells, and PSCs) with various energy storage solutions (for example, supercapacitors, lithium-ion batteries, and nickel metal hydride batteries). Connecting the energy conversion and storage units using wires is the simplest approach to integrate them (Xu et al., 2015; Yan et al., 2014). For example, a thin amorphous Si PV cell was coupled with a Li-ion battery of the iron phosphate type, for instance, according to Gibson and Kelly. Overall effectiveness of solar energy storage and conversion for the combined system was 14.5% (Gibson and Kelly, 2010). The PV voltage, initially low, was raised to almost 300 V through a DC–DC converter, leading to the charging of the high-voltage NiMH battery pack. This resulted in the creation of an integrated system with high efficiencies in PV -to-battery power storage (Liu et al., 2012). Recent advances in high-performance PSCs have increased opportunities for the creation of solar cells that are both highly efficient (up to 20%) and inexpensive, as well as for the incorporation of solar cells into IECSSs for use in real-world applications (Gurung and Qiao, 2018; Stroyuk, 2018).

The wire was attached with IECSSs that provide several benefits when choosing and assembling individual units. However, the considerable separation between the elements responsible for energy conversion and storage could potentially reduce the overall efficiency of energy storage. One way to get around this problem is to combine the energy conversion and storage components into a single unit, which could also improve space efficiency and boost the system's volume energy density (Wee et al., 2011).

8.2.1.2 Photocatalytic Charging System

Numerous solar energy conversion devices like DSSCs and photocatalytic reactions of solar-hydrogen synthesis have been thoroughly studied (Nazeeruddin et al., 2011). Due to their comparable construction and electrochemical nature, DSSCs and capacitor/battery systems have received a lot of attention during the past ten years (Gräetzel et al., 2012). Because of this, most photocatalytic charging systems are based on DSSC configurations. In a photocatalytic charging system, the photogenerated charges can typically be stored in a variety of ways, involving chemical reactions of the electrode materials, electrochemical interactions in the electrolytic solution, double-layer charges processes on the electrode–electrolyte interface, or a combination of these.

8.2.2 Bioenergy Technologies

Bioenergy is power produced from non-fossilized resources of a biological origin. Raw woody biomass and animal waste currently meet a significant portion of the world's primary energy requirements, with the potential for transformation into various gaseous or liquid fuels.

8.2.2.1 Biogas

Biogas is a concoction of gases produced when organic matter breaks down without oxygen. It can be used to heat or cook food, power internal combustion engines, or be synthesized into synthetic petrol to create superior fuels or chemicals. In order to convert the solid biomass into greenhouse gases (GHGs), it gets heated in without the presence of air, which is subsequently cleaned and filtered to remove undesirable compounds (Uriarte, 2010). Another method for converting organic matter into renewable energy is anaerobic digestion, which is a type of fermentation. Anaerobic biogas consists of 60% methane, akin to landfill gas, with the remaining 40% comprised of carbon dioxide.

8.2.2.2 Biofuels

Examples of liquid biofuels include fuel oil, biogasoline, and pure plant oil, also referred to as straight vegetable oil. The main sources of ethanol production are starch-rich crops like maize, sugar, and corn. Long-chain alkyl esters found in biodiesels are diesel alternatives made from vegetable or animal fats that are typically created by reacting fatty substances with alcohol (Islam et al., 2019). Regular diesel engines can run on biodiesel in its pure form or in any ratio of biodiesel to petroleum diesel; for example, B100 stands for 100% pure biodiesel, B20 for 20% biodiesel mixed with 80% petroleum diesel, and so on (Ackermann, 2005).

8.2.3 Ocean Energy Technology

The vast energy present in ocean waves and tides can be harnessed for practical applications. Additionally, the shift in salinity at river mouths and the temperature difference between cold ocean depths and hot ocean waters can also be faced for the generation of electricity (Adimazoya and Doelle, 2021). The following are the current and future ocean or marine energy technologies.

8.2.3.1 Ocean Thermal Energy Conversion

Ocean Thermal Energy Conversion (OTEC) technology generates electricity by harnessing the temperature contrast between the warmer surface water of a marine lake and the colder waters beneath it (Hamedi and Sadeghzadeh, 2017). It has been found that operative energy can be produced even with a temperature differential of 20°C.

Claude cycle (open), Anderson cycle (closed), and hybrid cycle OTEC plants are the three different types. The Claude cycle uses hot water on the surface from tropical oceans as a heat transfer fluid (HTF). In a vessel that has been partially evacuated, hot water is permitted to expand swiftly and turn into steam. As the generated steam undergoes further expansion, it propels a low-pressure steam turbine. The vapour is condensed into freshwater with no salt at the turbine outflow using cool seawater, which can serve as a source of drinking water for neighbouring settlements. In the Anderson cycle, Freon, propane, or ammonia serves as the active HTF, and the entire system, including the turbine, is fully enclosed or sealed. The fluid is heated through a heat exchanger using hot surface water, and the vapour produced by this process is expanded over the blades of a turbine, which in turn powers a generator. The vapour is fed back to the evaporator after condensing with cool deep ocean water at the turbine exit. In hybrid cycles, heated water from closed-cycle OTECs is injected into an open-cycle OTEC system to increase power output rather than being discharged to the ocean (Islam et al., 2019).

8.2.3.2 Wave Energy

Ocean waves derive their energy from the combination of kinetic and gravitational potential energy, generated when wind moves across the water's surface. Wave energy availability is closely linked to wave height, and during storms, peak wave power can surpass 200 kW/m. Due to their unpredictability and the potential hazards they pose, wave resources are challenging and unsafe for navigation or manipulation (Reddy, 2017).

The energy from ocean waves can be captured by employing a water column in oscillation, which includes an air chamber interfacing with the ocean and a turbine. The oscillation of the water column induces the flow of air into and out of the column, facilitating the collection of wave energy (Islam et al., 2019). To transform wave oscillation motion into linear motion, well compressors with identical airfoil structures edges are counterbalanced by flywheels.

8.2.3.3 Tidal Power

Massive amounts of potential energy are present during tidal movements, and this potential energy can be used to drive a turbogenerator's input or outflow (or both) of seawater (Dye, 2012). A dam is built to separate tidal basins from the seafloor, creating in the water level disparity between the two. At high tide, water surges from the sea into the basin, propelling via a turbine to generate electricity. Conversely, at low tide, water exists of the tidal basin into the sea, causing the turbine to rotate in the different direction and generate power (Islam et al., 2019).

8.3 LIFE CYCLE ANALYSIS

Upon understanding the function of energy conversion and storage system, the next step is to understand their impact to the environment from the creation to the end of life of the devices or systems. A method known as life cycle analysis (LCA) is used to measure the environmental effects of the manufacturing, disposal, and consumption of a product or process. Creating up-to-date environmentally friendly choices can benefit greatly from LCA: (1) products are contrasted according to categories of environmental impact that can be conceptualized by actual environmental harm, (2) it is possible to identify unforeseen environmental trade-offs between effect categories, and (3) a standardized methodology enables life cycle assessments from various studies to be utilized to compare various product options (Guinee, 2002).

One of the methodologies currently available to examine the environmental impacts of construction processes is life cycle assessment analysis. It is a method for assessing how well a thing, activity, or procedure performs from the standpoint of how the environment affects it at each stage. The numerous evaluation processes are carried out by the LCA under ISO 14040's authorization and supervision.

The Society for Environmental Toxicology and Chemistry (SETAC) defines LCA as an evaluation procedure intended to (Anon, n.d.):

- Evaluate the materials and energy used and emitted into the environment to determine environmental issues;
- Identify the environmental impacts of these resources and energy; and
- Find potential environmental development strategies.

For example, LCA for PV devices can be generally summarized as in Tables 8.2–8.4. Though the tables do not provide details breakdown of the LCA information, they present the gist of the findings in terms of advantages and disadvantages. Discussion on LCA of solar cell devices is presented in the subsequent sub-sections.

8.3.1 Life Cycle Analysis of Batteries

An assessment methodology for the entire lifecycle is used to evaluate the battery life cycle studies that were evaluated below.

Specifically, the assessment includes studies that offer comprehensive presentations of process-specific data, particularly those featuring structured process flows or references to them. Figure 8.4 provides an example of the essential flows required to define a unit process. For example, the production of a PbA battery necessitates unit processes for generating lead, acid, battery casings, stakes, barriers, copper, and other components. Additionally, one or more processes are required to assemble these components into a finished product ready for sale (Sullivan and Gaines, 2010). Moreover, diverse unit processes are essential for material manufacturing. To illustrate, mines, beneficiation, ore preparation, casting, and refining represent the required unit operations for lead production. Figure 8.5 illustrates the full life cycle of products (batteries).

TABLE 8.2
The Earliest Photovoltaic Cells

Solar Cells Based on Silicon, or First-Generation PV Cells (–Si)

Due to the superior electrical, chemical, and mechanical qualities, Silicon remains the predominant material utilized in photovoltaic (PV) modules. Adapted from Aberle (2006). Some of the most developed solar technologies rely on this semiconductor.

Type	Efficiency	Advantages	Disadvantages
sc-Si	• The maximum efficiency when compared to other solar technologies was between 25% and 27% in a laboratory (Sørensen, 2017). • The band gap is 1.11–1.15 eV. • The commercial efficiency ranges from 16% to 22%.	The most efficient and pure sort of solar panel is this one.	The manufacturing process (CZ process) requires a lot of resources and energy.
mc-Si	• 15%–18% (Hossam-Eldin et al., 2015) • 1.11 eV serves as the band gap	A good substitute to lower the price of PV modules.	Ineffective compared to sc-Si cells.

TABLE 8.3
PV Cells of the Second Generation

Thin-Film Solar Cells (TFSCs), A Second Generation of Photovoltaic cells			
The thickness of film layers can be measured in nanometres (nm) or micrometres. Because of the minimal semiconductor material required for each cell, this method boasts an economical manufacturing cost and low material expenses. In comparison to first-generation PV, TFSC are flexible, lighter, and have less drag thanks to their low thickness. By using sputtering equipment to deposit tiny layers of certain materials on glass, plastic, or stainless-steel substrates, TFSCs are created. A few of these technologies are still in the development stage (Muteri et al., 2020).			
Type	**Efficiency**	**Advantages**	**Disadvantages**
a-Si	• 4%–8% for modules commercialized • small cells that are tested in the lab can reach 12% (Hossam-Eldin et al., 2015)	Least expensive	As the layers are thinner and contain less material for absorbing solar radiation compared to single-crystal silicon (sc-Si) and multicrystalline silicon (mc-Si) solar modules, they yield lower electricity production (15%–35%) under sunlight exposure (Ratner, 2004).
CdTe	• 1.45 eV is the band gap value • 21% record efficiency (Ratner, 2004) • 10%–15%	Compared to Si cells, CdTe cells can use a wider wavelength spectrum that is similar to the organic spectrum. Low prices because cadmium is plentiful and produced as a result of the extraction of crucial industrial minerals like zinc.	One of the most dangerous substances is cadmium, and cadmium telluride also has certain hazardous qualities.
CIGS	• 20% (in some circumstances) • The energy band gap is 1.68 electron volts (eV)	The process consumes less energy than the production of solar cells made of crystalline silicon and exhibits excellent heat resistance.	• CIGS solar cells are less effective than Si solar cells and use hazardous chemicals. • Currently, it costs a lot to produce.
GaAs	• In a laboratory, 29% efficiency set a record (Ratner, 2004) • 1.43 eV is the band gap value	Greater effectiveness and thinner than silicon ones.	Cost is high.
CIS	10%–13%	Methodologies use less energy than crystalline technology fabrication, and they have better heat resistance than silicon-based modules (Gul et al., 2016).	Comparatively pricey because of the materials used.

TABLE 8.4
PV Cells of the Third Generation

Third-Generation Solar Cells

The majority of the novel technologies used in third-generation solar panels are still in the research and development stages. Some cells use natural, quasi-organic, or inorganic substrates to produce energy. They also use hybrid systems or novel technological methods built on components at the nanometre and molecular scale (Muteri et al., 2020).

Type	Efficiency	Advantages	Disadvantages
DSSC	Roughly 10% (Muteri et al., 2020)	• Flexibility, lack of pollution, and simplicity of recycling. • Low cost because of the straightforward manufacturing method. • DSSC function even in low light. • Good efficiency also at high temperatures.	• The electrolyte comprises organic solvents with volatility and requires secure sealing due to its susceptibility to freezing at low temperatures, leading to diminished power production and potential physical damage.
QDs	Roughly 1.9%	Simple to produce and prepare.	Low efficiency
PSC	19%–22% (Ratner, 2004)	• Good performance and room for development. • Producing perovskite is less expensive than silicon.	• When exposed to heat, snow, moisture, etc., perovskite degrades quickly. • The existence of lead in perovskite is hotly contested due to toxicity issues.

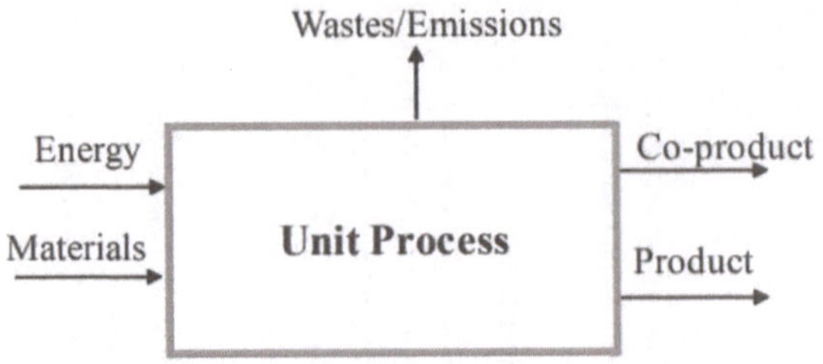

FIGURE 8.4 General unit process. Adapted from Sullivan and Gaines (2010).

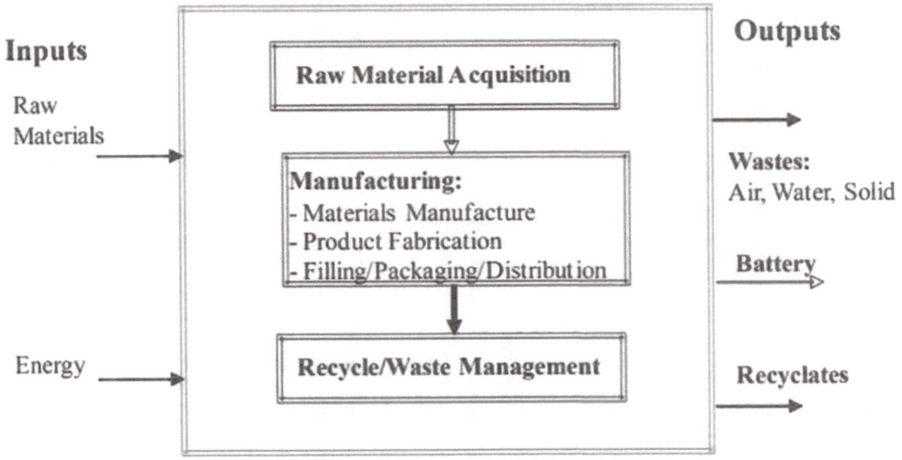

FIGURE 8.5 Limitations considered for cradle-to-gate study assessment. Adapted from Sullivan and Gaines (2010).

8.3.1.1 Lead-Acid Batteries

The basic battery materials must be transformed into the materials needed for the battery, which requires a significant amount of energy during battery manufacturing. In actuality, the manufacturing processes involve the production of grids, paste, plates, plastic moulding, and assembly. Lead peroxide and sponge lead are produced through electrolysis processing the lead oxide described first into a paste. The products from paste processing are then placed on lead grids, which are made with energy as well. About 9.2 MJ (30%) of the 31 MJ of energy (Ectg) necessary to produce a kilogram of PbAs battery is used to create the battery (Hittman Associates, 1980).

In his literature, Rantik elucidates the methods required for battery production, covering aspects such as grid creation, lead oxide synthesis, and paste formulation, as well as pasting, drying, curing, and formatting processes. As an illustrative example, he employs an Emnf value of 16.6 MJ/kg (Sullivan and Gaines, 2012).

8.3.1.2 Nickel–Cadmium Batteries

The following manufacturing procedures are necessary to create these batteries (Rantik, 1999). (1) to create sponge nickel, deposit and burn carbonyl Ni powder onto the cathode metal film in a reduction furnace; (2) $Ni(NO_3)_2$ should next be infused into the resultant cathode strip to produce $Ni(OH)_2$; (3) the anode base (nickel fibre grid) is coated with plastic-bonded (PTFE) Cd that has been mixed with a portion of graphite to increase porosity and, consequently, conductivity; (4) create the separators (three thin polymer strips made of nylon, polypropylene, and nylon); (5) the electrodes are charged with surplus electrolyte; (6) arrange separator strips between layers of cathode and anode that are alternately oriented; (7) shape the thermoplastic case; and (8) add the elements to the polypropylene case, such as the electrolyte and seal. For either crystalline or jelly-roll-designed batteries, this set of procedures are used. The Emnf range given by Rydh and Sanden (2005) is 46–63 MJ/kg of battery. They resisted talking about the manufacturing process, though.

8.3.1.3 Lithium-Ion Batteries

These batteries are made using a variety of procedures, which include: (1) manufacture of cathode pastes and cathodes using LiMexOy (Me = Ni, Co, Fe, Mn), binding agents, aluminium sheets, and solvent bought lithium metal oxides; (2) the creation of anodes using strips of copper and graphite pastes; (3) arrangement of separator sheets separating the anodes and cathodes; (4) Electrolyte addition, cell charging, and final assembly come next (Rydh, 2019). Ishihara et al. (2002) in their review found Li-ion Emnf values are around 30 MJ/kg and a high set greater than 100 MJ/kg.

8.3.1.4 Sodium-Sulphur Batteries

These batteries function at elevated temperatures and necessitate insulation to efficiently isolate the high-temperature interior from the surrounding environment, such as through the use of hollow crystal spheres. According to the Hittman estimate, Emnf is responsible for 25% of the overall production energy (Dell'Era et al., 2015).

In conclusion, life cycle energy values for some of the four battery types covered in this article can be found. PbA and NiCd batteries have values that are quite constant within every generation according to the literature. Na/S battery values are in uncertainty. For Li-ion on the other hand, a wide variety of Emnf values are discovered. To sum up, insufficient quantitative data on the energy and material flow during the construction of various batteries, particularly complex battery systems, raises doubts about the reliability of their respective Emnf values and additional life cycle burdens (Bmnf) (Sullivan and Gaines, 2010).

8.3.2 Life Cycle Analysis of Solar Cells

One layer of light-absorbing materials or several layers of light absorbing can be used to make solar cells. The first scenario involves single junction cells, which are straightforward in their production but not as efficient; in contrast, many configurations (multi-junctions) feature varied charge separation and absorption methods, making them more complicated and effective (Brutschy, 2000).

Cells made from single-crystalline silicon (sc-Si) and multicrystalline silicon (mc-Si), as well as conventional panels with crystalline silicon (c-Si) bases, are included in the first-generation.

The amorphous silicon type (a-Si), the cadmium telluride (CdTe), the cadmium sulphide (CdS), and copper indium gallium selenide (CIGS)/copper indium selenide (CIS) modules based on silicon are among the thin-film solar cells used in the second generation.

The third generation, sometimes referred to as the "next generation," includes novel quasi-silicon-based technologies and innovative devices, including biodynamic/semi-organic photovoltaic panels, PSCs, dye-sensitized solar cells (DSSCs), and quantum dot (QD) cells.

8.3.2.1 Silicon Solar Cells

A metalloid called silicon was found in 1824 (Bagher et al., 2015). Most glasses, high-tech devices, semiconductor components for computers, and auto parts are all made from this metalloid, which is the most abundant semiconductor in the world. PV modules made of crystalline silicon are created using a number of methods.

In an arc furnace, silicon dioxide, also known as SiO_2 or quartz sand silica, undergoes conversion into metallurgical-grade silicon (MG-Si) (Pappas, 2016). Additionally, silicon must undergo refinement to achieve silicon of solar-grade purity (>99.99%) can be produced through diverse methods, including the Siemens process and its modified versions. Operating at high temperatures (usually ranging between 1,100°C and 1,200°C) in the reaction chamber, this process consumes a substantial amount of energy to produce highly pure silicon (Peng et al., 2013).

With Japan's solar irradiance measuring 1,427 kWh/m^2/year, Kato et al. conducted an LCA on a 3-kWp home PV system installed on a rooftop utilizing off-grid monocrystalline silicon (mono-Si), multicrystalline silicon (multi-Si), and amorphous silicon (a-Si) (Wong et al., 2016). Based on the findings, mono-Si modules demonstrate an Energy Payback Time (EPBT) of 11.8 years and a GHG emissions rate of 60.1 g CO_2-eq/kWh for an annual production of 10 MW/year. In comparison, multi-Si modules and a-Si modules, both with an equivalent annual cell production of 10 MW/year, were projected to have shorter EPBTs of 2.4 years and 2.1 years, respectively (Figure 8.6).

Kannan et al. explored the EPBT and GHG emissions of a 2.7 kWp mono-Si PV system under Singapore's solar irradiation of 1,635 kWh/m^2/year. The EPBT and GHG emissions determined through LCA were identified as 6.74 years and 217 g CO_2-eq/kWh, respectively (Kato et al., 2017). Moreover, in a comparative LCA conducted by Wei Luo et al. (2018), three roof-integrated multicrystalline PV systems—AL-BSF (aluminium back surface field), PERC (passivated emitter and rear cell), and PERC solar cells with the frameless double-glass module structure—were investigated. The results revealed EPBTs of 1.11, 1.08, and 1.01 years, along with associated GHG emissions rates of 30.2, 29.2, and 20.9 g CO_2-eq/kWh for the respective PV systems mentioned (Kannan et al., 2006). Frameless double-glass PV module designs have been found to significantly lower EPBT and perform better in terms of GHG emissions.

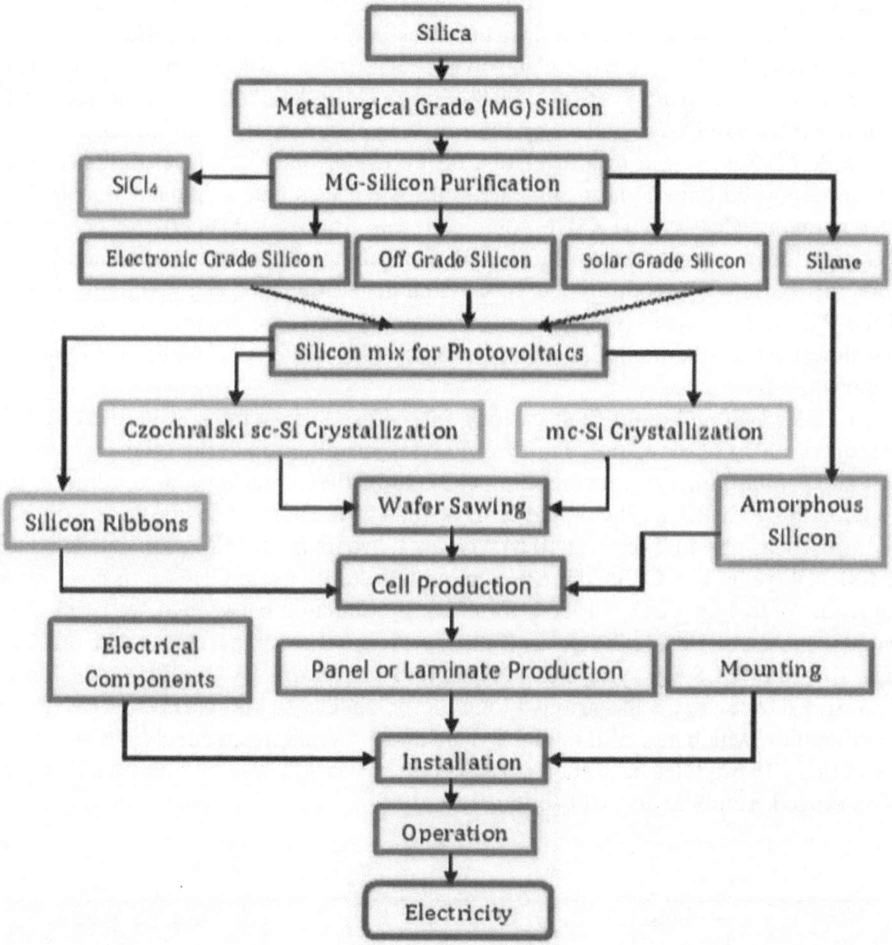

FIGURE 8.6 Stages of silicon-based PV components' life cycles. Adapted from Peng et al. (2013).

8.3.2.2 Thin-Film Solar Cells

The first LCA on these kinds of PVs was carried out in 2005 using 9% efficient CdTe modules. For the first solar facility in the US, LCI is carried out while accounting for the processes from raw materials including the BOS elements' removal to PV installation. The GHG emissions are 23.6 gCO_2-eq/kW h, and the EPBT is 1.2 years. Compared to silicon panels, this type has a smaller ecological footprint (Luo et al., 2018). Fthenakis outlined the phases of the life cycle of CdTe PV, specifically during thin-film production utilizing the cadmium telluride (CdTe) method (Nayeripour and Kheshti, 2011).

Most of the tellurium (Te) was extracted from the residue, known as slime, remaining after the electrolysis-based refining of copper. This residue contained copper

(Cu), selenium (Se), and other metals. The extraction of cadmium (Cd) from zinc ores (80%) and tellurium from copper ores was initially performed (Fthenakis et al., 2008; Alami, 2023). Following electrolytic purification, evaporating or distillation under vacuum were used to create >99.99% pure Cd and Te, which was subsequently used to synthesize CdTe. Following the initial application of a transparent conducting oxide (TCO) layer, a thin layer of CdS and subsequently a layer of CdTe were vapour-deposited onto a glass substrate. Subsequently, a heat treatment was applied after spraying $CdCl_2$. The CdTe solar cell was finalized through the deposition of a metal layer using vacuum arc-based methods to form the rear contact (Singh et al., 2013). Before becoming operational and generating electricity, the installation of the PV module, inverter, electrical components, and support systems took place. The flowchart for the manufacture of thin-film PV modules is shown in Figure 8.7 (Pappas, 2016).

To assess a-Si and multi-Si at various annual production rates, Kato et al. (2001) conducted an LCA on CdS/CdTe PV modules installed on residential rooftops to estimate primary energy consumption, CO_2, and EPBT emissions. According to the study, CdS/CdTe PV modules had EPBTs that were lower than those of a-Si and multi-Si, measuring 1.7 years at 10 MW/year, 1.4 years at 30 MW/year, and 1.1 years at 100 MW/year. The CdS/CdTe PV modules have life cycle GHG emissions ranging from 8.9 to 14.0 g CO_2-eq/kWh for annual production between 10 and 100 MW/year. In the Gobi Desert, Ito et al. (2008) performed a comparative LCA research for a 100-MW VLS-PV system employing four alternative PV module types: multi-Si, a-Si, CdTe, and CIS at various module efficiencies. In comparison to CdTe and a-Si modules, which had EPBTs of 1.9 years and 2.5 years, respectively, the scientists found that CIS modules have shorter EPBTs of 1.6 years. These modules' GHG emissions ranged from 9 to 16 g CO_2-eq/kWh.

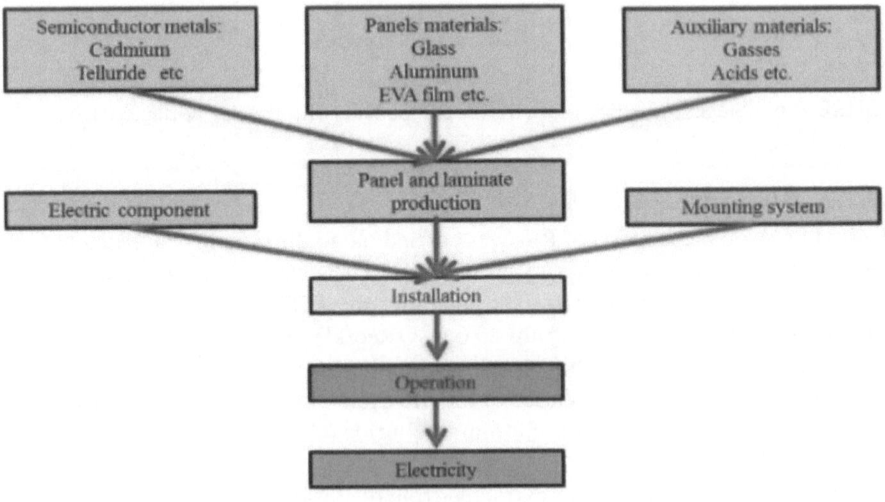

FIGURE 8.7 Stages of thin-film PV components' life cycles. Adapted from Pappas (2016).

In a 33 kW rooftop installation utilizing KC120 (multi-Si) and PV136 (a-Si) modules in Ann Arbor, United States, with solar irradiation levels of 1,359 kWh/m^2/year, Pacca et al. (2007) analysed three LCA parameters, namely, Net Energy Ratio (NER), EPBT, and GHG emissions. As per the study, the EPBT for a-Si is lower than that of multi-Si (3.2 years vs. 7.5 years). Yet, with potential improvements in conversion efficiency in the future, the anticipated EPBT for a-Si and multi-Si could be 1.6 and 5.7 years, respectively. The GHG emission rate follows a similar trend, with the a-Si GHG emission rate at 34.3 g CO_2-eq/kWh, considerably less than the multi-Si GHG emission rate of 72.4 g CO_2-eq/kWh (Ludin et al., 2018).

8.3.2.3 Dye-Sensitized Solar Cell (DSSC)

Due to their potential for having a lesser economic and environmental impact than standard PV technologies, DSSCs have some differentiation advantages. Initially, TCO coatings are applied to glass substrates, nanoporous titanium dioxide (TiO_2), which serves as a dye transporter—typically fluorine-doped tin oxide (FTO)—in the production of DSSCs (Ĺuque and Hegedus, 2003). After the deposition, the photoelectrode undergoes heating in a furnace to sinter the TiO_2, and the TiO_2 surface is subsequently coloured using dye sensitizer molecules, such as natural dye (Gong et al., 2012). The electrolyte solution, occupying the dye, contains I^-/$I3^-$ redox ions along with minimal-viscosity solvents that modify ion conductivity, influencing the cell's performance (Ĺuque and Hegedus, 2003). Platinum-coated counter-electrodes on other indium tin oxide glass sheets facilitate electron collection, completing the DSSC cell (Gong et al., 2012). Following this, DSSC cells are transformed into modules, subsequently mounted, and wired into the grid. The system boundaries for DSSC LCA experiments are depicted in Figure 8.8.

An environmental evaluation at laboratory scale was done on the production of DSSC PV in Southern Europe by Parisi et al. (2014). This analysis made the following assumptions about the DSSC module: an efficiency of 8%, a lifetime of 20 years, a performance ratio (PR) of 0.75, and the usage of the European Union for Coordination of Transmission Electricity as the module's energy source. The NER for the DSSC module stood at 12.67, with corresponding EPBT and CO_2 emissions of 1.58 years and 22.38 g CO_2-eq/kWh, respectively, according to the authors' findings. The LCA of the DSSC study was then compared to other chemical and biological thin-film technologies (Parisi et al., 2014). In a subsequent LCA on DSSCs from earlier investigations, Parisi et al. (2014) compared the DSSC with alternative thin-film technologies, including polymers, a-Si, CdTe, and CIS (de Wild-Scholten and Velkamp, 2007). GHG emissions rate ranged from 277 to 365 MJ/m^2, 0.6 to 1.8 years, and 9.8 to 120.0 g CO_2-eq/kWh, respectively. Notably, there was a strong correlation between DSSC GHG emissions and module functional lifetime (de Wild-Scholten and Velkamp, 2007). Longer lives and higher conversion efficiencies could lead to the DSSC's optimum performance.

8.3.2.4 Perovskite Solar Cell (PSC)

PSCs have gained considerable attention in recent years as one of the newly developed PV technologies that have excellent productivity with minimal cell production costs. In earlier LCA analyses of PSCs, Figure 8.9 depicts the system boundaries

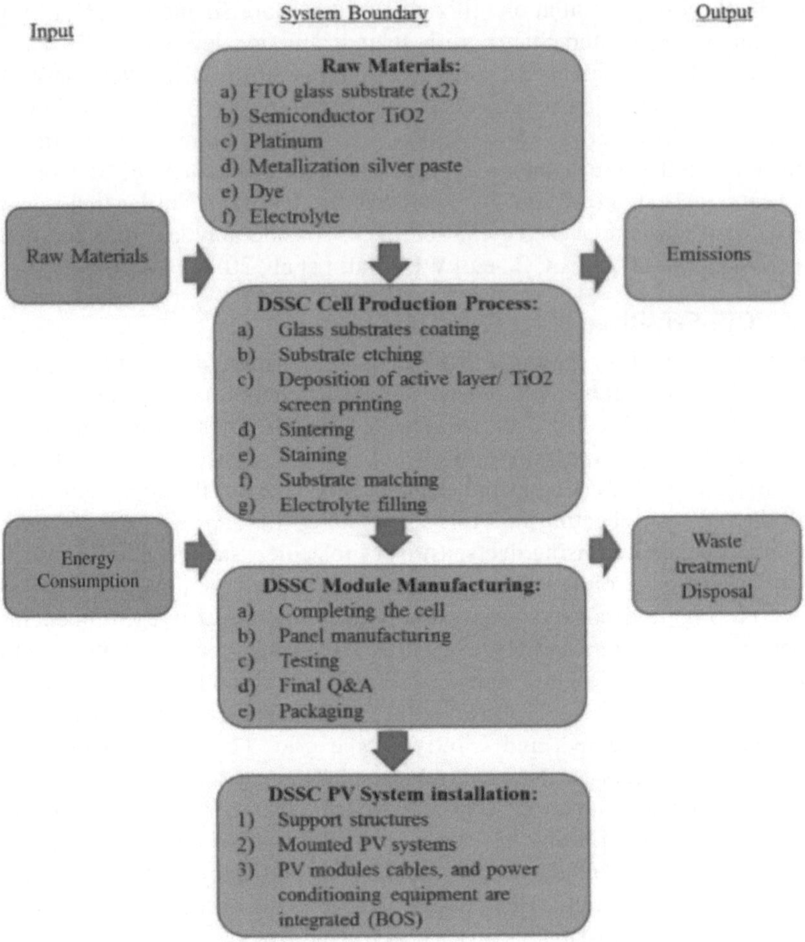

FIGURE 8.8 LCA from cradle to gate for the DSSC procedure. Adapted from Ludin et al. (2018).

taking into account a cradle-to-grave approach, encompassing the production of the individual modules, their use, and their disposal (Gong et al., 2015).

The initial phase commences with the extraction of raw materials required for manufacturing components essential in the subsequent stage. These materials include FTO glass, BL-TiO_2 ink, PbI_2, gold, and PET production. There are multiple phases involved in making a perovskite PV module, beginning with FTO glass manufacturing, substrates, blocking TiO_2 layer formation, followed by the process involves depositing an electron transport layer, a perovskite layer, and a hole transport layer on the substrate, followed by the thermal evaporation of gold (Gong and Sumathi, 2015). In the usage phase, the perovskite modules were used to produce electricity after being put together. In the disposal stage, the waste modules were finally dumped in the ground.

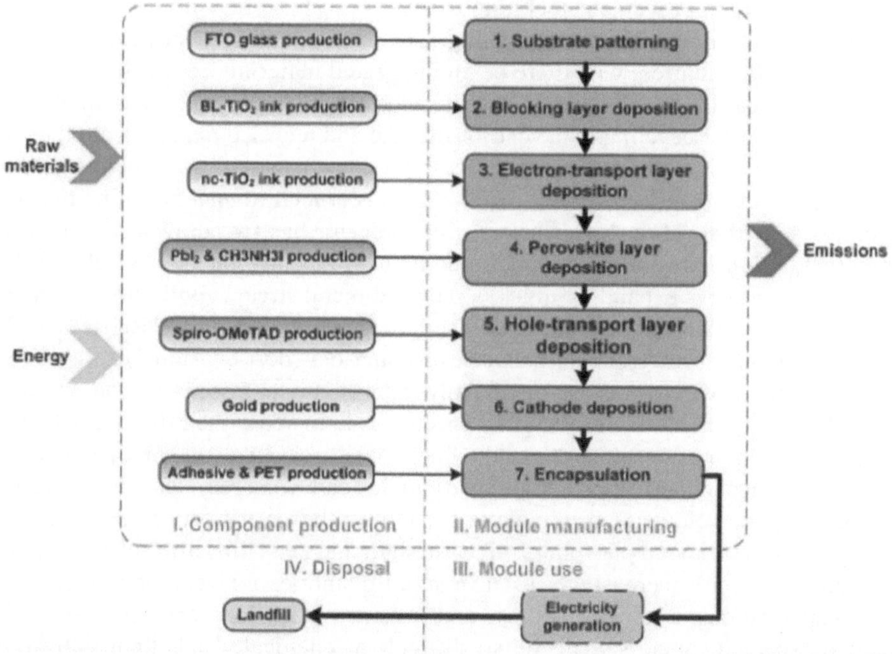

FIGURE 8.9 TiO_2 system boundary in perovskite solar module production. Adapted from Ludin et al. (2018).

Assuming a 15% module efficiency, a PR of 0.75, a 15-year lifespan, and 65% active area under solar irradiation of 1,700 kWh/m^2/year, Celik et al. assessed the environmental impact of three PSCs structures: void, solution, and hole transfer layer-free devices (Celik et al., 2016). The findings indicate that, in comparison to advancements in crystalline and thin-layer technologies, the PSCs exhibited EPBT and GHG emissions ranging from 1.05 to 1.54 years and 100–150 g CO_2-eq/kWh, respectively.

The EPBT spanned from 0.2 to 5.4 years, while the GHG emissions rate varied from 56.65 to 497.2 g CO_2-eq/kWh. Notably, PSCs demonstrated significantly lower energy consumption compared to silicon and lightweight innovations, showcasing competitive environmental advantages and an EPBT that paves the way for future production (Ludin et al., 2018).

8.4 RECYCLING AS SUSTAINABLE USAGE OF MATERIALS

The fast growth of the developer's material palette over the past few decades has hampered the recycling of items in the "occasionally abandoned" or "owned by someone else" categories (Zhang et al., 2010; Rozendal et al., 2006). As a result, many goods are now more durable and efficient than before. Recycling has become much more complicated and difficult to accomplish, which was not the desired outcome. In recent years, several reviews of metal recycling have been published. They explore important topics like reuse and recycling of technological devices, financial constraints, and improvement techniques (Ayres et al., 2006). Some unanswered queries are as follows: What's happening, and what are the recent developments? What are the boundaries? Can there be an integrated materials economy?

How successful is the globe recycling the wide range of components found in contemporary goods? Recycling at the end-of-life rate and recycled materials are the two indicators that best address this issue. Recycled content specifies the percentage of scrap used in the manufacturing of metal, which is crucial to understanding the size of the reclaimed supplier. Nevertheless, this indicator has two drawbacks. Initially given that the lifespans of metal-containing items frequently extend several decades and that metal use is expanding quickly, recycled metal streams will only be able to satisfy a small percentage of demand for a very long period (Logan and Rabaey, 2012). Second, it does not differentiate between new and old (post-consumer) scrap as an input material, leaving it open to artificially increased rates based only on pre-consumer sources (fabricators may be given incentives to increase their discard output to meet supplementary consumer demand, which makes recycled content a justification for shortcomings in fabrication and production). Instead, what recycled content means to promote is the volume of used materials that meet the criteria for regeneration.

Some devices like red phosphors, high-strength magnets, thin-film solar cells, and computer chips, are increasingly used in small quantities for very precise technological applications. Recovery in those applications, which frequently involve finely blended "specialty metals," can be so difficult mechanically and financially that attempts are rarely performed.

8.4.1 Acquisition, Preliminary Processing, and Finishing Off

Depending on cost, administrative tasks, and other considerations, acquisition rates for various waste streams might be very varied. Contrarily, despite regulatory efforts, acquisition rates for waste electrical and electronic equipment are frequently quite low. The pre-processing process for blog-consumer metal begins with collecting and includes continuous screening (like manual labour, conductive, or visual methods) and both chemical and physical separation (Cheng et al., 2009).

Scale-related issues are crucial. Production of virgin materials often takes place on a massive scale and is supported by historically cheap energy rates. Recycling, on the other hand, is frequently more regional, difficult to implement, and small scale. The financial benefits in such a situation are frequently insufficient to support the acquisition of contemporary the "impression and kind" technology, and a great deal of otherwise recoverable material is wasted (Logan and Rabaey, 2012).

8.4.2 Leveraging the Recycling Process

While segregation and organizing efficiency are tied to recyclable technological advances, acquisition efficiencies are connected to societal and governmental variables. The situation is unfavourable from a materials standpoint that, due to scale and budgetary constraints, more sophisticated technologies (like laser beam, near-infrared, or x-ray sorting) are typically only applied to specific recycling streams. Examples of more basic technologies include splitting, compression, and magnetic sorting. The design of products frequently makes it difficult to disassemble and liberate materials (for example, constructed magnetic strips in desktops) (Nam and Logan, 2011).

Though there are significant examples of cutting-edge repurposing innovations, many of which are still in the demonstration stage, much more focus needs to be placed on transforming and improving existing generic systems if efficiency ratings are to increase from their current levels (Dahmus and Gutowski, 2007). Such technological advancement might coincide with a globalization of labour, as is typical in production operations. For electronics, the finest-of-both-worlds strategy suggests utilizing the cheap labour costs in poor nations for manual disintegration and the high productivity of specialized refineries, which are normally found in industrialized nations for halt-processing (Cheng et al., 2009).

8.4.3 Future Recycling Concerns to Be Addressed

The activity that has the greatest chance to promote the recycling of metal is collecting, despite the fact that it initially appears insignificant. It is mostly a matter of behavioural norms and encouragement, as well as programmes like the requirement of recycling deposits on products for consumption, to collect discards with high effectiveness and with due care (to avoid mixing that would impede subsequent treatment). A further obstacle to the collection and rehabilitation of many elements is the worldwide commerce in used goods, which sends complicated products to nations with insufficient recycling infrastructure (Speers and Reguera, 2012; Stoytcheva, 2011).

The next endeavour is to involve upcoming manufacturers in selecting the right combinations of materials with reusing in mind after taking care of the collection. The current tendency of more material mixing can only be reversed by designers, yet the contemporary designs are actually harder to recycle than they were a few generations prior. The next endeavour is to involve upcoming manufacturers in selecting the right combinations of materials with reusing in mind after taking care of the collection. The current tendency of more material mixing can only be reversed by designers, yet the contemporary designs are actually harder to recycle than they were a few generations prior (Steinbusch et al., 2009; Heidrich et al., 2011).

8.5 NEW OPTIONS OF SUSTAINABLE USAGE IN ENERGY CONVERSION

As discussed briefly above, solar cell devices are the straightforward option of sustainable energy conversion. Other new options include fuel cells and rechargeable batteries.

8.5.1 Fuel Cells: The Conversion of Hydrogen

For the conversion of hydrogen, various technologies have already been established. For instance, hydrogen can be utilized in both fuel cells and engines. Fuel cells are electrical devices that convert chemical energy, typically in the form of hydrogen, into electricity. Engines, on the other hand, can combust hydrogen similarly to the way they burn gasoline or natural gas (Sherif et al., 2005). In a hydrogen fuel cell, the elements hydrogen and oxygen mix electrochemically (reverse electrolysis) without burning, creating electrically charged energy. The primary obstacles to overcome are the high production costs brought on by the pricey components necessary to make electrodes (bipolar plates), electrolytes, membranes, and catalysts (particularly given the price of platinum).

The automotive sector has historically been interested in PEMFC fuel cells for zero-emission cars, as well as the power production industries for both residential and electronic uses. They are intended to be efficient and clean power sources, but their commercial success is hindered by two key factors: the high cost of Pt/support electrodes and the membrane's poor performance, especially at elevated temperatures (Serrano et al., 2009).

By greatly reducing the quantity and endurance of Pt catalyst required for the application, it is possible to significantly lower the overall expense of the electrodes. Additionally, there is a clear connection between this cost and the support's impact on catalyst performance. For polymer electrolyte membrane fuel cells (PEMFC), carbon nanotube–based electrodes have been created by Wang et al. (2004). They discovered that MWNT can be employed as the support for the catalyst rather than carbon powder, allowing for a 2/3 reduction in Pt usage when compared to conventional PEMFC fuel cells.

Other methods rely on creating metal-based nanostructured catalyst employing a variety of later removed self-organized templates as configuration-directing elements. In the context of serving as anodes for Direct Methanol Fuel Cells (DMFC), Choi et al. employed a template based on crystalline structures with nanosized holes. Within these holes, they created Pt-Ru nanoscale systems (Hossain, 2011).

The application of nanotechnology hydrophilic inorganic substances in electrolytes and an inorganic lithium salt as additives to improve the hydrogen ion conductivity of a membrane at high temperatures is one of the contributions of nanotechnology (the inorganic materials present high affinity to water and the salts work in the right temperature range). Other methods rely on the use of capillary fuel cells rather than flat cell membranes, which are significantly less expensive than the latter. As a result, their more compact construction results in a three to six times higher power density and enables the preservation of the electrodes' bipolar plates (Serrano et al., 2009).

8.5.2 Battery Rechargeables

Renewable lithium batteries are the current subject of most ongoing research in this area. The Li-ion chemistry results in a rise of 100%–150% in the capacity to store of energy for each unit of mass and volume when compared with hydrophilic batteries. Despite this, there are significant drawbacks, including low power and energy density, a big shift in volume on consequences, security, and expense. The drawbacks can be (or are being) eliminated through the use of nanomaterials in the production of lithium-ion batteries (Serrano et al., 2009). Silicon, alumina, or chromium nanoparticles can boost the conductivity of non-aqueous electrolytic solutions by a factor of six. Solid polymer electrolytes (SPE) and solid-state electrolytes have received the majority of attention. The PEO-based SPE has drawn the greatest attention since PEO is secure, environmentally friendly, and produces flexible films. However, polymers commonly exhibit low conductivities at room temperature, and depending on the SPE designs, they may have insufficient interfacial mobility and stable mechanical properties. In addressing these challenges, the utilization of nanocomposite polymer electrolytes has the potential to contribute to the development of highly efficient, secure, and environmentally friendly batteries (Agrawal and Pandey, 2008).

Graphite granules and nanotubes made of carbon were used in place of LiC6 electrodes, which had a capacity to store of 340 mAh/g, in certain efficient methods for nanostructuring the anode for lithium-ion rechargeable batteries. The lithium deposition as well as certain safety issues were avoided by replacing these with nanosized metallic oxides, primarily in the form of nanotubes and nanowires, such as titanium, aluminium, vanadium, cobalt, tin, and silicon oxides. This has allowed for an amount of storage that is comparable to that of conventional graphite electrodes (Nam et al., 2006; Chan et al., 2007). A study by Chan et al. (2007) achieved a charge capacity that was around 3,000 mAh/g for 10 cycles following the first one, shedding roughly one third of capacity in the first cycle, by synthesizing silica nanowires that are further fixed to the substrate functioning as current collector.

Additionally, $LiFePO_4$, $LiMn_2O_4$, and vanadium oxide (V_2O_5) have been successfully used to examine the nanostructuration of the cathode. Manganese oxides are a good option for both the environment and the economy. $LiMn_2O_4$ performs the best when used in large quantities. For this, many nanostructures have been examined. Since nanoparticles need to be coated to prevent sinterization and increase conductivity, they are less desired (Serrano et al., 2009).

8.6 CONCLUSION

We must continue to create, transform, and use energy responsibly while minimizing our negative environmental effects because the world's population is growing and has an increasing need for energy. The development of novel multifunctional materials from the vast and interdisciplinary discipline known as nanotechnology today is essential for overcoming least of the technological constraints of the many non-renewable energy options. PV solar cells' efficiency is rising as their production and electricity production costs are falling at a pace never seen before due to improved nanomaterials. Numerous technological advances are required to transition from an energy system based on carbon to one that is more sustainable, including those in the production of energy, shipping, alteration, storage, and final use. We face considerable scientific and engineering obstacles across all these processes. The materials we currently have are frequently unable to offer an alternative at the level of effectiveness necessary at a workable cost.

ACKNOWLEDGEMENT

The research was supported by the Ministry of Higher Education (MoHE), through the Fundamental Research Grant Scheme (FRGS/1/2022/TK08/UTAR/02/7).

REFERENCES

Aberle, A.G. (2006). Fabrication and characterisation of crystalline silicon thin-film materials for solar cells. *Thin Solid Films* 511–512, 26–34.

Ackermann, T. (2005). *Wind Power in Power Systems*, John Wiley and Sons, Hoboken, pp. 58–110.

Adeney, K. (2002). *Ethnic Conflict and Civic Life: Hindus and Muslims in India, Ashutosh Varshney*, Yale University Press, New Haven, pp. 157–159.

Adimazoya, T.N., and Doelle, M. (2021). Regulating Wave, Tidal and Ocean Thermal EnergyIn: *Elgar Encyclopedia of Environmental Law*, ed. Michael F., pp. 546–557.

Agrawal, R.C., and Pandey, G.P. (2008). Solid polymer electrolytes: Materials designing and all-solid-state battery applications: An overview. *Journal of Physics D: Applied Physics* 41(22), 223001.

Alami, A.H. (2023). *PV Technology and Manufacturing*. Springer Nature, New York.

Anon, Trove. Trove. https://trove.nla.gov.au/work/15414012 (accessed: Oct 2023)

Ayres, R.U., Gleich, A., and Von Gößling-Reisemann, S. (2007). *Sustainable Metals Management : Securing our Future-Steps Towards a Closed Loop Economy, Eco-Efficiency in Industry and Science, XVI*, 610 p.

Bagher, A. M., Vahid, M. M. A., and Mohsen, M. (2015). Types of solar cells and application. *American Journal of optics and Photonics*, 3(5), 94–113.

Basosi, R., and Parisi, M.L. (2015). Environmental life cycle analysis of nonconventional thin-film photovoltaics: The case of dye-sensitized solar devices. In: *Energy Security and Development: The Global Context and Indian Perspectives* ed. Reddy, B., Ulgiati, S.. Springer, India, pp. 193–210.

Brutschy, B. (2000). Structure of microsolvated benzene derivatives and the role of aromatic substituents. *Chemical Reviews* 100, 3891–3920.

Celik, I., Song, Z., Cimaroli, A.J., Yan, Y., Heben, M.J., and Apul, D. (2016). Life cycle assessment (LCA) of perovskite PV cells projected from lab to fab. *Solar Energy Materials and Solar Cells* 156, 157–169.

Chan, C.K., Peng, H., Twesten, R.D., Jarausch, K., Zhang, X.F., and Cui, Y. (2007). Fast, completely reversible Li insertion in vanadium pentoxide nanoribbons. *Nano Letters* 7, 490–495.

Chan, C.K., Peng, H., Liu, G., McIlwrath, K., Zhang, X.F., Huggins, R.A., and Cui, Y. (2008). High-performance lithium battery anodes using silicon nanowires. *Nature Nanotechnology* 3, 31–35.

Cheng, S., Xing, D., Call, D.F., and Logan, B.E. (2009). Direct biological conversion of electrical current into methane by electromethanogenesis. *Environmental Science and Technology* 43, 3953–3958.

Costanza, R., D'Arge, R., De Groot, R., Farber, S., Grasso, M., Hannon, B., Limburg, K., Naeem, S., O'Neill, R.V., Paruelo, J., et al. (1997). The value of the world's ecosystem services and natural capital. *Nature* 387, 253–260.

Dahmus, J.B., and Gutowski, T.G. (2007). What gets recycled: An information theory based model for product recycling. *Environmental Science and Technology* 41, 7543–7550.

de Wild-Scholten, M.J., and Veltkamp, A.C. (2007). Environmental life cycle analysis of dye sensitized solar devices: Status and outlook. In: *Presented at: 22nd European Photovoltaic Solar Energy Conference and Exhibition, Milan, Italy* vol. 3, pp. 1–5.

Dell'Era, A., Bocci, E., Sbordone, D., Villarini, M., Boeri, F., Di Carlo, A., and Zuccari, F. (2015). Experimental tests to recover the photovoltaic power by battery system. *Energy Procedia*, 81,. 548–557.

Dye, S.T. (2012). Geoneutrinos and the radioactive power of the Earth. *Reviews of Geophysics* 50, 19.

Fthenakis, V.M., Hyung, C.K., and Alsema, E. (2008). Emissions from photovoltaic life cycles. *Environmental Science and Technology* 42, 2168–2174.

Gibson, T.L., and Kelly, N.A. (2010). Solar photovoltaic charging of lithium-ion batteries. *Journal of Power Sources* 195, 3928–3932.

Graedel, T.E., Allwood, J., Birat, J.P., Buchert, M., Hagelüken, C., Reck, B.K., Sibley, S.F., and Sonnemann, G. (2011). What do we know about metal recycling rates? *Journal of Industrial Ecology* 15, 355–366.

Graetzel, M., Janssen, R.A.J., Mitzi, D.B., and Sargent, E.H. (2012). Materials interface engineering for solution-processed photovoltaics. *Nature* 488, 304–312.

Gong, J., Liang, J., and Sumathy, K. (2012). Review on dye-sensitized solar cells (DSSCs): Fundamental concepts and novel materials. *Renewable and Sustainable Energy Reviews* 16, 5848–5860.

Gong, J., Darling, S.B., and You, F. (2015). Perovskite photovoltaics: Life-cycle assessment of energy and environmental impacts. *Energy and Environmental Science* 8, 1953–1968.

Green, M.L., Espinal, L., Traversa, E., and Amis, E.J. (2012). Materials for sustainable development. *MRS Bulletin* 37, 303–308.

Guinee, J.B. (2002). Handbook on life cycle assessment operational guide to the ISO standards. *The International Journal of Life Cycle Assessment* 7, 311-313.

Gurung, A., and Qiao, Q. (2018). Solar charging batteries: Advances, challenges, and opportunities. *Joule* 2, 1217–1230.

Gul, M., Kotak, Y., and Muneer, T. (2016). Review on recent trend of solar photovoltaic technology. *Energy Exploration and Exploitation* 34, 485–526.

Hamedi, A.S., and Sadeghzadeh, S. (2017). Conceptual design of a 5 MW OTEC power plant in the Oman Sea. *Journal of Marine Engineering and Technology* 16, 94–102.

Heidrich, E.S., Curtis, T.P., and Dolfing, J. (2011). Determination of the internal chemical energy of wastewater. *Environmental Science and Technology* 45, 827–832.

Hittman Associates. (1980). Life cycle energy analysis of electric vehicle storage batteries. H-1008/001-80-964, submitted to the U.S. Department of Energy, contract number DE-AC02-79ET25420.A000

Hossam-Eldin, A. & Refaey, M., and Farghly, A. (2015). A review on photovoltaic solar energy technology and its efficiency. In: *Proceedings of the Conference: 17th International Middle-East Power System Conference (MEPCON'15)*, Mansoura, Egypt, 15–17 December 2015.

Hossain, M.M. (2012). *Heat and Mass Transfer: Modeling and Simulation*, InTech, Rijeka.

Ishihara, K., and Kihira, N. (2002). Environmental burdens of large lithium-ion batteries developed in a Japanese national project. *Power Industry*, Japan 11.

Islam, M.M., Hasanuzzaman, M., Pandey, A.K., and Rahim, N.A. (2019). Modern energy conversion technologies. In: *Energy for Sustainable Development: Demand, Supply, Conversion and Management*, ed. Hasanuzzaman, MD., and Rahim, N.A., Elsevier, Amsterdam, pp. 19–39.

Ito, M., Kato, K., Komoto, K., Kichimi, T., and Kurokawa, K. (2008). A comparative study on cost and life-cycle analysis for 100 MW very large-scale PV (VLS-PV) systems in deserts using m-Si, a-Si, CdTe, and CIS modules. *Progress in Photovoltaics: Research and Applications* 16, 17–30.

Kato, K., Murata, A., and Sakuta, K. (1998). Energy pay-back time and life-cycle CO_2 emission of residential PV power system with silicon PV module. *Progress in Photovoltaics: Research and Applications* 6, 105–115.

Kato, K., Hibino, T., Komoto, K., Ihara, S., Yamamoto, S., and Fujihara, H. (2001). Life-cycle analysis on thin-film CdS/CdTe PV modules. *Solar Energy Materials and Solar Cells* 67, 279–287.

Kannan, R., Leong, K.C., Osman, R., Ho, H.K., and Tso, C.P. (2006). Life cycle assessment study of solar PV systems: An example of a 2.7 kWp distributed solar PV system in Singapore. *Solar Energy* 80, 555–563.

Landi, D., Marconi, M., and Pietroni, G. (2022). Comparative life cycle assessment of two different battery technologies: Lithium iron phosphate and sodium-sulfur. *Procedia CIRP*, 105, 482–488.

Lawton, J.H. (1998). *Nature's Services: Societal Dependence on Natural Ecosystems*, Island Press, Washington, D.C.

LeSar, R., Chen, K., and Apelian, D. (2012). Teaching sustainable development in materials science and engineering. *MRS Bulletin* 37, 449–454.

Liu, P., Yang, H.X., Ai, X.P., Li, G.R., and Gao, X.P. (2012). A solar rechargeable battery based on polymeric charge storage electrodes. *Electrochemistry Communications* 16, 69–72.

Logan, B.E., and Rabaey, K. (2012). Conversion of wastes into bioelectricity and chemicals by using microbial electrochemical technologies. *Science* 337, 686–690.

Ludin, N.A., Mustafa, N.I., Hanafiah, M.M., Ibrahim, M.A., Asri Mat Teridi, M., Sepeai, S., Zaharim, A., and Sopian, K. (2018). Prospects of life cycle assessment of renewable energy from solar photovoltaic technologies: A review. *Renewable and Sustainable Energy Reviews* 96, 11–28.

Luo, B., Ye, D., and Wang, L. (2017). Recent progress on integrated energy conversion and storage systems. *Advanced Science* 4, 1700104.

Luo, W., Khoo, Y.S., Kumar, A., Low, J.S.C., Li, Y., Tan, Y.S., Wang, Y., Aberle, A.G., and Ramakrishna, S. (2018). A comparative life-cycle assessment of photovoltaic electricity generation in Singapore by multicrystalline silicon technologies. *Solar Energy Materials and Solar Cells* 174, 157–162.

Luque, A. and Hegedus, S. (2003). *Handbook of Photovoltaic Science and Engineering*. Wiley, New York.

Muehlebach, A. (2001). "Making place" at the United Nations: Indigenous cultural politics at the U.N. *Working Group on Indigenous Populations. Cultural Anthropology: Journal of the Society for Cultural Anthropology* 16, 415–448.

Muteri, V., Cellura, M., Curto, D., Franzitta, V., Longo, S., Mistretta, M., and Parisi, M.L. (2020). Review on life cycle assessment of solar photovoltaic panels. *Energies* 13, 252.

Nam, K.T., Kim, D.W., Yoo, P.J., Chiang, C.Y., Meethong, N., Hammond, P.T., Chiang, Y.M., and Belcher, A.M. (2006). Virus-enabled synthesis and assembly of nanowires for lithium ion battery electrodes. *Science* 312, 885–888.

Nam, J.Y., and Logan, B.E. (2011). Enhanced hydrogen generation using a saline catholyte in a two chamber microbial electrolysis cell. *International Journal of Hydrogen Energy* 36, 15105–15110.

Nayeripour, M. and Kheshti, M. (2012). *Sustainable Growth and Applications in Renewable Energy Sources*. InTech, Rijeka.

Nazeeruddin, M.K., Baranoff, E., and Grätzel, M. (2011). Dye-sensitized solar cells: A brief overview. *Solar Energy* 85, 1172–1178.

Ness, D. (2020). Growth in floor area: The blind spot in cutting carbon. *Emerald Open Research* 2, 2.

Pacca, S., Sivaraman, D., and Keoleian, G.A. (2007). Parameters affecting the life cycle performance of PV technologies and systems. *Energy Policy* 35, 3316–3326.

Pappas S. (2016). Facts about silicon. https://www.livescience.com/28893-silicon.html (accessed: Oct 2023).

Parisi, M.L., Maranghi, S., and Basosi, R. (2014). The evolution of the dye sensitized solar cells from Grätzel prototype to up-scaled solar applications: A life cycle assessment approach. *Renewable and Sustainable Energy Reviews* 39, 124–138.

Park, S.K. (2018). Social bonds for sustainable development: A human rights perspective on impact investing. *Business and Human Rights Journal* 3, 233–255.

Parris, T.M., and Kates, R.W. (2003). Characterizing and measuring sustainable development. *Annual Review of Environment and Resources* 28, 559–586.

Peng, J., Lu, L., and Yang, H. (2013). Review on life cycle assessment of energy payback and greenhouse gas emission of solar photovoltaic systems. *Renewable and Sustainable Energy Reviews* 19, 255–274.

Rantik, M. (1999). Life cycle assessment of five batteries for electric vehicles under different charging regimes. *Chalmers University of Technology*, Goteborg, Sweden

Ratner, M. (2004). The physics of solar cells ; third generation photovoltaics: Advanced solar energy conversion. *Physics Today* 57, 71–72.

Reddy, T. A. (2017). *Energy Conversion*, 2nd edn, CRC Press, Boca Raton.

Rozendal, R.A., Hamelers, H.V.M., and Buisman, C.J.N. (2006). Effects of membrane cation transport on pH and microbial fuel cell performance. *Environmental Science and Technology* 40, 5206–5211.

Rydh, C.J., and Sandén, B.A. (2005). Energy analysis of batteries in photovoltaic systems. Part I: Performance and energy requirements. *Energy Conversion and Management* 46, 1957–1979.

Rydh, C.J. (2019). Environmental assessment of NiCd: Battery manufacturing -past and present trends. Linnaeus Eco-Tech 461-470.

Serrano, E., Rus, G., and García-Martínez, J. (2009). Nanotechnology for sustainable energy. *Renewable and Sustainable Energy Reviews* 13, 2373–2384.

Sherif, S.A., Barbir, F., and Veziroglu, T.N. (2005). Wind energy and the hydrogen economy-review of the technology. *Solar Energy* 78, 647–660.

Singh, A., Olsen, S.I., and Pant, D. (2013). Importance of life cycle assessment of renewable energy sources. In: *Green Energy and Technology*, ed. Singh, A., Pant,D., and Olsen, S.I., Springer, New York, pp. 1–11.

Solow, R. (1993). An almost practical step toward sustainability. *Resources Policy* 19, 162–172.

Song, J.H., Murphy, R.J., Narayan, R., and Davies, G.B.H. (2009). Biodegradable and compostable alternatives to conventional plastics. *Philosophical Transactions of the Royal Society B: Biological Sciences* 364, 2127–2139.

Sørensen, B. (2017). *Renewable Energy: Physics, Engineering, Environmental Impacts, Economics and Planning*, 5th edn, Elsevier, Amsterdam.

Speers, A.M., and Reguera, G. (2012). Electron donors supporting growth and electroactivity of geobacter sulfurreducens anode biofilms. *Applied and Environmental Microbiology* 78, 437–444.

Steinbusch, K.J.J., Hamelers, H.V.M., Schaap, J.D., Kampman, C., and Buisman, C.J.N. (2010). Bioelectrochemical ethanol production through mediated acetate reduction by mixed cultures. *Environmental Science and Technology* 44, 513–517.

Stroyuk, O. (2018). Lead-free hybrid perovskites for photovoltaics. *Beilstein Journal of Nanotechnology* 9, 2209–2235.

Stoytcheva, M. (2012). *Pesticides: Formulations, Effects, Fate.* InTech, Rijeka.

Sullivan, J L, Gaines, L, and Energy Systems. (2010). A review of battery life-cycle analysis: State of knowledge and critical needs. US Department of Energy, Oak Ridge.

Sullivan, J.L., and Gaines, L. (2012). Status of life cycle inventories for batteries. *Energy Conversion and Management* 58, 134–148.

Uriarte, F.A. (2010). *Biofuels from Plant Oils: A Book for Practitioners and Professionals Involved in Biofuels, to Promote a Better and More Accurate Understanding of the Nature, Production, and Use of Biofuels from Plant Oils*. ASEAN Foundation, Jakarta.

Wang, C., Waje, M., Wang, X., Tang, J.M., Haddon, R.C., and Yan, Y. (2004). Proton exchange membrane fuel cells with carbon nanotube based electrodes. *Nano Letters* 4, 345–348.

Wee, G., Salim, T., Lam, Y.M., Mhaisalkar, S.G., and Srinivasan, M. (2011). Printable photo-supercapacitor using single-walled carbon nanotubes. *Energy and Environmental Science* 4, 413–416.

Wong, J.H., Royapoor, M., and Chan, C.W. (2016). Review of life cycle analyses and embodied energy requirements of single-crystalline and multi-crystalline silicon photovoltaic systems. *Renewable and Sustainable Energy Reviews* 58, 608–618.

Xu, J., Chen, Y., and Dai, L. (2015). Efficiently photo-charging lithium-ion battery by perovskite solar cell. *Nature Communications* 6, 8103.

Yan, N.F., Li, G.R., and Gao, X.P. (2014). Electroactive organic compounds as anode-active materials for solar rechargeable redox flow battery in dual-phase electrolytes. *Journal of The Electrochemical Society* 161, A736–A741.

Zhang, X., Cheng, S., Huang, X., and Logan, B.E. (2010). The use of nylon and glass fiber filter separators with different pore sizes in air-cathode single-chamber microbial fuel cells. *Energy and Environmental Science* 3, 659–664.

Index

MODERN BIOTECHNOLOGY IN HEALTHCARE

Advances and Applications

Edited by

Sheikh Umar Ahmad, PhD

AAP | APPLE ACADEMIC PRESS

First edition published 2023

Apple Academic Press Inc.
1265 Goldenrod Circle, NE,
Palm Bay, FL 32905 USA

760 Laurentian Drive, Unit 19,
Burlington, ON L7N 0A4, CANADA

CRC Press
6000 Broken Sound Parkway NW,
Suite 300, Boca Raton, FL 33487-2742 USA

4 Park Square, Milton Park,
Abingdon, Oxon, OX14 4RN UK

Library and Archives Canada Cataloguing in Publication

Title: Modern biotechnology in healthcare : advances and applications / edited by Sheikh Umar Ahmad, PhD.
Names: Ahmad, Sheikh Umar, editor.
Description: First edition. | Includes bibliographical references and index.
Identifiers: Canadiana (print) 20230171265 | Canadiana (ebook) 2023017132X | ISBN 9781774913628 (hardcover) | ISBN 9781774913635 (softcover) | ISBN 9781003371298 (ebook)
Subjects: LCSH: Biotechnology. | LCSH: Diagnosis. | LCSH: Medical care.
Classification: LCC TP248.2 .M63 2023 | DDC 660.6—dc23

Library of Congress Cataloging-in-Publication Data

CIP data on file with US Library of Congress

ISBN: 978-1-77491-362-8 (hbk)
ISBN: 978-1-77491-363-5 (pbk)
ISBN: 978-1-00337-129-8 (ebk)

About the Editor

Sheikh Umar Ahmad, PhD

Postdoctoral Research Associate, Department of Dermatology, Wisconsin Institutes for Medical Research – School of Medicine and Public Health, University of Wisconsin–Madison, United States

Sheikh Umar Ahmad, PhD, is currently working as a Postdoctoral Research Associate in the Department of Dermatology, Wisconsin Institutes for Medical Research – School of Medicine and Public Health, University of Wisconsin–Madison, United States. He is involved in research projects aimed at understanding the "Developmental Genetics of Planar Cell Polarity Control in Melanoma Development and Progression." Dr. Ahmad completed his BS (Biological Sciences – majors in Biotechnology) from the University of Kashmir Srinagar, Jammu and Kashmir in 2012 and MS in Biotechnology from BGSB University Rajouri, Jammu & Kashmir in 2015. Dr. Ahmad did his doctoral research studies at the CSIR–Indian Institute of Integrative Medicine, Jammu, India, where his research interests were focused on "Understanding the Interplay of Autophagy and DNA Damage Response Mechanism in the Regulation of Ultraviolet-(B)-Induced Skin Photodamage." Dr. Ahmad's thesis work mechanistically elucidated the role of interventional autophagy as a promising therapeutic strategy against skin photodamage disorders. Dr. Ahmad published his research findings in various international journals of repute and presented his work at several scientific conferences. He was previously awarded for his research works and has received the Best Research Paper Award at the Annual Awards presented by CSIR–Indian Institute of Integrative Medicine, Jammu, India. In addition, he has also received numerous awards for his academic and research distinction, including Visiting Fellowship Award from the University of Alabama at Birmingham, United States; the DST-INSPIRE Fellowship Award from the Ministry of Science and Technology, Government of India; Junior Research Fellowship by University Grants Commission, Government of India, Gold Medal of Excellence for 1st class first in master's degree from BGSB University Rajouri, India, besides qualifying a number of nationalized competitive examinations like CSIR-UGC NET-JRF, GATE (Biotechnology). Dr. Ahmad serves both as editor and reviewer for various international journals and holds membership in several scientific organizations, including the American Association for Cancer Research (AACR), Society of Toxicology (SOT), European Society for Photobiology, International Society for Development and Sustainability (ISDS), National Postdoctoral Association, United States, Health Physics Society (HPS). Outside the lab, Dr. Umar enjoys playing cricket and traveling on long expeditions. He is a nature lover.

Dedication

Dedicated to my beloved parents, Wali Mohd Sheikh and Azee Begum, who against all odds ensured that I completed my education, and what I am today is because of their selfless and sincere efforts. I pray to almighty Allah that he accepts their efforts and keep them steadfast on Iman. May they live a happy, blessed, and long life, Ameen.

Contents

Contributors

Sheikh Tasduq Abdullah
Academy of Scientific and Innovative Research (AcSIR), Ghaziabad, Uttar Pradesh, India;
CSIR–Indian Institute of Integrative Medicine Canal Road, Jammu, Jammu and Kashmir, India

Afreen Anwar
Department of Biotechnology, Indian Institute of Technology Roorkee, Roorkee, Uttarakhand, India

Suresh B. Arakera
Department of Applied Genetics, Karnatak University, Dharwad, Karnataka, India

Aalim Maqsood Bhat
Academy of Scientific and Innovative Research (AcSIR), Ghaziabad, Uttar Pradesh, India;
CSIR–Indian Institute of Integrative Medicine Canal Road, Jammu, Jammu and Kashmir, India

Basharat Ahmad Bhat
Department of Bioresources, School of Biological Sciences, University of Kashmir, Srinagar, Jammu and Kashmir, India

Neelofar Maqsood Bhat
Department of Botany, University of Kashmir, Jammu and Kashmir, India

Bhavin Bhatt
Shree Ramkrishna Institute of Computer Education and Applied Sciences, Faculty of Science, Sarvajanik University, Surat, Gujarat, India

Naresh Butani
Shree Ramkrishna Institute of Computer Education and Applied Sciences, Faculty of Science, Sarvajanik University, Surat, Gujarat, India

Aijaz A. John
Division of Rheumatology, Department of Medicine, University of Massachusetts Medical School, Worcester, MA, USA

Taruneet Kaur
Animal Biochemistry Division, National Dairy Research Institute, Karnal, Haryana, India

Jonaid Ahmad Malik
DBME, Indian Institute of Technology (IIT), Ropar, Punjab, India

Manzoor Ahmad Mir
Associate Professor, Department of Bioresources, School of Biological Sciences, University of Kashmir, Srinagar, Jammu and Kashmir, India

Shahid Hussain Naikoo
Academy of Scientific and Innovative Research (AcSIR), Ghaziabad, Uttar Pradesh, India;
CSIR–Indian Institute of Integrative Medicine Canal Road, Jammu, Jammu and Kashmir, India

Hiral R. Patel
M. B. Patel Applied Science College, Mogri, Gujarat, India

Rahul Raghavan
Hospital for Special Surgery, Weill Cornell Medicine, New York, USA

Supriya Ray
Department of Applied Genetics, Karnatak University, Dharwad, Karnataka, India

Bashir Ahmad Sheikh
Department of Bioresources, School of Biological Sciences, University of Kashmir, Srinagar, Jammu and Kashmir, India

Jae-Hyuck Shim
Division of Rheumatology, Department of Medicine, University of Massachusetts Medical School, Worcester, USA

Achuit K. Singh
Sr. Scientist, Crop Improvement Division, ICAR–Indian Institute of Vegetable Research, Varanasi, Uttar Pradesh, India

Bhawana Singh
Department of Medicine, Institute of Medical Sciences, Banaras Hindu University, Varanasi, Uttar Pradesh, India

Aashiq Sultan
Department of Zoology, School of Biological Sciences, University of Kashmir, Srinagar, Jammu and Kashmir, India

Viralkumar V. Surati
Shri A. N. Patel P. G. Institute of Science and Research, Anand, Gujarat, India

Vimal Verma
Department of Medicine, Institute of Medical Sciences, Banaras Hindu University, Varanasi, Uttar Pradesh, India

Abbreviations

3'UTR	3'-untranslated region
5-FC	5-fluorocytosine
5-FU	5-fluorouracil
AA	aplastic anemia
AACR	American Association for Cancer Research
AAT	alpha-1 antitrypsin
AAV	adeno-associated viruses
AB	*Acinetobacter baumannii*
AC	*Acinetobacter calcoaceticu*
ACC	acetyltransferases
ACT	*Actinobacillus*
ACTH	adrenocorticotropic hormone
ACTm	*Actinomyces*
Ad	adenoviruses
AD	Alzheimer's disease
AER	*Aeromicrobium*
AGO	Argonaute
AHSCT	allogeneic hematopoietic stem cell transplantation
AIET	autologous immune enhancement therapy
AMO	anti-miRNA oligonucleotides
AMP	adenosine monophosphate
AMP	antimicrobial peptides
ANT	aminoglycoside nucleotidyltransferase
APH	aminoglycoside phosphotransferases
APP	amyloid precursor protein
ASC	adult stem cells
AT	agrobacterium tumefacien
ATP	adenosine triphosphates
AuNP	gold nanoparticles
Aβ	β-amyloid
BAC	*Bacillus*
BACT	*Bacteroides*
BC	*Bacillus clausii*

BM	bone marrow
BMI	body mass index
BMSC	bone marrow stem cells
BP	*Bacillus pumilus*
BP	*Burkholderia pseudomallei*
BRM	biological response modifiers
BS	*Bacillus subtilis*
CAMP	childhood asthma management program
CAR	chimeric antigen receptor
CARMEN	combinatorial arrayed reactions for multiplexed evaluation of nucleic acids
CAT	catalases
CAT	chloramphenicol acetyltransferases
CB	*Clostridium butyricum*
CC	*Campylobacter coli*
CD	*Clostridium difficile*
CD	cytosine deaminase
CDI	*Clostridium difficile* infection
CHIP	chromatin immunoprecipitation
CHR	*Chryseobacterium* spp.
CIFN	consensus IFN
CIT	*Citrobacter*
CLGP	cysteinyl leukotriene genetic pathway
CLPP	cysteinyl leukotriene pharmacological pathway
CLR-1	cysteinyl leukotriene receptor 1
ClyA	cytolysin A
CML	chronic myelogenous leukemia
CMV	cytomegalovirus
COR	*Corynebacterium*
CP	*Clostridium perfringens*
CPP	cell-penetrating peptide
CPT	chloramphenicol 3-O-phosphotransferase
CRISPR	clustered regularly interspaced short palindromic repeats
CRISPRa	CRISPR activation
CRISPR-Cas	clustered regularly interspaced short palindromic sequences–CRISPR-associated protein
CRISPRi	CRISPR interference

crRNA	CRISPR RNA
CSC	cancer stem cell
dCas9	dead Cas9
DHFR	dihydrofolate reductase
DOTAP	1,2-dioleoyloxy-3-trimethylammoniumpropane
DPP	diabetes prevention program
DPP4	dipeptidyl peptidase inhibitors-4
DPSC	dental pulp-derived stem cells
DSB	double-strand break
dsDNA	double-stranded DNA
DSS	dextran sodium sulfate
EA	*Enterobacter aerogene*
EBV	Epstein Barr virus
EC	*Escherichia coli*
EcN	*E. coli* Nissle 1917
ECV	extracellular vesicles
EF	*Enterococcus faecalis*
EGCG	epigallocatechin-3-gallate
ENT	*Enterobacter* spp.
ENT	*Enterococcus*
ENTB	*Enterobacter*
EntC	*Enterobacter cloacae*
ESA	erythropoietic-stimulating agents
ESC	embryonic stem cell
ESC	*Escherichia*
EUB	*Eubacterium*
FDA	Food and Drug Administration
FEV_1	forced expiratory volume in 1 second
FQ	fluorophore quencher
FUS	*Fusobacterium*
GAR	*Gardnerella*
GCK	glucokinase
gFET	graphene-based field-effect transistor
GLP	glucagon-like peptide
GPP	glucocorticoid pharmacological pathway
gRNA	guideRNA
GVHD	graft-versus-host disease
GWAS	genome-wide association studies

H_2O_2	hydrogen peroxide
HA	hemagglutinin
HAE	*Haemophilus*
HBV	hepatitis B infection
HCV	hepatitis C virus
HDR	homology-directed repair
HEPN	higher eukaryotic and prokaryotic nucleotide
HER2	human epidermal growth factor receptor 2
hESC	human embryonic stem cells
HGP	human genome project
HI	*Haemophilus influenza*
HLA	human leucocyte antigens
HMP	human microbiome project
HNF1-α	hepatic nuclear factor 1-α
HSCs	hematopoietic stem cells
HSCT	hematopoietic stem cell transplantation
IBD	inflammatory bowel disease
ICS	inhaled corticosteroids
IFN-α	interferon-α
IKTC	independent Korean trial cohort
ILs	interleukins
IPRs	IP rights
iPSCs	induced pluripotent stem cells
IRIS	immune reconstitution inflammatory syndrome
IST	immunosuppressive therapy
ITRs	inverted terminal repeats
KLE	*Klebsiella*
KLs	*Kluyvera* spp.
KO	knockout
KP	*Klebsiella pneumonia*
LA	*Listonella anguillarum*
LABA	long-acting beta-agonist
LAC	*Lactobacillus*
LAK	lymphokine-activated killer
LFR	lung function response
LL	*Lactococcus lactis*
LM	*Listeria monocytogenes*
LMR	leukotriene modifier response

LNA	locked nucleic acid
LO	lipoxygenase
LOD	limit of detection
LOI	leukotriene inhibitor
LPS	lipopolysaccharide
LR	*Lactobacillus reuteri*
LRA	leukotriene receptor antagonist
LRM	leukotriene receptor modifiers
LTA4H	leukotriene A_4 hydrolase
LTC4S	leukotriene C_4 synthase
MEAC	multiethnic asthma cohorts
MFS transporters	major facilitator superfamily transporters
MHC	histocompatibility complex
MIC	*Micromonospora*
MICC	*Micrococcus*
miRNA	microRNA
MM	*Morganella morganii*
MODY	maturity-onset diabetes of the young
MRI	magnetic resonance imaging
mRNA	messenger RNA
MRP1	multiple drug-resistant proteins 1
MRSA	methicillin-resistant *Staphylococcus aureus*
MSC	mesenchymal stem cells
MT	*Mannheimia taxon* 10
MV	*Mannheimia varigena*
NAE	N-acylethanolamines
NAG	N-any nucleotide
NAPE	N-acylphosphatidylethanolamines
NASBA	nucleic acid sequence-based amplification
NASBACC	nucleic acid sequence-based amplification–CRISPR cleavage
nCas9	nickase Cas9
NCI	US National Cancer Institute
ncRNA	non-coding RNA
NEI	*Neisseria*
NG	*Neisseria gonorrhoeae*
NGS	next-generation sequencing
NHEJ	non-homologous end joining

NM	*Neisseria meningitides*
NPC	neural precursor cells
NSC	neural stem cells
NSP	non-starch polysaccharide
O_2^-	superoxide radical
OH	hydroxyl radical
PA	*Pseudomonas aeruginosa*
PABA	*p*-aminobenzoic acid
PAM	protospacer adjacent motif
PAMAM	poly(amidoamine)
PBPs	penicillin-binding proteins
PD	*photobacterium damselae* subsp. *Piscicida*
PDE	phosphodiesterase inhibitors
PDX-1	pancreatic and duodenal homeobox gene 1
PED	*Pediococcus*
PEFR	peak expiratory flow rate
PEG	polyethylene glycol
pegRNA	prime editing guideRNA
PEI	polyethyleneimine
PEP	*Peptosteptococcus*
PLGA	poly(lactide-co-glycolide)
PM	*Pasteurella multocida*
PM	personalized medicine
PMb	*Proteus mirabilis*
PNA	peptide nucleic acid
POR	*Porphyromonas*
POR	*Proteus*
PP	*Pseudomonas putida*
PRE	*Prevotella*
PSCs	perinatal stem cells
PSEN-1	presenilin-1
QRDR	quinolone resistance determining the region
rAAV	recombinant AAV
RCC	renal cell carcinoma
RES	reticuloendothelial system
RND	resistance nodulation division
RNP	ribonucleoprotein
ROS	reactive oxygen species

RPR	regulatory promoter region
SA	*Staphylococcus aureus*
SABA	short-acting beta-agonist
SAC	*Streptomyces acrimycini*
SAg	*Streptococcus agalactiae*
SAV	self-amplifying vaccines
SCCmec	*Staphylococcal cassette* chromosome mec
SCFA	short-chain fatty acids
SCI	spinal cord injury
SE	*Salmonella enteritidis*
SEL	*Selenomonas*
SF	*Shigella flexneri*
sgRNA	single chimeric guideRNA
SH	*Staphylococcus haemolyticus*
SHERLOCK	specific high sensitivity enzymatic reporter UnLOCKing
SHINE	streamlined highlighting of infections to navigate epidemics
shRNAs	short-hairpin RNAs
SI	*Staphylococcus intermedius*
SLE	systemic lupus erythematosus
SM	*Serratia marcescens*
SNP	short nucleotide polymorphism
SNV	single nucleotide variants
SOD	superoxide dismutase
SP	*Streptococcus pyogenes*
SS	*Streptococcus sius*
SSC	somatic stem cells
ST	*Salmonella typhi*
STA	*Staphylococcus aureus*
STIP1	heat shock organizing protein gene
STM	*Salmonella typhimurium*
StM	*Stenotrophomonas malyophilia*
STP	*Staphylococcus*
STR	*Streptococcus*
STRM	*Streptomyces*
SUR1	sulfonylureas receptor subunit 1
T1D	type 1 diabetes

T2D	type 2 diabetes
TALENs	transcription activator-like effector nucleases
TFFs	trefoil factors
tgRNA	tuned guideRNA
TILs	transfer of tumor-infiltrating T-cells
TPMT	thiopurine methyltransferase
tracrRNA	trans-activating CRISPR RNA
TRE	*Treponema*
TRP	tandem repeat polymorphism
TSPSC	tissue-specific progenitor stem cells
UE	uncultured eubacterium
ULABA	ultra-long-acting beta-agonists
VC	*Vibrio cholerae*
VEI	*Veillonella*
VL	visceral leishmaniasis
WOL	*Wolinella*
ZFNs	zinc-finger nucleases

Preface

Biotechnology, as a fast-developing multidisciplinary modern technology at the interface of biology and engineering and as an advanced science, has already shown its impacts on different aspects of public health, including on many other fields of particular interest to humanity. Biotechnology holds tremendous potential in using fundamental biological knowledge in developing new tools and techniques that can in turn be used in the synthesis and production of novel biological and medicinal entities for the benefit of humankind. This advanced field encompasses a wide variety of highly multidisciplinary subjects and a series of enabling technologies involving the practical application of organisms or their components in manufacturing biological materials, environmental management, and in value addition to food and products of medicinal value. Historically, biotechnology used to be more of an artisan skill rather than a field of science involved in the preparation of wines, beers, and cheese, through traditional approaches and techniques without knowing the actual biological processes and mechanisms involved. As the knowledge and the scientific basis of these traditional biotechnological processes evolved through interdisciplinary research-based approach(es), it has resulted in efficient manufacturing from a commercial point of view and enhanced the rewards of biotechnology viz-a-viz its role in medicine, public health, industry, and agriculture. Interventions in modern biotechnology have led to a range of novel lifesaving biological products and techniques significantly impacting the modern healthcare sector. Modern advanced technology has further evolved in designing molecular tools and techniques allowing desired molecular changes that can be made to living organisms, now heralding into a new field and age of biotechnology called genetic engineering, with a special focus in the areas of environment, medicine, health, and food sustainability. It seems clear that biotechnology will be playing a pivotal role as an enabling technology in the 21st century, having the potential to ensure food and environmental sustainability in addition to its role in revolutionizing the healthcare sector.

Considering its huge impact on the economy, inter- and intra-governmental agencies, including stakeholders from the private sector, have already taken up numerous measures despite many challenges and are building competent human resources in promoting research and development in this field. However, with the advent of new and enabling modern biotechnological tools, it will be of particular interest to see how advanced biological research using the latest and innovative technologies will revolutionize the public health sector with particular emphasis on disease diagnosis, detection, and treatment.

Keeping these diverse applications of this interdisciplinary field in mind, this book, in a very simple and lucid manner, aims to cover many, if not all, aspects of modern biotechnology and its applications with special attention given to how modern biotechnology will advance the healthcare sector. Reading this book will give readers an idea of the intricate relationship between biotechnology and healthcare and will also explain different biotechnological processes and their applications in medicine.

This volume presents readers with some research trends and areas of improvement to better understand its prospects involving modern biotechnology-based tools and techniques. This volume also sheds light on how biotechnology can be used in disease diagnosis and treatment that will change the perspective of modern biotechnology and its role in healthcare.

This volume:

- Provides an overview of MicroRNAs as next-generation therapeutic and diagnostic agents.
- Discusses the evolving role of CRISPR-Cas-based diagnostics and its reliability.
- Discusses engineered gut microbiome in treating diseases and its applications in modern biotechnology.
- Discusses antibiotics and plant-derived antimicrobials as alternative sources to control infections.
- Sheds light on how understanding the individual pathophysiological intricacies can lead to personalized medicine.
- Discusses stem cell technology (SCT) and regenerative medicine as an emerging therapeutic intervention against a diverse set of diseases.
- Discusses the process methodology and potential role of CRISPR-Cas genome-editing tool as a prospective application to study human diseases.

CHAPTER 1

MicroRNAs: Next-Generation Therapeutic and Diagnostic Agents

AIJAZ A. JOHN,[1] AFREEN ANWAR,[2] TARUNEET KAUR,[3] and JAE-HYUCK SHIM[1]

[1]*Division of Rheumatology, Department of Medicine, University of Massachusetts Medical School, Worcester, USA*

[2]*Department of Biotechnology, Indian Institute of Technology Roorkee, Roorkee, Uttarakhand, India*

[3]*Animal Biochemistry Division, National Dairy Research Institute, Karnal, Haryana, India*

ABSTRACT

microRNAs (miRNAs) are a new class of endogenous regulators for gene expression at the post-transcriptional level whose deregulation is associated with numerous diseases. These small, noncoding molecules have gained a lot of interest in the scientific community across the globe as therapeutic and diagnostic agents. Their prominent presence in biopsies and body fluids makes them highly sensitive and specific circulating biomarkers in nature. Some diagnostic miRNAs are already available to clinicians and sheltered by major insurance companies. On the other hand, many therapeutic miRNAs are in clinical and preclinical trials, but none of them have been used in the market yet. Although there is a large number of obstacles

Modern Biotechnology in Healthcare: Advances and Applications. Sheikh Umar Ahmad (Ed.)

to translating the findings of miRNA therapeutics from bench to bedside, but the day is not far when such obstacles will be overcome, and this class of therapeutic drugs will enter into the clinic as next-generation medicine. This chapter summarizes the importance of the chemical modification of oligonucleotides and their delivery systems for the development of therapeutic microRNAs. Further, it also highlights the efforts of various pharmaceutical companies toward the development of therapeutic and diagnostic microRNAs. Lastly, the chapter delivers insights into the major barrier associated with the clinical application of therapeutic miRNAs.

1.1 INTRODUCTION

In the last few decades, exponential progress in the field of biological sciences like metabolomics, systems biology, and proteomics in conjunction with bioinformatics have generated a huge amount of data to address and understand the biological mechanisms associated with diseases and discover therapeutic interventions. It has been well-established that only 1.5% (approximately) of all human genes encode for the proteins that make up our cells and tissues. The rest of the genome is made of noncoding DNA (Li et al., 2001). The noncoding DNA is essential for the regulation and expression of the coding regions (Gloss & Dinger, 2018). microRNAs (miRNAs) are a new class of post-transcriptional regulators of endogenous gene expression. These molecules are short noncoding RNAs (ncRNAs) of ~22 nucleotides that down-regulate expression of their target genes by either mRNA degradation or translational inhibition by guiding Argonaute (AGO) proteins to target sites in the 3'-untranslated region (3'UTR) of messenger RNAs (mRNAs) (Huntzinger & Izaurralde, 2011; Swarts et al., 2014). miRNAs are involved in virtually every cellular process or biological process, including developmental timing, cellular differentiation, proliferation, apoptosis, gene regulation, insulin secretion, cholesterol biosynthesis, and cancer development (Ambros, 2003; Xu et al., 2003). The latest issue of the miRBase database (v22) contains 48,860 different mature microRNA sequences produced from 38,589 hairpin precursor microRNAs entries from 271 organisms. This issue represents 1,917 annotated hairpin precursors and 2,654 mature sequences from the human genome (Kozomara et al., 2019).

The increased research on miRNAs leads to a huge number of studies revealing that miRNA deregulation is associated with numerous diseases

(Esteller, 2011), predominantly cancer (Bracken et al., 2016). These small, noncoding, endogenous molecules have gained a lot of interest in the scientific community across the globe as therapeutic and diagnostic agents. microRNAs are highly sensitive and specific circulating biomarker as these can be easily detected in biopsies and body fluids (De Guire et al., 2013). The potential importance of these small molecules as therapeutics and diagnostics is apparent by the fact that nearly 11,000 studies have been published depicting their huge potentials as future diagnostic agents whereas 3,500 studies reveal their prospective as therapeutic agents (Bonneau et al., 2019).

The entry of Miravirsen (or SPC3649), an antagomiR targeting miR-122 into clinical trial has fueled hope into the scientific community across the globe that these disease modifying agents can be used as future medicines. Further, current advancement in RNA chemistry and technologies to deliver RNA molecules *in vivo* has made it possible for the success of microRNA-based therapeutics (Rupaimoole & Slack, 2017). In the United States, the biotechnology industry revenue was just $20 billion in 1996 and increased up to $70.1 billion in 2008. The major portion of this revenue comes from the pharmaceutical industries. Since 1983, a larger number of pharmaceutical companies have been established which are actively investing in the development of miRNA-based therapeutic molecules (Table 1.1) (Chakraborty et al., 2017).

In this chapter, we will discuss about the chemical modifications of miRNA-based therapeutics, advances in technologies to deliver such therapeutic molecules and microRNAs that are currently in clinical trials. Furthermore, we will discuss about the future prospects of microRNAs as therapeutic and diagnostic agents.

1.2 CHEMICAL MODIFICATIONS OF MIRNA-BASED THERAPEUTICS

Therapeutic microRNAs have emerged as an important biopharmaceutical that are in commercial space as future medicine. The miRNA-based therapeutics approach is based on the fact to restore normal miRNA expression levels either by downregulating high levels of miRNAs using anti-miRs (innovative molecules designed for endogenous miRNA degradation) or by upregulating low levels of miRNAs using mimics (molecules which mimic endogenous miRNAs) in a specific disease stage or pathological

TABLE 1.1 List of Pharmaceutical Companies Involved in miRNA and siRNA Therapeutics

Biopharmaceutical Company	Country	Year of Establishment
Marina Biotech	USA	1983
Quark Pharmaceuticals	USA	1993
Rosetta Genomics	Israel	2000
Alnylam Pharmaceuticals	USA	2002
Santaris Pharma	Denmark	2003
RXi Pharmaceuticals	USA	2003
Asuragen	USA	2006
Mirage Therapeutics	USA	2007
Mirna Therapeutics	USA	2007
Regulus Therapeutics	USA	2007
InteRNA Technologies	Netherlands	2008

condition. MicroRNA mimic is a synthetic RNA duplexes which is identical to the miRNA of interest with chemical changes in order to improve its stability and cellular uptake (Bader et al., 2011; Thorsen et al., 2012) whereas anti-miR is a RNA molecule having a complementary binding site to a particular miRNA or a family of miRNA in order to inhibit their function. It is commonly known as anti-miRNA oligonucleotides (AMOs) (Ebert & Sharp, 2010). The first generation of mimics and AMOs employed were unmodified and they were quickly degraded by ribonucleases present in serum and in the intracellular environment. Scientists quickly realized that chemical modification of these synthetic oligonucleotides is important to prevent their degradation by ribonucleases present in the blood and from phagocytosis carried out by reticuloendothelial system (RES) (Zhang & Farwell, 2008). Current chemical modifications are not carried out just to confer nuclease resistance to synthetic oligonucleotides but are also carried out to increase binding affinity and to increase the cellular uptake of these synthetic oligonucleotides. Other chemical modifications are made to alter the ability to trigger a response by the innate immune system (Lennox & Behlke, 2011).

A range of chemical alterations have been incorporated into these synthetic oligonucleotides particularly the AMOs. These modifications can be carried out in the phosphate moiety, e.g., phosphorothioate-containing oligonucleotides or in sugar group, e.g., 2'-O-methyl ribose containing oligonucleotides. The details of these chemical modifications are described in Table 1.2.

TABLE 1.2 Various Chemical Modifications of Oligonucleotides Improve Stability, Biodistribution, Cellular Uptake, and Delivery Efficiency, and Their Function

Oligonucleotides	Chemical Modification	Structure	References
Phosphorothioate-containing oligonucleotides	• α-oxygen of the phosphate replaced by sulfur • Improves stability	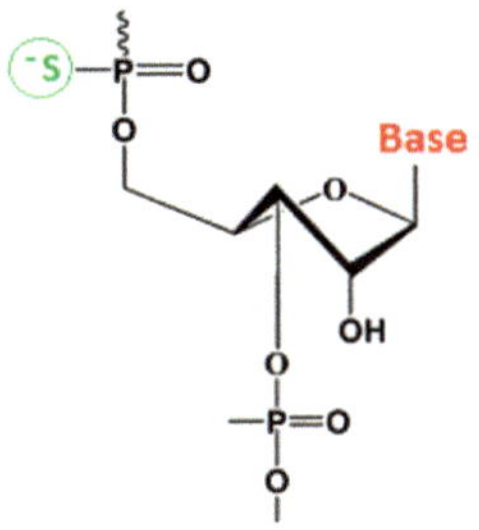	Sharma & Watts (2015)
Methylphosphonate	• α-oxygen of the phosphate replaced by methyl (CH_3) group • Improves stability	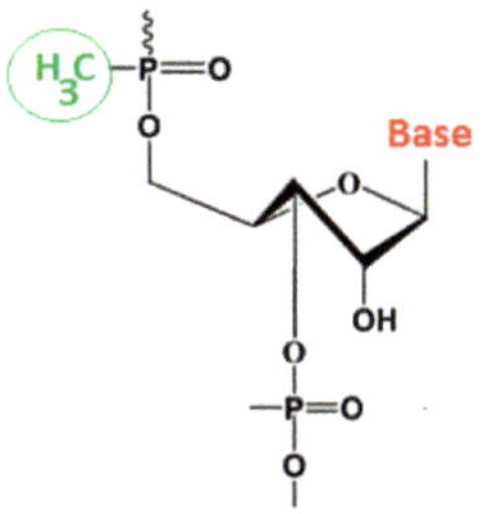	
2'-O-methyl-(2'-O-Me)	• 2'-O-methyl group added to 2'-oxygen of the ribose • Improves binding stability and protects from nuclease degradation	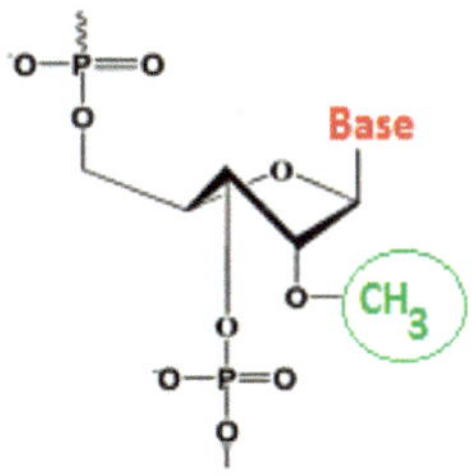	Bernardo et al. (2015)
2'-Fluoro oligonucleotides	• Fluoro group added to 2'-oxygen of the ribose • Improves binding stability and protects from nuclease degradation	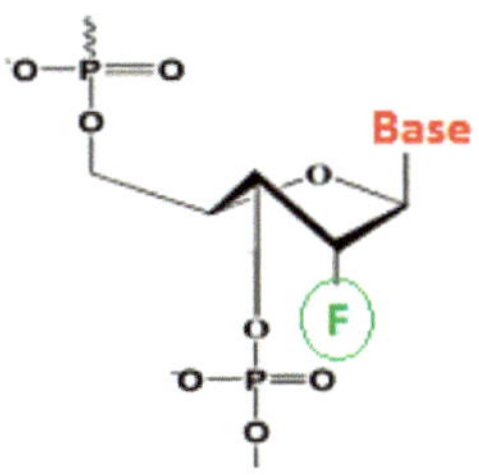	

TABLE 1.2 *(Continued)*

Oligonucleotides	Chemical Modification	Structure	References
Locked nucleic acid (LNAs)	• 2',4' methylene link in the ribose to form a bicyclic nucleotide • Improves sensitivity and specificity	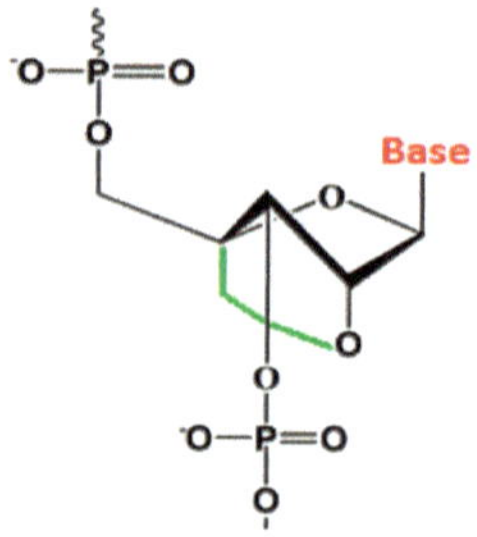	Chabot et al. (2012)
Terminally modified oligonucleotides	• Contains Cy3- or cholesterol or biotin at the terminal phosphate group • Improves stability and their tracer function in *in-vivo* delivery	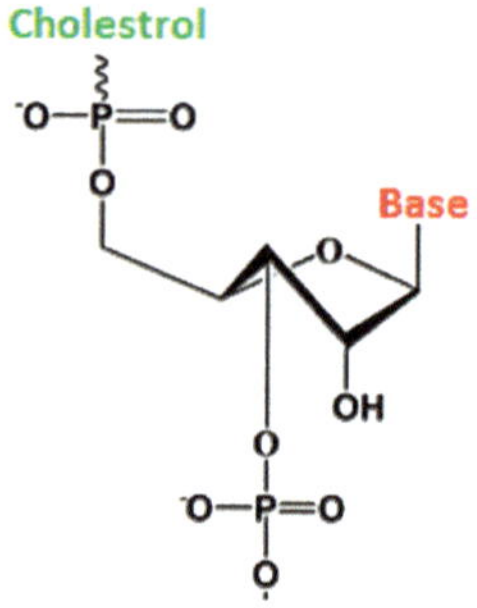	Krützfeldt et al. (2007)

1.2.1 ADVANCES IN MIRNA DELIVERY SYSTEMS

We all know that the success of microRNA-based therapeutics is directly dependent on the advancement in RNA chemistry and technologies to deliver RNA molecules *in vivo*. Usually, chemically modified oligonucleotides show poor cellular uptake and inadequate tissue-specific delivery when delivered in the absence of a carrier. To overcome such problems, miRNA delivery systems are used, which can be broadly divided into two categories, i.e., viral and nonviral miRNA delivery systems.

1.2.2 VIRAL MIRNA DELIVERY SYSTEMS

The basic idea behind the use of viruses as delivery systems is to remove or replace pathogenic genes with therapeutic gene(s). The Viral based delivery systems are useful for the long-term expression of miRNA (to restore normal miRNA expression) or anti-miRNA (to inhibit upregulated miRNA) into different tissues or organs. The viruses for such oligonucleotide delivery include adenovirus vectors, adeno-associated virus vectors (AAV), retroviral vectors, and lentivirus vectors (Fu et al., 2019).

Although these viral vectors have high delivery efficiency but there are few disadvantages, including low loading capacity, high toxicity, and strong immune response in the host (Buchbinder et al., 2008; McElrath, De Rosa, et al., 2008).

Adenoviruses (Ad) are nonenveloped viruses that contain an icosahedral nucleocapsid and double-stranded DNA (36 kb) with two inverted terminal repeats (ITRs) at its ends (Douglas, 2007). These viruses are highly attractive for scientists as delivery systems because these viruses can infect a broad range of cells without integrating into the host genome. Furthermore, these viruses can be manipulated easily to get the desired transduction efficacy in proliferating as well as quiescent cells (Cao et al., 2004). Currently, researchers use Next generation "gutless" adenoviral vectors with the packaging capacity of ~37 kb compared to the initial adenoviral vectors with a maximum 8 kb. Most of the genes in these Next generation "gutless" adenoviral vectors are completely removed and are left with the genes encoding for packaging and regulatory signals. These manipulations not only increase the loading capacity of Ad but also reduce their immunogenicity (Chen et al., 1997; Amalfitano, 1999). Next-generation "gutless" adenoviral vectors have advantages over initial adenoviral vectors, but these vectors need a helper virus for their propagation (Crettaz et al., 2009).

Despite many advantages as a delivery system, Ad has the disadvantage to induce an immune response (adaptive and innate), resulting in cell toxicity when repeatedly administered (Marshall, 1999).

Adeno-associated viruses (AAV) are nonenveloped viruses having a single-stranded DNA genome (4.7 kb) flanked by ITRs. AAV requires helper virus (adenovirus or herpes simplex virus) for its replication and in the absence of helper virus it releases its genome inside the nucleus where it remains as an episomal form or rarely integrates its DNA into the AAVS1 region of chromosome 19 (Schultz & Chamberlain, 2008). The integration of the AAV genome in the host cell is mainly carried out by the Rep protein. This Rep protein is deleted in recombinant AAV (rAAV), hence making its integration less efficient than wild-type AAV (Herrera-Carrillo et al., 2017). Around 11 nonpathogenic (absence of pathogenicity in humans) serotypes of AAV have been discovered and each of them shows a unique host's cell/tissue tropism with varying immunological properties. This cell/tissue-specific tropism makes AAV a favorite delivery system *in vivo* (Wu et al., 2006). Additionally, AAV shows a long duration of transgene expression in various organs of mice, including brain, liver, lung, eyes, and muscles (Nakai et al., 1998; Kaplitt et al., 1994; Grant et al., 1997).

Retroviruses are enveloped virion particles with single-stranded RNA (approx. 10 kb) (Pages & Bru, 2004). Once the virus infects the host cell, double-stranded DNA is formed from single-stranded RNA by the activity of reverse transcriptase enzyme. The double-stranded DNA then randomly integrates into one of the host chromosomes, which can allow its sustained expression (Liu & Berkhout, 2011). Although the durable expression of integrated fragment is a favorable characteristic, it can also result in high levels of insertional mutagenesis (Herrera-Carrillo et al., 2017).

Lentiviruses are very similar to Retroviruses as both of them belong to the Retroviridae family. Both of these viruses lead to stable integration of a transgene into the host genome, resulting in sustained gene expression in the host. In contrast to Retroviruses, useful for diving cells, Lentiviruses can transduce both dividing and nondividing cells, which makes Lentiviruses is more preferable than Retroviruses for stable gene expression in the cells (Greber & Fassati, 2003). Finally, Lentiviruses usually integrate within active transcriptional units, which significantly reduces the incidence of insertional oncogenesis (Montini et al., 2009).

1.2.3 NONVIRAL MIRNA DELIVERY SYSTEMS

Nonviral delivery systems also serve as essential platforms for the supply of miRNA-based therapeutics to target sites. These platforms are generally regarded as less immunogenic and toxic in nature. Besides, these delivery strategies do not have any size constraint of the packaged nucleic acid although insert size is an important determinant of transfection efficacy (Yang, 2015). An efficient nonviral delivery system transports exogenous synthetic miRNA or miRNA vectors without their degradation from cellular nucleases (Fu et al., 2019). These platforms encompass physical methods as well as chemical methods. Physical approaches, such as electroporation, gene gun, ultrasound, hydrodynamic-, and laser-dependent energy, apply forces from outside for imparting cell permeability towards the gene therapy (Yang, 2015) and mainly used in *in vitro* culture system and intermittently applied in the miRNA delivery systems (Balacescu et al., 2018). However, these delivery methods occasionally cause damages to cellular integrity. Whereas Chemical methods entail lipid-driven, polymer-driven, and inorganic carriers (Wang et al., 2018).

Lipid-based methodologies exploit the lipid/nucleic acid multiplexes, called lipoplexes or liposome as transporting systems. Cationic, anionic, or neutral liposomes exist, and they are made of a membrane-like exterior, with the miRNA or nucleic acids captured inside (Sforzi et al., 2020). Cationic lipoplexes are frequently employed in nonviral miRNA supplies due to their nonpathogenic and non-immunogenic properties, and high affinity towards cellular membranes. For example, liposome-mediated delivery of miR-300 to the osteoblast cultures resulted in precise overexpression for miR300, followed by a decrease in osteogenic mineralization (Kaur et al., 2020). Several cationic lipoplex (DharmaFECT®, SilentFect™, Lipofectamine® RNAi-MAX, SiPORT™) complexes are commercially available for miRNA delivery (Yoshizuka et al., 2016). However, the most acute difficulty for cationic liposome distribution schemes is their short half-lives in serum. Due to their short half-life and nonspecific interactions with blood proteins, anionic and neutral liposomes are instead utilized for miRNA delivery. For example, the anionic and neutral lipoplexes were effective to the delivery of miR-29b and miR-34a to acute myeloid leukemia and lymphoma, respectively (Garzon et al., 2009). Conjugation of polyethylene glycol (PEG) with the lipids improves steadiness with a longer half-life (up to 72 hours) (Pasut & Veronese, 2009). Pramanik and workers assembled lipid nanoparticles using DOTAP and PEG to deliver miR-143/145 cluster and miR-34a to a xenograft mouse model for a pancreatic cancer study. These assembled particles further established an amplified nanoparticle gathering at tumor sites with reduced tumor size (Kent et al., 2010).

Polymer dependent methods comprise of poly(lactide co glycolide) (PLGA), poly(amidoamine) (PAMAMs) dendrimers, polyethylenimine (PEI) or cell-penetrating peptide (CPP) as delivery transporters. Low molecular weight PEIs display minor cell membrane damage and related cellular toxicity in comparison with high molecular weight PEIs. Treatment of a xenograft mouse model of colon cancer with miR-33a and miR-145 using low molecular weight PEIs diminished tumor development and improved apoptosis (Ibrahim et al., 2011). Nevertheless, the transfection competence is indeed low and associated with limited biodegradability (Yang, 2015).

One crucial Food and Drug Administration (FDA) approved delivery vehicle, namely PLGA, has been used in miRNA-based therapeutic interventions against several critical pathological conditions (Christopher et al., 2016). However, the hydrophobic trait associated with PLGA reduces its transfection efficiency.

PAMAMs polymers, being positively charged in nature, exhibits high transfection ability compared to other polymers. Anti-miR21-PAMAM dendrimers have been reported to suppress the progression of glioblastoma cells (Játiva & Ceña, 2017). The chief shortcoming of PAMAM dendrimers is the polymer buildup in the liver (Vu et al., 2019). Besides these synthetic polymers, natural polymers viz. CPPs are employed for miRNA delivery. CPP (arginine-rich) from natural protamine efficiently transported miR-29b into osteo-stem cells (Suh et al., 2013). Additionally, Natural polymers CPPs are less toxic but disposed to degradation in serum in comparison to synthetic polymers.

In comparison to lipids and polymers, the importance of inorganic materials as miRNA gene carriers has been reported. Current inorganic miRNA vectors embrace gold nanoparticles (AuNPs), silica nanoparticles and Fe_3O_4-based nanoparticles (Wang et al., 2013). Among these AuNPs are the most habitually used. AuNPs have been stated to deliver miR-130b efficaciously (Bakhshinejad et al., 2014) and anti-miR29b to the surfaces of tumor cells (Roma-Rodrigues et al., 2017). The inorganic delivery platforms exhibit high *in vivo* stability and are devoid of microbial attack. Though, the binding affinity between these carriers and nucleic acids is generally less. To this end, inorganic and organic hybrid materials are innovated and hence employed in molecular delivery systems. AuNP10 conjugated to PEG0.5 delivers miR-1 into cancer cells, depicting higher transfection competence, longer half-lives, and lower toxicity in contrast to lipofection (Yang, 2015).

Viral and nonviral miRNA delivery systems have both advantages and disadvantages. Viral vectors show higher transfection efficiency but result in toxicity with immunogenic response, while nonviral carriers display reduced delivery efficiency but are relatively safe. Consequently, operative and nontoxic miRNA delivery systems are urgent need of the hour. The construct for future delivery systems must combine the benefits from both systems resulting in effective miRNA therapies from bench to clinics.

1.3 THERAPEUTIC MIRNAS IN CLINICAL TRIALS

It is entrenched that miRNAs are restorative for molecular therapy and gene silencing. With the help of emerging pharmaceutical firms, nucleic acid-based gene therapies are thereby entering the commercial sector. miRNAs are either overexpressed for the therapeutic purpose by miRNA mimics or silenced in the target tissue/cell by AMOs. These oligonucleotides are particularly responsive to nuclease-mediated decay in live cells and chemical changes are normally made prior to their distribution in the body (Lennox et al., 2013). In addition, several biomedical companies, including Rosetta Genomics, Alnylam, and Regulus Therapeutics concentrate on the RNAi driven pharmaceuticals in treatment for myriad of diseases for the societal benefit (Haussecker, 2012). Lately, Onpattro (FDA approved), a medication for the polyneuropathy treatment, has identified the reasons for the drug demand of RNAi technology-driven therapies (Akinc et al., 2019; Kaur et al., 2020). RNA biodrug(s) has been a significant business among numerous pharmaceutical companies from the last 20 years, performing miRNA clinical trials that are functional against several disorders. The majority of pharmaceutical companies are working on the pharmacokinetics and pharmacodynamics for miRNAs in the preclinical and clinical stages of development (Table 1.3).

1.3.1 MIRAVIRSEN

Miravirsen is an antisense phosphorothioate miRNA, modified by LNA (locked nucleic acid) inhibiting miR-122. It has been tagged as the first miRNA that entered phase II clinical trials, funded by Santaris Pharma (Roche Innovation Center), in order to clarify its efficiency and admissibility in patients. Clinical trials for its use against HCV infection have been performed in the US. Besides, phase II clinical trials are currently being conducted in other countries as well (Chakraborty et al., 2020; Gebert et al., 2014).

1.3.2 RG-101 AND RG-012

RG-101, a N-acetylgalactosamine (GalNAc)-conjugated inhibitor that restricts action of liver-specific miRNA-122 for the treatment of HCV, has been developed by Regulus Therapeutics. RG-012, another miRNA silencer targeting miRNA-21, developed by Regulus Therapeutics is currently in a clinical pipeline. It executes an imperative part in fibrogenic ailments associated with distinct organs, specifically the kidneys. Hence the drug is majorly employed in the prevention of Alport nephropathy (Ali et al., 2016; Chakraborty et al., 2017).

1.3.3 ANTI-MIRS

Unconventional RNA targeting therapy, has been developed by U.S.-based biopharmaceutical corporation miRagen with a special focus on miRNAs for unaddressed human healthcare needs. Several miRNA-dependent therapeutics have been designed from miRagen therapeutics, including but not limited to, MGN-2677, MGN-4220, MGN-1374, MGN-6114, MGN-4893 MGN-8107, and MGN-9103. MGN-9103 was proposed to play a vital role in diabetes and obesity. Besides, miRagen therapeutics is conducting clinical studies on MRG-106 prohibiting miR155. While, for red blood cell expansion, miR-451 is needed and suppression of miR-451 using anti-miR-451 (MGN-4893) in mice showed inhibition for erythrocyte's differentiation. This was proved beneficial against diseases involving anomalous red blood cell count (Chakraborty et al., 2017, 2020).

1.3.4 MRX34

MRX34, double-stranded miRNA mimic supplied in a liposome formulation initially developed by miRNA Therapeutics, which is now under the Synlogic Inc. Preclinical studies demonstrated therapeutic effects of MRX34 on several cancers (renal, hepatocellular, etc.) (Beg et al., 2017).

1.4 DIAGNOSTIC MIRNAS CURRENTLY IN PRECLINICAL AND CLINICAL TRIALS

There is a plethora of studies indicating miRNA as a biomarker in the diagnosis of various disorders (Table 1.4). Rosetta Genomics was the first company that focused on the arena of diagnostic miRNAs. The company launched RossettaGX Reveal (2016), used to distinguish between benign and indeterminate thyroid nucleus, which was authenticated via. multi-center cohort (189) FNA smears (Chakraborty et al., 2020).

Interpace diagnostics (NASDAQ: IDXG), a diagnostic firm that assimilated a combination of miRNA classifier, ThyraMIR®, and oncogene panel ThyGeNEXT®, for the stratification of thyroid cancer. The ThyraMIR® compromises of miR-31-5p, miR-29b-1-5p, miR-146b-5p miR-139-5p, miR-204-5p, miR-155, miR-375, miR-222-3p, miR-138-1-3p, and miR-551b-3p (Partyka et al., 2018).

Further, Hummingbird Diagnostics (Comprehensive Biomarker Center), generated novel miRNA candidates in discovery of human disorders, such as cancer (melanoma, non-small-cell lung carcinoma, breast cancer), neurodegenerative (Alzheimer, multiple sclerosis, Parkinson) and inflammatory disease (Keller et al., 2017). Clinical validation of these miRNA signatures as biomarkers is in progress, supported by three European FP7-funded consortia (BestAgeing, RiskyCAD, and EURenOmics) (Bonneau et al., 2019).

DiamiR has developed the application of miRNAs as biomarkers in various neurodegenerative and neurodevelopmental pathologies. CogniMIR™, panel for early determination of Alzheimer's disease (AD) which formerly is in clinical trials and testified by DiamiR (Chakraborty et al., 2020). Moreover, TAmiRNA developed another miRNA-based detection panel associated with clinical utility for osteoporosis in the form of their first product OsteomiR™ (Ladang et al., 2020). Interestingly, TAmiRNA also offered a panel of 11 miRNAs ThrombomiR™, for evaluating the platelet function and a panel of 19 miRNAs, ToxomiR™ for assessing the toxicity associated with certain tissues (Mussbacher et al., 2020).

Mirnext biomedical firm recognized miR123 5p as a suitable biomarker in heart failure after validation studies in a large number of clinical samples. Additionally, DestiNA developed a patented PCR-free technology for the miRNA discovery and estimation from human samples. The studies demonstrated that the associated assays are capable of detecting miR-122 providing an edge over prevailing protein biomarkers to evaluate liver toxicity (Chakraborty et al., 2020).

1.5 FUTURE PROSPECTS

It has been well established that RNA-based therapeutics are developing rapidly along with enormous potentials for the treatment of various diseases.

TABLE 1.3 Diagnostic miRNAs Currently in the Preclinical and Clinical Trials of Development

SL. No.	Product	miRNA	Companies	Disease	Development Progress
1.	miRview Mets	miRNA library	Rosetta Genomics	Cancer	Available
2.	OsteomiR	Panel of 19 miRNAs	TAmiRNA	Osteoporosis	Available
3.	CogniMIR	Panel (unknown)	DiamiR	Alzheimer	Phase 1
4.	Simoa	miR-122	Quanterix	Liver toxicity	Preclinical
5.	ThyraMIR/ ThyGENX	miR-29b-1-5p, miR-31-5p, miR-138-1-3p, miR-139-5p miR-146b-5p, miR-155, miR-204-5p, miR-222-3p, miR-375 and miR-551b-3p	Interpace Diagnostics	Thyroid and pancreatic cancer	Available
6.	Mirnext	Panel with miR 423-5p	Mirnext	Heart failure	Development

Source: Chakraborty et al. (2020); Bonneau et al. (2019); Gebert et al. (2014); Ladang et al. (2020); Mussbacher et al. (2020).

TABLE 1.4 Therapeutic miRNAs Currently in the Preclinical and Clinical Trials of Development

Product	miRNA	Disease	Development Phase	Pharmaceutical Company
RG-101	miR-122	HCV	Phase 2	Regulus Therapeutics
RG-012	miR-21	Alport syndrome	Phase 1	
RG-125	miR-103	NASH	Phase 1	
Mesomir	Mesomir	Mesothelioma	Phase 2	ENGeneIC
MRG-107	miR-155	ALS	Preclinical	MiRagen Therapeutics
MRG-106	miR-155	Lymphoma, leukemia	Phase 1 and Phase 2	
MRG-110	miR-92	Ischemia	Phase 1	
Miravirsen	miR-122	HCV	Phase 2	Roche/Santaris
ABX464	miR-124	IBD	Phase 2	Abivax
MRX34	miR-3	Hepatocellular cancer	Phase I clinical trial	miRNA Therapeutics
–	miR-208	Heart failure	Preclinical	miRagen
–	Let-7	Cancer	Preclinical	miRNA Therapeutics

Source: Chakraborty et al. (2020); Titze-De-Almeida et al. (2020); Nana-Sinkam & Croce (2013).

The discovery of these disease modifying agents is the most exciting drug development, IP rights (IPRs), and the people who are interested in pharmaceutical business. Additionally, the discovery of diagnostic miRNAs is as advanced as the development of therapeutic miRNAs. Several diagnostic miRNAs are already available to clinicians and sheltered by major insurance companies, while the therapeutic miRNAs are yet to enter the market despite their huge potential as therapeutic agents. A major barrier associated with the clinical application of therapeutic miRNAs is to develop a carrier with tissue/cell-specific tropisms to minimize side effects. As this field of therapeutics is new and continues to grow, a better understanding of miRNA biogenesis and function along with the development of efficient vehicles for the targeted delivery to specific cells, tissues, and organs will guide future endeavors in miRNA-based therapeutics. Many obstacles may lie due to safety, target selection, delivery technologies, clinical trial design or commercial considerations. Proper optimization of such factors will minimize the risk of product failure and a high rate of successful drug production. Taken together, miRNA therapeutics is a new and emerging field with large number of obstacles, but the day is not far when such obstacles will be overcome, and this class of therapeutic drugs will enter into the clinic as next-generation medicine.

CONFLICT OF INTEREST

JHS is a scientific co-founder of AAVAA therapeutics and holds equity in the company. Other authors declare no conflict of interest.

KEYWORDS

- **adeno-associated virus vectors**
- **human diseases**
- **microRNAs**
- **molecular diagnostics**
- **next generation therapeutic and diagnostic agents**
- **reticuloendothelial system**

REFERENCES

Akinc, A., Maier, M. A., Manoharan, M., Fitzgerald, K., Jayaraman, M., Barros, S., et al., (2019). The Onpattro story and the clinical translation of nanomedicines containing nucleic acid-based drugs. *Nature Nanotechnology, 14*, 1084–1087.

Ali, S. S., Kala, C., Abid, M., Ahmad, N., Sharma, U. S., & Khan, N. A., (2016). Pathological microRNAs in acute cardiovascular diseases and microRNA therapeutics. *Journal of Acute Disease, 5*, 9–15.

Amalfitano, A., (1999). Next-generation adenoviral vectors: New and improved. *Gene Therapy, 6*, 1643–1645.

Ambros, V., (2003). MicroRNA pathways in flies and worms: Growth, death, fat, stress, and timing. *Cell, 113*, 673–676.

Bader, A., Brown, D., Stoudemire, J., & Lammers, P., (2011). Developing therapeutic microRNAs for cancer. *Gene Therapy, 18*, 1121–1126.

Bakhshinejad, B., Javidi, M. A., Babashah, S., & Babashah, S., (2014). Nanocarriers and microRNA-based scenarios for cancer therapy. *MicroRNAs: Key Regulators of Oncogenesis* (pp. 387–411). Springer.

Balacescu, O., Visan, S., Baldasici, O., Balacescu, L., Vlad, C., & Achimas-Cadariu, P., (2018). MiRNA-based therapeutics in oncology, realities, and challenges. *Antisense Therapy*. IntechOpen.

Beg, M. S., Brenner, A. J., Sachdev, J., Borad, M., Kang, Y. K., Stoudemire, J., et al., (2017). Phase I study of MRX34, a liposomal miR-34a mimic, administered twice weekly in patients with advanced solid tumors. *Investigational New Drugs, 35*, 180–188.

Bernardo, B. C., Ooi, J. Y., Lin, R. C., & McMullen, J. R., (2015). miRNA therapeutics: A new class of drugs with potential therapeutic applications in the heart. *Future Medicinal Chemistry, 7*, 1771–1792.

Bonneau, E., Neveu, B., Kostantin, E., Tsongalis, G., De Guire, V., (2019). How close are miRNAs from clinical practice? A perspective on the diagnostic and therapeutic market. *Ejifcc, 30*, 114.

Bracken, C. P., Scott, H. S., & Goodall, G. J., (2016). A network-biology perspective of microRNA function and dysfunction in cancer. *Nature Reviews Genetics, 17*, 719–732.

Buchbinder, S. P., Mehrotra, D. V., Duerr, A., Fitzgerald, D. W., Mogg, R., Li, D., et al., (2008). Efficacy assessment of a cell-mediated immunity HIV-1 vaccine (the step study): A double-blind, randomized, placebo-controlled, test-of-concept trial. *The Lancet, 372*, 1881–1893.

Cao, H., Koehler, D. R., & Hu, J., (2004). Adenoviral vectors for gene replacement therapy. *Viral Immunology, 17*, 327–333.

Chabot, S., Orio, J., Castanier, R., Bellard, E., Nielsen, S. J., Golzio, M., et al., (2012). LNA-based oligonucleotide electrotransfer for miRNA inhibition. *Molecular Therapy, 20*, 1590–1598.

Chakraborty, C., Sharma, A. R., & Sharma, G., (2020). Therapeutic advances of miRNAs: A preclinical and clinical update. *Journal of Advanced Research*.

Chakraborty, C., Sharma, A. R., Sharma, G., Doss, C. G. P., & Lee, S. S., (2017). Therapeutic miRNA and siRNA: Moving from bench to clinic as next generation medicine. *Molecular Therapy-Nucleic Acids, 8*, 132–143.

Chen, H. H., Mack, L. M., Kelly, R., Ontell, M., Kochanek, S., & Clemens, P. R., (1997). Persistence in muscle of an adenoviral vector that lacks all viral genes. *Proceedings of the National Academy of Sciences, 94*, 1645–1650.

Christopher, A. F., Kaur, R. P., Kaur, G., Kaur, A., Gupta, V., & Bansal, P., (2016). MicroRNA therapeutics: Discovering novel targets and developing specific therapy. *Perspectives in Clinical Research, 7,* 68.

Crettaz, J., Otano, I., Ochoa, L., Benito, A., Paneda, A., Aurrekoetxea, I., et al., (2009). Treatment of chronic viral hepatitis in woodchucks by prolonged intrahepatic expression of interleukin-12. *Journal of Virology, 83*, 2663–2674.

Dastmalchi, N., Safaralizadeh, R., Baradaran, B., Hosseinpourfeizi, M., & Baghbanzadeh, A., (2020). An update review of deregulated tumor suppressive microRNAs and their contribution in various molecular subtypes of breast cancer. *Gene, 729*, 144301.

De Guire, V., Robitaille, R., Tetreault, N., Guerin, R., Menard, C., Bambace, N., et al., (2013). Circulating miRNAs as sensitive and specific biomarkers for the diagnosis and monitoring of human diseases: Promises and challenges. *Clinical Biochemistry, 46*, 846–860.

Douglas, J. T., (2007). Adenoviral vectors for gene therapy. *Molecular Biotechnology, 36*, 71–80.

Ebert, M. S., & Sharp, P. A., (2010). MicroRNA sponges: Progress and possibilities. *RNA, 16*, 2043–2050.

Esteller, M., (2011). Noncoding RNAs in human disease. *Nature Reviews Genetics, 12*, 861–874.

Fu, Y., Chen, J., & Huang, Z., (2019). Recent progress in microRNA-based delivery systems for the treatment of human disease. *ExRNA, 1*, 1–14.

Garzon, R., Heaphy, C. E., Havelange, V., Fabbri, M., Volinia, S., Tsao, T., et al., (2009). MicroRNA 29b functions in acute myeloid leukemia. *Blood, 114*, 5331–5341.

Gebert, L. F. R., Rebhan, M. A. E., Crivelli, S. E. M., Denzler, R., Stoffel, M., & Hall, J., (2014). Miravirsen (SPC3649) can inhibit the biogenesis of miR-122. *Nucleic Acids Research, 42*, 609–621.

Gloss, B. S., & Dinger, M. E., (2018). Realizing the significance of noncoding functionality in clinical genomics. *Experimental & Molecular Medicine, 50*, 1–8.

Grant, C. A., Ponnazhagan, S., Wang, X. S., Srivastava, A., & Li, T., (1997). Evaluation of recombinant adeno-associated virus as a gene transfer vector for the retina. *Current Eye Research, 16*, 949–956.

Greber, U. F., & Fassati, A., (2003). Nuclear import of viral DNA genomes. *Traffic, 4*, 136–143.

Haussecker, D., (2012). The business of RNAi therapeutics in 2012. *Molecular Therapy-Nucleic Acids*, 1.

Herrera-Carrillo, E., Liu, Y. P., & Berkhout, B., (2017). Improving miRNA delivery by optimizing miRNA expression cassettes in diverse virus vectors. *Human Gene Therapy Methods, 28*, 177–190.

Huntzinger, E., & Izaurralde, E., (2011). Gene silencing by microRNAs: Contributions of translational repression and mRNA decay. *Nature Reviews Genetics, 12*, 99–110.

Ibrahim, A. F., Weirauch, U., Thomas, M., Grünweller, A., Hartmann, R. K., & Aigner, A., (2011). MicroRNA replacement therapy for miR-145 and miR-33a is efficacious in a model of colon carcinoma. *Cancer Research, 71*, 5214–5224.

Játiva, P., & Ceña, V., (2017). Use of nanoparticles for glioblastoma treatment: A new approach. *Nanomedicine, 12*, 2533–2554.

Kaplitt, M. G., Leone, P., Samulski, R. J., Xiao, X., Pfaff, D. W., O'Malley, K. L., et al., (1994). Long-term gene expression and phenotypic correction using adeno-associated virus vectors in the mammalian brain. *Nature Genetics, 8*, 148–154.

Kaur, T., John, A. A., Sharma, C., Vashisht, N., Singh, D., Kapila, R., et al., (2020). miR300 intervenes Smad3/β-catenin/RunX2 crosstalk for therapy with an alternate function as indicative biomarker in osteoporosis. *Bone*, 115603.

Kaur, T., Kapila, R., & Kapila, S., (2020). MicroRNAs as next generation therapeutics in osteoporosis. *Bone Regeneration*. IntechOpen.

Keller, A., Beier, M., Meese, E., Leidinger, P., & Wendschlag, A., (2017). *Complex Sets of miRNAs as Non-Invasive Biomarkers for Kidney Cancer.* Google Patents.

Kent, O. A., Chivukula, R. R., Mullendore, M., Wentzel, E. A., Feldmann, G., Lee, K. H., et al., (2010). Repression of the miR-143/145 cluster by oncogenic RAS initiates a tumor-promoting feed-forward pathway. *Genes & Development, 24*, 2754–2759.

Kozomara, A., Birgaoanu, M., & Griffiths-Jones, S., (2019). miRBase: From microRNA sequences to function. *Nucleic Acids Research, 47*, D155–D62.

Krützfeldt, J., Kuwajima, S., Braich, R., Rajeev, K. G., Pena, J., Tuschl, T., et al., (2007). Specificity, duplex degradation and subcellular localization of antagomirs. *Nucleic Acids Research, 35*, 2885–2892.

Ladang, A., Beaudart, C., Locquet, M., Reginster, J. Y., Bruyère, O., & Cavalier, E., (2020). Evaluation of a panel of microRNAs that predicts fragility fracture risk: A pilot study. *Calcified Tissue International, 106*, 239–247.

Lennox, K. A., Owczarzy, R., Thomas, D. M., Walder, J. A., & Behlke, M. A., (2013). Improved performance of anti-miRNA oligonucleotides using a novel non-nucleotide modifier. *Molecular Therapy-Nucleic Acids, 2*, e117.

Lennox, K., & Behlke, M., (2011). Chemical modification and design of anti-miRNA oligonucleotides. *Gene Therapy, 18*, 1111–1120.

Li, W. H., Gu, Z., Wang, H., & Nekrutenko, A., (2001). Evolutionary analyses of the human genome. *Nature, 409*, 84784–84789.

Liu, Y. P., & Berkhout, B., (2011). miRNA cassettes in viral vectors: Problems and solutions. *Biochimica et Biophysica Acta (BBA)-Gene Regulatory Mechanisms, 1809*, 732–745.

Marshall, E., (1999). Gene therapy death prompts review of adenovirus vector. *Science, 286*, 2244, 2245.

McElrath, M. J., De Rosa, S. C., Moodie, Z., Dubey, S., Kierstead, L., Janes, H., et al., (2008). HIV-1 vaccine-induced immunity in the test-of-concept step study: A case–cohort analysis. *The Lancet, 372*, 1894–1905.

Montini, E., Cesana, D., Schmidt, M., Sanvito, F., Bartholomae, C. C., Ranzani, M., et al., (2009). The genotoxic potential of retroviral vectors is strongly modulated by vector design and integration site selection in a mouse model of HSC gene therapy. *The Journal of Clinical Investigation, 119*, 964–975.

Mussbacher, M., Krammer, T. L., Heber, S., Schrottmaier, W. C., Zeibig, S., Holthoff, H. P., et al., (2020). Impact of anticoagulation and sample processing on the quantification of human blood-derived microRNA signatures. *Cells, 9*, 1915.

Nakai, H., Herzog, R. W., Hagstrom, J. N., Walter, J., Kung, S. H., Yang, E. Y., et al., (1998). Adeno-associated viral vector-mediated gene transfer of human blood coagulation factor IX into mouse liver. *Blood, The Journal of the American Society of Hematology, 91*, 4600–4607.

Nana-Sinkam, S., & Croce, C., (2013). Clinical applications for microRNAs in cancer. *Clinical Pharmacology & Therapeutics, 93*, 98–104.

Pages, J., & Bru, T., (2004). Toolbox for retrovectorologists. *The Journal of Gene Medicine: A Cross-Disciplinary Journal for Research on the Science of Gene Transfer and Its Clinical Applications, 6*, S67–S82.

Partyka, K. L., Randolph, M. L., Lawrence, K. A., Cramer, H., & Wu, H. H., (2018). Utilization of direct smears of thyroid fine-needle aspirates for ancillary molecular testing: A comparison of two proprietary testing platforms. *Diagnostic Cytopathology, 46*, 320–325.

Pasut, G., & Veronese, F. M., (2009). PEGylation for improving the effectiveness of therapeutic biomolecules. *Drugs of Today (Barcelona, Spain: 1998), 45*, 687–695.

Roma-Rodrigues, C., Raposo, L. R., Cabral, R., Paradinha, F., Baptista, P. V., & Fernandes, A. R., (2017). Tumor microenvironment modulation via gold nanoparticles targeting malicious exosomes: Implications for cancer diagnostics and therapy. *International Journal of Molecular Sciences, 18*, 162.

Rupaimoole, R., & Slack, F. J., (2017). MicroRNA therapeutics: Towards a new era for the management of cancer and other diseases. *Nature Reviews Drug Discovery, 16*, 203.

Schultz, B. R., & Chamberlain, J. S., (2008). Recombinant adeno-associated virus transduction and integration. *Molecular Therapy, 16*, 1189–1199.

Sforzi, J., Palagi, L., & Aime, S., (2020). Liposome-based bioassays. *Biology, 9*, 202.

Sharma, V. K., & Watts, J. K., (2015). Oligonucleotide therapeutics: Chemistry, delivery and clinical progress. *Future Medicinal Chemistry, 7*, 2221–2242.

Suh, J. S., Lee, J. Y., Choi, Y. S., Chong, P. C., & Park, Y. J., (2013). Peptide-mediated intracellular delivery of miRNA-29b for osteogenic stem cell differentiation. *Biomaterials, 34*, 4347–4359.

Swarts, D. C., Makarova, K., Wang, Y., Nakanishi, K., Ketting, R. F., Koonin, E. V., et al., (2014). The evolutionary journey of Argonaute proteins. *Nature Structural & Molecular Biology, 21*, 743.

Thorsen, S. B., Obad, S., Jensen, N. F., Stenvang, J., & Kauppinen, S., (2012). The therapeutic potential of microRNAs in cancer. *The Cancer Journal, 18*, 275–284.

Titze-de-Almeida, S. S., Soto-Sánchez, C., Fernandez, E., Koprich, J. B., Brotchie, J. M., & Titze-de-Almeida, R., (2020). The promise and challenges of developing miRNA-based therapeutics for Parkinson's disease. *Cells, 9*, 841.

Vu, M. T., Bach, L. G., Nguyen, D. C., Ho, M. N., Nguyen, N. H., Tran, N. Q., et al., (2019). Modified carboxyl-terminated PAMAM dendrimers as great cytocompatible nano-based drug delivery system. *International Journal of Molecular Sciences, 20*, 2016.

Wang, H., Liu, S., Jia, L., Chu, F., Zhou, Y., He, Z., et al., (2018). Nanostructured lipid carriers for microRNA delivery in tumor gene therapy. *Cancer Cell International, 18*, 101.

Wang, Z., Bai, Y., Wei, W., Xia, N., & Du, Y., (2013). Magnetic Fe_3O_4-based sandwich-type biosensor using modified gold nanoparticles as colorimetric probes for the detection of dopamine. *Materials, 6*, 5690–5699.

Wu, Z., Asokan, A., & Samulski, R. J., (2006). Adeno-associated virus serotypes: Vector toolkit for human gene therapy. *Molecular Therapy, 14*, 316–327.

Xu, P., Vernooy, S. Y., Guo, M., & Hay, B. A., (2003). The drosophila microRNA Mir-14 suppresses cell death and is required for normal fat metabolism. *Current Biology, 13*, 790–795.

Yang, N., (2015). An overview of viral and nonviral delivery systems for microRNA. *International Journal of Pharmaceutical Investigation, 5*, 179.

Yoshizuka, M., Nakasa, T., Kawanishi, Y., Hachisuka, S., Furuta, T., Miyaki, S., et al., (2016). Inhibition of microRNA-222 expression accelerates bone healing with enhancement of osteogenesis, chondrogenesis, and angiogenesis in a rat refractory fracture model. *Journal of Orthopaedic Science, 21*, 852–858.

Zhang, B., & Farwell, M., (2008). microRNAs: A new emerging class of players for disease diagnostics and gene therapy. *Journal of Cellular and Molecular Medicine, 12*, 3–21.

CHAPTER 2

Evolving CRISPR-Cas-Based Diagnostics and Its Reliability

RAHUL RAGHAVAN

Hospital for Special Surgery, Weill Cornell Medicine, New York, USA

ABSTRACT

Detection of genetic material is important for detecting the pathogen or progression of various diseases. The CRISPR-based diagnostic tools are cost-effective in many ways compared to PCR-based DNA detection, as it does not require a thermal cycler machine. Properties of the Cas system, such as detection of the specific protospacer adjacent motif (PAM), specific cleavage of double-strand DNA, and collateral activity on nonspecific sequences, have been utilized for developing diagnostics tools. This chapter explains the types CRISPR-Cas systems, their mechanism, CRISPR-Cas-based diagnostic tools, and their reliability.

2.1 INTRODUCTION

Detection of nucleic acid-based markers is considered highly accurate for determining infectious diseases. Finding nucleic acid-based markers are important for determining the progression of various pathological conditions, including cancer. The recent pandemic of SARS-CoV-2 has increased the need for developing fast nucleic acid detection tools.

Modern Biotechnology in Healthcare: Advances and Applications. Sheikh Umar Ahmad (Ed.)

Nucleic acid-based markers are usually detected using PCR-based techniques and often require a thermal cycle machine. The evolution of CRISPR-Cas (clustered regularly interspaced short palindromic sequences–CRISPR-associated protein)-based techniques have changed the face of diagnostic due to their specificity and reliability. CRISPR-based diagnostics has a limit of detection (LOD) in the picomolar range; hence, they can directly detect the target when the sample has a high concentration of DNA or RNA, e.g., Viral infections (Gootenberg et al., 2017; Ramachandran & Santiago, 2021). Ever since the first Cas9-based Zika virus DNA variant detection system was developed, Class 2 CRISPR-Cas systems have been sculptured for multiple diagnostic tools. Diagnostic tools based on the CRISPR-Cas system are extensively reviewed (Kaminski et al., 2021).

The CRISPR-based diagnostic tools are more cost-effective in many cases compared to PCR-based DNA detection, as it does not require thermal cycler machines, and preamplification can be done at isothermal conditions. Properties of Cas enzymes to specifically recognize proto-spacer adjacent motif (PAM), cleave double-strand DNA or single strand (nickase), and initiate collateral cleavage of nucleic acid has been utilized for developing CRISPR-Cas-based diagnostic. NASBACC (nucleic acid sequence-based amplification–CRISPR cleavage), a CRISPR-Cas 9 system in combination with a PCR-based technique, has also been developed to detect Zika virus DNA. A CRISPR Cas12 system-based diagnostic tool, CAT-SMelor, used Cas enzymes as a reporter. CAT-SMelor detects the nanomolar concentration of small molecules such as Uric acid and parahydroxybenzoic acid. The ability of the Cas12 enzyme to specifically cleave double-stranded DNA (dsDNA) and the property of bacterial allosteric transcription factor to detect small molecules were used in combination to develop CAT-SMelor. SHERLOCK one pot Cas 12-based SARS-CoV-2 detecting kit has shown 93.1% sensitivity and 98.5% specificity and could be used widely in low complexity labs (Joung et al., 2020). This chapter explains the types of CRISPR-Cas systems used in diagnostics, their mechanism, and the methods adopted to mold the CRISPR-Cas system to develop various diagnostic tools. Finally, the reliability of diagnostic tools needs to be proved. Various research groups have utilized next-generation sequencing to study the off-target effects of the CRISPR-Cas system. The above-mentioned studies were explored to explain the reliability of CRISPR-Cas-based diagnostic tools.

2.1.1 HISTORY OF CRISPR-CAS SYSTEM

The CRISPR-Cas gene editing technology is an adaption of the bacterial immune system. An unusual genetic structure composed of repeating DNA segments separated by non-repeating sequences was first observed by Japanese scientists (Ishino et al., 1987). It was later discovered that these genomic structures were part of the acquired bacterial immune system (Barrangou et al., 2007). The acquired immunity in bacteria against bacteriophage is achieved through three stages, adaption, expression, and interference (Makarova et al., 2011). In the adaptive stage, upon infection by bacteriophage, Cas enzymes in bacteria cut the phage DNA at a specific site (approximately 30 bp), recognized by PAM. Short segments of bacteriophage genomic DNA (protospacer) were inserted with a duplicated noncontiguous direct repeating segment added adjacent to the inserted segment at the CRISPR locus. The expression stage involves the transcription of cr-RNA, from the CRISPR locus of the bacterial genome. During the expression stage bacterial genome transcribes the spacer-containing region from Pre-crRNA. A Tracer RNA, which is constitutively expressed in a bacterial system, specifically detects and binds to the CRISPR repeat sequence adjacent to protospacers in Pre-crRNA. CRISPR-associated proteins such as RNAase III process the Pre-crRNA:Tracer RNA complex into guideRNA (gRNA). In the interference stage, Cas enzymes get activated upon loading the gRNA and specifically detect and cleaves the target DNA (Makarova et al., 2011). Thus, Cas enzymes loaded with gRNA can specifically damage the bacteriophage DNA and confers resistance.

2.1.2 TYPES OF CRISPR-CAS SYSTEMS

Based on the evolutionary relationship CRISPR Cas system constitute two classes, six types and subtypes (Kick et al., 2017; Xu & Li, 2020). Based on the composition of the ribonucleoprotein (RNP) effector complex (effector nuclease) CRISPR-Cas system is divided into two classes. Class 1 has multiple effector proteins involved in the formation of the effector complex. In the Class 2 systems, effector nuclease involves a single crRNA binding protein. Hence, Class 2 Cas enzymes are widely used in diagnostics as they are easy to reconstitute. Classification of Class 2 enzymes and their requirements for specifically cleaving the target are given in Table 2.1. Mechanisms of Cas enzymes such as Cas9, Cas12, and Cas13 are given in Figure 2.1.

TABLE 2.1 Classification of Class II Cas Enzymes and Requirements for Cleavage of Target

Class	Types	Cas Enzymes	Type of DNA Cleavage	Tracer RNA Requirement	cRNA	Collateral Activity	References
Class II	Type II	Cas9	Blunt cut, double-strand break	Required	Required	No	Jinek et al. (2012); Makarova et al. (2020)
	Type V	Cpf I (Cas12a) or C2c1 (Cas12b)	Generates 5' overhang, double-strand break	No for Cas12h and Cas12i; Yes for Cas12b and Cas12g.	Required	Cas12a, Cas12b-collateral ssDNA degradation	Chen et al. (2018); Rusk (2019)
	Type VI	C2c2 (Cas 13a), C2c6 (Cas 13b)	Targets single-strand RNA	No	Required	Cas13-collateral RNA degradation	Abudayyeh et al. (2016)

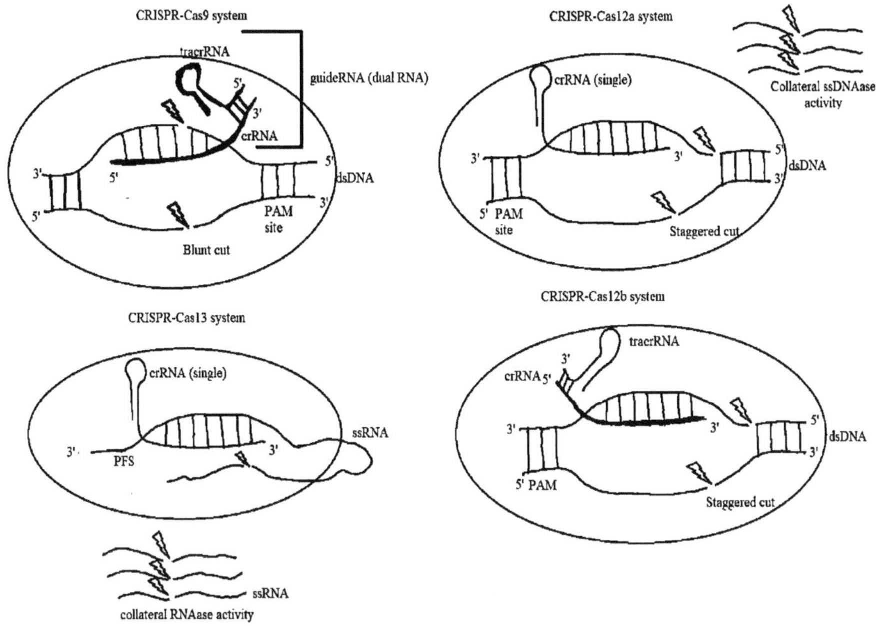

FIGURE 2.1 Mechanism of Cas9, Cas12, and Cas13 enzymes binding to target sequence.

2.2 CRISPR-CAS 9 SYSTEM AND ITS MODIFICATION FOR DIAGNOSTICS

Cas9 enzymes are best characterized Class II enzymes and have been widely used and modified for diagnostic purposes. The enzyme requires Tracer RNA for generating gRNA. The effector nuclease forms double-teamed blunt cuts on target DNA. Cas9 enzyme require PAM motif, NGG, and very rarely NAG (N-any nucleotide) to bind to the DNA. PAM motifs are located downstream to the binding site of Cas9. Cas9 enzymes cleave the DNA using two separate catalytic domains, a RuvC-like domain and C-terminal HNH-like domain. RuvC domain cleaves the noncomplementary strand and HNH domain cleaves the complementary strand binding to cRNA-gRNA (Jinek et al., 2012). Each catalytic domain cleaves one strand of DNA resulting in double-stranded blunt-end cleavage of DNA or with short 3 bp overhang (Vriend et al., 2014). Cas9 enzymes can be made catalytically dead by mutating both catalytic domains, but the mutated Cas9 can still bind to DNA at specific sites. This property of Cas9 has been put to use in visualizing genomic loci in living cells (Chen et al., 2013).

2.2.1 CAS9-NASBA-BASED DETECTION OF TARGET GENE

Cas9 system in combination with nucleic acid sequence-based amplification (NASBA) technique was used to differentiate between African and American Zika virus strains. After preamplification of virus RNA by NASBA technique, the sample will be then applied to a toehold RNA-regulated LacZ mRNA along with guideRNA and Cas9 enzyme. Cas9 enzyme specifically detects strain-specific PAM motif and cleaves the DNA. If strain Specific PAM motif is not present on target RNA, it will activate toehold RNA upstream to LacZ gene. Activation of LacZ gene leads to color production in the presence of substrate (Pardee et al., 2016). Figure 2.2 shows the NASBA-Cas9-based method of detection and differentiation of viral strains.

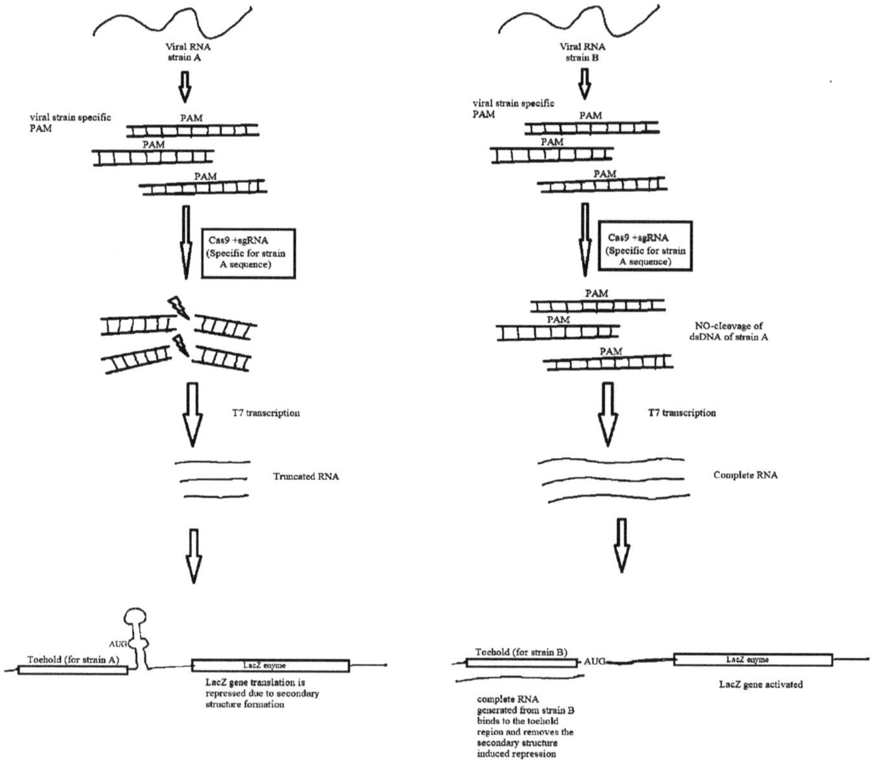

FIGURE 2.2 NASBA-Cas9-based method of detection and differentiation of viral strains.

Source: Pardee (2016).

2.2.2 CRISPR-CAS9 CHIP FOR DETECTION OF TARGET GENE

Catalytically deactivated Cas9 system directly immobilized on a graphene sheet to develop CRISPR-enhanced graphene-based field-effect transistor (gFET or CRISPR-chip). CRISPR-Chip is the best example for using the Cas9 system for preamplification-free detection of nucleic acid biomarkers. Catalytically deactivated Cas9 scan the whole genome sample binds to the target sequence. Hybridization of target DNA to CRISPR-gRNA complex will lead to change in electric characteristics of CRISPR-Chip (Hajian et al., 2019). Figure 2.3 shows the diagram of CRISPR-Cas9-based chip for detecting target DNA sequences.

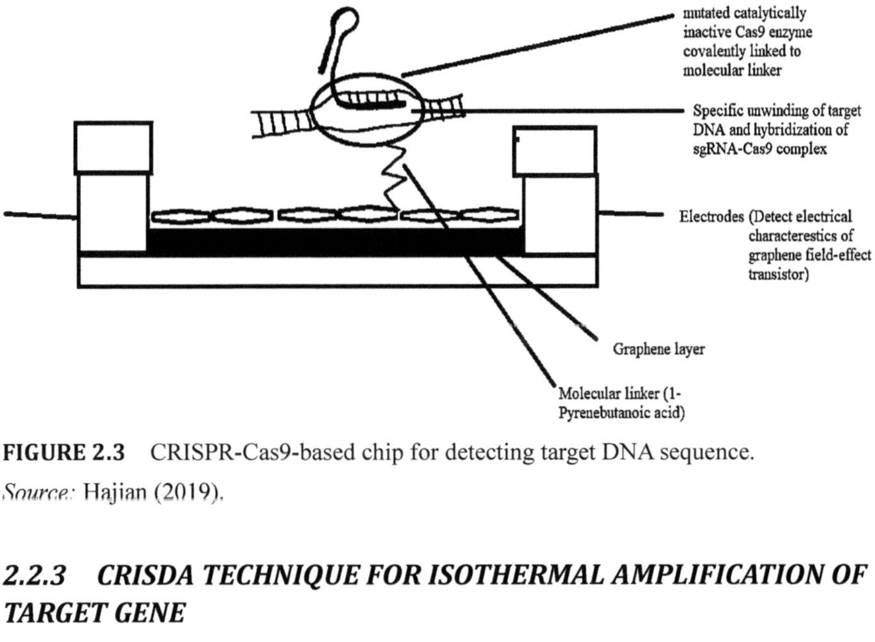

FIGURE 2.3 CRISPR-Cas9-based chip for detecting target DNA sequence.

Source: Hajian (2019).

2.2.3 CRISDA TECHNIQUE FOR ISOTHERMAL AMPLIFICATION OF TARGET GENE

Using PCR machine to amplify DNA is a major hurdle in onsite diagnostics for detecting nucleic acid biomarkers. CRISPR–Cas9-triggered nicking endonuclease-mediated Strand displacement amplification method (CRISDA) technique using Cas9H840A system could isothermally amplify and detect the target DNA. Cas9 enzymes with H840A mutation is a nickase, which cleaves only one of the strands of dsDNA (Nickases) while exposing the other strand. CRISDA method of isothermal amplification require Cas9H840A enzyme, Specific guideRNA, forward and reverse primer with 5' Nb.BbvCI endonuclease nicking site for initiating the amplification. Specific gRNA is designed to target the DNA at two sites (~186 bp). Upon reconstituting the enzyme and gRNAs, the enzyme will nick the DNA from both sides exposing the complementary strand for primer binding. Upon reconstituting the klenow DNA polymerase (exo-), DNA polymerization will be initiated from both ends leading to amplification of the target DNA. The targetDNA has active Nb.BbvCI nickase site at the 5' end. Nb.BbvCI nickase enzyme cleaves the DNA, Klenow polymerase (exo-) binds to the nicked site and add new nucleotide and displaces the other strand. Thus, a continuous cycle of nicking activity and strand displacement by klenow polymerase (exo-) leads to amplification of target DNA at isothermal conditions. A specific peptide nucleic acid (PNA) probes detects the amplified target DNA (Zhou et al., 2018). Figure 2.4 shows the diagrammatic view of CRISDA technique.

2.2.4 CAS-EXPAR-BASED ISOTHERMAL AMPLIFICATION OF TARGET GENE

Isothermal amplification of target gene was possible using Cas9 system. Cas9 enzymes cleave single-stranded DNA and single-stranded RNA when Cas9-sgRNA system is treated with separate PAM containing oligonucleotide nucleotide (PAMmer) (O'Connell et al., 2014). This property of Cas9 system was made used to achieve isothermal amplification of the target gene in CRISPR/Cas9 triggered exponential amplification method (CAS-EXPAR). Cas-Expar technique requires an exogenous single-stranded Expar template oligonucleotide containing Nickase enzyme recognition site (Nt.BstNBI) flanking two sgRNA target sequences. After specific cleavage of single-stranded DNA by Cas9-sgRNA-Pammer complex, the specifically cleaved single-stranded DNA will bind to the Expar template oligonucleotide. DNA polymerase enzymes further add nucleotide, and a complete dsDNA is generated. The generated new dsDNA has a nickase enzyme recognition site in the center, Nt.BstNBI creates a nick and vent (exo-) DNA polymerase binds to the nicked region and adds nucleotides, and displaces the complementary single-strand DNA. This released single-stranded DNA serves as a primer for amplifying the target DNA. Thus, cycle of Cas9 cleavage, binding of Expar template oligonucleotide, nicking by Nt.BstNBI, vent (exo-) DNA polymerase activity to release

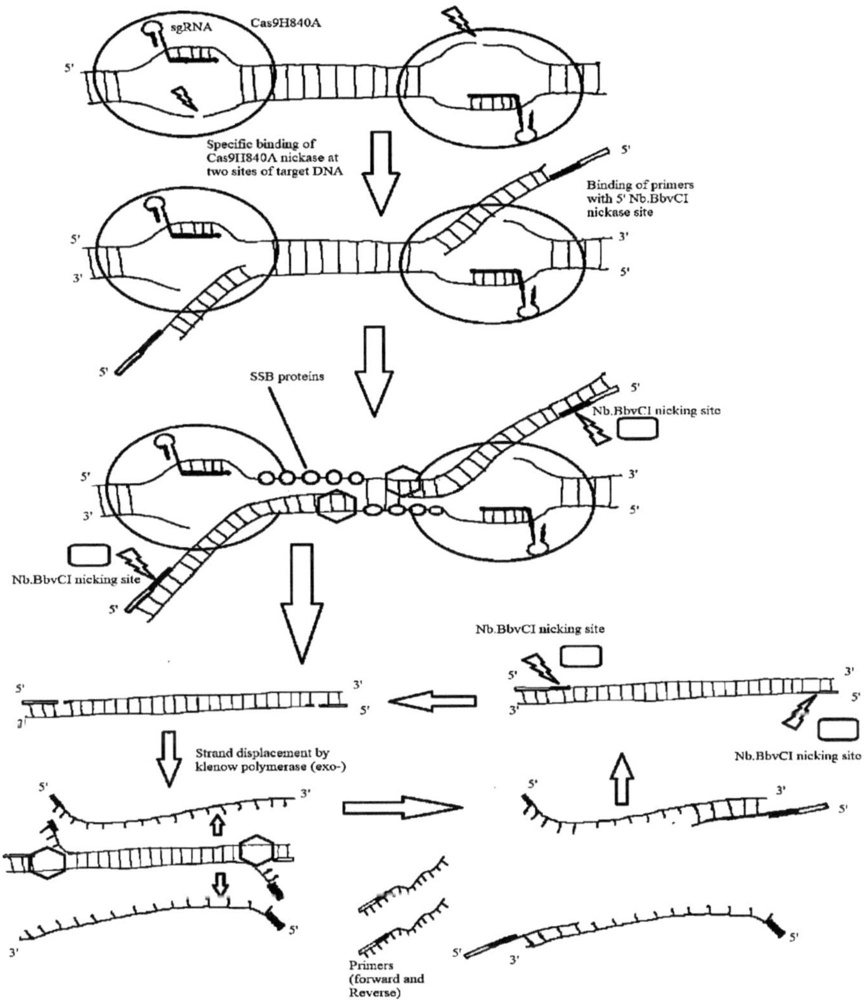

FIGURE 2.4 Diagrammatic view of CRISDA technique.

Source: Zhou et al. (2018).

a single strand leads to amplification of the target gene (Huang et al., 2018). Figure 2.5 shows the diagrammatic view of the Expar technique of isothermal amplification using CRISPR-Cas9 system.

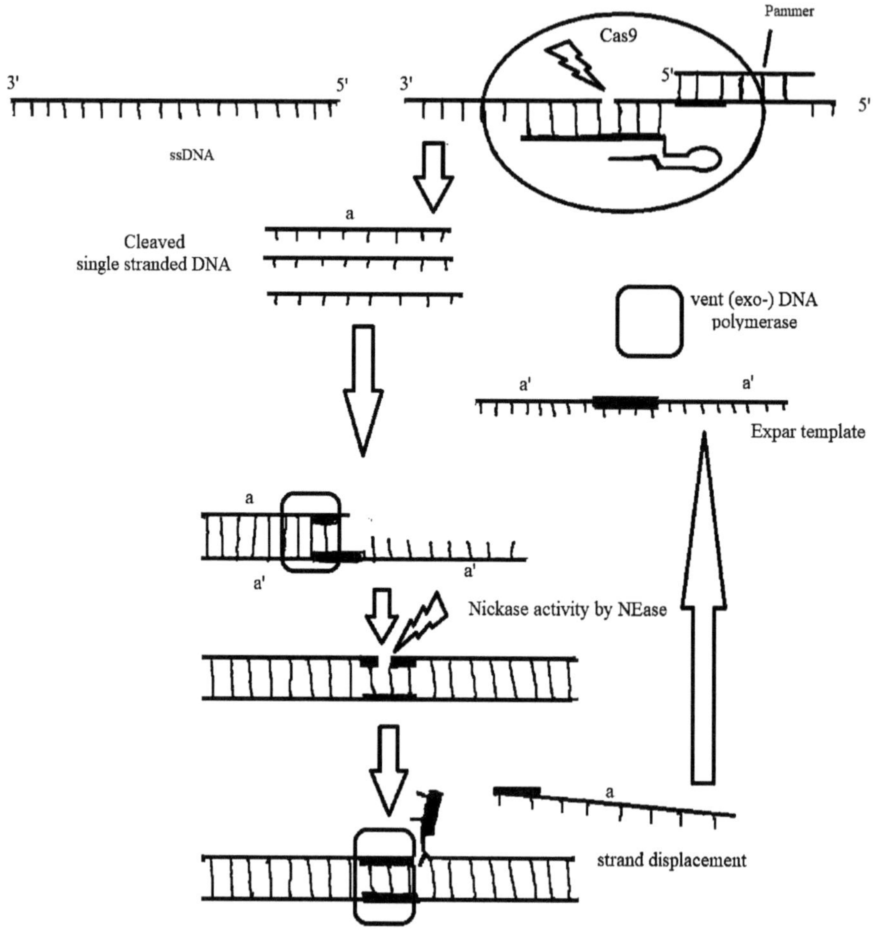

FIGURE 2.5 Cas-Expar technique of isothermal amplification using CRISPR-Cas9 system.

Source: Huang (2018).

2.2.5 RNA-GUIDED CAS9 SYSTEM FOR ISOTHERMAL AMPLIFICATION

Isothermal amplification was achieved using RNA-guided Cas9 system and (exo-) klenow polymerase. The target dsDNA is specifically nicked using nCas9 (D10A) and sgRNA. (exo-)klenow polymerase adds nucleotides and displaces the opposite single-stranded DNA. Specific primers containing protospacer and PAM on 5' end will bind to the released single-stranded Target DNA. The (exo-)klenow polymerase adds nucleotides and forms dsDNA. The dsDNA formed has protospacer and PAM motif at 5' end, nCas9-sgRNA complex target this region and again insert a nick. The (exo-)klenow polymerase binds to the nicked DNA region and add nucleotide and displaces the complementary single stranded DNA. The second set of primers containing protospacer and PAM on 5' end will bind to the released single-stranded Target DNA. Thus, the cycle of specific primers containing protospacer and PAM motif at 5' end specific nCas9 nicking leads to amplification of target at isothermal condition (Wang et al., 2019).

2.3 CRISPR/CAS12 SYSTEM AND DIAGNOSTICS

Cas12 are class II type V enzyme. It requires gRNA for specific target DNA detection and induces specific cleavage of target DNA. Cas12 enzymes, Cas12a requires single crRNA for activation, but Cas12b requires dual RNA, both tracer and gRNA for activation (Murugan et al., 2017). Cas12a enzyme process its own crRNA (Swarts et al., 2017; van der Oost et al., 2017). Cas12 enzyme specifically recognizes a T rich PAM. Cas12a enzyme specifically targets dsDNA and form a staggered cut. Cas12a also degrade nonspecific single-stranded DNA along with specific cleavage of dsDNA target. This property of the Cas12a enzyme is made used in CRISPR-Cas12-based diagnostics to detect viral genomes (Chen et al., 2018). Cas12a enzyme cleaves the specific double-stranded target DNA by RuvC catalytic domain (Stella et al., 2017). Nonspecific dsDNA degradation (collateral activity) is proposed to be happening in RuvC domain itself (Li et al., 2018).

2.3.1 CAS12-BASED DETECTION OF RNA-DETECTR

DETECTR is a Cas12-based diagnostic method to detect human papilloma virus directly from anal swabs. Crude DNA extracted from the anal swabs were treated with either Cas12a-crRNA specific to Human papillomavirus 16 or 18 (hypervariable loop region of HPV16 or 18). crRNA specifically detected the target and induced a double-stranded cleavage of target DNA. Collateral ssDNAse activity initiated immediately after cleavage of specific target DNA leads to the cleavage of the fluorescent quenched reporter (ssDNA). Fluorescent signal released indicates the presence of HPV16 or HPV18 (Chen et al., 2018).

2.3.2 CAS12-BASED DNA DETECTION-HOLMES

A Cas12a-based diagnostic HOLMES (a one-HOur Low-cost Multipurpose highly Efficient System) combines Rapid PCR-based amplification of target gene and then detection by Cas12a-crRNA complex. Similar to DETECTR technique This method also made use of the collateral ssDNA degradation activity of Cas12a enzyme to cleave the quenched fluorescent ssDNA reporter following binding to the target DNA. Amplification of the target DNA using PCR increased the specificity of detection by HOLMES. HOLMES could rapidly detect a gout risk-related short nucleotide polymorphism (SNP), rs101429 and was able to discriminate between PRV Ra classical strain and Bartha-K61 vaccine strain (Li et al., 2018).

2.3.3 CAS12B-BASED DNA DETECTION (CD DETECTION)

Compared to Cas12a enzyme Cas12b require both tracerRNA and gRNA for detection of specific target DNA. In the CD detection technique, tuned guideRNA (tgRNA)-Cas12b system was able to detect the dsDNA target specifically. CD detection could directly detect the human papilloma dsDNA concentration at attomolar level. CD detection with fine-tuned specific sgRNA could discriminate the human blood groups with high specificity. The tgRNA-Cas12b system also detected cancer-based SNPs such as *TP53* 856G>A, *BRCA1* gene (3232A>G and 3537A>G). Cas12b enzyme had preference for non-targeted ssDNA cleavage. Cas12b-sgRNA complex preferred to cleave poly-thyamine, poly-adenine, or poly-cytosine-based fluorophore quencher (FQ)-labeled homopolymer reporters (Teng et al., 2019).

2.4 CRISPR-CAS13 SYSTEM AND ITS MODIFICATION FOR DIAGNOSTICS

Cas13 enzymes are class II type VI enzyme. The Cas13 enzyme has RNAase activity and gRNA processing activity. The enzyme has both

RNAse and gRNA maturation activity. The RNAse activity is through higher eukaryotic and prokaryotic nucleotide (HEPN) binding domain and gRNA maturation activity is mediated by both HEPN2 and helical-1 domain (O'Connell, 2019). One of the important properties of Cas9 enzyme is its collateral activity, once the HEPN domain of Cas13 enzyme is activated by binding to specific gRNA it cleaves specific target RNA and in addition to the target it will also degrade the neighboring nonspecific RNA (Abudayyeh et al., 2016). This collateral degradation activity of Cas13 enzyme is put to use in CRISPR/Cas13-based diagnostics to amplify the signal after specific cleavage.

2.4.1 CAS13-BASED DIAGNOSTIC SPECIFIC HIGH SENSITIVITY ENZYMATIC REPORTER UNLOCKING (SHERLOCK)

Cas13-based diagnostic SHERLOCK could detect Zika virus and Dengue virus at attomolar sensitivity. In this method, the recombinase polymerase amplification (isothermal preamplification) technique was coupled with T7 Transcription to convert target DNA into RNA. Cas13 with sgRNA specifically detect the target DNA and induce cleavage of specific viral RNA. Collateral RNA degradation initiated by Cas13 after detecting specific RNA target degrades quenched RNA probes releasing the fluorescence. Appearance of fluorescence indicates the presence of viral RNA. SHERLOCK technique was also able to differentiate between related Zika virus and Dengue virus (Gootenberg et al., 2017). SHERLOCK one pot testing kit for detecting SARS-CoV-2 is a Cas13-based diagnostic tool, it showed 98.5% specificity and 93.1% sensitivity (Joung et al., 2020).

2.4.2 CAS13-BASED DIAGNOSTIC STREAMLINED HIGHLIGHTING OF INFECTIONS TO NAVIGATE EPIDEMICS (SHINE)

SARS-CoV-2 detection was done using Cas13-based diagnostic, SHINE. SHINE technique was able to detect 10 copies of viral RNA per microliter. SHINE is a modification of SHERLOCK method. This method combined various steps – reverse transcription of RNA to DNA, recombinase-based polymerization (isothermal amplification of target), T7 transcription and Cas13-based detection of target RNA into a single step. Upon detection of viral RNA, Cas13-gRNA cleaves the quenched fluorescent reporter tagged to ssDNA (Arizti-Sanz et al., 2020).

2.4.3 CAS13-BASED DETECTION OF VIRUS (CARMEN)

Combinatorial arrayed reactions for multiplexed evaluation of nucleic acids (CARMEN) is a Cas13-based platform for multiplexed pathogen detection (Ackerman et al., 2020). In this method, input samples are initially amplified by RPA method and then undergone T7 transcription, and then the samples are emulsified using fluorous oil to form emulsion droplets of 1 nanoliter size. Each sample for analysis is color coded for identification. Similarly, emulsion droplets containing Specific crRNA for target recognition, Cas13 enzyme and fluorescent quenched ssRNA reporters were prepared. All color-coded emulsions were mixed and loaded onto a microwell array chip, each well in microarray can hold two nanoemulsion droplets in random. Nanodroplets will be merged after application of electric field. Upon merging of nanodroplet, specific cRNA-Cas13 system will bind to the target and fluorescent reporter will be cleaved. Each droplet is identified and monitored by fluorescent microscopy. Figure 2.6 shows the diagrammatic view of the microwell array chip containing nanodroplet.

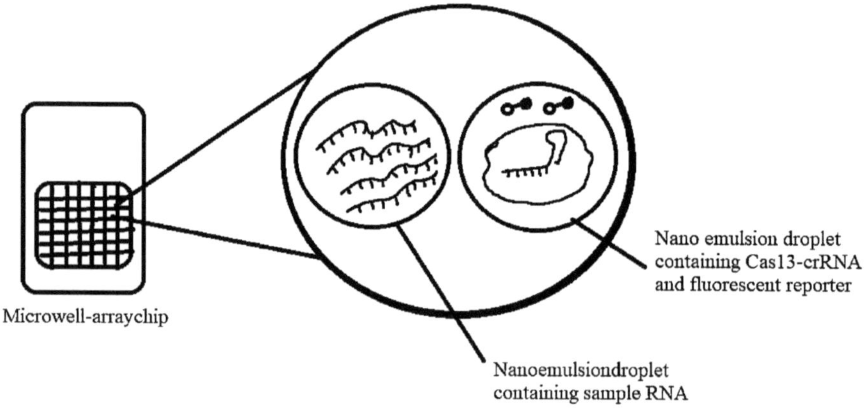

FIGURE 2.6 Diagrammatic view of microwell-array chip used in Cas13-based CARMEN.

Source: Ackerman (2020).

2.5 CRISPR-CAS SYSTEM AS REPORTER

Collateral activity of CRISPR-Cas system can be used as a method to amplify signals. Cat-SMelor technique combines the ability of a transcription factor to bind to specific sequence and the ability of CRISPR-Cas system to specifically bind to the target DNA (Liang et al., 2019). Cat-SMelor-based techniques were used to detect uric acid (25–500 nM) and p-hydrobenzoic acid (9–180 nM) in serum. In this method target DNA for both transcription factor and CRISPR-Cas system are same therefore, they compete for the target DNA sequence. In CaT-SMelor technique a bacterial transcription factor HucR fused to cellulose binding domain was immobilized on microcrystalline cellulose. The HucR-based CaT-SMelor technique was able to detect the levels of uric acid in serum. The bacterial transcription factor HucR constantly bind to the target DNA containing HucR binding motif and PAM sequence. HucR is a uric acid responsive protein, upon binding of uric acid HucR factors dissociate from target DNA, allowing the specific CRISPR-Cas12a system to bind and cleave the target dsDNA. Collateral nuclease activity initiated cleaves the fluorescently quenched ssDNA probes. The rate of cleavage of quenched fluorescent probe was depended on the concentration of uric acid. Figure 2.7 shows the diagrammatic view of CAT-SMelor technique.

2.6 CRISPR-BASED DIAGNOSTICS AND RELIABILITY

Recent whole genome sequencing studies following chromatin immunoprecipitation (CHIP) of a catalytically inactive Cas9 enzyme showed that only 1 to 5 bp of the immunoprecipitated DNA matched to the complimentary region in gRNA. These regions of gRNA are called seed sequence. This indicates that 1 to 5 bp adjacent to the PAM sequence or seed sequence on gRNA determines the specificity of Cas9 binding (Wu et al., 2014). The sequencing studies following CHIP only gave information of DNA bound to Cas9, but not the cleavage event (Duan et al., 2014; Zhang et al., 2015). Therefore, it is not sure if all these Cas9 bound sites are cleaved. Frequency of the seed sequence in the genome determines the specificity of binding. Therefore, CRISPR-Cas9-based diagnostics, which does not involve cleavage events such as direct observation of genomic loci could be interfered by off-target effects and therefore require optimization of guideRNA. The PAM sequence also affects the specificity

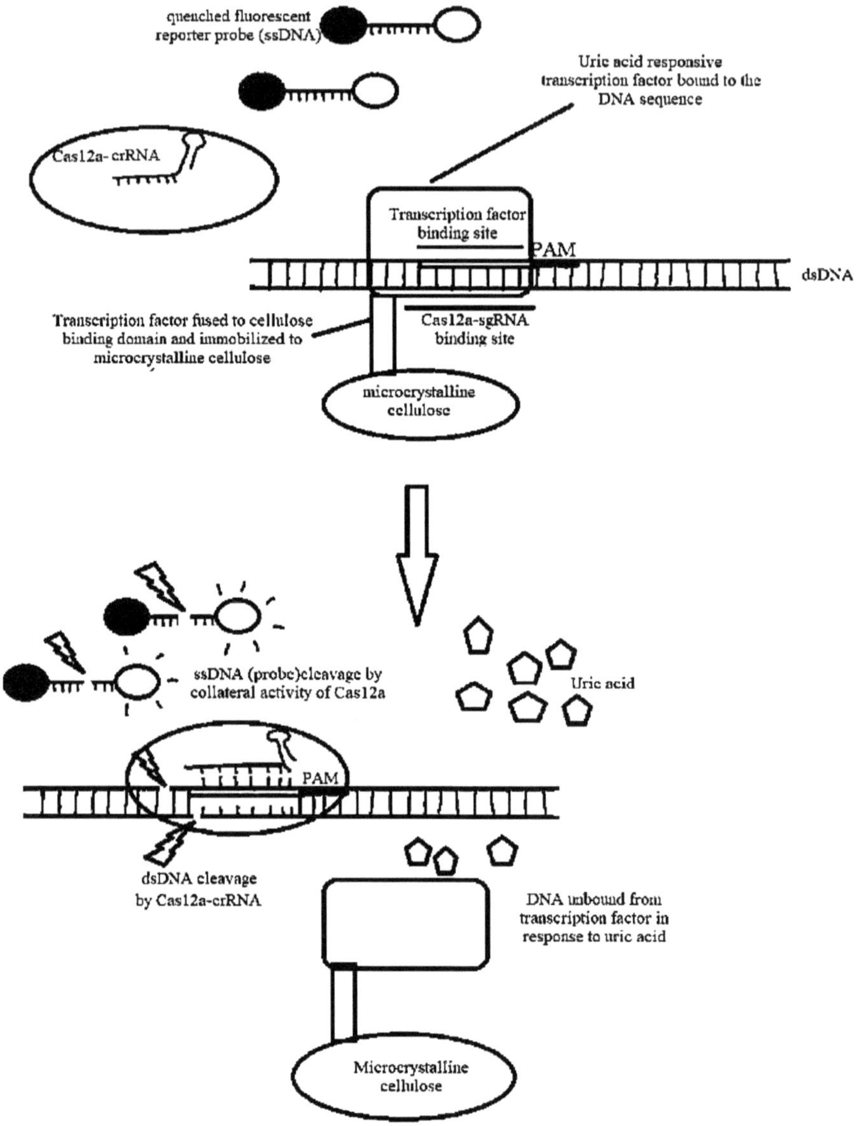

FIGURE 2.7 Diagrammatic view of CAT-SMelor technique.

Source: Liang (2019).

of binding, it was shown that Cas9 system uses the PAM sequence NGG (N for A, T, G or C) but Cas9 system also has less frequent preference for PAM sequence NRG (R for G or A) therefore diagnostics tools designed based on the Cas9 system should carefully consider for the presence of other less frequent PAM sequences in the system (Zhang et al., 2015). For improving the specificity of the CRISPR-Cas system, potential off target regions should be predicted and avoided. Target sequence with global similarity should be avoided (Figure 2.8).

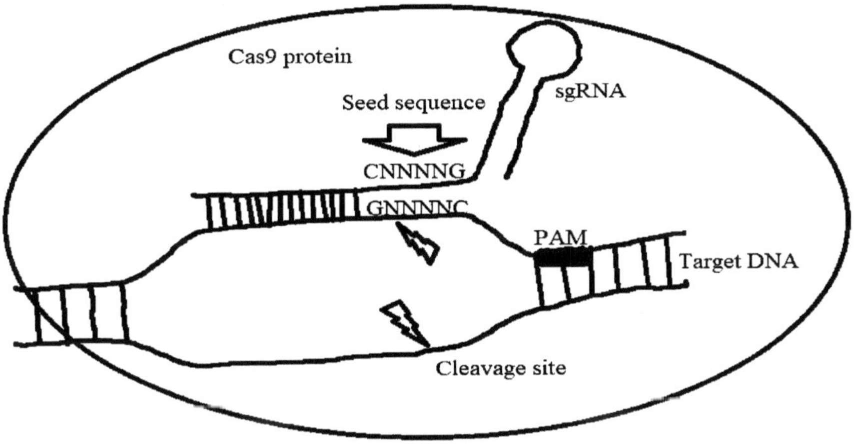

FIGURE 2.8 Seed sequence determines the specificity of Cas9 binding to target DNA.
Source: Zhang (2015).

2.7 CONCLUSION

Detecting genetic material of pathogens is the major requirement to confirm a disease condition. Detecting the pathogens at very low concentration could help in finding the disease at very early stages. Therefore, specificity and sensitivity of diagnostic tools need to be improved. In addition, diagnostics should be able to be performed without heavy machinery. The current pandemic has shown the requirement of portable diagnostic tools. CRISPR-Cas-based diagnostics tools are solution for all these problems. CRISPR diagnostic tools help in detecting DNA-based or RNA-based target DNA directly without machinery, such as thermal cycler. Different isothermal DNA amplification techniques have helped to improve the sensitivity of the CRISPR-Cas-based diagnostic tool. The recent development of various CRISPR-Cas-based diagnostic tools showed that they are very reliable. Collateral activity observed in Cas12 and Cas13 system largely helped in amplifying the signals in CRISPR-Cas-based diagnostics. Various studies show that CRISPR-Cas system-based diagnosis has very high specificity and sensitivity. SHERLOCK-CRISPR Cas13-based technique could detect SARS-CoV-2 RNA with 100% specificity and 100% sensitivity in fluorescent read-out method (Patchsung et al., 2020). Cas12b-based diagnostic tool could detect SARS-CoV-2 mRNA with 98.5% specificity and 93.1% sensitivity. The CRISPR-Cas system was modified to use as reporter, in CaT-SMelor. In CAT-Smelor it is used to detect the concentration of small molecules in serum (Liang et al., 2019). The CaT-SMelor technique was able to detect the concentration of small molecules in the serum. This study showed that collateral activity of Cas enzymes can be quantified by comparing the rate of cleavage of fluorescent probes.

KEYWORDS

- **CRISPR-Cas**
- **deoxyribonucleic acid**
- **disease diagnostics**
- **peptide nucleic acid**
- **protospacer adjacent motif**
- **ribonucleic acid**
- **short nucleotide polymorphism**

REFERENCES

Abudayyeh, O. O., et al., (2016). C2c2 is a single-component programmable RNA-guided RNA-targeting CRISPR effector. *Science, 353*(6299), aaf5573.

Ackerman, C. M., et al., (2020). Massively multiplexed nucleic acid detection with Cas13. *Nature, 582*(7811), 277–282.

Arizti-Sanz, J., et al., (2020). Streamlined inactivation, amplification, and Cas13-based detection of SARS-CoV-2. *Nature Communications, 11*(1), 1–9.

Barrangou, R., et al., (2007). CRISPR provides acquired resistance against viruses in prokaryotes. *Science, 315*(5819), 1709–1712.

Chen, B., et al., (2013). Dynamic imaging of genomic loci in living human cells by an optimized CRISPR/Cas system. *Cell, 155*(7), 1479–1491.

Chen, J. S., et al., (2018). CRISPR-Cas12a target binding unleashes indiscriminate single-stranded DNase activity. *Science, 360*(6387), 436–439.

Duan, J., et al., (2014). Genome-wide identification of CRISPR/Cas9 off-targets in human genome. *Cell Research, 24*(8), 1009–1012.

Gootenberg, J. S., et al., (2017). Nucleic acid detection with CRISPR-Cas13a/C2c2. *Science, 356*(6336), 438–442.

Hajian, R., et al., (2019). Detection of unamplified target genes via CRISPR–Cas9 immobilized on a graphene field-effect transistor. *Nature Biomedical Engineering 3*(6), 427–437.

Huang, M., et al., (2018). Clustered regularly interspaced short palindromic repeats/Cas9 triggered isothermal amplification for site-specific nucleic acid detection. *Analytical Chemistry, 90*(3), 2193–2200.

Ishino, Y., et al., (1987). Nucleotide sequence of the iap gene, responsible for alkaline phosphatase isozyme conversion in *Escherichia coli*, and identification of the gene product. *Journal of Bacteriology, 169*(12), 5429–5433.

Jinek, M., et al., (2012). A programmable dual-RNA–guided DNA endonuclease in adaptive bacterial immunity. *Science, 337*(6096), 816–821.

Joung, J., et al., (2020). Detection of SARS-CoV-2 with SHERLOCK one-pot testing. *New England Journal of Medicine, 383*(15), 1492–1494.

Kaminski, M. M., et al., (2021). CRISPR-based diagnostics. *Nature Biomedical Engineering 5*(7), 643–656.

Kick, L., et al., (2017). CRISPR-Cas9: From a bacterial immune system to genome-edited human cells in clinical trials. *Bioengineered, 8*(3), 280–286.

Li, S. Y., et al., (2018). CRISPR-Cas12a has both cis-and trans-cleavage activities on single-stranded DNA. *Cell Research, 28*(4), 491–493.

Li, S. Y., et al., (2018). CRISPR-Cas12a-assisted nucleic acid detection. *Cell Discovery, 4*(1), 1–4.

Liang, M., et al., (2019). A CRISPR-Cas12a-derived biosensing platform for the highly sensitive detection of diverse small molecules. *Nature Communications, 10*(1), 1–9.

Makarova, K. S., et al., (2011). Evolution and classification of the CRISPR–Cas systems. *Nature Reviews Microbiology, 9*(6), 467–477.

Makarova, K. S., et al., (2020). Evolutionary classification of CRISPR–Cas systems: A burst of class 2 and derived variants. *Nature Reviews Microbiology, 18*(2), 67–83.

Murugan, K., et al., (2017). The revolution continues: Newly discovered systems expand the CRISPR-Cas toolkit. *Molecular Cell, 68*(1), 15–25.

O'Connell, M. R., (2019). Molecular mechanisms of RNA targeting by Cas13-containing type VI CRISPR–Cas systems. *Journal of Molecular Biology, 431*(1), 66–87.

O'Connell, M. R., et al., (2014). Programmable RNA recognition and cleavage by CRISPR/Cas9. *Nature, 516*(7530), 263–266.

Pardee, K., et al., (2016). Rapid, low-cost detection of Zika virus using programmable biomolecular components. *Cell, 165*(5), 1255–1266.

Patchsung, M., et al., (2020). Clinical validation of a Cas13-based assay for the detection of SARS-CoV-2 RNA. *Nature Biomedical Engineering 4*(12), 1140–1149.

Ramachandran, A., & Santiago, J. G., (2021). CRISPR enzyme kinetics for molecular diagnostics. *Analytical Chemistry, 93*(20), 7456–7464.

Rusk, N., (2019). Spotlight on Cas12. *Nature Methods, 16*(3), 215–215.

Stella, S., et al., (2017). Structure of the Cpf1 endonuclease R-loop complex after target DNA cleavage. *Nature, 546*(7659), 559–563.

Swarts, D. C., et al., (2017). Structural basis for guide RNA processing and seed-dependent DNA targeting by CRISPR-Cas12a. *Molecular Cell, 66*(2), 221–233. e224.

Teng, F., et al., (2019). CDetection: CRISPR-Cas12b-based DNA detection with sub-attomolar sensitivity and single-base specificity. *Genome Biology, 20*(1), 1–7.

Vriend, L. E., et al., (2014). Assaying break and nick-induced homologous recombination in mammalian cells using the DR-GFP reporter and Cas9 nucleases. *Methods in Enzymology, 546*, 175–191. Elsevier.

Wang, T., et al., (2019). An RNA-guided Cas9 nickase-based method for universal isothermal DNA amplification. *Angewandte Chemie, 131*(16), 5436–5440.

Wu, X., et al., (2014). Genome-wide binding of the CRISPR endonuclease Cas9 in mammalian cells. *Nature Biotechnology, 32*(7), 670–676.

Xu, Y., &. Li, Z., (2020). CRISPR-Cas systems: Overview, innovations and applications in human disease research and gene therapy. *Computational and Structural Biotechnology Journal, 18*, 2401–2415.

Zhang, X. H., et al., (2015). Off-target effects in CRISPR/Cas9-mediated genome engineering. *Molecular Therapy-Nucleic Acids, 4*, e264.

Zhou, W., et al., (2018). A CRISPR–Cas9-triggered strand displacement amplification method for ultrasensitive DNA detection. *Nature Communications, 9*(1), 1–11.

CHAPTER 3

Engineered Gut Microbiome in Treating Diseases and Its Applications in Modern Biotechnology

SUPRIYA RAY and SURESH B. ARAKERA

Department of Applied Genetics, Karnatak University, Dharwad, Karnataka, India

ABSTRACT

Microbiomes are composed of trillions of diverse microbial communities that evolved with us and existed in all ecosystems. Any hindrance in the balance of the microbial communities leads to chronic diseases and creates an imbalance in the ecosystem. Gut microbes play a crucial role in health and disease and in maintaining the ecosystem. Manipulation of the commensal bacterial metabolites can be applied in treating diseases such as IBD, cancer, diabetes, metabolic diseases, and many more. Microbiome engineering thus holds a promising future in increasing agricultural sustainability and balance in the ecology. This chapter mainly focuses on the engineered bacterial strain to produce metabolites which then incorporate into the host microbiota for therapeutic treatment, and the current applications of microbiome engineering, mainly in humans, plants, animals, and soil reviewed in improving human health and agricultural productivity.

Modern Biotechnology in Healthcare: Advances and Applications. Sheikh Umar Ahmad (Ed.)

3.1 INTRODUCTION

Adaptation to microorganisms to form ecological communities (microbiomes) on earth occurred over millions of years of co-evolution before any existence of humans, which shows that humans have learned and co-evolved with them (Grice & Segre, 2012; de Vos & de Vos, 2012). These microbial communities exist in living hosts such as humans, plants, animals, oceans, soil, and air. They interact with their hosts and form synergistic relationships. These interactions in human determine the composition of the gut microbiome, which is associated with the physiology and physiological health of the host (Grice & Segre, 2012), and the microbiome composition of the soil influence the productivity of crops and ecosystem (Chaparro et al., 2012).

With the increase in recent advances in metagenomic analyzes, it has provided us with many information about the composition of the gut microbiota and their differences in healthy and diseased individuals. More than 25 diseases are linked to the alteration of the intestinal microbiota (de Vos & de Vos, 2012). The most studied diseases that are linked to the altered gut microbiota are metabolic syndrome, obesity, type II diabetes, inflammatory bowel diseases (IBDs), Crohn's disease, colorectal cancer, Alzheimer's disease (AD), autoimmune disease, and Parkinson's disease (Backhed et al., 2004). Studies showed that in few cases metabolites released from the gut microbiota directly affect the host (Feng et al., 2018; Bilotta & Cong, 2019). Such studies suggest manipulation of the gut microbial metabolites in improving host health and treating diseases. Approaches for alteration of gut microbial metabolites include use of medication, manipulation of diet, manipulation of gut microbiota by probiotics (Mukherjee et al., 2018; Choi & Cho, 2016). This chapter focuses on the genetically engineered specific bacterial species to produce beneficial metabolites which then incorporate in the host microbiota of target clinical condition and the current applications of microbiome engineering in improving human health and agricultural productivity.

Engineered bacteria offer more therapeutic benefit than localized bacteria because they can be manipulated to have additional features and also to produce higher yields of bacterial products (Pinero-Lambea et al., 2015). Till now, engineered products produced by therapeutic bacteria include interleukins (ILs), enzymes, vaccine antigens, and antibodies. Bacteria chosen for modification must be localized to the appropriate target region since the composition of the gut microbiota varies longitudinally and latitudinally (Sekirov et al., 2010). The commonly used genetically engineered bacterial strains are *Escherichia coli* and lactic acid bacteria as it can culture in large-scale, easy genetic manipulation and has low toxicity (Behnsen et al., 2013).

3.2 ENGINEERED LACTIC ACID BACTERIA IN INFLAMMATORY BOWEL DISEASE (IBD)

Lactic acid bacteria are gram-positive, non-pathogenic, non-spore forming, facultative anaerobes that are used in food fermentation. Studies showed that lactic acid bacteria like *Lactococcus lactis* is not a part of the human gut flora but can survive in the human gut environment (Klijn et al., 1995). In IBD patients, it has been observed that there is an overall decrease in microbial diversity and stability in the intestinal microbiota (Scaldaferri et al., 2013). The current therapies to treat IBD are based on anti-inflammatory drugs integrated with immunosuppressives (Kuhbacher & Folsch, 2007). Recombinant lactic acid bacteria such as *L. lactis* strain was used to treat and prevent colitis by producing and delivering an anti-inflammatory cytokine IL-10 in different mouse models *in situ* (Steidler et al., 2000). Colitis in mice induced by administration of dextran sodium sulfate (DSS) showed a decrease in 50% after daily mucosal administration of recombinant *L. lactis* secreting IL-10 (Dumot et al., 2001).

Lactic acid Bacteria producing trefoil factors (TFFs) to treat IBD has also been studied. TFFs are non-mitogenic peptides which plays an important role in repairing and protection of intestinal mucosa after damage (Playford et al., 1995), which is why it has become an interesting molecule to treat IBD. These molecules when administered by an oral route, adheres to the gut and then absorbed at the intestinal level. Interestingly, when recombinant *L. lactis* secreting TFF were administered intragastric, it expressed active peptides in the colon, which prevents and repairs the damage caused by acute colitis induced by DSS (Vandenbroucke et al., 2004). Another study showed that use of recombinant *L. lactis* secretes LcrV antigen, an inflammatory protein produced by *Yersinia pseudotuberculosis* is also used to treat colitis. Two colitis murine models, i.e., DSS and trinitrobenzene sulfonic acid was used to evaluate the therapeutic and protective potential of this strain (Foligne et al., 2007).

Another strategy to treat IBD is oral administration of antioxidant enzymes. Previous studies have shown that the inpouring of macrophages and neutrophils with the production of inflammatory mediators like cytokines, proteases, and reactive oxygen species (ROS) leads to inflammation in the gastrointestinal tract (Segui et al., 2005). ROS includes hydrogen peroxide (H_2O_2), superoxide radical (O_2^-), hydroxyl radical (·OH). Reactivity of these ROS towards DNA, proteins, and lipids causes mutagenic cellular damage and cytotoxicity (Grisham et al., 1990). To detoxify ROS, antioxidant enzymes like catalases (CAT) and superoxide dismutase (SOD) degrade H_2O_2 and O_2, respectively—thus preventing the formation of ·OH (Wardman, 2007). Many studies with this context have shown that use of recombinant strain of *Lactobacillus* spp. expressing either CAT or SOD can reduce GIT inflammation in mouse models (Rochat et al., 2007; Watterlot et al., 2010). A natural protease inhibitor elafin which is expressed in healthy intestinal mucosa is diminished in IBD patients, therefore, use of lactic acid bacteria secreting elafin in acute and chronic colitis showed protection against inflamed epithelium from increased intestinal permeability and from release of chemokines and cytokines (Motta et al., 2012).

It has been well established that recombinant lactic acid bacteria producing *L. lactis* strain is widely used in the production of heterologous protein and is considered as live delivery vector model and lactic acid bacteria model (Bermudez-Humaran, 2009). However, this bacterium has a shorter life span of about 24 hours in the human GIT thus leading to a reduced time of action. A study carried out by Hamady et al. chose *Bacteroides ovatus* bacterium for *in vivo* delivery of proteins due to its ability to utilize properties of xylan and colonize the colon (Hamady, 2013; Hamady et al., 2008). They developed a xylan-regulated delivery of human tumor growth factor TGF-b_1 in mice to treat colitis (Hamady et al., 2011) and human keratinocyte growth factor-2 to treat inflamed colon (Hamady et al., 2010). These promising developments confirm the potentiality for *in vivo* delivery of recombinant commensal.

3.3 ENGINEERED *E. COLI* IN OBESITY, DIABETES, CARDIOMETABOLIC DISEASE, AND CANCER

E. coli are gram-negative facultative anaerobe reside in the normal gut of mammals. Several *E. coli* are pathogenic which causes infectious diseases, but the majority belongs to commensals (Huang et al., 2012). Some common lab strains of *E. coli* used for engineering are BL21DE3 and MG1655 but *E. coli* Nissle 1917 (EcN), a probiotic strain that highly colonize, was used previously by Germany for treating intestinal infections, chronic constipation and ulcerative colitis (Behnsen et al., 2013). *E. coli* has been used as a model organism for several synthetic biology studies because of its rapid growth, high yield production, ease of genetic manipulation, biochemistry, and well-characterized genetics (Huang et al., 2012).

A recent study revealed that incorporation of EcN in gut microbiota can inhibit the development of obesity and can treat related metabolic disorders. N-acylphosphatidylethanolamines (NAPE) and N-acylethanolamines (NAE) produce bioactive lipids which regulates food intake but production of these is downregulated in fat diet. NAPE is a biosynthetic precursor of NAE and several studies are focusing on increasing its bacterial production (Gillum et al., 2008). Reduced inflammation, reduced pain and increased satiety are some of the beneficial effects that shown on increasing NAEs (Piomelli, 2013). In a study, C57BL/6J mice were maintained on a high-fat diet and administered with NAPE expressing EcN in the drinking water, showed an inhibition of the development of obesity. Due to colonization in the intestine, this anti-obesity effect persisted for 4 weeks after ending the treatment (Chen et al., 2014).

Engineered EcN have also been developed for diabetes treatment. Engineered EcN has revealed that it produces insulin topic protein GLP-1 or pancreatic and duodenal homeobox gene 1 (PDX-1) (Duan et al., 2008) which stimulate the intestinal epithelial cells to synthesize insulin. *In vitro* studies of GLP-1 (1–37) secreting bacteria are recently exploring whether by oral administration of these bacteria in rat models can enhance hyperglycemia in diabetes (Duan et al., 2015).

Increased risk of cardiometabolic diseases which includes diabetes, non-alcoholic fatty liver and cardiovascular diseases are associated with obesity. NAPE expressing EcN significantly reduced early signs of fibrosis, serum cholesterol, body weight, liver fat as well as inflammation, and some extent of necrosis in a mouse model of fatty liver disease and atherosclerosis (May-Zhang et al., 2019).

The use of engineered *E. coli* in treating cancer is similar to *Salmonella* as it efficiently targets the necrotic region of the solid tumors (Kocijancic, 2016; Weibel et al., 2008). Engineered *E. coli* expressing cytosine deaminase (CD) ameliorate conversion of 5-Fluorocytosine (5-FC) to cytotoxic 5-fluorouracil (5-FU) at the site of tumor (Lehouritis et al., 2013). For the treatment of tumors, several novel deliveries in *E. coli* have been engineered. CTNNB1 (catenin β-1), a gene mutated or over-expressed in colorectal cancer can be suppressed by engineered *E. coli* BL21DE3 through production of short-hairpin RNAs (shRNAs) against CTNNB1 (catenin β-1). Xiang et al. when administered these bacteria to nude mice xenografted with human colon cancer cells, showed significant suppression of CTNNB1 expression in those cancer cells (Xiang et al., 2006). Cytolysin A (ClyA) is a hemolytic bacterial protein that forms pore and has the ability to kill mammalian cells is expressed by engineered *E. coli* MG1655. Single-dose administration of ClyA slows down the rate of tumor growth in BALB/c mice implanted with CT26 colon cancer cells subcutaneously. Treatment with ClyA suppresses the lung metastases thus extending survival in mice injected with CT26 cancer cells intravenously (Jiang et al., 2010).

3.4 ENGINEERED *BIFIDOBACTERIUM* SPP., *BACILLUS* SPP., AND *CLOSTRIDIUM* SPP. IN VARIOUS DISEASES

Bifidobacterium spp. are non-pathogenic, non-spore forming gram-positive anaerobes present in mammals in the lower gastrointestinal tract. *Bifidobacterium* spp. provide many benefits such as preventing carcinogenesis and protection against viral infections (Hidaka et al., 2007). Engineered *Bifidobacterium* are used as a therapeutic for treating IBD. *Bifidobacterium longum* was engineered to produce α-melanocyte-stimulating hormone that has anti-inflammatory properties. This engineered bacterium inhibits DSS-induced ulcerative colitis by colonizing the intestine in mice (Wei et al., 2016). For the treatment of type 2 diabetes (T2D), Wei et al. engineered *Bifidobacterium longum* to express bioactive penetratin – GLP-1 fusion protein (Wei et al., 2015). Another therapeutic strategy with the concept of delivering engineered *B. longum* secreting human manganese superoxide dismutase (rhMnSOD) through transporter peptide PEP-1 penetratin fusion protein in the colon has also been studied (Liu et al., 2018). Oral administration of *Bifidobacterium* is more stable in the intestine and increases its efficiency and retention time of target drugs. *Bifidobacterium* are used as a model organism for treating cancers because similar to *E. coli* and *Salmonella, Bifidobacterium* grow in the hypoxic region of tumors. Engineered *Bifidobacterium* strains delivers cytosine deaminase enzyme to hypoxic region of solid tumors that inhibit tumor growth. Engineered *B. adolescentis* secreting endostatin inhibits angiogenesis therefore suppressing the growth of Heps liver cancer cells in induced BALB/c mice (Li et al., 2003).

Bacillus subtilis are gram-positive endospore forming aerobic bacterium (Buescher et al., 2012; Koo et al., 2017). It is found in the gastrointestinal tract of humans and in soil. *Bacillus subtilis* used as a probiotic for both humans and animals due to its GRAS status (Hong et al., 2005). *B. subtilis* production and secretion ability is high (van Dijl & Hecker, 2013) which makes it suitable to use it as a model organism for various genetics and metabolism studies (Buescher et al., 2012; Koo et al., 2017). *B. subtilis* produces vitamins, enzymes, sugars, and engineered *B. subtilis* developed biofilms known as smart living glues (Zhang et al., 2019) Killed *B. subtilis* spores express streptavidin which enable the targeting colon cancer cells (Nguyen et al., 2013). In 2017, *B. subtilis* with 168 strains were constructed with a reduction of genome by 36% (Reuß et al., 2017). In genetically engineered *B. subtilis*, insertion of two chimeric genes in two thymidylate synthase genes, i.e., *thy*A and *thy*B makes the spores more dependent on thymine or thymidine, thus unable to survive in thymine or thymidine deficient environment (Hosseini et al., 2018; Duan et al., 2015).

Clostridium spp. are gram-positive anaerobic bacterium, and it can grow in necrotic or hypoxic regions of solid tumors (Lehouritis et al., 2013). *C. histolyticum* was the first strain to induce lysis and tumor regression in mice. *Clostridia* are injected as spores since it is a spore forming bacteria that travel to the site of the tumor and proliferate only in anoxic areas (Parker et al., 1947). *Clostridium* are pathogenic strains that lead to intestinal infections (Lehouritis et al., 2013). One example of pathogenic strain is *C. tetani* that has found to grow in and shrunk the tumors which results in rapid death of those animals due to elevated toxicity (Malmgren & Flanigan, 1955). Therefore, it is clear that for safe use of treatments, *Clostridium* strains should be non-pathogenic. One such strain is *Clostridium novyi* which was genetically engineered to become non-pathogenic by removing the gene encoding for lethal α-toxin NT. This engineered strain then gets its ability to destroy tumors (Dang et al., 2001) and it also secretes liposomase which inflate the delivery of liposome encapsulated drugs within the tumors (Cheong et al., 2006).

Radio-induced promoters control the expression of gene temporally and spatially, therefore fusion of this promoter improves the gene specificity targeting to hypoxic tumors. Administration of cytosine deaminase expressing *C. sporogenes* incorporated with radio-induced promoters increases the specificity and anti-tumor response of rats induced with rhabdomyosarcoma (Nuyts et al., 2001). Cytosine deaminase expressing *Clostridium sporogenesis* delivers the enzyme to tumor cells in mice which results in anti-tumor effect through intravenous administration of 5-FC (Liu et al., 2002). Engineered *Clostridia* induced with radiation-inducible recA promoter secretes TNFα that directly gets delivered in tumor, thus amplifying the therapeutic use for cancer treatment (Nuyts et al., 2001). Therefore, incorporation of radiotherapy with *Clostridium* expressing protein delivery proves to be a new possibility for cancer therapy.

3.5 APPLICATIONS OF MICROBIOME ENGINEERING

With the advances in meta-omics tools, certain culture independent analysis of microbiome has been made possible to do, which showed a clear picture of how the composition of the microbiome can impact the ecosystem (Chaparro et al., 2012; Mueller & Sachs, 2015). As it becomes more clearer, there is an increasing demand in engineering the microbiomes for shaping the microbiota to change the ecosystem. Microbial communities as well as microorganisms and host within themselves have a complex network of interconnection between them which portray the idiosyncratic characteristics of each ecosystem (Waldor et al., 2015). Any perturbation to the balanced microbiomes leads to dyshomeostasis, which then cause detrimental effect to both host fitness, productivity, and soil fertility (Round & Mazmanian, 2009; Navarrete et al., 2015). Therefore, microbial compositions can be manipulated by microbiome engineering to improve health and ecosystem.

3.5.1 HUMAN MICROBIOME ENGINEERING

Microbiome engineering is mostly applied to human microbiome due to the possibilities of manipulating the gut microbiota for the treatment of various diseases (Grice & Segre, 2012). The largest and most diverse population of microbes inhabit in the gastrointestinal tract of human

(Zhao, 2013). The introduction of human microbiome project (HMP) led to characterize and identify the particular microbiota in human body such as skin, urogenital tract, oral, gut, nasal, and can differentiate the microbiota with healthy and diseased ones (Group et al., 2009). It is well established that any disturbance to the microbiome can lead to gastrointestinal diseases like IBD, ulcerative colitis, *Clostridium difficile* infection (CDI), also diabetes, metabolic diseases, obesity, and many more (Gough et al., 2011).

Fecal microbiota transplantation is a method of collecting fecal material from healthy donors and transplanting it to the gastrointestinal tract of the patients suffering from any disease (Jung Lee et al., 2015). Studies have shown that FMT holds a promising therapy in treating CDI. Previously, CDI was treated traditionally using antibiotics vancomycin and metronidazole to remove *Clostridium difficile,* but more use of antibiotic makes the pathogens resistance to antibiotic which leads to washing out of the beneficial bacterial population from the gut, thus causing the patients to get ill again. Through FMT, it restores beneficial bacteria like *Bacteroidetes* and *Firmicutes*, thus reviving a healthy microbiome (Russell et al., 2014). FMT has also shown its potential in treating IBD, diabetes, multiple sclerosis, rheumatoid arthritis, and autism (Rossen et al., 2015).

Other strategies besides FMT are the antimicrobial peptides (AMPs), i.e., thuricin CD (Rea et al., 2011), pyocin S5 (Saeidi et al., 2011) which by targeting the pathogens can alter the gut microbiota explored in animal model studies. Another strategy is using of prebiotics signaling inhibitor LED209 which prevents autophosphorylation of membrane which in turn activate the virulent factors that eliminate pathogenic strains (Rasko et al., 2008). Other prebiotics, for example, galactooligosaccharides increase beneficial *Bifidobacterium* and decrease pathogenic *Clostridium histolyticum,* which improves health (Costabile et al., 2016). All these studies show the potentiality of microbiome engineering in therapeutic applications performed in mice models, but to understand the efficacy of these strategies, clinical studies need to be performed in humans. Treatment of skin diseases such as psoriasis, atopic dermatitis is treated using probiotic *Lactobacillus* which enhance the growth of *Staphylococcus epidermidis* and hinder the growth of pathogenic bacteria in skin (Grice, 2014). Using AMP C16G2, oral microbiome has been engineered to eliminate *Streptococcus mutans* which causes tooth decay (Guo et al., 2015).

3.5.2 ANIMAL MICROBIOME ENGINEERING

Animals, like humans, also inhabit diverse and large population of microbial communities. Rats and mice were extensively studied but as an animal model for engineering human microbiome. More studies should focus on understanding the animal microbiome with respect to animals such as cows, rats, mice, broiler for improving the health and agricultural productivity. Prebiotics, probiotics, and feed enzymes usage can alter the microbial composition of livestock animals.

Studies shows that most dominant prebiotics such as fructooligosaccharides, xylooligosaccharides, inulin have been used to alter the microbial composition in animals. Administration of prebiotics leads to the production of: (i) antimicrobial factors; (ii) short chain fatty acids (SCFAs) that provides energy, regulate immune system and metabolism; (iii) prebiotic derivatives (Pourabedin & Zhao, 2015). Prebiotics benefits have also been noticed in broiler chicken (De Maesschalck et al., 2015). Delivery of prebiotic inulin in swine gut has shown a rise in probiotic and decline of pathogenic microorganisms (Samanta et al., 2015).

For engineering animal microbiomes, the administration of probiotic has proven to be efficacious. Administration of probiotic has been used to treat infectious diseases and to increase the growth performance as seen in *Bacillus subtilis* CH16. On administration of this probiotic in broiler chicken, gaining weight and reduction in food conversion rate have been observed. The most used dominant probiotic strains *Bacillus*, *Saccharomyces boulardii*, *Lactobacillus*, *Enterococcus*, and *Bifidobacteria* (Nguyen et al., 2015). Amalgamate of probiotics with prebiotics has been shown to increase the probiotic bacterial strain in the gastrointestinal tract (Tanner et al., 2015).

Feed enzymes improve gut health in swine by increasing prebiotics production from non-starch polysaccharide (NSP)-degrading enzymes and enhancing digestion of substrate. Feed enzymes include NSP-degrading enzymes, amylase, lysozyme, phytase, and proteases. So far feed enzymes have shown positive effects in inhibiting pathogens, reducing infectious diseases like salmonellosis, dysentery in piglets and in aiding animal growth (Kiarie et al., 2013).

3.5.3 SOIL MICROBIOME ENGINEERING

Soil microbiome engineering has led to create more microbial and diverse communities which improves soil health and plant fertility.

Soil microbiome is attained by implementing agricultural practices like organic farming (Chaparro et al., 2012). By practicing organic farming, it magnifies biological cycles, microbial diversity of soil, and biological activity with the slightest usage of pesticides and herbicides. Long-term use of organic fertilizers improves soil health by changing the structure of soil microbiome, reduces scattering and increases richness of soil (Hartmann et al., 2015). Other strategies for soil microbiome engineering, which can alter the structure and function of soil microbiome is to change of land utilization that has been observed in tropical forests in South-East Asia where the forests were converted into oil palms plantation (Tripathi et al., 2016). Besides, change of vegetation, other agricultural practices include cropping systems (Xiong et al., 2015), logging (Hartmann et al., 2014) and tillage (Souza et al., 2016) have shown to be effective in shifting the structure and function of soil microbiome that led to alter the microbial diversity and functions of soil. For decades, agricultural practices have been applied to improve crops.

3.5.4 PLANT MICROBIOME ENGINEERING

Microbial communities of plant are diverse, including fungal and bacterial taxa which reside in the root and rhizosphere hence defines as root associated microbiome. The structures of communities get affected due to microbe-microbe interaction, abiotic environment, and host genotypes (Agler et al., 2016). Therefore, by engineering the root associated microbiome it can alter the microbial composition of the plant and can improve the plant fertility and growth. It has been noted that, for engineering plant microbiome, synthetic microbiomes, microbiome transfer have been engaged. A recent approach has been adapted for plant disease management is to merge disease conducive soils and disease suppressive soils. This strategy has been applied to inhibit tobacco black root rot, potato common swab and sugar beet infection (Gopal et al., 2013). Another approach is the use of a plant model (*Arabidopsis thaliana*) for transferring microbiome from soil to plant to increase biomass and avoid drought condition (Zolla et al., 2013).

Soil that manufactures phenotypes in previous generation's rhizosphere microbiomes is engineered by generating duplication of *Arabidopsis thaliana* in the soil microbiome to alter the biomass and detoxification of the toxic compounds. Hence, usage of this host mediated artificial selection can alter ecology and develop plant microbiomes (Swenson et al., 2000; Edwards et al., 2015). Chemical molecules such as salicylic acid from *Arabidopsis thaliana* plant which is the regulator of plant immune system increases the microbial composition in root. Hence, this approach can be taken into consideration for improving microbial diversity, its productivity and increasing resistance to environmental stress (Lebeis et al., 2015). *Arabidopsis thaliana* is used extensively as a plant model for studying plant microbiome, this studies can then be extrapolated to study other plants. Extensive crop microbiome engineering studies are required for improving productivity, sustainability, drought, and disease resistance.

3.6 CONCLUSION AND FUTURE DIRECTIONS

It is not too distant in our future, when we will be able to manipulate our gut with genetically engineered bacteria that will detect and eliminate chronic diseases at its earlier stage. Intestinal microbiota is an integral part of the human body, and with its interaction, it can lead to long lasting hereditary programming. Engineered bacteria expressing therapeutic compounds which when administered, colonize the gut, shows a remarkable potentiality in treating various chronic diseases such as cancer, gastrointestinal diseases, metabolic diseases, and many more. The futile hunt for diseases like Crohn's disease has set off a new theory that all these lifestyle diseases may have been caused by the intestinal microbiota. Therefore, the approach of modifying or engineering the gut by administration of probiotics, prebiotics or FMT can prevent and treat chronic diseases.

Expression of various proteins by engineered bacteria seems to be a new tide of medicine. Despite all the success in animal study experiments, there is a lack of clinical trials in human which will prove the efficacy of therapeutic prospects of engineered bacteria. The evident roadblock for conducting human trials is the unsettlement of whether the pre-emptive measures are enough to ensure that the individual consented to treatment are exposed to the engineered bacteria. Therefore, to obtain the effectiveness of the therapeutic bacteria, public confidence and support, more rigorous and well controlled studies are needed urgently. Another roadblock for both conducting of human trials and acceptance of therapeutic engineered bacteria is the absence of antibiotic selection for production of large stocks of the recombinant therapeutic protein. Large scale fermentation approach without antibiotic selection for engineered bacteria needs to be developed fully, since large-scale fermentation strategy is well-rooted for unaltered probiotic strains.

Microbiome engineering in the field of synthetic biology holds great potential in improving agricultural productivity and sustainability of the ecosystem because of the advances in high throughput multi omics tools. With the help of synthetic biology smart microbes that have beneficial effects can be engineered. This strategy can be applied to eliminate the pathogenic strains from the microbial community for microbiome engineering. Despite development of omics tools in determining microbial composition and attaining knowledge of the microbiomes, kinetics of the microbiome as well as how it interacts to regulate the composition of microbiome still needs to be investigated. More research needs to be focused on other microbiomes located at other places of a body, such as oral cavity, lungs, urogenital tract, and skin because peripheral effects of microbiome far off from the site need to expand to understand the relevancy of therapeutic microbiome engineering. Advance studies of microbiomes for clinical trial are obligatory to understand the potential of microbiome engineering.

KEYWORDS

- **dextran sodium sulfate**
- **inflammatory bowel disease**
- **microbiome engineering**
- **microbiota**
- **N-acylethanolamines**
- **reactive oxygen species**
- **superoxide dismutase**

REFERENCES

Agler, M. T., Ruhe, J., Kroll, S., Morhenn, C., Kim, S. T., Weigel, D., & Kemen, E. M., (2016). Microbial hub taxa link host and abiotic factors to plant microbiome variation. *PLoS Biology, 14*(1), e1002352.

Backhed, F., Ding, H., Wang, T., Hooper, L. V., Koh, G. Y., Nagy, A., Semenkovich, C. F., & Gordon, J. I., (2004). The gut microbiota as an environmental factor that regulates fat storage. *PNAS, 101*(44), 15718–15723.

Behnsen, J., Deriu, E., Sassone-corsi, M., & Raffatellu, M., (2013). Probiotics: Properties, examples, and specific applications. *Cold Spring Harbor Perspectives in Medicine, 3*(3), 1–14.

Bermudez-Humaran, L. G., (2009). *Lactococcus lactis* as a live vector for mucosal delivery of therapeutic proteins. *Human Vaccines, 5*(4), 264–267.

Bilotta, A. J., & Cong, Y., (2019). Gut microbiota metabolite regulation of host defenses at mucosal surfaces: Implication in precision medicine. *Precision Clinical Medicine, 2*(2), 110–119.

Buescher, J. M., Liebermeister, W., Jules, M., Uhr, M., Muntel, J., Botella, E., Hessling, B., et al., (2012). Global network reorganization during dynamic adaptations of *Bacillus subtilis* metabolism. *Science, 335*(6072), 1099–1103.

Chaparro, J. M., Sheflin, A. M., Manter, D. K., & Vivanco, J. M., (2012). Manipulating the soil microbiome to increase soil health and plant fertility. *Biology and Fertility of Soils, 48*, 489–499.

Chen, Z., Guo, L., Zhang, Y., Walzem, R. L., Pendergast, J. S., Printz, R. L., Morris, L. C., et al., (2014). Incorporation of therapeutically modified bacteria into gut microbiota inhibits obesity. *The Journal of Clinical Investigation, 124*(8), 3391–3406.

Cheong, I., Huang, X., Bettegowda, C., Diaz, L. A., Kinzler, K. W., Zhou, S., & Vogelstein, B., (2006). A bacterial protein enhances the release and efficacy of liposomal cancer drugs. *Science, 314*(5803), 1308–1311.

Choi, H. H., & Cho, Y. S., (2016). Fecal microbiota transplantation: Current applications, effectiveness, and future perspectives. *Clinical Endoscopy, 49*(3), 257–265.

Costabile, A., Deaville, E. R., Morales, A. M., & Gibson, G. R., (2016). Prebiotic potential of a maize-based soluble fibre and impact of dose on the human gut microbiota. *PLoS One, 11*(1), e0144457.

Dang, L. H., Bettegowda, C., Huso, D. L., Kinzler, K. W., & Vogelstein, B., (2001). Combination bacteriolytic therapy for the treatment of experimental tumors. *PNAS., 98*(26), 15155–15160.

De Maesschalck, C., Eeckhaut, V., Maertens, L., De Lange, L., Marchal, L., Nezer, C., De Baere, S., et al., (2015). Effects of xylo-oligosaccharides on broiler chicken performance and microbiota. *Applied and Environmental Microbiology, 81*(17), 5880–5888.

De Vos, W. M., de Vos, E. A., (2012). Role of the intestinal microbiome in health and disease: From correlation to causation. *Nutrition Reviews, 70*(Suppl 1), S45–56.

Duan, F. F., Liu, J. H., & March, J. C., (2015). Engineered commensal bacteria reprogram intestinal cells into cells for the treatment of diabetes. *Diabetes, 64*(5), 1794–1803.

Duan, F., Curtis, K. L., & March, J. C., (2008). Secretion of insulinotropic proteins by commensal bacteria: Rewiring the gut to treat diabetes. *Applied and Environmental Microbiology, 74*(23), 7437, 7438.

Dumot, J. A., Conwell, D. L., Zuccaro, G., Vargo, J. J., Shay, S. S., Easley, K. A., & Ponsky, J. L., (2001). A randomized, double blind study of interleukin 10 for the prevention of ERCP-induced pancreatitis. *The American Journal of Gastroenterology, 96*, 2098–2102.

Edwards, J., Johnson, C., Santos-Medellin, C., Lurie, E., Podishetty, N. K., Bhatnagar, S., Eisen, J. A., & Sundaresan, V., (2015). Structure, variation, and assembly of the root-associated microbiomes of rice. *PNAS, 112*(8), E911–E920.

Feng, W., Ao, H., & Peng, C., (2018). Gut microbiota, short-chain fatty acids, and herbal medicines. *Frontiers in Pharmacology, 9*, 1354–1366.

Foligne, B., Dessein, R., Marceau, M., Poiret, S., Chamaillard, M., Pot, B., Simonet, M., & Daniel, C., (2007). Prevention and treatment of colitis with *Lactococcus lactis* secreting the immunomodulatory *Yersinia* LcrV protein. *Gastroenterology, 133*(3), 862–874.

Gillum, M. P., Zhang, D., Zhang, X. M., Erion, D. M., Jamison, R. A., Choi, C., Dong, J., et al., (2008). N-acylphosphatidylethanolamine, a gut-derived circulating factor induced by fat ingestion, inhibits food intake. *Cell, 135*, 813–824.

Gopal, M., Gupta, A., & Thomas, G. V., (2013). Bespoke microbiome therapy to manage plant diseases. *Frontiers in Microbiology, 4*, 355.

Gough, E., Shaikh, H., & Manges, A. R., (2011). Systematic review of intestinal microbiota transplantation (fecal bacteriotherapy) for recurrent clostridium difficile infection. *Clinical Infectious Diseases, 53*(10), 994–1002.

Grice, E. A., & Segre, J. A., (2012). The human microbiome: Our second genome. *Annual Review of Genomics Human Genetics, 13*, 151–170.

Grice, E. A., (2014). The skin microbiome: Potential for novel diagnostic and therapeutic approaches to cutaneous disease. *Seminars in Cutaneous Medicine and Surgery, 33*(2), 98–103.

Grisham, M. B., Gaginella, T. S., Von, R. C., Tamai, H., Be, R. M., & Granger, D. N., (1990). Effects of neutrophil-derived oxidants on intestinal permeability, electrolyte transport, and epithelial cell viability. *Inflammation, 14*(5), 531–542.

Group, N. H. W., Peterson, J., Garges, S., Giovanni, M., McInnes, P., Wang, L., Schloss, J. A., et al., (2009). The NIH human microbiome project. *Genome Research, 19*(12), 2317–2323.

Guo, L., McLean, J. S., Yang, Y., Eckert, R., Kaplan, C. W., Kyme, P., Sheikh, O., et al., (2015). Precision-guided antimicrobial peptide as a targeted modulator of human microbial ecology. *PNAS, 112*(24), 7569–7574.

Hamady, Z. Z., (2013). Novel xylan-controlled delivery of therapeutic proteins to inflamed colon by the human anaerobic commensal bacterium. *Annals of The Royal College of Surgeons of England, 95*(4), 235–240.

Hamady, Z. Z., Farrar, M. D., Whitehead, T. R., Holland, K. T., Lodge, J. P., & Carding, S. R., (2008). Identification and use of the putative *Bacteroides ovatus* xylanase promoter for the inducible production of recombinant human proteins. *Microbiology, 154*(Pt 10), 3165–3174.

Hamady, Z. Z., Scott, N., Farrar, M. D., Lodge, J. P., Holland, K. T., Whitehead, T., & Carding, S. R., (2010). Xylan-regulated delivery of human keratinocyte growth factor-2

to the inflamed colon by the human anaerobic commensal bacterium *Bacteroides ovatus. Gut, 59*(4), 461–469.
Hamady, Z. Z., Scott, N., Farrar, M. D., Wadhwa, M., Dilger, P., Whitehead, T. R., Thorpe, R., et al., (2011). Treatment of colitis with a commensal gut bacterium engineered to secrete human TGF-beta1 under the control of dietary xylan 1. *Inflammatory Bowel Diseases, 17*, 1925–1935.
Hartmann, M., Frey, B., Mayer, J., Mader, P., & Widmer, F., (2015). Distinct soil microbial diversity under long-term organic and conventional farming. *The ISME Journal, 9*(5), 1177–1194.
Hartmann, M., Niklaus, P. A., Zimmermann, S., Schmutz, S., Kremer, J., Abarenkov, K., Lüscher, P., Widmer, F., & Frey, B., (2014). Resistance and resilience of the forest soil microbiome to logging-associated compaction. *The ISME Journal, 8*(1), 226–244.
Hidaka, A., Hamaji, Y., Sasaki, T., Taniguchi, S., & Fujimori, M., (2007). Exogenous cytosine deaminase gene expression in Bifidobacterium breve I-53-8w for tumor-targeting enzyme/ prodrug therapy. *Bioscience, Biotechnology, and Biochemistry, 71*(12), 2921–2926.
Hong, H. A., Duc Le, H., & Cutting, S. M., (2005). The use of bacterial spore formers as probiotics. *FEMS Microbiology Reviews, 29*(4), 813–835.
Hosseini, S., Curilovs, A., & Cutting, S. M., (2018). Biological containment of genetically modified *Bacillus subtilis. Applied and Environmental Microbiology*, *84*(3), 1–15, e02334-17.
Huang, C. J., Lin, H., & Yang, X., (2012). Industrial production of recombinant therapeutics in *Escherichia coli* and its recent advancements. *Journal of Industrial Microbiology and Biotechnology, 39*(3), 383–399.
Jiang, S., Phan, T. X., Nam, T., Nguyen, V. H., Kim, H., Bom, H., Choy, H. E., et al., (2010). Inhibition of tumor growth and metastasis by a combination of *Escherichia coli*-mediated cytolytic therapy and radiotherapy. *Molecular Therapy, 18*(3), 635–642.
Jung, L. W., Lattimer, L. D., Stephen, S., Borum, M. L., & Doman, D. B., (2015). Fecal microbiota transplantation: A review of emerging indications beyond relapsing clostridium difficile toxin colitis. *Gastroenterology and Hepatology, 11*(1), 24–32.
Kiarie, E., Romero, L. F., & Nyachoti, C. M., (2013). The role of added feed enzymes in promoting gut health in swine and poultry. *Nutrition Research Reviews, 26*(1), 71–88.
Klijn, N., Weerkamp, A. H., & De Vos, W. M., (1995). Genetic marking of *Lactococcus lactis* shows its survival in the human gastrointestinal tract. *Applied and Environmental Microbiology, 61*(7), 2771–2774.
Kocijancic, D., (2016). Therapy of solid tumors using probiotic symbioflor-2-restraints and potential. *Oncotarget, 7*(16), 22605–22622.
Koo, B. M., Kritikos, G., Farelli, J. D., Todor, H., Tong, K., Kimsey, H., Wapinski, I., et al., (2017). Construction and analysis of two genome-scale deletion libraries for *Bacillus subtilis. Cell Systems, 4*(3), 291–305.
Kuhbacher, T., & Folsch, U. R., (2007). Practical guidelines for the treatment of inflammatory bowel disease. *World Journal of Gastroenterology, 13*(8), 1149–1155.
Lebeis, S. L., Paredes, S. H., Lundberg, D. S., Breakfield, N., Gehring, J., McDonald, M. S., et al., (2015). Salicylic acid modulates colonization of the root microbiome by specific bacterial taxa. *Science, 349*(6250), 860–864.

Lehouritis, P., Springer, C., & Tangney, M., (2013). Bacterial-directed enzyme prodrug therapy. *Journal of Controlled Release, 170*(1), 120–131.

Li, X., Fu, G. F., Fan, Y. R., Liu, W. H., Liu, X. J., Wang, J. J., & Xu, G. X., (2003). Bifidobacterium adolescentis as a delivery system of endostatin for cancer gene therapy: Selective inhibitor of angiogenesis and hypoxic tumor growth. *Cancer Gene Therapy, 10*(2), 105–111.

Liu, M., Li, S., Zhang, Q., Xu, Z., Wang, J., & Sun, H., (2018). Oral engineered *Bifidobacterium longum* expressing rhMnSOD to suppress experimental colitis. *International Immunopharmacology, 57*, 25–32.

Liu, S. C., Minton, N. P., Giaccia, A. J., & Brown, J. M., (2002). Anticancer efficacy of systemically delivered anaerobic bacteria as gene therapy vectors targeting tumor hypoxia/necrosis. *Gene Therapy, 9*(4), 291–296.

Malmgren, R. A., & Flanigan, C. C., (1955). Localization of the vegetative form of *Clostridium tetani* in mouse tumors following intravenous spore administration. *Cancer Research, 15*(7), 473–478.

May-Zhang, L. S., Chen, Z., Dosoky, N. S., Yancey, P. G., Boyd, K. L., Hasty, A. H., Linton, M. F., & Davies, S. S., (2019). Administration of N-acylphosphatidylethanolamine expressing bacteria to low density lipoprotein receptor (−/−) mice improves indices of cardiometabolic disease. *Scientific Reports, 9*(1), 420–433.

Motta, J. P., Bermudez-Humaran, L. G., Deraison, C., Martin, L., Rolland, C., Rousset, P., Boue, J., et al., (2012). Food-grade bacteria expressing elafin protect against inflammation and restore colon homeostasis. *Science Translational Medicine, 4*(158).

Mueller, U. G., & Sachs, J. L., (2015). Engineering microbiomes to improve plant and animal health. *Trends in Microbiology, 23*(10), 606–617.

Mukherjee, S., Joardar, N., Sengupta, S., & Babu, S. P. S., (2018). Gut microbes as future therapeutics in treating inflammatory and infectious diseases: Lessons from recent findings. *The Journal of Nutritional Biochemistry, 61*, 111–128.

Navarrete, A. A., Tsai, S. M., Mendes, L. W., Faust, K., De Hollander, M., Cassman, N. A., Raes, J., et al., (2015). Soil microbiome responses to the short-term effects of Amazonian deforestation. *Molecular Ecology, 24*(10), 2433–2448.

Nguyen, A. T., Nguyen, D. V., Tran, M. T., Nguyen, L. T., Nguyen, A. H., & Phan, T. N., (2015). Isolation and characterization of *Bacillus subtilis* CH16 strain from chicken gastrointestinal tracts for use as a feed supplement to promote weight gain in broilers. *Letters in Applied Microbiology, 60*(6), 580–588.

Nguyen, V. A., Huynh, H. A., Hoang, T. V., Ninh, N. T., Pham, A. T., Nguyen, H. A., Phan, T. N., & Cutting, S. M., (2013). Killed *Bacillus subtilis* spores expressing streptavidin: A novel carrier of drugs to target cancer cells. *Journal of Drug Targeting, 21*(6), 528–541.

Nuyts, S., Theys, J., Landuyt, W., vanMellaert, L., Lambin, P., & Anne, J., (2001). Increasing specificity of anti-tumor therapy: Cytotoxic protein delivery by non-pathogenic clostridia under regulation of radio-induced promoters. *Anticancer Research, 21*(2A), 857–861.

Nuyts, S., Van, M. L., Theys, J., Landuyt, W., Bosmans, E., Anne, J., & Lambin, P., (2001). Radio-responsive recA promoter significantly increases TNFalpha production in recombinant clostridia after 2 Gy irradiation. *Gene Therapy, 8*(15), 1197–1201.

Parker, R. C., Plummer, H. C., Siebenmann, C. O., & Chapman, M. G., (1947). Effect of histolyticus infection and toxin on transplantable mouse tumors. *Proceedings of the Society for Experimental Biology and Medicine, 66*(2), 461–467.

Pinero-Lambea, C., Ruano-Gallego, D., & Fernandez, L. A., (2015). Engineered bacteria as therapeutic agents. *Current Opinion in Biotechnology, 35*, 94–102.

Piomelli, D., (2013). A fatty gut feeling. *Trends in Endocrinology and Metabolism, 24*(7), 332–341.

Playford, R. J., Marchbank, T., Chinery, R., Evison, R., Pignatelli, M., Boulton, R. A., Thim, L., & Hanby, A. M., (1995). Human spasmolytic polypeptide is a cytoprotective agent that stimulates cell migration. *Gastroenterology, 108*, 108–116.

Pourabedin, M., & Zhao, X., (2015). Prebiotics and gut microbiota in chickens. *FEMS Microbiology. Letters, 362*(15), fnv122.

Rasko, D. A., Moreira, C. G., Li de, R., Reading, N. C., Ritchie, J. M., Waldor, M. K., Williams, N., et al., (2008). Targeting QseC signaling and virulence for antibiotic development. *Science, 321*(5892), 1078–1080.

Rea, M. C., Dobson, A., O'Sullivan, O., Crispie, F., Fouhy, F., Cotter, P. D., Shanahan, F., et al., (2011). Effect of broad- and narrow-spectrum antimicrobials on *Clostridium difficile* and microbial diversity in a model of the distal colon. *PNAS, 108*(Suppl. 1), 4639–4644.

Reuß, D. R., Altenbuchner, J., Mäder, U., Rath, H., Ischebeck, T., Sappa, P. K., Thürmer, A., et al., (2017). Large-scale reduction of the *Bacillus subtilis* genome: Consequences for the transcriptional network, resource allocation, and metabolism. *Genome Research, 27*(2), 289–299.

Rochat, T., Bermudez-Humaran, L., Gratadoux, J. J., Fourage, C., Hoebler, C., Corthier, G., & Langella, P., (2007). Anti-inflammatory effects of *Lactobacillus casei* BL23 producing or not a manganese-dependant catalase on DSS-induced colitis in mice. *Microbial Cell Factories, 6*(22).

Rossen, N. G., MacDonald, J. K., De Vries, E. M., D'Haens, G. R., De Vos, W. M., Zoetendal, E. G., & Ponsioen, C. Y., (2015). Fecal microbiota transplantation as novel therapy in gastroenterology: A systematic review. *World Journal of Gastroenterology, 21*(17), 5359–5371.

Round, J. L., & Mazmanian, S. K., (2009). The gut microbiota shapes intestinal immune responses during health and disease. *Nature Reviews Immunology, 9*(5), 313–323.

Russell, G. H., Kaplan, J. L., Youngster, I., Baril-Dore, M., Schindelar, L., Hohmann, E., & Winter, H. S., (2014). Fecal transplant for recurrent clostridium difficile infection in children with and without inflammatory bowel disease. *Journal of Pediatric Gastroenterology and Nutrition, 58*(5), 588–592.

Saeidi, N., Wong, C. K., Lo, T. M., Nguyen, H. X., Ling, H., Leong, S. S., Poh, C. L., & Chang, M. W., (2011). Engineering microbes to sense and eradicate *Pseudomonas aeruginosa*, a human pathogen. *Molecular Systems Biology, 7*(521), 1–11.

Samanta, A. K., Jayaram, C., Jayapal, N., Sondhi, N., Kolte, A. P., Senani, S., Sridhar, M., & Dhali, A., (2015). Assessment of fecal microflora changes in pigs supplemented with herbal residue and prebiotic. *PLoS One, 10*(7), e0132961.

Scaldaferri, F., Gerardi, V., Lopetuso, L. R., Del Zompo, F., Mangiola, F., Boškoski, I., Bruno, G., et al., (2013). Gut microbial flora, prebiotics, and probiotics in IBD: Their current usage and utility. *BioMed Research International, 2013*, 435268.

Segui, J., Gil, F., Gironella, M., Alvarez, M., Gimeno, M., Coronel, P., Closa, D., Pique, J. M., & Panes, J., (2005). Down-regulation of endothelial adhesion molecules and leukocyte adhesion by treatment with superoxide dismutase is beneficial in chronic immune experimental colitis. *Inflammatory Bowel Diseases, 11*(10), 872–882.

Sekirov, I., Russell, S. L., Antunes, L. C. M., & Finlay, B. B., (2010). Gut microbiota in health and disease. *Physiological Reviews, 90*(3), 859–904.

Souza, R. C., Mendes, I. C., Reis-Junior, F. B., Carvalho, F. M., Nogueira, M. A., Vasconcelos, A. T. R., Vicente, V. A., & Hungria, M., (2016). Shifts in taxonomic and functional microbial diversity with agriculture: How fragile is the Brazilian Cerrado? *BMC Microbiology, 16*(42).

Steidler, L., Hans, W., Schotte, L., Neirynck, S., Obermeier, F., Falk, W., Fiers, W., & Remaut, E., (2000). Treatment of murine colitis by *Lactococcus lactis* secreting interleukin-10. *Science, 289*(5483), 1352–1355.

Swenson, W., Wilson, D. S., & Elias, R., (2000). Artificial ecosystem selection. *PNAS, 97*(16), 9110–9114.

Tanner, S. A., Lacroix, C., Del'Homme, C., Jans, C., Zihler, B. A., Bernalier-Donadille, A., & Chassard, C., (2015). Effect of *Bifidobacterium thermophilum* RBL67 and fructo-oligosaccharides on the gut microbiota in Gottingen minipigs. *The British Journal of Nutrition, 114*(7), 746–755.

Tripathi, B. M., Edwards, D. P., Mendes, L. W., Kim, M., Dong, K., Kim, H., & Adams, J. M., (2016). The impact of tropical forest logging and oil palm agriculture on the soil microbiome. *Molecular Ecology, 25*(10), 2244–2257.

Van, D. J. M., & Hecker, M., (2013). *Bacillus subtilis*: From soil bacterium to super-secreting cell factory. *Microbial Cell Factories, 12*(3).

Vandenbroucke, K., Hans, W., Van, H. J., Neirynck, S., Demetter, P., Remaut, E., Rottiers, P., & Steidler, L., (2004). Active delivery of trefoil factors by genetically modified *Lactococcus lactis* prevents and heals acute colitis in mice. *Gastroenterology, 127*(2), 502–513.

Waldor, M. K., Tyson, G., Borenstein, E., Ochman, H., Moeller, A., Finlay, B. B., Kong, H. H., et al., (2015). Where next for microbiome research? *PLoS Biology, 13*(1), e1002050.

Wardman, P., (2007). Fluorescent and luminescent probes for measurement of oxidative and nitrosative species in cells and tissues: Progress, pitfalls, and prospects. *Free Radical Biology and Medicine, 43*(7), 995–1022.

Watterlot, L., Rochat, T., Sokol, H., Cherbuy, C., Bouloufa, I., Lefevre, F., Gratadoux, J. J., et al., (2010). Intragastric administration of a superoxide dismutase-producing recombinant *Lactobacillus casei* BL23 strain attenuates DSS colitis in mice. *International Journal of Food Microbiology, 144*(1), 35–41.

Wei, P., Yang, Y., Ding, Q., Li, X., Sun, H., Liu, Z., Huang, J., & Gong, Y., (2016). Oral delivery of *Bifidobacterium longum* expressing α-melanocyte-stimulating hormone to combat ulcerative colitis. *Journal of Medical Microbiology, 65*(2), 160–168.

Wei, P., Yang, Y., Li, T., Ding, Q., & Sun, H., (2015). A engineered *Bifidobacterium longum* secreting a bioactive penetratin-glucagonlike peptide 1 fusion protein enhances glucagon-like peptide 1 absorption in the intestine. *Journal of Microbiology and Biotechnology*.

Weibel, S., Stritzker, J., Eck, M., Goebel, W., & Szalay, A. A., (2008). Colonization of experimental murine breast tumors by *Escherichia coli* K-12 significantly alters the tumour microenvironment. *Cellular Microbiology, 10*(6), 1235–1248.

Xiang, S., Fruehauf, J., & Li, C. J., (2006). Short hairpin RNA-expressing bacteria elicit RNA interference in mammals. *Nature Biotechnology, 24*(6), 697–702.

Xiong, W., Li, Z., Liu, H., Xue, C., Zhang, R., Wu, H., Li, R., & Shen, Q., (2015). The effect of long-term continuous cropping of black pepper on soil bacterial communities as determined by 454 pyrosequencing. *PLoS One, 10*(8), e0136946.

Xiong, W., Zhao, Q., Zhao, J., Xun, W., Li, R., Zhang, R., Wu, H., & Shen, Q., (2015). Different continuous cropping spans significantly affect microbial community membership and structure in a vanilla-grown soil as revealed by deep pyrosequencing. *Microbial Ecology, 70*, 209–218.

Zhang, C., Huang, J., Zhang, J., Liu, S., Cui, M., An, B., Wang, X., et al., (2019). Engineered *Bacillus subtilis* biofilms as living glues. *Materials Today, 28*, 40–48.

Zhao, L., (2013). The gut microbiota and obesity: From correlation to causality. *Nature Reviews. Microbiology, 11*(9), 639–647.

Zolla, G., Badri, D. V., Bakker, M. G., Manter, D. K., & Vivanco, J. M., (2013). Soil microbiomes vary in their ability to confer drought tolerance to *Arabidopsis*. *Applied Soil Ecology, 68*, 1–9.

CHAPTER 4

Antibiotics and Plant-Derived Antimicrobials as an Alternative Source to Control Infections

VIRALKUMAR V. SURATI,[1] BHAVIN BHATT,[2] NARESH BUTANI,[2] HIRAL R. PATEL,[3] and ACHUIT K. SINGH[4]

[1]*Shri A. N. Patel P. G. Institute of Science and Research, Anand, Gujarat, India*

[2]*Shree Ramkrishna Institute of Computer Education and Applied Sciences, Faculty of Science, Sarvajanik University, Surat, Gujarat, India*

[3]*M. B. Patel Applied Science College, Mogri, Gujarat, India*

[4]*Sr. Scientist, Crop Improvement Division, ICAR–Indian Institute of Vegetable Research, Varanasi, Uttar Pradesh, India*

ABSTRACT

The discovery of penicillin started the era of antibiotics and was used for the treatment of infectious diseases to ensure the elimination of the major cause of human morbidity. However, microbial antibiotic resistance was soon observed after the introduction of antibiotics. Diverse mechanisms of antibiotic resistance have been revealed, including molecular mechanisms. Intrinsic mechanisms are attributed to a mutation in the host's chromosomal gene, whereas mutations in antibiotic-targeted genes and the transfer of resistance traits found on extrachromosomal elements like

Modern Biotechnology in Healthcare: Advances and Applications. Sheikh Umar Ahmad (Ed.)

plasmids, bacteriophages transposons, and integrons are responsible for bacterial antibiotic resistance. The development of drug resistance results in clinical failures in the treatment of infections. Hence, new antibiotics are required to combat pathogens. Additionally, the development pace of new antimicrobial drugs is not enough to counteract the fast emergence of resistant strains. Natural small molecules could be used as an alternative to antibiotics.

4.1 INTRODUCTION

The invention of antibiotics has been considered as one of the greatest achievements of biomedicine. Antibiotics are compounds that inhibit or kill microorganisms (Pathogens). The use of antibiotics has decreased mortality rate and morbidity rate of humans. Penicillin was the first antibiotic discovered by Fleming (1929). Nowadays, numerous different classes of antibiotics are known, and they are classified based on their mode of action. Mode of action of antibiotics includes inhibition of DNA replication, transcription, translation, inhibition of cell wall synthesis, and inhibition of energy metabolism (Walsh, 2000).

Broad and uncontrolled use of antibiotics leads to resistance development in pathogens. Resistance in bacteria is not new. It was found in antimicrobial compound-producing microorganisms. Other peripheral organisms also became resistance due to selective pressure exerted by the compound to which they are exposed (Walsh, 2000; Barbosa & Levy, 2007). Over the past 70 years, bacteria have been exposed to different antibiotics used to treat human diseases and adapted genetic traits to nullify the effect of the antibiotic (Walsh, 2000). This chapter will discuss the antibiotic mode of action and both intrinsic as well as acquired resistance mechanisms.

4.2 MOLECULAR MECHANISMS OF RESISTANCE

In response to antibiotics, bacteria adapt genetically and physiologically to combat the action of antibiotics (Barbosa & Levy, 2007). The mechanisms involved in resistance emergence are enzymatic inactivation of antibiotics, reduced cell permeability, efflux of the antibiotics by specific or non-specific pumps and modification of the antibiotic targets (Figure 4.1). There are two mechanisms through which bacteria develop resistance against antibiotics; one, intrinsic mechanism and second, acquired mechanism (Barbosa & Levy, 2007; Wright & Sutherland, 2007; Alekshun & Levy, 2007).

The intrinsic mechanism includes a mutation in the host's chromosomal gene that encodes antibiotic targets and overexpression of efflux systems. In acquired mechanisms, bacteria become resistant by mutations in antibiotic-targeted genes and the transfer of resistance traits found on plasmids (accomplished by conjugation), bacteriophages (accomplished by transduction), transposons, and integrons (Barbosa & Levy, 2007; Wright & Sutherland, 2007; Alekshun & Levy, 2007). For several classes of antibiotics, development of intrinsic and acquired resistance is described in Figure 4.1 along with its mode of action.

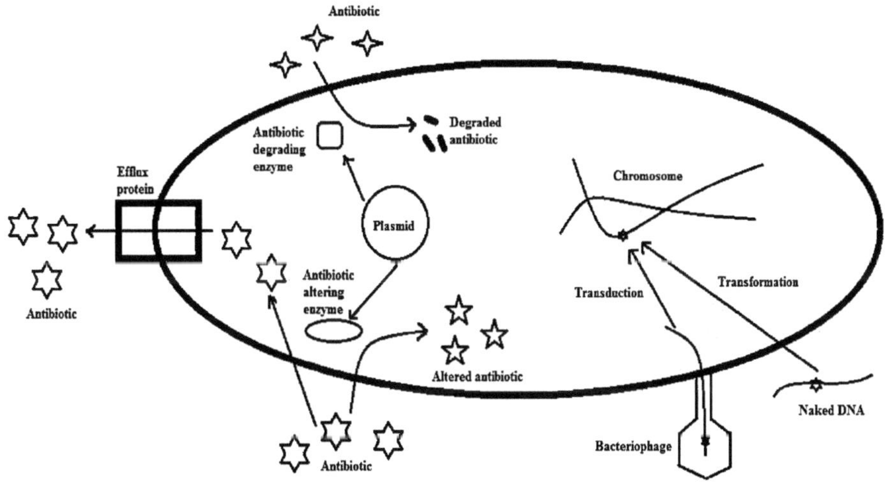

FIGURE 4.1 The antibiotic resistance mechanism of bacteria. Red star represents mutation.

4.2.1 AMINOGLYCOSIDES

4.2.1.1 HISTORY AND MODE OF ACTION

In 1943, the first aminoglycoside discovered was streptomycin (Figure 4.2), isolated from *Streptomyces griseus*. Streptomyces was clinically used to treat tuberculosis. Several years later, new aminoglycosides, such as neomycin (1949); gentamycin (1963); tobramycin (1967); and sisomycin

(1970), were developed to overcome the development of resistance against streptomycin. They show bactericidal activity against selective gram-positive bacteria as well as gram-negative bacteria. There is the rapid emergence of resistance against first-generation aminoglycosides, like dibekacin (1971); amikacin (1972); arbekacin (1973); isepamicin (1975); and netilmicin (1976) (Becker & Cooper, 2013).

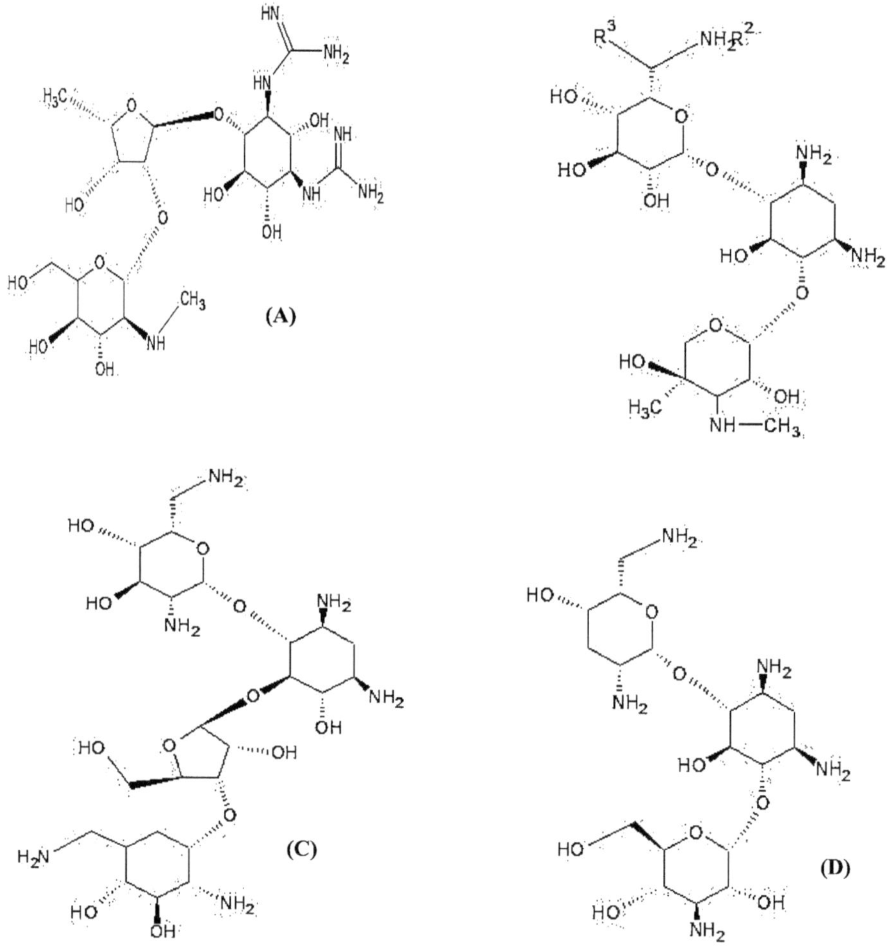

FIGURE 4.2 Aminoglycoside antibiotic structure: (A) streptomycin; (B) gentamycin; (C) neomycin; and (D) tobramycin.

Source: All chemical structures are drawn with Chem Draw Software, Perkin Elmer Informatics.

Aminoglycoside show dose dependent bactericidal activity and inhibit protein synthesis by specifically binding to A site of 16S ribosomal (rRNA). Thus, the presence of aminoglycoside in cytosol generally hampers the peptide elongation by inhibiting of translocation of ribosomes (Becker & Cooper, 2013; Kaul et al., 2004; Ramirez & Tolmasky, 2010).

4.2.1.2 RESISTANCE MECHANISMS

Intrinsic resistance mechanism against aminoglycoside includes efflux pumps, mutations in 16S rRNA and 16S rRNA methylases mediated resistance. Mutation in A1408 leads rearrangement of the molecular structure of 16S rRNA. This rearrangement causes resistance to neomycin and gentamycin (De et al., 1989). Ribosomal protein S12 (RpsL) is also the main target of streptomycin-mediated translational inhibition. Mutation in *the rpsL* gene affects binding of streptomycin, thereby leading to resistance (Springer et al., 2001).

16S rRNA methylase mediated methylation of specific nucleotides within the A-site of 16S RNA prevents binding of aminoglycosides to the 30S ribosomal subunits, initiating emergence of resistance against aminoglycoside such as amikacin, tobramycin, and gentamycin (Doi & Arakawa, 2007). It has been reported that self-transmissible plasmid *pIP1204* confers 16S rRNA methylase in *Klebsiella pneumonia*, isolated from urinary tract infection and became resistance to aminoglycoside (Galimand et al., 2003).

Resistance to aminoglycoside in *P. aeruginosa* strains was due to changes to the outer membrane lipopolysaccharide (LPS). In limited Mg^{2+} concentrations, the PhoP-PhoQ two-component regulatory systems induce the changes of outer membrane LPS level in the presence of polyamines. Ultimately, organisms reduce the net negative charge of the outer membrane (Macfarlane et al., 2000; Kwon & Lu, 2006; Wang & Quinn, 2010).

The efflux pumps of the family of resistance nodulation division (RND), which is mainly present in gram-negative bacteria, are able to transport aminoglycoside antibiotics. The RND family efflux pump-mediated resistance has been reported in *Pseudomonas*, *Acinetobacter*, *Brucella*, *Bukholderia*, *Enterobacter*, *Escherichia*, *Helicobacter*, and *Stenotrophomonas* species (Li & Nikaido, 2009).

Acquired resistance mechanism involves modification of aminoglycoside by enzymes. These enzymes can be classified into three families based on the type of modification: aminoglycoside acetyltransferases (ACCs), aminoglycoside phosphotransferases (APHs) and aminoglycoside nucleotidyltransferase (ANTs). Aminoglycoside acetyltransferase transfers acetyl group to amino group of aminoglycosides. Aminoglycoside phosphotransferase catalyzes phosphorylation reaction to the hydroxyl group of aminoglycosides while ANTs transfers adenosine monophosphate (AMP) group from adenosine triphosphate (ATP) to a hydroxyl group of the aminoglycoside. There are four acetyltransferases: AAC(1), AAC(2'), AAC(3), and AAC(6'); seven phosphotransferases: APH(2"), APH(3'), APH(4), APH(6), APH(7"), and APH(9) and five nucleotidyltransferases: ANT(2"), ANT(3"), ANT(4"), ANT(6), and ANT(9) have been reported till date (Table 4.1) (Becker & Cooper, 2013; Ramirez & Tolmasky, 2010).

TABLE 4.1 Aminoglycoside Modifying Enzymes and Their Antibiotic Substrates and Sources (Ramirez & Tolmasky, 2010)

Enzyme	Classification	Type	Antibiotic Substrate	Bacteria Source
N-Acetyltransferases	AAC(1)	–	Apramycin Butirosin Lividomycin Paromomycin	*Escherichia coli*, *Campylobacter* spp.
–	–	–	Butirosin Lividomycin Paromomycin	*Actinomycete*
–	AAC(3)	I	Gentamicin Sisomicin	Gram-negative *Enterobacteriaceae*
		Ia	Fortimicin	*Serratia marscens*
–	–	Ie	–	*Proteus vulgaris*, *Pseudomonas aeruginosa*, *Salmonella enterica* subsp. *enterica*
–	–	IIa	Gentamicin Netilmicin Tobramycin Sisomicin	*Acinetobacter* spp., *Enterobacter* spp., *Escherichia* spp., *Klebsiella* spp., *Pseudomonas* spp., *Salmonella* spp.
–	–	IIb	–	–
–	–	IIc	–	–
		III	–	*Pseudomonas* spp., *Klebsiella* spp., *Pneumoniae* spp.

TABLE 4.1 *(Continued)*

Enzyme	Classification	Type	Antibiotic Substrate	Bacteria Source
		IV	–	*Escherichia coli*, *Campylobacter jejuni*, *Pseudomonas stutzeri*
		Via	–	*Escherichia coli*, other *Enterobacteriaceae*, *Salmonella enterica*
		VII	Kanamycin	*Actinomycetes*
		VIII	Dibekacin	
		IX	Arbekacin Amikacin	
		X		
	AAC(2)	Ia	Gentamicin	*Providencia stuartii*
		Ib	Tobramycin Dibekacin	*Mycobacterium fortuitum*, *Acinetobacter baumannii*
		Ic	Kanamycin Netilmicin	*Mycobacterium tuberculosis*, *Mycobacterium bovis*
		Id		*Mycobacterium smegmatis*
		Ie		*Mycobacterium leprae*
–	AAC(6')	I	Amikacin Gentamicin C1a, C2	Gram-positive, gram-negative
		II	Gentamicin C1, C1a, C2	
O-Nucleotidyltransferase	ANT(6)	I	Streptothricin	*Enterococcus faecalis*, *Campylobacter jejuni*, *Staphylococci* spp., *Enterococci* spp., *Bacillus subtilis*, *Campylobacter fetus* subsp. *fetus*
	ANT(9)	Ia	Streptomycin	*Enterococcus avium*, *Enterococcus faecalis*, *Enterococcus faecium*
		Ib		*Enterococcus faecalis*
	ANT(4')	Ia	Dibekacin Tobramycin Amikacin Isepamicin	*Staphylococci* spp., *Enterococci* spp., *Bacillus* spp.
		IIa	Tobramycin Amikacin	*Pseudomonas* spp., *Enterobacteriaceae*
		IIb	Isepamicin	*Pseudomonas aeruginosa*

TABLE 4.1 *(Continued)*

Enzyme	Classification	Type	Antibiotic Substrate	Bacteria Source
	ANT(2”)	Ia	Gentamicin Tobramycin Dibekacin Sisomicin Kanamycin	*Enterobacteriaceae* Gram-negative
–	ANT(3”)	Ia	Spectinomycin Streptomycin	*Pseudomonas aeruginosa*
O-Phosphotransferase	APH(4)	Ia	Hygromycin	–
	APH(6)	Ia	Streptomycin	*Streptomyces griseus*
		Ib		*Streptomyces glaucescens*
		Ic		Gram-negative
		Id		Gram-negative Gram-positive
	APH(9)	Ia	Spectinomycin	*Legionella pneumophila*
		Ib		*Streptomyces flavopersicus*, *Streptomyces spectabilis*
	APH(3’)	I	Kanamycin Neomycin Paromomycin Ribostamycin Lividomycin	*Corynebacterium* Gram-negative
		IIa IIb IIc	Kanamycin Neomycin Paromomycin Ribostamycin Butirosin	*Pseudomonas aeruginosa* *Salmonella maltophilia*
–	–	IIIa	Kanamycin Neomycin Lividomycin Paromomycin Livostamycin Butirosin Amikacin Isepamicin	Gram-positive
		Iva	Neomycin Paromomycin Ribostamycin	*Bacillus circulans*

TABLE 4.1 *(Continued)*

Enzyme	Classification	Type	Antibiotic Substrate	Bacteria Source
		Va	Neomycin	*Actinomycetes*
		Via	Paromomycin	*Acinetobacter baumannii*
		VIb	Butirosin Amikacin Isepamicin Kanamycin Ribostomycin	*Klebsiella pneumonia, Serratia marcesans*
		VIIa	Kanamycin Neomycin	*Campylobacter jejuni*
	APH(2")	I	Gentamicin	Gram-positive
		II		–
	APH(3")	Ia	Streptomycin	*Streptomyces griseus, Mycobacterium fortuitum*
		Ib		*Vibrio cholarae*
		Ic		*Mycobacterium fortuitum*
–	APH(7")	Ia	Hygromycin	*Streptomyces hygroscopicus*

There are two different nomenclature systems in use for the aminoglycosides modifying enzymes. One is from an enzymatic perspective and consists of a three-letter code to identify the activity (APH, ANT, and AAC), followed by a number in parentheses that identifies the site of modification, then a Roman numeral that describes a particular resistance profile that is emerged in the organism (Subclass), and lastly a lower case letter as an individual identifier. The parentheses and the subclass are separated by a hyphen, AAC(3)-IIa, for example. The other nomenclature system has a genetic perspective, with three lower case letters in italics for the type of activity (*aph, aac*, and *aad*), a capital letter for the site of modification and a number as a unique identifier of the individual genes (Becker & Cooper, 2013; Ramirez & Tolmasky, 2010).

4.2.2 B-LACTAM ANTIBIOTICS

4.2.2.1 HISTORY AND MODE OF ACTION

The β-lactam antibiotics (Figure 4.3), one of the largest classes of antibiotics, contain a β-lactam nucleus in their molecular structure. Penicillin was the first *β*-lactam antibiotic discovered in 1929 by Alexander Fleming (1929). The *β*-lactam antibiotic family includes penicillin and its derivatives, cephalosporins, carbapenems, and monobactams. Early *Penicillins* were less antibacterial against gram-negative pathogens. Aminopenicillin was developed later that showed antibacterial activity against *E. coli*, *Shigella*, and *Salmonella* but not against *P. aeruginosa* and *Klebsiella* species. Carboxypenicillin was generated by replacing an amino group of aminopenicillin with a carboxy group. Carboxypenicillin was active against *P. aeruginosa* (Bodey, 1990).

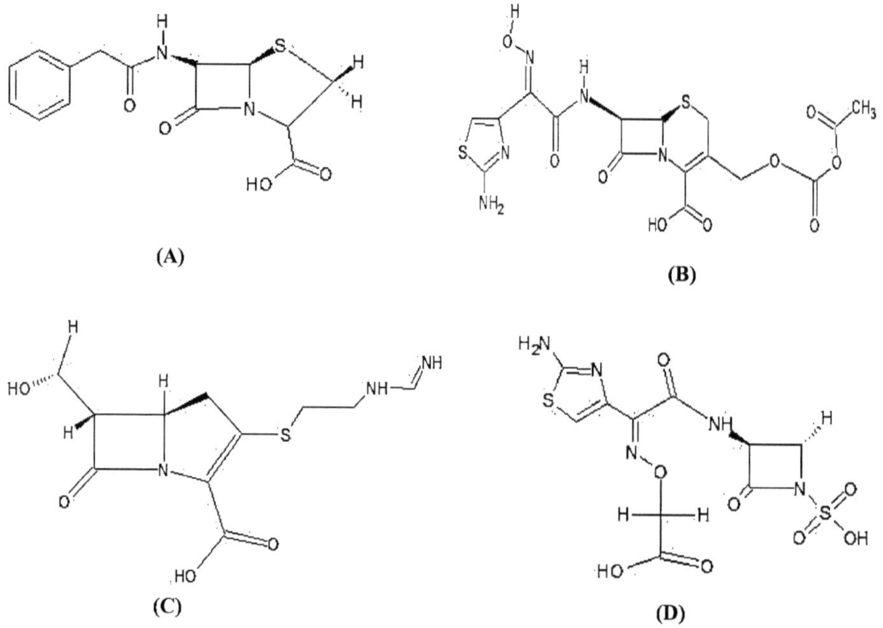

FIGURE 4.3 β-lactam antibiotic structure: (A) benzylpenicillin; (B) cefotoxamine; (C) imipenam; and (D) aztreonam.

In the late 1940s, Cephalosporin was developed to treat *β*-lactamase producing staphylococcal infections. According to their spectrum of activity and time of introduction, cephalosporins can be categorized into first, second, third, and fourth generations. The first generation shows good gram-positive activity and modest activity against gram-negative bacteria; second generation cephalosporin has increased gram-negative activity and less gram-positive activity while the third generation shows improved gram-negative and variable gram-positive activity. The fourth generation has broad spectrum activity against both gram-negative as well as gram-positive bacteria (Harrison & Bratcher, 2008). Imipenem, first carbapenems introduced in 1985, shows broad spectrum activity as they diffuse easily in the bacteria. The discovery of monobactams provides therapeutic options for bacterial infection. Aztreonam is one of the monobactam introduced in 1986 (Bodey, 1990).

The β-lactams antibiotics inhibit the cell wall synthesis by binding to penicillin binding proteins (PBPs) in bacteria, which catalyzes the cross-linking of peptidoglycan. Inhibition of transpeptidation reaction results in less integrity of cell wall and finally leads to cytolysis due to osmotic pressure (Bodey, 1990; Tipper & Strominger, 1965).

4.2.2.2 RESISTANCE MECHANISMS

The β-lactam antibiotic resistance is increased at a significant rate in bacteria. Hence it constitutes one of the most common problems for public health care. Resistance to β-lactam antibiotics occurs through two mechanisms: (i) the production of β-lactamase, which hydrolyzes β-lactam antibiotics; and (ii) production of altered PBPs with lower affinity for most β-lactam antibiotics.

By late 2009, more than 890 β-lactamases were reported (Bush et al., 1995). Most of the β-lactamases are encoded by genes located on plasmids and transposons, whereas others are chromosomally encoded. β-Lactamases can be classified on the basis of either the functional or molecular characteristics of the enzymes. Based on the molecular characteristic, β-lactamases can be divided into four classes. This classification is also known as "Ambler" classification. Classes A, C, and D β-lactamases contains serine in their active sites and catalyzes the reaction by forming an acyl enzyme whereas class B β-lactamases are metalloenzymes who require zinc as a metal cofactor for their catalytic activity (Bush, 1989). Based on the functional characteristic, β-lactamases can also be classified into 1 to 4 groups, also called Bush Jacoby-Medeiros groups (Figure 4.4). AmpC is the first β-lactamases encoded from the chromosome. AmpC found in many species of *Enterobacteriaceae* and *P. aeruginosa*, is a member of class C β-lactamases. In 2005, it has been reported that plasmid-borne *ampC* genes encode AmpC protein. Class D enzymes, also called oxacillinases, have been found in *P. aeruginosa*, *Acinetobacter* ssp., and *Aeromonas* spp. (Drawz & Bonomo, 2010).

Based on their potent activity to hydrolyze a small number or a variety of β-lactam, the enzymes can be further subdivided into narrow, moderate, broad, and extended-spectrum β-lactamases. The broad spectrum β-lactamases can hydrolyze penicillins and cephalosporins, but they are susceptible to β-lactamase inhibitors such as clavulanic acid and tazobactam. The extended-spectrum β-lactamases enable to provide resistance to the penicillins in the first, second, and third generation cephalosporins and aztreonam, while they are susceptible to carbapenems and β-lactamase inhibitors (Bradford, 2001).

Another mechanism of β-lactam resistance is altered PBPs. The first methicillin resistance *Staphylococcus aureus* was identified in 1961, soon after two years of the introduction of methicillin. The Staphylococci emerged as methicillin resistant by acquiring gene encoding for PBPs. The altered PBPs of methicillin resistant *S. aureus* are conferred by *mecA*. The *mecA* gene is regulated by *mecR1-mecI*. The *mecA* is transformed as mobile genetic elements, called staphylococcal cassette chromosome *mec* (SCC*mec*). The *mecA* operon encodes the site-specific recombinases responsible for transformation (Fuda et al., 2005).

During treatment of cystic fibrosis, *Pseudomonas aeruginosa* developed resistance to β-lactam antibiotics by altered PBPs, which have a low affinity for penicillin G (Godfrey et al., 1981). Clinically isolated *S. pneumonia* from the United States showed a decrease in the affinity of PBP group 1 and 2 for Penicillin G. The most resistant strain showed a decrease in the amount of PBP2b (Hakenbeck et al., 1980).

4.2.3 CHLORAMPHENICOL

4.2.3.1 HISTORY AND MODE OF ACTION

Chloramphenicol (Figure 4.5), initially known as chloromycetin, was isolated from *Streptomyces venezuelae*. Production of chloramphenicol was originally observed by Ehrlich in 1947 (Mclean et al., 1949; Izard & Jacqualine, 2000; Schwarz et al., 2004). Three derivatives of chloramphenicol have been developed by substitution of functional groups. Azidamfenicol (Figure 4.5) is developed by replacing two chlorine atoms

(Cl_2) with an azide group. In thiamphenicol (Figure 4.4), the nitro group ($–NO_2$) is substituted by a methyl sulfonyl residue ($–SO_2CH_3$) while florfenicol (Figure 4.4) is developed by substituting the nitro group (NO_2) and hydroxyl group (–OH) with a methyl sulfonyl group and fluorine (–F), respectively (Schwarz et al., 2004).

Chloramphenicol inhibits protein synthesis by preventing peptide chain elongation. It binds specifically to the peptidyl transferase center at the 50S ribosomal subunit of 70S ribosomes. It has broad spectrum activity as it is active against gram-positive, gram-negative, aerobic, and anaerobic bacteria (Izard & Jacqualine, 2000; Schwarz et al., 2004).

Ambler class	A							B	C	D
Bush Jacoby-Medeiros class	2a	2b	2be	2br	2c	2e	2f	3	1	2d
Substrates	Penicillins	Penicillin Narrow spectrum cephalosprins	Penicillins Narrow and wide spectrum Cephalosporin	Penicillins	Penicillins Carbenicillin	Broad spectrum cephalosporin	Penicillins, Cephalosporins, Carbapenems	Most β-lactams including carbapenems	Cephalosporins	Penicillins Cloxacillin
Inhibition by Clavulanic acid	+	+	+	-	+	+	±	-	-	±
Bacteria	KP, EC, BP, PA, AB, HI, NG, KLs, EntC, SM							PA, KP, AB, BACs, CHRs, StM	ENTs, CITs, PA	AB, PA

FIGURE 4.4 β-lactamase classification schemes (Drawz & Bonomo, 2010).

Abbreviations: AB: *Acinetobacter baumannii*; BACs: *Bacillus* spp.; BP: *Burkholderia pseudomallei*; CHRs: *Chryseobacterium* spp.; CITs: *Citrobacter* spp.; EC: *Escherichia coli*; EntC: *Enterobacter cloacae*; ENTs: *Enterobacter* spp.; HI: *Haemophilus influenza*; KLs: *Kluyvera* spp.; KP: *Klebsiella pneumonia*; NG: *Neisseria gonorrhoeae*; PA: *Pseudomonas aeruginosa*; SM: *Serratia marcescens*; StM: *Stenotrophomonas malyophilia.*

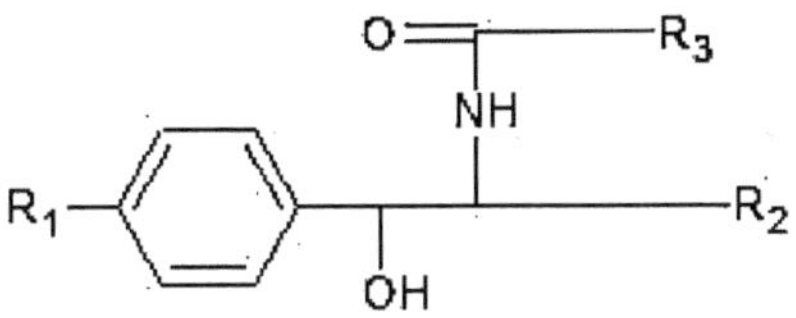

FIGURE 4.5 The structure of chloramphenicol and related antibiotics.

Name of Antibiotic	R_1	R_2	R_3
Chloramphenicol	$-NO_2$	–OH	$=Cl_2$
Azidamfenicol	$-NO_2$	–OH	Azide group
Thiamphenicol	$-SO_2CH_3$	–OH	$=Cl_2$
Florfenicol	$-SO_2CH_3$	–F	$=Cl_2$

4.2.3.2 RESISTANCE MECHANISMS

The mechanisms of bacterial resistance to chloramphenicol involved enzymatic inactivation, efflux system, mutation, and permeability of barriers. Chloramphenicol acetyltransferases (CATs) modifies the molecular structure of chloramphenicol and azidamfenicol by replacing hydroxyl group (-OH) with an acetyl group. However, the hydroxyl group is absent in florfenicol and hence, it is resistant to CATs inactivation. Two types of genes encoding CATs, which distinctly differ in their structure, have been reported: the classical CATs and the Xenobiotic CATs (Figure 4.6) (Schwarz et al., 2004). All CATs are heterotrimer of 24 to 26 kDa (Murray & Shaw, 1997).

Chloramphenicol 3-O-phosphotransferase (CPT) has a polypeptide chain of 19 kDa that inactivates the chloramphenicol by phosphorylation. The CPT is active as a homodimer. CPT-mediated phosphorylation requires ATP as phosphoryl donor to transfer the γ-phosphate to the (C-3) hydroxyl of chloramphenicol (Izard & Jacqualine, 2000).

Specific exporters play a key role in chloramphenicol resistance (Schwarz et al., 2004). The first non-enzymatic chloramphenicol resistance was observed in *P. aeruginosa* having Inc P plasmid R26 and transposon Tn1696 on the Inc P plasmid R1033. Both contain the same gene (*cmlA1*) that encodes transmembrane transport proteins of the major facilitator superfamily (George & Hall, 2002). Currently, eight different groups of specific exporters are known (Table 4.2).

4.2.4 GLYCOPEPTIDE

4.2.4.1 HISTORY AND MODE OF ACTION

Vancomycin, the first glycopeptide, was isolated as secondary metabolites from a soil bacterium; *Streptomyces orientalis* (now renamed *Amycolatopsis orientalis*). Teicoplanin is the second glycopeptide, used against gram-positive bacterial infections (Pootoolal et al., 2002).

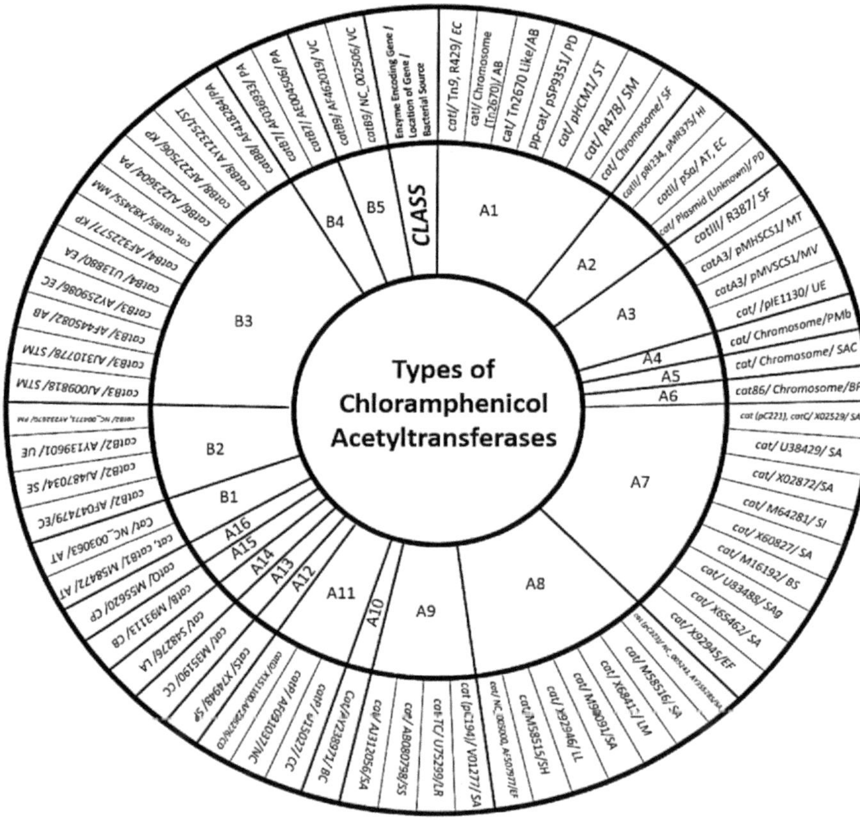

FIGURE 4.6 Types of chloramphenicol acetyltransferases (Schwarz et al., 2004).

Note: Pattern of reading in outermost concentric circle: Name of enzyme encoding gene/location of gene/bacterial source.

Abbreviations: AB: *Acinetobacter baumannii*; AC: *Acinetobacter calcoaceticus*; AT: *Agrobacterium tumefaciens*; BC: *Bacillus clausii*; BP: *Bacillus pumilus*; BS: *Bacillus subtilis*; CB: *Clostridium butyricum*; CC: *Campylobacter coli*; CD: *Clostridium difficile*; CP: *Clostridium perfringens*; EA: *Enterobacter aerogenes*; EC: *Escherichia coli*; EF: *Enterococcus faecalis*; HI: *Haemophilus influenza*; KP: *Klebsiella pneumonia*; LA: *Listonella anguillarum*; LL: *Lactococcus lactis*; LM: *Listeria monocytogenes*; LR: *Lactobacillus reuteri*; MM: *Morganella morganii*; MT: *Mannheimia taxon* 10; MV: *Mannheimia varigena*; NM: *Neisseria meningitides*; PA: *Pseudomonas aeruginosa*; PD: *Photobacterium damselae* subsp. *Piscicida*; PM: *Pasteurella multocida*; PMb: *Proteus mirabilis*; PP: *Pseudomonas putida*; SA: *Staphylococcus aureus*; SAC: *Streptomyces acrimycini*; SAg: *Streptococcus agalactiae*; SE: *Salmonella enteritidis*; SF: *Shigella flexneri*; SH: *Staphylococcus haemolyticus*; SI: *Staphylococcus intermedius*; SM: *Serratia marcescens*; SP: *Streptococcus pyogenes*; SS: *Streptococcus sius*; ST: *Salmonella typhi*; STM: *Salmonella typhimurium*; UE: Uncultured eubacterium; VC: *Vibrio cholera.*

TABLE 4.2 The Specific Exporters Involved in Chloramphenicol or Chloramphenicol/Florfenicol Resistance (Schwarz et al., 2004)

Group	Gene Designation	Plasmid/ Transposon/ Chromosome	Database Accession No.	Bacterial Source
E1	*cmlB, cmlA2*	pIP833	AF034958	*Enterobacter aerogenes*
	cmlA	Plasmid	AJ487033	*Salmonella typhimurium*
	cmlA5	R751 (Tn2000)	AF205943	*Escherichia coli*
	cmlA1	pILT-3	AF458080	*Klebsiella pneumoniae*
	cmlA1	RPL11 (Tn1403)	AF313472	*Pseudomonas aeruginosa*
	cmlA4	pTK1	AF156486	*Klebsiella pneumoniae*
	cmlA5	pSp1	AY115475	*Uncultured bacterium*
	cmlA6	Plasmid	AF294653	*Pseudomonas aeruginosa*
	cmlA7	Chromosome	AJ511268	*Pseudomonas aeruginosa*
	cmlA, cmlA1	pR1033:Tn1696	U12338 M64556 AF078527	*Pseudomonas aeruginosa*
E2	*Cml*	R26	M22614	*Escherichia coli*
E3	*cmlA-like*	Chromosome	AF071555	*Salmonella typhimurium DT104*
	floR	Chromosome	AF118107	*Salmonella typhimurium DT104*
	Flo	Chromosome	AJ251806	*Salmonella typhimurium DT104*
	floR	Chromosome	AF261825	*Salmonella typhimurium DT104*
	floR	Chromosome	AY339985	*Salmonella typhimurium DT104*
	Flo	Plasmid	AF252855	*Escherichia coli*
	floR	Plasmid	AF231986	*Escherichia coli*
	floR	pMBSF1	AJ518835	*Escherichia coli*
	floR	R55	AF332662	*Klebsiella pneumoniae*
	floR	Chromosome (SXT element)	AY034138	*Vibrio cholerae*
	floR	Chromosome (SXT element)	AY055428	*Vibrio cholerae*
	pp-flo	pSP92088	D37826	*Photobacterium damselae* subsp. *piscicida*

TABLE 4.2 *(Continued)*

Group	Gene Designation	Plasmid/ Transposon/ Chromosome	Database Accession No.	Bacterial Source
E4	*fexA*	pSCFS2	AJ549214	*Staphylococcus lentus*
E5	*Cml*	Chromosome	X59968	*Streptomyces lividans*
E6	*Cmlv*	Chromosome	U09991	*Streptomyces venezuelae*
E7	*cmrA*	Tn5561	AF015087	*Rhodococcus rhodochrous*
	Cmr	pRF2	Z12001	*Rhodococcus fascians*
E8	*Cmr*	pXZ10145	U85507	*Corynebacterium glutamicum*
	Cmx	pTP10:Tn5564	AF024666	*Corynebacterium striatum*

Haemophilus influenza, which lacks CAT activity, developed chloramphenicol resistance due to the loss of an outer membrane protein (Burns et al., 1985). The same mechanism is also observed in *P. cepacia* from a patient with cystic fibrosis (Burns et al., 1989). The clinically isolated *Salmonella typhi* developed as it lacks porin (OmpF) (Toro et al., 1990). *E. coli* and *B. subtilis* have developed chloramphenicol due to a mutation in the major ribosomal protein gene cluster (Baugman & Fahnestock, 1979; Anderson et al., 1984). Mutation in 23S rRNA in *E. coli* is also responsible for chloramphenicol resistance emergence (Ettayebi et al., 1985).

Glycopeptide inhibits peptidoglycan synthesis. It inhibits transpeptidase and transcarboxylase activity by specific interaction with D-Ala-D-Ala terminus of the cell wall peptidoglycan precursors (Pootoolal et al., 2002).

4.2.4.2 RESISTANCE MECHANISM

Six different phenotypes (VanA, VanB, VanC, VanD VanE and VanG) are responsible for the emergence of glycopeptide resistance in enterococci. *Enterococcus faecium* acquired Tn1546 contains three genes of seven genes which are responsible for Vancomycin resistance: *vanH*, *vanA*, and *vanX*. The *vanH*, *vanA*, and *vanX* conferred α-keto acid reductase, D-Ala-D-Ala ligase, and a Zn binding metallodipeptidase, respectively. The α-keto acid reductase generates the D-isomer of lactate. Dipeptidase cleaves D-Ala-D-Ala, whereas D-Ala-D-Ala ligase ligates the D-Ala-D-Lact (Figure 4.7). Thus, cells having transposon become resistant to a change in the antibiotic target structure (Pootoolal et al., 2002).

The cells having VanB and VanD phenotype produce D-Ala-D-Lact terminating peptidoglycan. The production of D-Ala-D-Lact terminating peptidoglycan requires VanHB, VanAB, and VanXB enzymes, having similar role enzyme produced in cells having VanA phenotype.

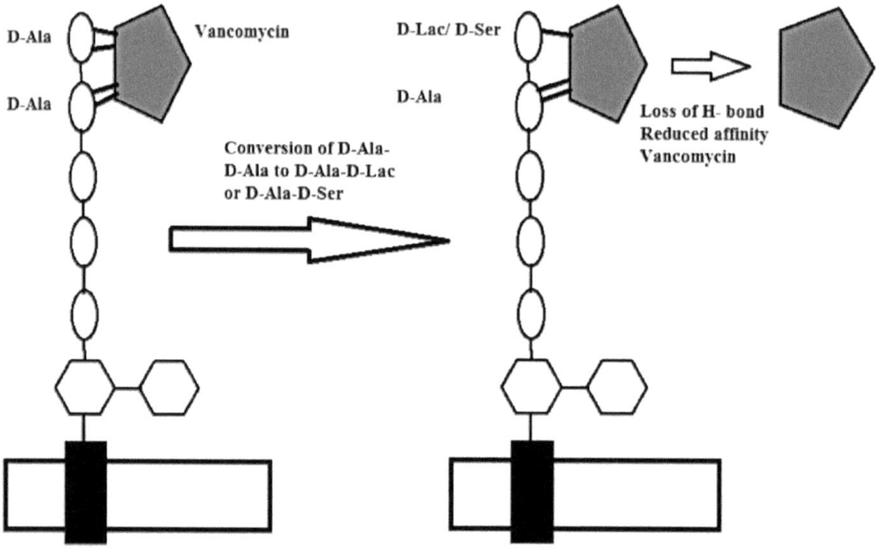

FIGURE 4.7 Vancomycin resistance. Conversion of D-Ala-D-Ala to D-Ala-D-Lac or D-Ala-D-Ser results in loss of one hydrogen bond, which reduces vancomycin affinity towards peptidoglycan.

VanD phenotype is encoded by chromosome and responsible for vancomycin and teicoplanin resistance. The VanC, VanE, and VanG phenotypes produce D-Ala-D-Ser terminating peptidoglycan that lowers the affinity for glycopeptides (Figure 4.7) (Pootoolal et al., 2002).

Staphylococcus aureus developed vancomycin resistance by thickening the cell wall, increased peptidoglycan synthesis and altered PBPs. Mutation in VncS, a sensor His kinase of a two-component regulatory system, leads autolysis in response to vancomycin. Phosphorylation of VncR represses autolysin gene expression under normal conditions. VncS has both kinase and phosphatase activity. In the presence of vancomycin, VncR dephosphorylates VncR, which leads to autolysin production and cause cell lysis. Inactivation of VncS inactivates the phosphatase activity (Pootoolal et al., 2002).

4.2.5 MACROLIDE-LINCOSMAIDE-STREPTOGRAMIN B (MLS ANTIBIOTICS)

4.2.5.1 HISTORY AND MODE OF ACTION

Structurally, macrolides contain a large lactone ring to which amino acids or neutral sugars are attached by glycosidic bonds. The first macrolide, erythromycin, was discovered in 1952, isolated from *Saccharopolyspora erythraea*. The new 14, 15, and 16-membered ring macrolides, such as clarithromycin and azithromycin, have been developed to mimic the chemical instability, poor absorbance, and bitter taste of erythromycin (Roberts et al., 1999). Although lincosamide, streptogramin, and macrolide are structurally unrelated to each other, they have a similar mode of action. However, many macrolides resistance genes code for the resistance of two or all three of the macrolide-lincosamide-streptogramin B (MLSB) (Roberts, 2002).

MLSB binds to the 50S ribosomal subunit and stimulates dissociation of the peptidyl-tRNA from the ribosome during elongation, and thus inhibiting the protein synthesis (Roberts et al., 1999; Roberts, 2002; Weisblum, 1998).

4.2.5.2 RESISTANCE MECHANISMS

MLS resistance mechanisms involve enzymatic modification, mutation of ribosomal RNA (rRNA) target site and efflux system. The most common mechanism of MLS resistance is methylation of 23S rRNA by rRNA methylases, which is encoded by the *erm* gene (*erm* tends erythromycin–ribosome methylation) (Figure 4.8). The rRNA methylases methylate at the site of the peptidyl transferase circle of the 23S rRNA domain. It has been observed that A2058 of 23S rRNA methylates post-transcriptionally by 23S rRNA methylase (Roberts et al., 1999; Roberts, 2002).

Two esterases, two hydrolases, seven transferases and three phosphorylases have been reported to inactivate the MLS antibiotics (Figure 4.8). Mutation in 23S rRNA domain V and domain II results in MLS resistance. Mutational alteration of 23S rRNA at A2058 was first reported in clinically isolated *Mycobacterium intracellulare*. Mutation in domain II of the 23S rRNA contains a Shine Dalgarno sequence (GGAGG), which destabilizes the secondary structure of Domain II.

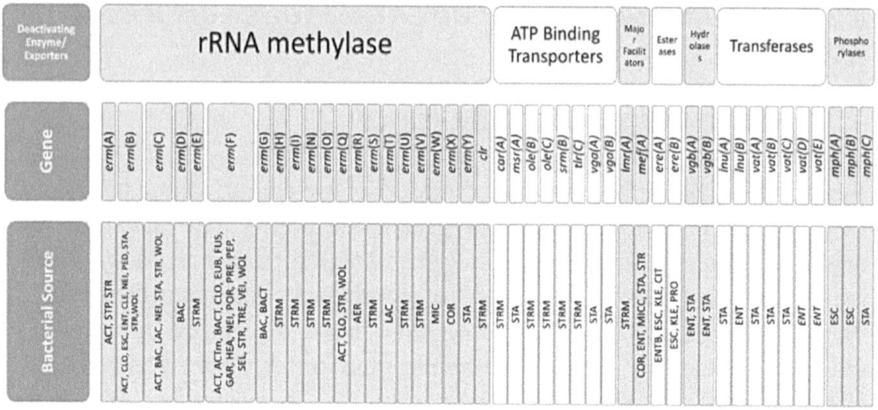

FIGURE 4.8 Inactivating enzymes and exporters genes, which confer resistance to MLS antibiotics.

Abbreviations: ACT: *Actinobacillus*; ACTm: *Actinomyces*; AER: *Aeromicrobium*; BAC: *Bacillus*; BACT: *Bacteroides*; CIT: *Citrobacter*; CLO, CLE: *Clostridium*; COR: *Corynebacterium*; ENT: *Enterococcus*; ENTB: *Enterobacter*; ESC: *Escherichia*; EUB: *Eubacterium*; FUS: *Fusobacterium*; GAR: Gardnerella; HAE: *Haemophilus*; KLE: *Klebsiella*; LAC: *Lactobacillus*; MIC: Micromonospora; MICC: *Micrococcus*; NEI: *Neisseria*; PED: *Pediococcus*; PEP: *Peptosteptococcus*; POR: *Porphyromonas*; PRE: *Prevotella*; POR: *Proteus*; SEL: *Selenomonas*; STA: *Staphylococcus aureus*; STP: *Staphylococcus*; STR: *Streptococcus*; STRM: *Streptomyces*; TRE: *Treponema*; VEI: *Veillonella*; WOL: *Wolinella.*

Erythromycin resistance is regulated by either transcriptional or translational attenuation (Weisblum, 1998). Instead of efflux of the drug out of the cell or the cellular membrane, the efflux protein pumps keep drug intracellular concentration lower so that the ribosome can function normally. ATP transporters and major facilitator super family transporters (MFS transporters) play a crucial role in MLS resistance emergence. Eight different ATP transporter genes and two different MFS transporters have been reported in gram-positive and gram-negative bacteria (Figure 4.8). The *msr* and *vga* genes are found in *Enterococcus* and *Staphylococcus* ssp. The *lmr* and *mef* genes encode MFS transporter proteins and confer lincosamide resistance. The *lme* gene is reported as present in *Streptomyces*, whereas the *mef* gene is present in a gram-positive and gram-negative bacterium (Roberts et al., 1999; Roberts, 2002; Weisblum, 1998).

4.2.6 QUINOLONE

4.2.6.1 HISTORY AND MODE OF ACTION

Nalidixic acid, a quinolone derivative was first introduced in clinical care in 1962, as it has antibacterial activity against the gram-negative bacteria. Nalidixic acid was produced during the synthesis of chloroquine (Lesher et al., 1962). Later, fluoroquinolone was generated by adding fluoride atom at the 6th position of quinolone molecule. The addition of fluoride atom to quinolone enhances its antibacterial activity. The second and third generations of quinolone were introduced in health care in the 1980s and 1990s, respectively. The second generation of quinolone includes ciprofloxacin, norfloxacin, and ofloxacin, while the third generation comprises levofloxacin and sparfloxacin. These generations of quinolone show wide spectrum antibacterial activity against gram-positive as well as gram-negative (Hooper, 2000; King et al., 2000).

They kill the bacteria directly by inhibiting DNA replication (Hooper, 2001). They trap the DNA-DNA gyrase or topoisomerase IV complex and forms barrier to DNA replication and transcription (Hiasa et al., 1996; Willmott et al., 1994; Shea & Hiasa, 1999). Both are type 2 topoisomerase, having tetramer subunits (GyrA and GyrB subunits of DNA gyrase, ParC, and ParE subunits of topoisomerase IV).

4.2.6.2 RESISTANCE MECHANISM

Two main mechanisms of quinolone resistance have been reported: alterations in the targets of quinolones and decreased intracellular concentration due to impermeability of the membrane and/or over-expression of efflux pump systems. Both of the mechanisms are intrinsic. Furthermore, the *qnr* genes are found on a plasmid, which confers quinolone resistance (Ruiz, 2003).

Quinolone inhibits the action of the type II topoisomerases, DNA gyrase and topoisomerase IV. Mutation in *gyrA*/*gyrB* and *parC*/*parE* leads in an amino acid substitution that has been reported in *Escherichia coli*. Mutations in *gyrA* are predominantly found in the quinolone resistance determining the region (QRDR). Mutation at the 67th and 106th positions in the QRDR results in the substitution of serine in place of alanine and arginine/histidine in place of glutamine. Additionally, mutation is found also in the *parC* gene (Ruiz, 2003; Yoshida et al., 1990).

Decreased intracellular concentration of quinolone may be due to an increase in the bacterial impermeability or the overexpression of efflux pumps. Hydrophobic quinolones are reported to diffuse through the porins. Thus, alteration in the composition of LPS and expression of porins may lead to the creation of quinolone resistance. Decreased expression of OmpF increases the resistance level in *E. coli* to quinolone. Expression of OmpF is regulated by chromosome loci such as MarRAB and SoxRS. The MarRAB contains three genes: *marR* that encodes repressor protein, *marA* that encodes activator protein and *marB* that encodes an unknown protein. SoxRS operon contains two genes, *soxR* encoding regulatory protein and *soxS* encoding activator protein (Alekshun and Levy., 1997; Oethinger et al., 1998).

In *E. coli*, the MarRAB and SoxRS operons also reported regulating the expression level of efflux pump AcrAB. The disruption or inhibition of AcrAB expression increase susceptibility to ciprofloxacin, enrofloxacin, and marbofloxacin in *S. typhimurium*. Three different efflux systems such as MexAB-OprM, MexCD-OprJ or MexEF-OprN have been reported in *P. aeruginosa*. The fourth Mex XY efflux system has also been described, but no downstream open reading frame for an outer membrane protein has been found. In *S. aureus*, NorA, an ATP dependent efflux pump, is reported to pump out hydrophilic quinolones like enoxacin or norfloxacin, but not sparfloxacin (Hydrophobic quinolone). The over-expression of Bmr and Blt, the NorA related efflux pumps in *B. subtilis,* provide quinolone resistance (Hiroshi, 1996).

Plasmid contains a *qnr* gene that encodes a protein of 218 amino acids, belonging to the pentapeptide repeat family. This protein protects the DNA gyrase from quinolone inhibition, although its effect on topoisomerase IV is unknown. In 1998, plasmid-mediated quinolone resistance was first reported in *Klebsiella pneumoniae* (Strahilevitz et al., 2009; Tran & Jacoby, 2002).

4.2.7 SULFONAMIDES

4.2.7.1 HISTORY AND MODE OF ACTION

Sulfonamides were the first synthetic drugs with a selective effect on bacteria. They were first used to treat *Streptococcus pyogenes* infection in mice by Domagk in 1932 (Skold, 2000). To combat the rapid development of sulfonamide resistance after their introduction, sulfonamide was combined with trimethoprim due to the synergistic bactericidal effect resulted from the combination of both drugs (Huovinen et al., 1995).

Sulfonamide, structural analog to *p*-aminobenzoic acid (PABA), completely inhibits the dihydropteroate synthase (DHPS). DHPS catalyzes the condensation of PABA and 7,8-dihydro-6-hydroxymethylpterin pyrophosphate to form dihydropteroic acid, which is the final step in the folate biosynthetic pathway. Folate biosynthesis is required for thymine production and bacterial cell growth (Skold, 2000).

4.2.7.2 RESISTANCE MECHANISM

Sulfonamide resistant *Escherichia coli* have been isolated under selective pressure of sulfonamide. The mutant *E. coli* produces altered DHPS that shows lower affinity to sulfonamide, compared to the parent strain of *E. coli* (Pato & Brown, 1963). The sequencing of a *folP* gene that encodes DHPS revealed that the mutation results in the substitution of leucine codon instead of phenylalanine codon.

Plasmid borne sulfonamide resistance has been observed in *E. coli*. They produce altered DHPS enzymes (Skold, 1976; Wise & Abou-Donia, 1975). The plasmid mediated genes are characterized as *sul1* and *sul2*. The *sul1* gene is normally found on the *Tn21* type integrin, while *sul2* is usually located on *IncQ* family plasmid and *pBP1* (Swedberg & Skold, 1983).

4.2.8 TETRACYCLINE

4.2.8.1 HISTORY AND MODE OF ACTION

The first members of the tetracycline group were characterized in the late 1940s as Chlortetracycline and Oxytetracycline. These molecules were produced by *Streptomyces aureofaciens* and *Streptomyces rimosus*, respectively. Later, other tetracyclines were discovered either as naturally occurring molecules mostly in *Streptomyces* spp. or as semi-synthetic products such as methacycline, doxycycline, and minocycline (Swedberg & Skold, 1991).

Tetracycline enters in bacterial cells through porin channels and prevents the interaction of aminoacyl tRNA with the bacterial ribosome. They mainly bind to the 30S subunit of the ribosome in the protein synthesis machinery (Swedberg & Skold, 1991).

4.2.8.2 RESISTANCE MECHANISMS

The bacteria use three strategies to become resistant to tetracycline: efflux systems, ribosomal protection proteins and enzymatic inactivation. Acquisition of tetracycline resistance (*tet*) and oxytetracylcine (*otr*) gene is responsible for the development of tetracycline resistance in most bacteria. Approximately 38 different *tet* and *otr* genes have been reported so far. Out of the 38 different *tet* and *otr* genes, 23 genes encode energy dependent efflux proteins, 11 genes encode for ribosome protection proteins and 3 genes encode an inactivating enzyme (Table 4.3) (Roberts, 2005).

TABLE 4.3 Distribution of *tet* Genes in Bacteria (Roberts, 2005)

Gene	Mechanism	Bacterial Source
tet(A)	Efflux	*Acinetobacter, Haemophilus, Veillonella*
tet(B)	Efflux	*Acinetobacter, Brevundimonsa, Neisseria, Photobacterium, Pseudomonas*
tet(C)	Efflux	*Aeromonas, Chlamydia*
tet(D)	Efflux	*Alteromonas*
tet(E)	Efflux	–
tet(G)	Efflux	*Escherichia, Providencia*
tet(H)	Efflux	*Actinobacillus, Acinetobacter, Moraxella, Pasteurella*
tet(J)	Efflux	–
tet(K)	Efflux	*Lactobacillus, Norcardia, Streptomyces*
tet(L)	Efflux	*Actinobacillus, Morganella, Norcardia, Salmonella, Veillonella*
tet A(P)	Efflux	–
tet(V)	Efflux	–
tet(Y)	Efflux	–
tet(Z)	Efflux	–
tet(30)	Efflux	–
tet(31)	Efflux	–
tet(33)	Efflux	*Corynebacterium*
tet(35)	Efflux	*Stenotrophomonas, Vibrio*

TABLE 4.3 *(Continued)*

Gene	Mechanism	Bacterial Source
tet(38)	Efflux	*Staphylococcus*
tet(39)	Efflux	*Acinetobacter*
tcr3	Efflux	–
Otr(B)	Efflux	–
Otr(C)	Efflux	–
tet(M)	Ribosomal protection	*Acinetobacter, Afipia, Enterobacter, Erysipelothrix, Escherichia, Klebsiella, Lactobacillus, Lactococcus, Microbacterium, Mitsuokella, Mycobacterium, Neisseria, Prevotella, Porphyromonas, Ralstonia, Photobacterium, Pseudomonas, Selenomonas, Streptomyces, Vibrio*
tet(O)	Ribosomal protection	*Megasphaera, Neisseria*
tet(S)	Ribosomal protection	*Lactococcus, Lactobacillus, Veillonella*
tet(T)	Ribosomal protection	–
tetB (P)	Ribosomal protection	–
tet(Q)	Ribosomal protection	*Neisseria*
tet(W)	Ribosomal protection	*Actinomyces, Arcanobacterium, Bacillus, Butyrivibrio*
–	–	*Clostridium, Lactobacillus, Mitsuokella, Megasphaera, Neisseria, Prevotella, Porphyromonas, Roseburia, Selenomonas, Staphylococcus, Streptococcus, Streptomyces, Veillonella*
tet(32)	Ribosomal protection	*Clostridium*
tet(36)	Ribosomal protection	*Bacteroides, Clostridium, Lactobacillus*
Tet	Ribosomal protection	–
Otr(A)	Ribosomal protection	–
tet(X)	Enzymatic inactivation	–
tet(34)	Enzymatic inactivation	*Aeromonas, Pseudomonas, Serratia, Vibrio*
tet(X37)	Enzymatic inactivation	*Unknown*
tet(U)	Unknown	*Staphylococcus*

Energy-dependent and membrane-associated proteins, encoded by *tet* and *otr* genes, decrease intracellular concentration of tetracycline. Thereby it protects the bacterial ribosomes *in vivo* (Roberts, 2005). Upon binding of tetracycline, a conformational change occurs in the ribosome, which disrupts the elongation cycle and protein synthesis stops. The ribosomal protection proteins interact with the ribosome and cause an allosteric change of tetracycline binding site, which results in the release of tetracycline molecule from the ribosome. Thus, ribosome regains its active conformation and starts protein synthesis. The ribosomal protection protein requires GTP to be functional. Tet(O) and Tet(M) are extensively studied ribosomal protection proteins (Connell et al., 2003).

An NADPH-dependent oxidoreductase, encoded by *tet(X)*, inactivates tetracycline in the presence of oxygen and NADPH. The *tet(X)* gene is found only in strict anaerobe, bacteroides (Chopra & Roberts, 2001). The *tet(34)* in *Vibreo* spp. encodes xanthine-guanine phosphoribosyl transferases, which inactivates tetracycline (Nonaka & Suzuki, 2002).

4.2.9 TRIMETHOPRIM

4.2.9.1 HISTORY AND MODE OF ACTION

Trimethoprim, 2,4-diamino-5-(3,4,5-trimethoxy-benzyl)-pyrimidine was synthesized in 1962 and being explored for its antibacterial activity (Barbara et al., 1962). All the later agents have been developed from older antibiotics, which belong to families of agents. Like sulfonamides, trimethoprim is also a synthetic antibiotic that belongs to the diaminopyrimidine group of compounds (Skold, 2001).

Trimethoprim binds to dihydrofolate reductase (DHFR) and inhibits its activity. The DHFR catalyzes the reduction of dihydrofolate to tetrahydrofolate. Thus, it is also known as antifolate (Skold, 2001).

4.2.9.2 RESISTANCE MECHANISMS

There are three types of intrinsic trimethoprim resistance: one, chromosomally located in Tn7 transposon, second, thymine auxotrophic mutation in bacteria and third, mutational changes in the chromosomally encoded DHFR. It has been reported that Tn7 transposon is responsible for trimethoprim resistance. The Tn7 specifically and efficiently transposes into the *Escherichia coli* chromosome (Lichtenstein & Brenner, 1981). Under selective pressure of trimethoprim, the mutation occurs in the gene coding for thymidilate synthetase. The *thy*-mutants require thymine exogenously to synthesize DNA and cause trimethoprim resistance (Maskell et al., 1978). The mutation in *dhfr* gene leads to the overproduction of altered enzyme and trimethoprim resistance (de Groot et al., 1988; Powell et al., 1991). Similarly, several 100-fold overproductions of the chromosomally encoded DHFR are reported in highly trimethoprim resistant *E. coli*. The enzyme overproduction is due to a mutational substitution of a glycine for a tryptophan at position 30 (Flensburg & Skold, 1987).

In 1972, plasmid mediated trimethoprim resistance was first reported in *E. coli* and *K. aerogenes*, isolated from infected urines (Fleming et al., 1972). During investigations on the R-plasmids mediated trimethoprim resistant *E. coli* and *Citrobacter*, it has been found that two types of R-plasmids DHFRs were identified. Both were different from the chromosomally encoded enzyme in terms of binding of dihydrofolate, NADPH, folate, and 2,4 diamino pyrimidine (Pattishall et al., 1977). Based on phylogenetic analysis, plasmid borne DHFR can be classified into two families. The family 1DHFR includes enzyme type I, V, VI, VII, and Ib. They are 64 to 88% identical. They are a homodimer of pentapeptide containing 157 amino acids. They confer a very high level of trimethoprim resistance. The family 2 DHFRs includes type IIa, IIb, and IIc.

They are homotetramer proteins of 78 amino acids (Huovinen et al., 1995). The Tn5086 transposon, 15.3 kb trimethoprim resistant, was found in two distinct plasmids. They contain the *dhfrVII* gene that encodes DHFR which has got glutamate instead of aspartic acid residue at the 27th position of the chromosomally encoded enzyme from *E. coli* (Sunstrom et al., 1993). In trimethoprim resistant *E. coli* and *Shigella* spp., *dhfrXII* gene has been found in the Tn21 like element (Heikkila et al., 1993).

4.3 NATURAL SMALL MOLECULES AS A SOURCE OF BIOACTIVE AND THERAPEUTIC AGENT

Antibiotics derived from natural sources, semi-synthetic processes, and synthetic processes, are commonly used for the treatment of ailments. Researchers believed that the infectious diseases would be soon eradicated from this world. However, nature has given strength to every living organism to fight for survival under adverse conditions. With the introduction of novel antibiotics in clinical usage, resistance to them also evolved in microorganisms. Over the past 70 years, bacteria have been exposed to different antibiotics during treatment of human diseases, and many adapted genetic traits to nullify the effect of antibiotics. Since past two decades, several reports indicating the development of resistance in bacteria not only to one antibiotic but multiple antibiotics have been reported (Walsh, 2000), which disturbs the scientist and doctors alike.

Spreading of resistance to antibiotics has fascinated the researcher to develop new antibiotics. However, it is predicted from history that new antimicrobials will have a short life. Thus, resistance to new drugs leads shortage of antimicrobials as the discovery of new antimicrobial drug is not enough to cope with the fast emergence of resistant strains (Levy & Marshall, 2004). This requires new potent antimicrobial compounds as an alternative to existing antibiotics (Cowan, 1999). Natural molecules are a more promising source of drugs. Natural molecules have diverse chemical structure and biochemical specificity. Biologically active natural molecules usually have low molecular weight; they are easily absorbed and metabolized in the body. Major resources of natural molecules include plants, marine organisms, and microbes (Harvey, 2008). The plant produces innumerable secondary metabolites with several biological activities. Hence, it is worthwhile to explore the potential of plants for the development of new therapeutic agents (Lahlou, 2013).

4.3.1 PHYTOCHEMICAL AS AN ANTIMICROBIAL AGENTS

The plant produces secondary metabolites which help in plant defense mechanisms against predation by microorganisms, insects, and herbivores. Some metabolites are responsible for odor and pigments production. These metabolites are known as phytochemicals. Depending on species and climate, they have a wide range of biological activity (Cowan, 1999). Phytochemicals can be separated into several categories: terpenes, phenolics, and nitrogen containing compounds (Chinou, 2008). Each plant species, genus, and family produce a mixture of these metabolites.

4.3.1.1 TERPENES

Terpene is a simple hydrocarbon molecule which contains isoprene units. Terpenes are the most structurally diverse plant molecule. The isoprene unit contains five carbon molecules. Therefore, isoprene is the most basic class of terpenes, which is known as hemiterpenes (C5). Monoterpene (C10) contains two isoprene units. Diterpenes (C20), triterpenes (C30) and tetraterpenes (C40) have two, three, and four terpene units, respectively. Sesquiterpenes (C15) contain three isoprene units. When oxygen is added to terpenes, they termed as terpenoids (Cowan, 1999; Taiz & Zeiger, 2010). Sterols are triterpenes which contain an alcoholic group. Sterol acts as a precursor for the synthesis of cholesterol (Taiz & Zeiger, 2010).

The Essential oils commonly comprise monoterpenes and sesquiterpenes. They have pharmaceutical properties like antifungal, antimalarial, etc. The isolated monoterpenes, diterpenoids, sesquiterpenes, and triterpenoids have been reviewed recently for their antibacterial activity (Kurek et al., 2011). Sesquiterpenes (Diterpenoids) isolated from diverse medicinal plants showed antibacterial activity against gram-positive bacteria and also have growth inhibition activity against *Mycobacterium tuberculosis* (Kurek et al., 2011; Garcia et al., 2012). Terpenoids are effective against bacteria and fungi. However, the mode of action of terpenoids is not identified yet. It is assumed that they interact with a lipophilic compound of the membrane and disrupt the membrane. It has been reported that essential oil of plant origin showed inhibitory effect on *Listeria monocytogenes* (Aureli et al., 1992). Essential oil of tea tree having a composition of terpinene 4-ol, α pinene, linalool, and α-terpineol have antibacterial activity and reported to disrupt the cell membrane of bacteria (Gustafson et al., 1998; Carson et al., 2002).

4.3.1.2 PHENOLICS

Plants produce a large number of secondary metabolites that have phenolic compounds. Many phenolic compounds protect plants from herbivores and plant pathogens. Some compounds act as an attractant of pollinator. Plants secrete many phenolic compounds to inhibit the growth of other plants near its periphery. Others act as a shield to protect from ultraviolet radiation damage (Taiz & Zeiger, 2010).

Polyphenolic compounds have potential antioxidant activity and antimicrobial activity. Simple phenolic acids, ferulic acid and gallic acids are effective against bacteria by rupturing or making pores in the cell membrane (Borges et al., 2013). Catechol and pyrogallol are the allelochemicals which are produced by plants. It is believed that they are bio-synthesized by plants to control plant interaction with pathogens. Both compounds are toxic to bacteria (Kocacaliskan et al., 2006). Quinones are also classified as a phenolic compound having potential antimicrobial properties. They make a stable complex with amino acids in microbial proteins and leads to the loss of functions (Cowan, 1999; Saleem et al., 2010).

Flavonoids are phenolic compounds, and most of them have flavone nucleus with two side aromatic rings. These compounds play a key role in the protection of plant from pathogens as well as predators and also play a major role in physiological functions (Cowan, 1999; Taiz & Zeiger, 2010). Antibacterial activity of flavonoids has been extensively studied. Olive leaves showed antibacterial activity and found to have luteolin 7-O-glucoside, rutin, apigenin 7-O-glucoside and luteolin 4'-O-glucoside (Pereira et al., 2007). Seven pure flavonoids were isolated and evaluated for their antibacterial activity. Isolated flavonoids show more activity on gram-negative bacteria (Basile et al., 1999). Epigallocatechin-3-gallate (EGCG) is a flavonoid compound found in high concentrations in green (unfermented) leaves of *Camellia sinensis* (tea). It has been found that consumption of green tea has a variety of health benefits. It also showed anticancer and antimicrobial properties. Gram-positive, gram-negative, and fungal pathogens, including methicillin-resistant *Staphylococcus aureus* (MRSA) and MDR *Acinetobacter baumannii,* were sensitive to EGCG (Taylor et al., 2005).

Tannin is high molecular weight phenolic phytochemical which is produced by diverse medicinal plants. Tannins can be categorized into two categories: hydrolysable and condensed tannins. Condensed tannins are a polymer of flavonoid units. They can be easily converted into anthocyanidins in the presence of strong acids. Hydrolyzed tannins are a polymer of phenolic acids like gallic acids and simple sugars (Cowan, 1999; Taiz & Zeiger, 2010). Like other phenolic compounds, tannins also protect from herbivores and plant pathogens. They act as a repellent and force animals away from the plant (Taiz & Zeiger, 2010). However, it has been reported that the ingestion of beverages having a substantial amount of tannin, especially green teas, and red wines, can provide immunity against ailments in humans (Cowan, 1999; Taiz & Zeiger, 2010). The duramen has a higher concentration of tannin that inhibits decaying activity initiated by bacterial as well as fungal growth (Taiz & Zeiger, 2010). Thus, it also has antimicrobial activity. Gallotannins were isolated from Mango (*Mangifera indica* L.) Kernels and showed gram-positive bacterial growth inhibition activity. However, bacterial growth was regained by the addition of iron to the medium. Thus, it has been proved that the inhibitory effects are due to their iron-binding properties (Engels et al., 2009).

4.3.1.3 NITROGEN-CONTAINING COMPOUNDS

Structurally, alkaloids are a vastly distinct group of compounds that comprise an aromatic ring structure and a nitrogen atom. Alkaloids are produced by a wide variety of medicinal plants. Alkaloids have the excellent property to act as either hydrogen acceptor or hydrogen-donor for hydrogen bonding which plays a crucial role in the interaction of the molecule with enzymes, proteins, and receptors (Kittakoop et al., 2014). Berberine is isoquinoline alkaloids. It usually occurs in the root, rhizome, and stem bark of different group of plants like *Berberis aquifolium* (Oregon grape), *Berberis vulgaris* (Barberry), *Berberis aristata* (Tree Turmeric), etc. Berberine showed antibacterial activity against enterovirulent multidrug resistance *Escherichia coli* (Bandyopadhyay et al., 2013). Two new benzophenanthrene alkaloids, namely 8-acetonyldihydronitidine and 8-acetonyldihydroavicine, were purified from *Zanthoxylum tetraspermum* and *Z. caudatum* that possessed note-worthy antibacterial activity (Nissanka et al., 2001). The indole alkaloid found present in the ethanol extract of *Tabernaemontana catharinensis* root bark which showed an antimicrobial activity (Medeiros et al., 2011).

4.4 CONCLUSION

Antimicrobial agents are essential for our healthcare needs as they are lifesaving arsenals. However, uncontrolled use of them has emerged antimicrobial resistance as soon as they were introduced. Increased development of drug resistance in pathogens has become a viable problem for the healthcare system. Antimicrobial-resistant microorganisms have higher morbidity and mortality, and in turn, it leads to higher healthcare costs. Thus, it is very important that discovery and production pace of new active molecules should be maintained on a scale to fulfill increased demand for antibiotics. The pharmaceutical companies are in search of novel antimicrobial agents, but they are non-productive. However, it indicates that health care will face an antibiotic crisis. To deal with the global crisis of antibiotics, coordinated efforts should be made to implement new policies, and research efforts on novel antimicrobial drugs. Natural small molecules, which are produced by plants, animals, and microbes, could be the alternative of source of antimicrobial agents.

KEYWORDS

- **acetyltransferases**
- **adenosine monophosphate**
- **antibiotics**
- **bacteriophages**
- **integrons**
- **lipopolysaccharide**
- **morbidity**
- **plasmids**
- **resistance transposons**

REFERENCES

Alekshun, M. N., & Levy, S. B., (1997). Regulation of chromosomally mediated multiple antibiotic resistance: The mar regulon. *Antimicrobial Agents and Chemotherapy, 41*, 2067–2075.

Alekshun, M. N., & Levy, S. B., (2007). Molecular mechanisms of antibacterial multidrug resistance. *Cell, 128*, 1037–1050.

Alexander, F., (1929). On the antibactrial action of cultures of a penicillium, with special reference to their use in the isolation of *B. influenzae*. *The British Journal of Experimental Pathology, 10*, 226–236.

Anderson, L. M., Henkin, T. M., Chamliss, G. H., & Bott, K. F., (1984). New chloramphenicol resistance locus in *B. subtilis*. *Journal of Bacteriology, 158*, 1386–1388.

Aureli, P., Costantinf, A., & Zolea, S., (1992). Antimicrobial activity of some plant essential oils against *Listeria monocytogenes*. *Journal of Food Protection, 55*, 344–348.

Bandyopadhyay, S., Patra, P. H., Mahanti, A., Mondal, D. K., Dandapat, P., Bandyopadhyay, S., Samanta, I., et al., (2013). Potential antibacterial activity of berberine against multi drug resistant enterovirulent *Escherichia coli* isolated from yaks (*Poephagus grunniens*) with haemorrhagic diarrhea. *Asian Pacific Journal of Tropical Medicine*, 315–319.

Barbara, R. B., Falco, E. A., & Hitchings, G. H., (1962). 5-Benzyl-2,4-diaminopyrimidines as antimicrobial agents. I. Synthesis and antimicrobial activity *in vitro*. *Journal of Medicinal Chemistry, 5,* 1103–1123.

Barbosa, T. M., & Levy, S. B., (2000). The impact of antibiotic use on resistance development and persistance. *Drug Resistance Updates, 3*, 303–311.

Basile, A., Giordanoa, S., LoÂpez-SaÂez, J. A., & Cobianchi, R. C., (1999). Antibacterial activity of pure flavonoids isolated from mosses. *Phytochemistry, 52*, 1479–1482.

Baugman, G. A., & Fahnestock, S. R., (1979). Chloramphenicol resistance mutation in *E. coli* which maps in the major ribosomal protein gene cluster. *Journal of Bacteriology, 137*, 1315–1323.

Becker, B., & Cooper, M. A., (2013). Aminoglycoside antibiotics in the 21st century. *ACS Chemical Biology, 8*, 105–115.

Bodey, G. P., (1990). Penicillins, monobatams and carbapenems. *Texas Heart Institute Journal, 17*, 315–329.

Borges, A., Ferreira, C., Saavedra, M. J., & Simoes, M., (2013). Antibacterial activity and mode of action of ferulic and gallic acids against pathogenic bacteria. *Microbial Drug Resistance*, 1–10.

Bradford, P. A., (2001). Extended spectrum *β*-lactamases in the 21st century: Characterization, epidemiology and detection of this important resistance threat. *Clinical Microbiology Review*, *14*, 933–951.

Burns, J. L., Hedin, L. A., & Lien, D. M., (1989). Chloramphenicol resistance in *Pseudomonas cepacia* because of decreased permeability. *Antimicrobial Agents and Chemotherapy, 33*, 136–141.

Burns, J. L., Mendelman, P. M., Levy, J., Stull, T. L., & Smith, A. L., (1985). A permeabilty barrier as a mechanism of chloramphenicol resistance in *Haemophilus influenzae*. *Antimicrobial Agents and Chemotherapy, 27*, 46–54.

Bush, K., (1989). Characterization of *β*-lactamases. *Antimicrobial Agents and Chemotherapy, 33*, 259–263.

Bush, K., Jacoby, G. A., & Medeiros, A. A., (1995). A functional classification scheme for *β*-lactamases and its correlation with molecular structure. *Antimicrobial Agents and Chemotherapy, 39*, 1211–1233.

Carson, C. F., Mee, B. J., & Riley, T. V., (2002). Mechanism of action of *Melaleuca alternifolia* (tea tree) oil on *Staphylococcus aureus* determined by time-kill, lysis, leakage and salt tolerance assays and electron microscopy. *Antimicrobial Agents and Chemotherapy, 46*, 1914–1920.

Chinou, I., (2008). Primary and secondary metabolites and their biological activity. In: Waksmundzka-Hajnos, M., Sherma, J., & Kowalska, T., (eds.), *Thin Layer Chromatography in Phytochemistry*. CRC Press, Boca Raton.

Chopra, I., & Roberts, M., (2001). Tetracycline antibiotics: Mode of action, applications, molecular biology and epidomology of bacterial resistance. *Microbiology and Molecular Biology Reviews, 65*, 232–260.

Connell, S. R., Tracz, D. M., Nierhaus, K. H., & Taylor, D. E., (2003). Ribosomal protection proteins and their mechanisms of tetracycline resistance. *Antimicrobial Agents and Chemotherapy, 47*, 3675–3681.

Cowan, M. M., (1999). Plant products as antimicrobial agents. *Clinical Microbiology Review, 12*, 564–582.

De Groot, R., Campos, J., Moseley, S. L., & Smith, A. L., (1988). Molecular cloning and mechanism of trimethoprim resistance in *Haemophilus influenzae*. *Antimicrobial Agents and Chemotherapy, 32,* 477–484.

De Stasio, E. A., Moazed, D., Noller, H. F., & Dahlberg, A. E., (1989). Mutations in 16S ribosomal RNA disrupt antibiotic-RNA interactions. *The EMBO Journal, 8*, 1213–1216.

Doi, Y., & Arakawa, Y., (2007). 16S ribosomal RNA methylation: Emerging resistance mechanism against aminoglycosides. *Clinical Infectious Diseases, 45*, 88–94.

Drawz, S. M., & Bonomo, R. A., (2010). Three decades of beta lactamase inhibitors. *Clinical Microbiology Reviews, 23*, 160–201.

Engels, C., Knodler, M., Zhao, Y., Carle, R., Ganzle, M. G., & Schieber, A., (2009). Antimicrobial activity of gallotannins isolated from mango (*Mangifera indica* L.) Kernels. *Journal of Agricultural Food Chemistry, 57*, 7712–7718.

Ettayebi, M., Prasad, S. M., & Morgan, E. A., (1985). Chloramphenicol-erythromycin resistance mutation in a 23S rRNA gene of *E. coli*. *Journal of Bacteriology, 162*, 551–557.

Fleming, M. P., Datta, N., & Grunberg, R. N., (1972). Trimethoprim resistance determined by R-factors. *British Medical Journal, 1*, 726–728.

Flensburg, J., & Skold, O., (1987). Massive overproduction of dihydrofolate reductase in bacteria as a response to the use of trimethoprim. *European Journal of Biochemistry, 162*, 473–476.

Fuda, C. C., Fisher, J. F., & Mobashery, S., (2005). Beta-lactam resistance in *Staphylococcus aureus*: The adaptive resistance of a plastic genome. *Cellular and Molecular Life Sciences, 62*, 2617–2633.

Galimand, M., Courvalin, P., & Lambert, T., (2003). Plasmid mediated high level resistance to aminoglycosides in *Enterobacteriaceae* due to 16S rRNA methylation. *Antimicrobial Agents and Chemotherapy, 47*, 2565–2571.

Garcia, A., Bocanegra-Garcia, V., Palma-Nicolas, J. P., & Rivera, G., (2012). Recent advances in antitubercular natural products. *European Journal of Medicinal Chemistry, 49*, 1–23.

George, A. M., & Hall, R. M., (2002). Efflux of chloramphenicol by the CmlA1 protein. *FEMS Microbiology Letter, 209*, 209–213.

Godfrey, A. J., Bryan, L. E., & Rabin, H. R., (1981). Beta-lactam resistant *P. aeruginosa* with modified penicillin binding proteins emerging during cystic fibrosis treatment. *Antimicrobial Agents and Chemotherapy, 19*, 705–711.

Gustafson, J. E., Liew, Y. C., Chew, S., Markham, J., Bell, H. C., Wyllie, S. G., & Warmington, J. R., (1998). Effects of tea tree oil on *Escherichia coli*. *Letters in Applied Microbiology, 26*, 194–198.

Hakenbeck, R., Tarpay, M., & Tamasz, A., (1980). Multiple changes of penicillin binding proteins in penicillin resistant clinical isolates of *Streptococcus pneumoniae*. *Antimicrobial Agents and Chemotherapy, 17*, 364–371.

Harrison, C. J., & Bratcher, D., (2008). Cephalosporins: Review. *Pediatrics Review, 29*, 264–273.

Harvey, L. H., (2008). Natural products in drug discovery. *Drug Discovery Today, 13*, 894–901.

Heikkila, E., Surnik, M., Sundstrom, L., & Huovinen, P., (1993). A novel dihydrofolate reductase cassette inserted in an integron borne on a Tn21 like element. *Antimicrobial Agents and Chemotherapy, 37*, 1297–1304.

Hiasa, H., Yousef, D. O., & Marians, K. J., (1996). DNA strand cleavage is required for replication fork arrest by a frozen topoisomerase-quinolone-DNA ternary complex. *Journal of Biological Chemistry, 271*, 26424–26429.

Hiroshi, N., (1996). Multidrug efflux pumps of gram-negative bacteria. *Journal of Bacteriology, 178*, 5853–5859.

Hooper, D. C., (2000). Mechanisms of action and resistance of older and newer fluoroquinolones. *Clinical Infectious Diseases, 31*, S24–S28.

Hooper, D. C., (2001). Emerging mechanisms of fluoroquinolone resistance. *Emerging Infectioius Diseases, 7,* 337–341.

Huovinen, P., Sundstrom, L., Swedberg, G., & Skold, O., (1995). Trimethoprim and sulfonamide resistance. *Antimicrobial Agents and Chemotherapy, 39*, 279–289.

Izard, T., & Jacqualine, E., (2000). The crystal structures of chloramphenicol phosphotransferase reveal a novel inactivation mechanism. *The EMBO Journal, 19*, 2690–2700.

Kaul, M., Barbieri, C. M., & Pilch, D. S., (2004). Fluorescence based approach for detecting and characterizing antibiotic induced conformational changes in ribosomal RNA: Comparing aminoglycoside binding to prokaryotic and eukaryotic ribosomal RNA sequences. *Journal of American Chemical Society, 126*, 3447–3453.

King, D. E., Malone, R., & Lilley, S. H., (2000). New classification and update on the quinolone antibiotics. *American Family Physician, 61*, 2741–2748.

Kittakoop, P., Mahidol, C., & Ruchirawat, S., (2014). Alkaloids as important scaffolds in therapeutic drugs for the treatments of cancer, tuberculosis, and smoking cessation. *Current Topics in Medicinal Chemistry, 14*, 239–252.

Kocacaliskan, I., Talan, I., & Terzi, I., (2006). Antimicrobial activity of catechol and pyrogallol as allelochemicals. *Z Naturforsch C.*, 639–642.

Kurek, A., Grudniak, A. M., Kraczkiewicz-Dowjat, A., & Wolska, K. I., (2011). New antibacterial therapeutics and strategies. *Polish Journal of Microbiology, 60*, 3–12.

Kwon, D., & Lu, C. D., (2006). Polyamines induce resistance to cationic peptide, aminoglycoside and quinolone antibiotics in *Pseudomonas aeruginosa PAO1*. *Antimicrobial Agents and Chemotherapy, 50*, 1615–1622.

Lahlou, M., (2013). The success of natural products in drug discovery. *Pharmacology and Pharmacy, 4*, 17–31.

Lesher, G. Y., Froelich, E. J., Gruett, M. D., Bailey, J. H., & Brundage, R. P., (1962). 1,8-Naphthridine derivatives: A new class of chemotherapy agents. *Journal of Medicinal Chemistry, 5*, 1063–1065.

Levy, S. B., & Marshall, B., (2004). Antibacterial resistance worldwide: Causes, challenges and responses. *Nature Medicine, 10*, S122–S129.

Li, X. Z., & Nikaido, H., (2009). Efflux mediated drug resistance in bacteria: An update. *Drugs, 69*, 1555–1623.

Lichtenstein, C., & Brenner, S., (1981). Site specific properties of Tn7 transposition of into *Escherichia coli* chromosome. *Molecular and General Genetics, 183*, 380–387.

Macfarlane, E. L. A., Kwasnicka, A., & Hancock, R. E. W., (2000). Role of *Pseudomonas aeruginosa* PhoP–PhoQ in resistance to antimicrobial cationic peptides and aminoglycosides. *Microbiology, 146*, 2543–2554.

Maskell, R., Okubadejo, O. A., Payne, R. H., & Pead, L., (1978). Human infections with thymine requiring bacteria. *Journal of Medical Microbiology, 11*, 33–46.

Mclean, J. W., Schwab, J. L., Hillegas, A. B., & Schlingman, A. S., (1949). Susceptibiltiy of microorganisms to chloramphenicol (chloromycetin). *The Journal of Clinical Investigation, 28*, 953–963.

Medeiros, M. R., Prado, L. A., Fernandes, V. C., Figueiredo, S. S., Coppede, J., Martins, J., Fiori, G. M., et al., (2011). Antimicrobial activities of indole alkaloids from tabernaemontana catharinensis. *Natural Product Communications, 6*, 193–196.

Murray, J. A., & Shaw, W. V., (1997). *O*-Acetyltransferases for chloramphenicol and other natural products. *Antimicrobial Agents and Chemotherapy, 41*, 1–6.

Nissanka, A. P., Karunaratne, V., Bandara, B. M., Kumar, V., Nakanishi, T., Nishi, M., Inada, A., et al., (2001). Antimicrobial alkaloids from zanthoxylum tetraspermum and caudatum. *Phytochemistry, 56*, 857–861.

Nonaka, L., & Suzuki, S., (2002). New Mg^{2+} dependent oxytetracycline resistance determinant tet34 in *Vibreo* isolates from marine fish intestinal contents. *Antimicrobial Agents and Chemotherapy, 46*, 1550–1552.

Oethinger, M., Podglajen, I., Kern, W. V., & Levy, S. B., (1998). Overexpression of the *marA* or *soxS* regulatory gene in clinical topoisomerase mutants of *Escherichia coli. Antimicrobial Agents and Chemotherapy, 42*, 2089–2094.

Pato, M. L., & Brown, G. M., (1963). Mechanisms of resistance of *E. coli* to sulfonamides. *Archieves of Biochemistry and Physics, 103*, 443–448.

Pattishall, K. H., Acar, J., Burchall, J. J., Goldstein, F. W., & Harvey, R. J., (1977). Two distinct types of trimethoprim resistant dihydrofolate reductase specified by R-plasmids of different compatibility groups. *The Journal of Biological Chemistry, 252*, 2319–2323.

Pereira, A. P., Ferreira, I. C. F. R., Marcelino, F., Valentão, P., Andrade, P. B., Seabra, R., Estevinho, L., et al., (2007). Phenolic compounds and antimicrobial activity of olive (*Olea europaea* L. Cv. Cobrancosa) leaves. *Molecules, 12*, 1153–1162.

Pootoolal, J., Neu, J., & Wright, G. D., (2002). Glycopeptide antibiotic resistance. *Annual Review Pharmacology and Toxicology, 42*, 381–408.

Powell, M., Hu, Y., & Livemore, D. M., (1991). Resistance to trimethoprim in *Haemophilus influenzae. Infection, 19*, 174–177.

Ramirez, M. S., & Tolmasky, M. E., (2010). Aminoglycoside modifying enzymes. *Drug Resistance Updates, 13*, 151–171.

Roberts, M. C., (2002). Resistance to tetracycline, macrolide–licosamide–streptogramin, trimethoprim and sulfonamide drug class. *Molecular Biotechnology, 20*, 261–284.

Roberts, M. C., (2005). Update on acquired tetracycline resistance gene. *FEMS Microbiology Letter, 245*, 195–203.

Roberts, M. C., Sutcliffe, J., Courvalin, P., Jensen, L. B., Rood, J., & Seppala, H., (1999). Nomenclature for macrolide and macrolide–lincosamide–strptogramin-B resistant determinants. *Antimicrobial Agents and Chemotherapy, 43*, 2823–2830.

Ruiz, J., (2003). Mechanisms of resistance to quinolones: Target alterations, decreased accumulation and DNA gyrase protection. *Journal of Antimicrobial Chemotherapy, 51*, 1109–1117.

Saleem, M., Nazir, M., Ali, M. S., Hussain, H., Lee, Y. S., Riaz, N., & Jabbar, A., (2010). Antimicrobial natural products: An update on future antibiotic drug candidates. *Natural Product Reports, 27*, 238–254.

Schwarz, S., Kehrenberg, C., Doublet, B., & Cloeckarert, A., (2004). Molecular basis of bacterial resistance to chloramphenicol and florfenicol. *FEMS Microbiology Review, 28*, 519–542.

Shea, M. E., & Hiasa, H., (1999). Interactions between DNA helicases and frozen topoisomerase IV-quinolone-DNA ternary complexes. *Journal of Biological Chemistry, 274*, 22747–22754.

Skold, O., (1976). R-factor mediated resistance to sulfonamides by a plasmid borne, drug resistant dihydropteroate synthase. *Antimicrobial Agents and Chemotherapy, 9*, 49–54.

Skold, O., (2000). Sulfonamide resistance: Mechanisms and trends. *Drug Resistance Updates, 3*, 155–160.

Skold, O., (2001). Resistance to trimethoprim and sulfonamides. *Veterinary Research, 32*, 261–273.

Springer, B., Kidan, Y. G., Prammananan, T., Ellrott, K., Erik, C., Bottger, C. E., & Sander, P., (2001). Mechanisms of streptomycin resistance: Selection of mutations in the 16S rRNA gene conferring resistance. *Antimicrobial Agents and Chemotherapy, 45*, 2877–2884.

Strahilevitz, J., Jacoby, G. A., Hooper, D. C., & Robicsek, A., (2009). Plasmid mediated quinolone resistance: A multifaceted threat. *Clinical Microbiology Reviews, 22*, 664–689.

Sunstrom, L., Swedberg, G., & Skold, O., (1993). Characterization of transposon Tn5086, carrying the site specifically inserted gene *dhfrVII* mediating trimethoprim resistance. *Journal of Bacteriology, 175*, 1796–1805.

Swedberg, G., & Skold, O., (1983). Plasmid borne sulfonamide resistance determinants studied by restriction enzyme analysis. *Journal of Bacteriology, 153*, 1228–1237.

Swedberg, G., & Skold, O., (1991). Genetic analysis of sulfonamide resistance and its dissemination in gram negative bacteria illustrate new aspects of R-plasmid evolution. *Antimicrobial Agents and Chemotherapy, 35*, 1840–1848.

Taiz, L., & Zeiger, E., (2010). Secondary metabolites and plant defense. In: *Plant Physiology* (5th edn, pp. 283–308). Sinauer Associates: Sunderland, USA.

Taylor, P. W., Hamilton-Miller, J. M. T., Paul, D., & Stapleton, P. D., (2005). Antimicrobial properties of green tea catechins. *Food Science Technology Bulletin, 2*, 71–81.

Tipper, D. J., & Strominger, J. L., (1965). Mechanism of action of penicillins: A proposal based on their structural similarity to acyl-D-alanyl-D-alanine. *Proceedings of the National Academy of Sciences, 54*, 1133–1141.

Toro, C. S., Lobos, S. R., Calderon, I., Rodriguez, M., & Mora, G. C., (1990). Clinical isolate of a porinless *Salmonella typhi* resistant to high levels of chloramphenicol. *Antimicrobial Agents and Chemotherapy, 34*, 1715–1719.

Tran, J. H., & Jacoby, G. A., (2002). Mechanism of plasmid-mediated quinolone resistance. *Proceedings of the National Academy of Sciences, 99*, 5638–5642.
Walsh, C., (2000). Molecular mechanisms that confer antibacterial drug resistance. *Nature, 406*, 775–781.
Wang, X., & Quinn, P. J., (2010). Lipopolysaccharide: Biosynthetic pathway and structure modification. *Progress in Lipid Research, 49*, 97–107.
Weisblum, B., (1998). Macrolide resistance. *Drug Resistance Updates, 1*, 29–41.
Willmott, C. J., Critchlow, S. E., Eperon, I. C., & Maxwel, A., (1994). The complex of DNA gyrase and quinolone drugs with DNA forms a barrier to transcription by RNA polymerase. *Journal of Molecular Biology, 242*, 351–363.
Wise, E. M., & Abou-Donia, M. M., (1975). Sulfonamide resistance mechanism in *Escherichia coli*: R-plasmids can determine sulfonamide resistant dihydropteroate synthases. *Proceedings of The National Academy of Sciences, 72*, 2621–2625.
Wright, G. D., & Sutherland, A. D., (2007). New strategies for combating multidrug resistant bacteria. *Trends in Molecular Medicine, 13*, 260–267.
Yoshida, H., Bogaki, M., Nakamura, M., & Nakamura, S., (1990). Quinolone resistance-determining region in the DNA Gyrase gyrA Gene of *Escherichia coli. Antimicrobial Agents and Chemotherapy, 34,* 1271, 1272.

CHAPTER 5

Personalized Medicine: Understanding the Individual Intricacies

JONAID AHMAD MALIK

DBME, Indian Institute of Technology (IIT), Ropar, Punjab, India

ABSTRACT

Personalize medicine/precision medicine/stratified medicine/targeted medicine/pharmacogenomics, as the name suggests, refers to the medicines that are designed and formulated based on specific characteristics of a particular person, which in simple terms can be defined as the right drug providing for the right patient at the right dose at the right time. The concept of personalized medicine (PM) came when the sequencing of the entire human genome began in the 21st century. The developments in biotechnology have revolutionized the science in the modern world and allowed scientists to elucidate the role of genetics in behaving with chemotherapy which in turn gave birth to pharmacogenomics. PM has the potential to match therapy with a therapeutic response with the lowest toxicity to ensure better patient care. PM promises to eradicate health-related problems by enabling early diagnosis, risk assessments, and optimal treatments for each particular patient (Vogenberg et al., 2010). Personalized medicine provides a platform to explore the agents that are targeted to patient groups that do not respond to conventional therapies and where the therapies have failed. The most important in Personalize medicine is to elucidate the variations in DNA and RNA characteristics

Modern Biotechnology in Healthcare: Advances and Applications. Sheikh Umar Ahmad (Ed.)

with the drug response, seeking to understand the expression of mRNA, which affects the body's response to the medications.

5.1 INTRODUCTION

The best patient care is achieved by having the best response and highest safety margin which becomes possible by the potential of matching therapy with patient characteristics through personalized medicine (PM), and it allows the particular patient to be looked for early-stage diagnosis, risk assessments, and proper treatment. The lowering of economic costs with an overall improvement in health is the main characteristic promise that has been made by it (Vogenberg et al., 2010). For medical applications like patient care, including disease markers for targeted chemotherapy and early disease diagnosis and treatment matching then, how individual genes have to be studied is defined by medical genetics, which is rapidly revolutionizing healthcare, making it more personalized than ever (Offit, 2011). In the National Institute of General Medical Sciences, 2013 the first human genome was completely mapped down and cataloged all the genes in a human (Blix, 2014). The Human Genome Project (HGP) was completed in 2003 and approximately 21,000 genes in a human were cataloged, respectively (Stratton et al., 2009). Researchers in medical genetics have found that not only single genes predict disease but there are other factors which are equally important. Also medical genomics is a broad term which includes the study of individual genes as well as factors (Epigenetic and environmental factors) that have active participation in inheritance patterns and how health or medical care can be matched to the individual genetic makeup (Blix, 2014). The nurses need to be stay informed about the health/medical care to be delivered in which PMs represent a major change like the clinical usage, pharmacogenomics, and the social impact, respectively (Blix, 2014). Genetics plays a great role fundamentally in cancer so, advances in genetics have a prominent impact on cancer therapeutics, and nurses as well as other members of the health care team will be looked at by the patients more importantly (Riley et al., 2012). For obtaining approval of PMs from regulatory agencies like the Food and Drug Administration (FDA) there have been many challenges (Goetz & Schork, 2018). Physicians, health care executives, insurance companies, and finally patients have many issues in accepting

PMs (Goetz & Schork, 2018). Our understanding of Human diseases has advanced more significantly with more focus on the sequencing of the human genome, particularly over the past 10 years (Tremblay & Hamet, 2013). Advances in the technological process in analyzing intergenic variants like SNP (single nucleotide polymorphisms), millions of genes and copy number variations per person have enhanced our understanding of inter person differences in the overall genetic profile (Tremblay & Hamet, 2013). The common genetic differences associated with many diseases have been successfully identified by GWAS (genome-wide association studies) (Tremblay & Hamet, 2013). There occurs a change in the way of new drugs that are being discovered from all the knowledge that we have gained so far, particularly to treat cancer (Tremblay & Hamet, 2013). Diagnosis, prognosis, risk assessment, and treatment are the main application of genomics in PM (Tremblay & Hamet, 2013).

The classical examples in personalized medicines of the genomic investigation areas:

i. Fragile X syndrome, the severity of which is predicted by the number of tri-nucleotide repeats (Sherman et al., 2005).
ii. Risk assessment of breast cancer through BRC1/2 (Tremblay & Hamet, 2013).
iii. Diagnosis of breast cancer subtypes via gene expression analysis (Sørlie et al., 2001).
iv. The women possessing HER2-positive breast cancer, Herceptin® also called trastuzumab administration is confined based on diagnosis (Piccart-Gebhart et al., 2005; Romond et al., 2005).
v. Diminished therapeutic response due to CYP2C19 variants (Shuldiner et al., 2009).

5.1.1 NEED FOR PERSONALIZED MEDICINE (PM)? WHY?

- Subjects having similar symptoms but distinct disorders;
- Genetics will help in medical intervention more specifically;
- In some subjects, drugs work while not in others;
- To reduce allergic and adverse events;
- Design best target based therapies;
- Need for evidence-based medicines;

- A good of percentage of drugs available are not efficacious in treating ailments;
- Some subjects respond to less dose, some to more dose, some to normal and some do not respond at all (Figure 5.1).

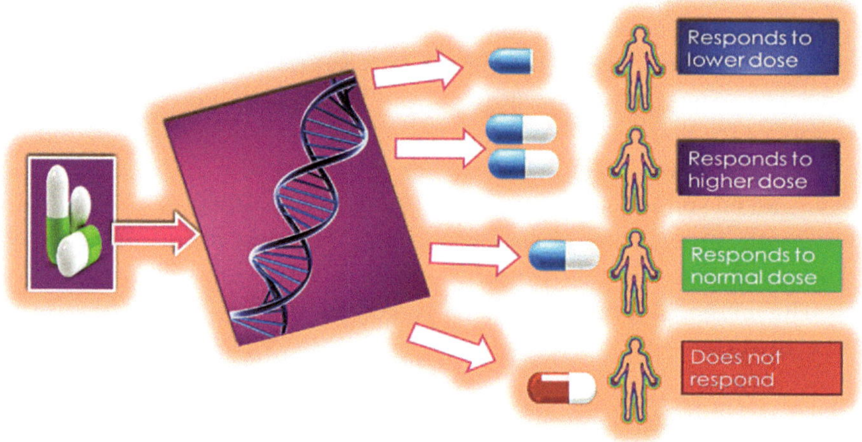

FIGURE 5.1 Pictorial overview of drug responses with different genetic profiles.

5.2 ENVIRONMENTAL FACTORS INFLUENCING GENOMICS

Environmental influences many biological pathways and are responsive to environmental stimuli in the most complex and polygenic diseases, like diabetes, hypertension and atherosclerosis, which are heritable (Hunter, 2005). Cigarette smoking constitutes one of the big environmental risk factors for atherosclerosis (Tremblay & Hamet, 2013). The carrier state of the ApoE epsilon 4 allele with cigarette smoking is having a synergistic effect and increases the atherosclerosis risk, respectively (Mertens, 2010). A group of genomic markers that were related to smoking cessation was identified by GWAS, allowing to develop a quit-success genotype score a prototype by its authors (Rose et al., 2010). Now the data revealed provide ways of smoking stoppage which may be based on genotype score, nicotine dependence, or carbon monoxide observance (Tremblay & Hamet, 2013). A personalized and adaptive approach to smoking cessation cure needs support as the study has not been replicated in the independent cohorts and based on genotypic as well as phenotypic characteristics of a particular smoker, PM offers smoking cessation ways (Tremblay & Hamet, 2013). Computational modeling needs to be developed to investigate how polygenes talk with factors associated with the environment to perturb the phenotype expression as the gene environmental interactions are complex in nature (Tremblay & Hamet, 2013). Genes play a varying role in both females as well as males in the same families when the interaction between current and post smoking with cardiovascular responses to stress, hypertension, obesity, and BMI (body mass index) modulated by synaptic plasticity was recently described (Nikpay et al., 2012).

5.3 GENOMIC STUDIES: INSIGHTS

Genetic makeup makes humans distinct from one another, simultaneously it also plays a role in the susceptibility to various disorders/diseases (Jiang & Wang, 2010). New technological advances in genomic sequencing have revolutionized the studies on patterns of gene expression, miRNA expression, DNA copy number variations, single nucleotide polymorphisms, and factors assessed by utilizing multiple pathological and clinical parameters, respectively (Panel, 2001; Crivellari et al., 2003). Two gene prognosis signatures panels for the prognosis of breast cancer were developed independently and were validated by two teams respectively one gene signature consisting of 76 genes and the other was comprised of 70 genes (Van De Vijver et al., 2002; Wang et al., 2005). Prediction success was better in both gene signature panels for distant relapse risk than that of adjuvant online software (Jiang & Wang, 2010). In terms of prediction ability of distant relapse risk of breast cancer both gene signature panels show similarities and in the two-breast cancer prognostic gene signature panels there were few genes that overlapped (Jiang & Wang, 2010). Due to differences in the characteristics of the patient population, there was a small overlap. The patients included were of under 55 years of age in 70-gene identification study and patients ages were under 40–70 years old with an average of 54 years in Wang et al. study (Panel, 2001; Wang et al., 2005).

The reference or standard human genome sequence completion has facilitated the cataloging and discovery of differences/variations in the genome sequences among subjects including both diseased as well healthy and in different populations (Ginsburg & Willard, 2009) (Figure 5.2).

About 10–15 million genomic sequence variants have been estimated and are of sufficient frequency in one or more populations to be considered genetically polymorphic in humans (Ginsburg & Willard, 2009). In a single or few subjects, countless rare variants exist and will be accessible via direct genome sequencing (Geoffrey & Huntington, 2009). A key question is to what extent genome impact the likelihood of disease onset, alarm the natural history of the disease, or gives relevant clues about the management of the disease if we talk in the context genomic PM (Geoffrey & Huntington, 2009).

5.4 PHARMACOGENOMICS IN PERSONALIZED MEDICINE (PM)

Pharmacogenomics allows the use of genetics to optimize the drug discovery and development process. The main focus is too tailor-made drugs in which right drugs are developed for the right patient and we can say that it is just a marriage between molecular pharmacology and functional genomics (Norton, 2001; Fakruddin & Chowdhury, 2013). Pharmacogenomics aims to discover the correlations of subject's genotype with their therapeutic response to tailor the specific genetic makeup (Torres et al., 2003). The overall process involves genes encoding proteins are identified that can be used as potential drug targets and to analyze or understand the variations in the genes (Fakruddin & Chowdhury, 2013). The growing evidence that the genetic makeup is a key predictor of how effective drugs will behave has gained the interest and funding (Fakruddin & Chowdhury, 2013) (Jiang & Wang, 2010). One of the best examples is a breast cancer treatment in which gene expression analysis data has been exploited clinically for the determination of the prognosis of a breast cancer patient. Luminal, HER2 positive, and basal-like breast cancers were classified as three major subtypes via gene expression profiling studies which are based on the differentiation status and tumor grades, expression levels of the HER2 (human epidermal growth factor receptor 2), positive or negative lymph node status (Jiang & Wang, 2010). Patients classified into subtypes can help in the prediction of treatment outcomes (Jiang & Wang, 2010). The patients who present the same pathological and clinical manifestation cannot predict the patient's treatment accurately based on traditional prognostics.

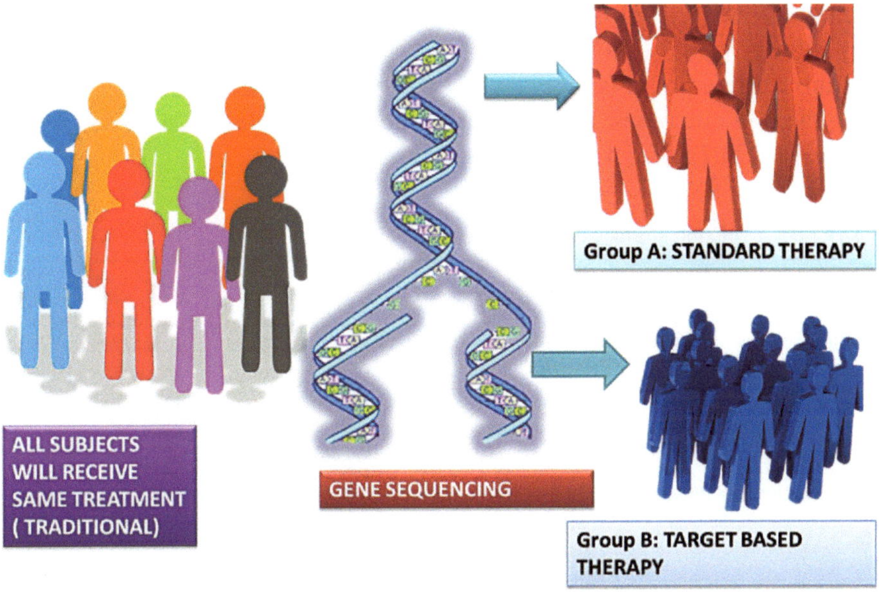

FIGURE 5.2 The overview of pharmacogenomics where the main focus remains to tailor-made right drugs for the right patients.

5.4.1 APPLICATIONS IN PHARMACOGENOMICS

The DNA variations in genes that code for CYP 450 (cytochrome P) family of enzymes which metabolizes 30 different classes of drugs can influence their ability to metabolize certain drugs and the enzymes which are unable to metabolize can cause drug overdose problems in patients, for example, less active enzymes/inactive enzymes are unable to break down some drugs which ultimately lead to drug overdose (Association et al., 1999; Fakruddin & Chowdhury, 2013). Nowadays, genetic tests are being performed in clinical trials by researchers for checking variations in CYP 450 genes for the purpose of screening and monitoring patients (Fakruddin & Chowdhury, 2013) and chemical compounds developed in pharmaceutical companies are screened with variants of CYP enzymes to elucidate how they are getting metabolized. TPMT (thiopurine methyltransferase), an enzyme that breaks down thiopurines, a class of therapeutic compounds, but in case of Caucasians a small percentage of the population have variations in genetic makeup that make them unable to produce an active form of this TPMT protein which results in the accumulation of thiopurines leading to toxicity (Pistoi, 2002). The pharmacogenomics mainly focuses on three key goals, including reducing risk, increasing efficacy, and developing diagnostic tools that can impact therapeutic decisions and improve the clinical outcomes. The various clinical pharmacogenomics studies are described in Table 5.1.

TABLE 5.1 Clinical Pharmacogenomics Studies

Disorder	Drug	Genetic Polymorphism	Result	References
Schizophrenia	Clozapine	5-HT2A protein C102 allele	The homozygotes of C102 response to atypical antipsychotics like clozapine	Arranz et al. (1995); Masellis et al. (1998); Joober et al. (1999)
Atherosclerosis	Pravastatin	TaqB1alleles of B1 and B2 site in CETP (cholesteryl-ester transfer protein), LDL (lipoprotein lipase), and β-fibrinogen.	B1B2 heterozygotes have less better response as compared to homozygotes with pravastatin	Kuivenhoven et al. (1998)
Alzheimer	Tacrine	ApoE4 protein	APOE4 in subjects having other than homozygous alleles show better response.	Poirier et al. (1995); Richard et al. (1997); Rigaud et al. (2000)
Asthma	Montelukast and Zileuton	ALOX5 genotype	In heterozygotes, there is reduced response	Drazen et al. (1999)

5.5 THE CONCEPT OF BIOBANKS IN PERSONALIZED MEDICINE (PM)

In PM, based on different biomarkers in the patient's blood and tissue therapies are selected (Hewitt, 2011). The bio-specimens required for research purposes to detect useful biomarkers are provided by biobanks, PM shortly will focus on the translation of fast-growing knowledge about disease processes into useful therapies. The path of developing a medical product is highly challenging which is inefficient and costly in today's world, and only a concerted effort in modernizing the critical path of developing a new medical product will succeed (Hewitt, 2011). The FDA critical path initiative identified a better evaluation tool like a biomarker for the concerted effort of developing medical products (Hewitt, 2011). The recommendation of biomarkers came from AACR–FDA (American Association for Cancer Research), NCI (US National Cancer Institute), and the aim was given on the necessity of the need for bio-banking services as well as bio-specimen quality control, respectively (Khleif et al., 2010). Concerning the cancer research area, in particular, accelerating PM the improvement of bio-banking services is high on goal (Hewitt, 2011). The term biobank has been described in several ways, one of the definitions describe it like an organized collection of a human biological sample and related information stored for one or more research purposes (Cambon-Thomsen et al., 2007; Kauffmann & Cambon-Thomsen, 2008). The collection of non-human samples like a plant, microbe, environmental, and animal may also be explained as biobanks (Hewitt, 2011). Biobanks of human material fall into many categories, depending on design and purpose and it includes population-based biobanks, twin cohort studies and disease-oriented biobanks (Gottweis & Zatloukal, 2007; Hewitt, 2011). The disease-oriented biobanks are the most frequent hospital-based and consist of tumor banks, collection of blood, and other samples from different diseases together with normal control groups (Hainaut et al., 2009; Bevilacqua et al., 2010). The ethical discussion in bio-banking has focused specifically on the validity of general consent (Cambon-Thomsen et al., 2007). The information covers all aspects related specifically to the personal choice then the consent of the person is appropriately informed (Forsberg et al., 2009). The bio-banking collection model a move towards general consent for future research use is considered as a trend in bio-banking ethics (Cambon-Thomsen et al., 2007; Forsberg et al., 2009).

5.6 NEXT-GENERATION SEQUENCING (NGS): IMPLICATIONS IN PERSONALIZED MEDICINE (PM) AND PHARMACOGENOMICS

The recent trends in genetics have dramatically enhanced efforts to unravel the mechanism of human diseases (Rabbani et al., 2016). In many cases, the phenotypic consequences of various disorders are affected by the genotype and searching the casual variants may lead to the development of novel paths of medical intervention. The pharmacogenetics research involvement in the clinical practice could significantly change in the way of medications are consumed or in another way, it can improve the health care globally and will help in the administration of drugs (Farberov et al., 2013; Goodhead et al., 2013; Girault et al., 2014; Rabbani et al., 2016). The international scientific research effort called HGP aimed to sequence and map down the human genome by using the Sanger sequencing method and the sequence of the human genome was published in 2003 (Cheung et al., 2001; Consortium, 2001; Sequencing, 2004). Generally, the Sanger approach is having excellent accuracy and reasonable read length but low throughput and expensive to sequence down all the genes; however, it's a robust method for clinical application (Rabbani et al., 2016). Next-generation sequencing (NGS) platforms from the last 10 years now provide affordable and high throughput sequencing for the evaluation of functional variations in DNA in many disorders consisting of both monogenic as well as polygenic phenotypes like diabetes, cancer, neurological, and cardiovascular disorders or the regulation of normal human body functions like height, BMI, blood pressure (Metzker, 2010; Bamshad et al., 2011; Kingsmore & Saunders, 2011; Diaz-Horta et al., 2012; Rabbani et al., 2012, 2014). Nowadays, there is increasing demand to apply personal genome sequencing for predicting risk, medical care, lifelong well-being, and therapeutic response in the era of PM (Rabbani et al., 2016). Millions of DNA variants in different populations have been identified by DNA sequencing, mainly SNV (single nucleotide variants) (Altshuler et al., 2012; Consortium, 2010). About 38 million single-nucleotide variants, 14,000 large deletions, and 14 million bi-allelic deletions or insertions have been discovered (Altshuler et al., 2012). The methodology of GWAS was established by the use of a haplotype map of the human genome in HapMap and with the help of GWAS several SNVs associated with cancer, type 2 diabetes (T2D), asthma, and common traits like fat mass, height can be determined (Easton et al., 2007; Moffatt et al., 2007; Sladek et al., 2007; Stacey et al., 2007; Todd et al., 2007; Sanna et al., 2008).

5.7 TRENDS OF PM WITH REFERENCE TO DIABETES

Diabetes is a metabolic disorder characterized by increased blood glucose levels. Approximately 382 million subjects are suffering from diabetic disorder globally and it's an estimate in the near future that it will increase by 50% in the next two decades (Kleinberger & Pollin, 2015). The implementation of PM in Diabetes intervention is very small. The American Diabetes Association classifies diabetes into four classes:

1. **Type 1 Diabetes Mellitus (T1DM):** It refers to an auto-immune disease in which autoantibodies against pancreatic beta cells results in mass cell destruction causing a decrease in insulin production, reduction in weight, polyuria, polydipsia, and make the subject dependent on external insulin (Chiang et al., 2014; Kleinberger & Pollin, 2015). The complex causative factors responsible for T1DM has involved about 40 genes with unknown clinical implications (Chiang et al., 2014).
2. **Type 2 Diabetes Mellitus (T2DM):** It generally refers to a metabolic disorder with a reduction in insulin secretion or increased resistance to insulin (Association, 2017). About 80 susceptible genetic locations have been detected to understand the genetic cause of T2DM (Hara et al., 2014). Genetic testing can revolutionize the diagnosis process in subjects having diabetes or pre-diabetes and can help to intervene in lifestyle change or therapeutic regimen to prevent it at an early stage.
3. **Monogenic Diabetes:** The third category of diabetes has clinical presentation the same as T1DM and T2DM. The clinical presentations are almost the same as T1DM as well as T2DM and are having known genetic and non-genetic causative factors. Monogenic diabetes with genetic testing has proven the ability for enhanced treatment; however, diagnosis is not made yet (Shepherd et al., 2009; Kleinberger & Pollin, 2015).
4. **Gestational Diabetes:** The causative factors may be the same as the other three diabetes subtypes described and may designate expression of elemental vulnerability certified by pregnancy prompted insulin resistance (2014). There is an immense need for development in the field of personalized diabetes in order to curb the growing concern. Almost 13 types of MODY (maturity-onset diabetes of the young) are categorized based on the non-functional gene causing the phenotype of which is one well-elucidated form of monogenic diabetes. Before the age of twenty-five MODY extant in skinny subjects and is inheritable; however, the subject has yet signed of β cell function (Turner et al., 1999). In the UK (United Kingdom) MODY3 is the most prevalent form of MODY consisting of 52% of total percentage however the presence of it changes by ethnicity as well as area (Shields et al., 2010). HNF1A gene which encodes HNF1-α (transcription factor hepatic nuclear

factor 1-α) which is responsible for transcription of many genes that are associated with insulin release, breakdown of glucose and synthesis of insulin. HNF1-α has 55% resemblance with HNF4-α which is mutated in MODY1 and MODY1 comprising 10% MODY in UK (Shields et al., 2010). The HNF1-α as well as HNF4-α have been demonstrated to talk in a way to suppress each other's effect (Boj et al., 2010). The diagnosis of subjects suffering from MODY1 and MODY3 is very crucial for therapeutics because they are more sensitive that is hypersensitivity to sulfonylureas (Pearson et al., 2000; Ješić et al., 2008; Shepherd et al., 2009). The reduction in HNF1-α and HNF4-α genes expression in the liver is responsible for hypersensitivity to sulfonylureas which leads to minimal uptake and causes the persistent enhancement of its blood levels (Boileau et al., 2002). So, the subjects will require only 1/10 dose of sulfonylurea and makes sulfonylureas as the drugs of choice for MODY3 and MODY1. In the above discussion, it's obvious how the genetic makeup plays a role in patient response towards drugs so the application of PM will assist in deciding the right drug for the right patient at the right time and with the right dose.

The other contributing factor is mutations in glucokinase (GCK) gene encoding GCK enzyme which is critical for glucose metabolism leading to MODY2. The mutations are responsible for a reduction in the activity of GCK enzyme that is important for pancreatic β cell supervision at circulating glucose levels. Subjects suffering from MODY2 usually have a mild increase in blood sugar levels and don't proceed to complications like micro-vascular or macro-vascular, So MODY2 subjects don't require any treatment (Ajjan & Owen, 2014). Although certain examinations have demonstrated that a mild increase in blood glucose levels can result in insulin resistance (Clement et al., 1996) and regarding the carriers the gestational diabetes mellitus is a prime example of having GSK gene mutations (Ellard et al., 2000; Shehadeh et al., 2005).

5.7.1 PHARMACOGENETICS OF SOME T2DM MEDICATIONS

Subjects suffering from T2DM contribute to the highest percentage of cases with diabetes mellitus. Metformin being the drug of choice for

T2DM is considered the first-line drug which operates by reducing gluconeogenesis in the liver. The class of drugs like sulfonylureas, meglitinide, glucagon-like peptide (GLP-1), thiazolidinediones, and DPP4 inhibitors (dipeptidyl peptidase inhibitors-4) are considered as second-line classes of drugs (2014). The examinations of the genetic variations that could change the therapeutic response to anti-diabetic drugs have demonstrated moderate effects, and few are totally opposite/contradicting. These examinations describe a judgment that may give us a clue about the upcoming pharmacogenetic recommendations for oral anti-diabetic drugs.

5.7.1.1 METFORMIN

The mechanism of metformin is poorly understood, and its target receptors have not been elucidated for pharmacogenetic variations. The GWAS elucidated that SNP rs11212617 close to ATM locus was connected with a decrease in HbA1c in response to metformin (van Leeuwen et al., 2012). The phosphatidylinositol 3-kinase family having a member namely ataxia telangiectasia mutated gene encoded by ATM is very crucial for cell cycle regulation and repairing of DNA. One of the meta-analysis replications of this very study has shown that the correlation was accredited however 1 of the 3 cohorts demonstrated no correlation (Zhou et al., 2011). The DPP (Diabetes Prevention Program) revealed that there was no correlation between rs11212617 as well as progression from impaired sugar tolerance (Florez et al., 2012). The single nucleotide polymorphism requires more accreditation and exploring more pathways for confirmation and upcoming studies into metformin therapeutic mode of action (a mechanism).

5.7.1.2 SULFONYLUREAS

The sulfonylureas act by ligating with sulfonylureas receptor subunit 1 (SUR1) (which is encoded by ABCC8 gene) near to ATP sensitive K^+ inward rectifying channel responsible for depolarization which is then followed by Ca^{++} influx which results in insulin secretion. The other subunit of the potassium ATP sensitive channel is kir6.2 (which is encoded by KCNJ11) very near to the ABCC8 gene having chromosome number 11 (Inagaki et al., 1995). The high dose of sulfonylurea is given to treat neonatal diabetes mellitus caused by stimulating mutations in KCNJ11 as well as ABCC8 although the low dose is administered as the first line or drugs of choice for the subjects suffering from MODY1 and MODY3. Investigations of varying shapes of these genes have revealed a familiar haplotype of E23K in KCNJ11 and S1369A which are affiliated to T2D mellitus (Florez et al., 2007). Path-clamp technique has revealed that sulfonylureas are less responsive to this haplotype (E23K in KCNJ11 and S1369A) (Lang et al., 2012). The polymorphisms have contradicted correlation with T2D mellitus and sulfonylurea therapeutic effectiveness (Sesti et al., 2006; Sato et al., 2010).

5.7.1.3 MEGLITINIDES

The meglitinides are oral anti-diabetic drugs that usually act by inhibiting potassium sensitive ATP channels causing depolarization leading to insulin release. They are metabolized in the liver (McLeod, 2004) which is the principal organ for drug metabolomics. Meglitinides are transported to the liver by a protein SLCO1B1 and the c.521T>C polymorphism has demonstrated a reduction in the rate of breakdown of meglitinides, but the changed pharmaco-kinetics have a less biological effect on sugar levels (Phillips et al., 2001; Cheng et al., 2013). Repaglinide which is metabolized by CYP2C8 and CYP3A4 differs from Nateglinide which s metabolized by CYP2C9. The genetic polymorphism in metabolic proteins may change the pharmacokinetics of the drugs but it doesn't demonstrate any significant effect on glucose levels of subjects (Kirchheiner et al., 2004).

5.8 THE OPPORTUNITIES FOR PHARMACOGENOMICS IN T2DM

T2D mellitus promotes to consume the pharmacogenomics for the production and development of new drug therapies in addition to how the drug action is affected by genetic polymorphisms. A specific genetic variation could give enough information about the pathophysiology of a disease. The most important prime discovery in pharmacogenomics is the nonsense mutation in SLC30A8 that is a safeguard contrary to T2D mellitus that could be consumed for drug development. So, by shooting SLC30A8 for prevention of up-regulation could assure T2D mellitus nonsense variants. The other prime genetic variant example is APOC3 nonsense variant R19X that could bring the potential upcoming treatment ways. The nonsense variants reduce the blood levels of APOC3, resulting in a reduction of serum triglycerides levels and safeguarding from coronary disaster (Pollin et al., 2008; TG, HDL Working Group of the Exome Sequencing Project et al., 2014). Nowadays the APOC3 has become the shoot point target for the therapeutic inhibition to safeguard from increased triglycerides (hypertriglyceridemia). The common complication of diabetes mellitus is coronary artery disease due to growing hypertriglyceridemia this can be an opportunistic direction for drug development (Fitzgerald et al., 2014). Pharmacogenomics can bring more dramatic changes in the upcoming future to eradicate the conventional means of drug development and will allow scientists to focus more on the target-based approach by elucidating the entire genomic functions.

5.9 TRENDS OF PM WITH REFERENCE TO ASTHMA

Asthma is one of the inflammatory disorders that affects airways by about 300 million people globally. There are many classes of drugs that are being employed to treat asthma targeting distinct biological pathways consisting of leukotriene antagonists, glucocorticoids, anti-cholinergic, beta-2 agonists, phosphodiesterase inhibitors (PDE) and monoclonal antibodies against IgE. Despite the availability of a large number of drug classes against asthma still, there are refractory asthmatic patients who do not respond to either of the treatments (Chan et al., 1998; Society, 2000). The classes of drugs targeting many pharmacological pathways have adverse and sometimes fatal adverse threatening reactions. The genes which regu late the multiple pharmacological pathways that utilize by anti-asthmatic drugs have many variations that in-turn alters the therapeutic efficacy and form the basis of development of genetic profiles for PM (Castle et al., 1993; Nelson et al., 2006; Salpeter et al., 2006). The genetic investigations of susceptibilities, severity, and therapeutic responses to anti-asthmatics can result in pathways to build up the genetic profiles (Ortega et al., 2015). The ideal genetic profile should be consisting of genetic variants that can predict the drug is choice with maximal therapeutic efficacy, the severity of diseases, the progression of the disease, and predict therapies avoidable for a particular subject. In the case of asthma genetics, the environmental effects intervene in the pharmacological therapy and alter in phenotypic therapeutic response (Figure 5.3). In twin-based clinical trial investigations in which the heritable pharmacological responses were monitored demonstrated that the therapeutic responses to different therapies within the same subjects vary much less than inter-subject therapeutic responsiveness (Kalow et al., 1998; Drazen et al., 2000). The pharmacogenetic investigations are focused on the genes which regulate the pharmacological pathways utilized by leukotriene, β_2-adrenergic agonists and the glucocorticoids and many genetic locations associated with therapeutic responses in asthma have been identified (Tables 5.2–5.4).

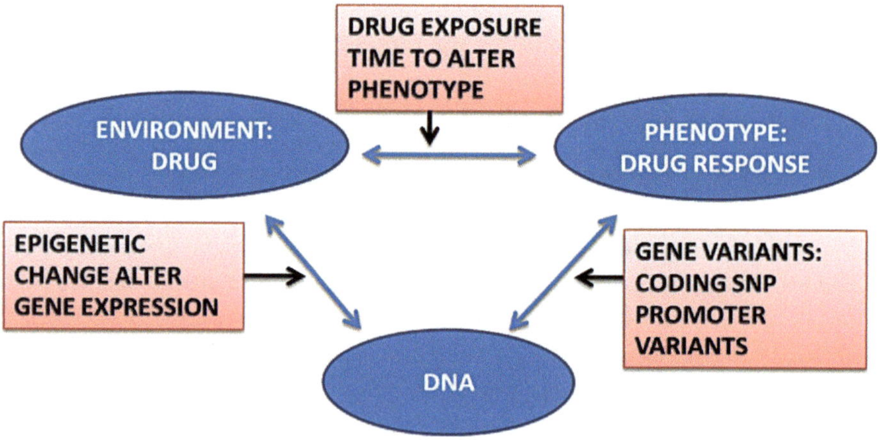

FIGURE 5.3 Pharmacogenomics research in complex disorders. For example, in asthma genetics, the environmental effects intervene in the pharmacological therapy and alter in phenotypic therapeutic response. (Modified from Ortega et al., 2015.)

In the case of lung cancer and cystic fibrosis, the predictive genetic profile is becoming reality in the therapeutic management however not yet in case management of asthma (Ramsey et al., 2011; Rosell et al., 2013). The pharmacogenetic research has identified genes that can affect therapeutic response to the drugs currently under development that shoot specifically pharmacological pathways.

5.9.1 GLUCOCORTICOID PHARMACOLOGICAL PATHWAY (GPP)

The glucocorticoids administered both orally or intravenously are the most common drugs used in inflammatory disorders and are one of the first-line therapies for asthma management along with inhaled

TABLE 5.2 Pharmacogenetic Subject Genes for Glucocorticoid Response in Asthmatics

Class of Drug	Design of Study	Gene	Loci Associated	Phenotype Response	References
Glucocorticoids (inhaled)	Subject gene study	CRHR1	rs242941, rs1876828	Forced expiratory volume in 1 second (FEV_1) response	Tantisira et al. (2004)
		STIP1	rs2236647, rs6591838, rs1011219		Hawkins et al. (2009)
		ADCY9	rs2230739 ($Met^{772}Ile$)		Tantisira et al. (2005)
	Genome-wide association studies	*GLCCI1*	rs37972=rs37973		Tantisira et al. (2011)
		T GENE	rs6456042, rs3099266, rs2305089, rs3127412		–
–	Subject gene study	CYP3A4	CYP3A4*22 allele	Symptom control	Stockmann et al. (2013)
	Subject gene study	TBX21	rs2240017 ($His^{33}Gln$)	Broncho-protection	Tantisira et al. (2004); Ye et al. (2009)

TABLE 5.3 Pharmacogenetic Subject Genes for β_2-Adrenergic Agonists Response in Asthmatics

β_2-Adrenergic Agonists	Design of Study	Gene	Loci Associated	Phenotype Response	References
The short-acting beta-agonists example albuterol	Subject gene study	CRHR2	rs7793837	Acute FEV_1 broncho-dilation	Poon et al. (2008); Himes et al. (2012)
		ADCY9	rs2230739 ($Ile^{772}Met$)		Tantisira et al. (2005); Himes et al. (2012)
		ARG1	rs2781659, rs2781667		Litonjua et al. (2008); Hattersley et al. (2009); Duan et al. (2011); Himes et al. (2012); Iordanidou et al. (2012)
		ARG2	rs7140310, rs10483801		Vonk et al. (2010); Iordanidou et al. (2012)
		NOS3	rs1799983 ($Asp^{298}Glu$)		Iordanidou et al. (2012)
		THRB	rs892940		Duan et al. (2013)
	Subject gene study and	*ADRB2*	rs1042713 ($Gly^{16}Arg$)	Acute FEV_1 bronchodilation	Martinez et al. (1997); Taylor et al. (2000); Israel et al. (2004); Choudhry et al. (2005); Duan et al. (2013)
	Genotype-stratified			Long-term peak expiratory flow rate response	
	Subject gene study			Long-term peak expiratory flow rate response	
	Genome-wide association studies	*SLC24A4*	rs77441273 ($Arg^{585}Gln$)	Acute FEV_1 bronchodilation	Drake et al. (2014)
		SPATS2L	rs295137		Himes et al. (2012); Meyers et al. (2014)
	Admixture mapping	*SLC22A15*	rs1281748, rs1281743		Drake et al. (2014); Flannick et al. (2014)

TABLE 5.3 *(Continued)*

β_2-Adrenergic Agonists	Design of Study	Gene	Loci Associated	Phenotype Response	References
The long-acting beta-agonists examples are salmeterol and formoterol	Subject gene study	*ADCY9*	rs2230739 ($Met^{772}Ile$)	Long-term FEV_1 response	Kim et al. (2011)
	Subject gene study and genotype-stratified	*ADRB2*	rs1042713 ($Gly^{16}Arg$)	Long-term peak expiratory flow rate response No effect on peak expiratory flow rate response Bronchoprotection Preference for montelukast or long-acting beta-agonist as an add-on to inhaled corticosteroids	Taylor et al. (2000); Wechsler et al. (2006, 2009); Kleinberger & Pollin (2015); Bleecker et al. (2006, 2007, 2010); Lee et al. (2004); Lipworth et al. (2013)

corticosteroids (ICS) (Peters et al., 2007; Sorkness et al., 2007). A large number of subjects suffering from asthma demonstrate better response with improving their lung functions against ICS and only a few subjects show negative responses (Szefler et al., 2005). The subjects which do not respond to ICS are said to be steroid-resistant or refractory asthmatics which continuously exacerbate the symptoms with no improvement in lung functions (Chan et al., 1998, 2000). The pharma-cogenetic investigations were focused on glucocorticoid pathways based on the biologic subject gene coding the corticosteroid biosynthetic signaling pathway which includes hetero-complex and chaperone proteins (Table 5.2). The biological subject gene investigation of the CRHR1 in 1,117 subjects suffering from asthma randomized to inhalational corticosteroid therapy from the clinical trial cohorts detected 2 CRHRI SNPs associated with lung function response (LFR) (Tantisira et al., 2004). Another subject gene investigation of one of these cohorts assessing different genes coding the glucocorticoid hetero-complex also detected 3 SNPs within STIP1 (heat shock organizing protein gene) associated with LFR amidst inhalational therapy (Vogenberg et al., 2010). The gene variations demonstrated while GPP talks with biological pathways that impact the therapeutic responses against inhalation corticosteroids alone or in combination with SABA or LABA. The ADCY9 gene encodes an enzyme namely Adenylyl cyclase type 9 with coding SNP, $Met^{772}Ile$ (rs2230739), associated with therapeutic bronchodilation effect against SABA, albuterol only in patient's treated with inhalational corticosteroid from CAMP cohort (Tantisira et al., 2005). The ADCY9 gene pathway talk was replicated in an IKTC (independent Korean trial cohort) treated with LABA and formoterol in combination with inhalational corticosteroids (Kim et al., 2011). The gene TBX21 encoding T-box expressed in T-cell transcription which controls native T cell development outside GPP consists a coding single nucleotide polymorphism $His^{33}Glu$ (rs2240017), was involved in the enhancement of bronchial hyper-responsiveness or bronchial-protection during inhalational corticosteroid therapy in the Childhood Asthma Management Program (CAMP) cohort and replicated in IKTC (Tantisira et al., 2004; Ye et al., 2009). An investigation of the genes *CYP3A4*, *CYP3A5*, and *CYP3A7* in 413 inhalational corticosteroid-treated children suffering from asthma identified a specific *CYP3A4* genotype in twenty children's involvement in the enhancement of asthmatic symptom control proposing that the genetic loci are critical for inhalational corticosteroid metabolism may also serve as biomarkers for therapeutic response (Stockmann et al., 2013).

5.9.2 CYSTEINYL LEUKOTRIENE PHARMACOLOGICAL PATHWAY (CLPP)

The cysteinyl leukotriene-altering drugs are also widely used in the management of asthma; however, they are less effective when compared to inhalational corticosteroids. The cysteinyl leukotriene exists as 5-LO blockers and CLR-1 (cysteinyl leukotriene receptor 1) blockers. The CLGP (cysteinyl leukotriene genetic pathway) is launched and limited by 5-LO encoded by ALOX5 gene followed by a group of proteins that biosynthesize distinct leukotrienes like LTA4H and LTC4S and MRP1 (multiple drug-resistant protein 1) that ship leukotrienes to extra-cellular area to ligate and stimulate CLR (CYSLTR1 and CYSLTR2) (Ortega et al., 2015). The leuckotreine altering drugs used in asthma have been subjected to little replicated pharmacogenetic subject gene investigations mainly due to small patient number although these investigations are beneficial and showed that the variability in pharmaco-therapeutic responsiveness to these drugs in at least fragmentary determined by gene variation (Table 5.4). One of the 1st investigations assessed a TRP (tandem repeat polymorphism) in the RPR (regulatory promoter region) of ALOX5 and demonstrated that this promoter variant was involved with LFR in 144 subjects suffering from asthma treated with 5-LO blocker and ABT-761 (Drazen et al., 1999).

The larger subject gene investigations have demonstrated that ALOX5 promoter variant and other ALOX5 SNPs may also affect the response to LRA (leukotriene receptor antagonist) (Lima et al., 2006; Klotsman et al., 2007; Telleria et al., 2008; Tantisira et al., 2009). The variants in LTC4S and MRP1 have been associated with LFR to another 5-LOI (leukotriene inhibitor) and LRA (Lima et al., 2006; Tantisira et al., 2009). The LPC gene investigations also involved SNPs in LTA4H and CYSLTR1 genes and have demonstrated contradicting results associated with sample sizes (n=61 to n=577) underpowered to consistently identify and replicate pharmacogenetic correlations (Lima et al., 2006; Tantisira et al., 2009).

TABLE 5.4 Pharmacogenetic Subject Genes for LMR in Asthma

Class	Gene	Loci Associated	Design of Study	Phenotype Response	References
LRM: 5-LO inhibitors (ABT-761 and zileuton)	ALOX5	Promoter repeat, rs892690, rs2029253	Subject gene study	Forced expiratory volume in 1-second	Drazen et al. (1999); Telleria et al. (2008)
–	–	rs2115819	–	Response	–
	LTC4S	rs272431			Tantisira et al. (2009)
	MRP1	rs215066, 119774			Tantisira et al. (2009)
Cysteinyl leukotriene inhibitors (montelukast)	*ALOX5*	Promoter repeat, rs2115819			Lima et al.(2006); Klotsman et al. (2007)
	MRP1	rs119774			Lima et al.(2006)
	LTA4H	RS266845		Risk of exacerbation	Lima et al.(2006)
	LTC4S	rs730012			Lima et al. (2006)
	SLCO2B1	rs12422149 ($Arg^{312}Gln$)		Symptom control and drug levels	Mougey et al. (2009)

5.9.3 B2-ADRENERGIC PHARMACOLOGICAL PATHWAY

The inhaled β_2-receptor agonists are the oldest and most widely used in asthma exacerbations which act through G-protein coupled receptor pathways. There are about three dug classes in this category which consist of SABA, LABA, and ULABA (ultra-long-acting beta-agonists). The asthma death epidemics were associated with a high dose of SABA which includes isoproterenol in the era of 1960s and non-selective fenoterol in the era of 1970s (Crane et al., 1995; Pearce et al., 1997, 1990; Stolley, 1972; Grainger et al., 1991). These death toxicities lead to the withdrawal of drugs from the global market.

The β_2-adrenergic receptor gene ADRB2 I not having any introns; however, a polymorphic gene with more than 49 genetic variants in the MEAC (multiethnic asthma cohorts) evaluated to date (Drysdale et al., 2000; Hawkins et al., 2006; Ortega et al., 2014). The most extensively investigated coding variant AAP 16, $Gly^{16}Arg$, which is associated with *in-vitro* altered receptor down-regulation. The earlier pharmacogenetic investigation of the ADRB2 gene have shown that GLY^{16} homozygotes have a lesser acute response to SABA bronchodilation compared to ARG^{16} homozygotes which has been verified in extra asthmatic subjects (Lima et al., 1999; Choudhry et al., 2005).

5.10 CONCLUSION

The PM has become popular day by day and in the coming years, it will help to ease the diagnostics as well as therapeutic targeting. The reasons for its popularity are because of the following reasons like:

- Reduce the burden of disease;
- Focuses on prevention;
- Diminishes the duration and severity of illness;
- Decreases health care costs;
- Enhance benefits and reduces risks.

PM has opened the door to integrating genomic-based medicine into a clinical perspective. If efforts, like complemented by well-orchestrated collaborative efforts ranging from stakeholders pursuing the activities designed to accelerate a shared vision of PM remarkable benefits in genome-informed clinical care, maybe favorable near these 10 years. The healthcare team, patients, and industry should club around meaningful opportunities like patient counseling, concordance in coverage policy, and the appropriate guidelines to facilitate wider adoption of PM.

KEYWORDS

- **biotechnology developments**
- **diagnosis**
- **glucocorticoid pharmacological pathway**
- **next-generation sequencing**
- **optimal treatments**
- **personalized medicine**
- **phosphodiesterase inhibitors**
- **risk assessment**

REFERENCES

Alan, R. S., Jeffrey, R. O., Kevin, P. B., Amish, G., Kathleen, R., Richard, B. H., Coleen, M. D., et al., (2009). Association of cytochrome P450 2c19 genotype with the antiplatelet effect and clinical efficacy of clopidogrel therapy. *JAMA, 302*, 849–857.

Altshuler, D. M., Durbin, R. M., Abecasis, G. R., Bentley, D. R., Chakravarti, A., & AG Clark, Genomes Project, C., (2012). An integrated map of genetic variation from 1,092 human genomes. *Nature, 491*, 56–65.

Altshuler, D. M., Gibbs, R. A., Peltonen, L., Dermitzakis, E., Schaffner, S. F., Yu, F., Bonnen, P. E., De Bakker, P. I., & Deloukas, P., (2010). Integrating common and rare genetic variation in diverse human populations; International HapMap 3 Consortium. *Nature, 467*, 52–58.

American Diabetes Association, (2014). Diagnosis and classification of diabetes mellitus. *Diabetes Care*, *37*(1), S81–S90.

American Diabetes Association, (2014). Standards of medical care in diabetes-2014. *Diabetes Care*, *37*, S14–S80.

American Diabetes Association, (2017). Classification and diagnosis of diabetes. *Diabetes Care, 40*, S11, S24.

American Thoracic Society, (2000). Proceedings of the ATS workshop on refractory asthma: Current understanding, recommendations, and unanswered questions. *Am. J. Respir. Crit. Care Med., 162*, 2341–2351.

Andrew, B., (2014). Personalized medicine, genomics, and pharmacogenomics: A primer for nurses. *Clinical Journal of Oncology Nursing, 18*, 437–441.

Andrew, H., Jan, B., Julian, S., Pal, N., & Kim, C. D., (2009). The diagnosis and management of monogenic diabetes in children and adolescents. *Pediatric Diabetes, 10*, 33–42.

Anne Cambon-Thomsen, Rial-Sebbag, E., & Bartha, M. K., (2007). Trends in ethical and legal frameworks for the use of human biobanks. *European Respiratory Journal, 30*, 373–382.

Arranz, M. J., Collier, D. A., Monsheel, S., David, B., Roberts, G. W., Sham, P., Kerwin, R., & Price, J., (1995). Association between clozapine response and allelic variation in 5-Ht2a receptor gene. *The Lancet, 346*, 281, 282.

Audrey, H. P., Kelan, G. T., Augusto, A. L., Ross, L., Jingsong, X., Jessica Lasky-Su, John, J. L., et al., (2008). Association of corticotropin releasing hormone receptor 2 (CRHR2) genetic variants with acute bronchodilator response in asthma. *Pharmacogenetics and Genomics, 18*, 373.

Augusto, A. L., Lasky-Su, J., Kady, S., Kelan, G. T., Ross, L., Barbara, K., John, J. L., ET AL., (2008). Arg1 is a novel bronchodilator response gene: Screening and replication in four asthma cohorts. *American Journal of Respiratory and Critical Care Medicine, 178*, 688–694.

Bahareh, R., Hirofumi, N., Shahin, A., Mustafa, T., & Nejat, M., (2016). Next generation sequencing: Implications in personalized medicine and pharmacogenomics. *Molecular BioSystems, 12*, 1818–1830.

Bahareh, R., Mustafa, T., & Nejat, M., (2014). The promise of whole-exome sequencing in medical genetics. *Journal of Human Genetics, 59*, 5–15.

Bahareh, R., Nejat, M., Kazuyoshi, H., Hirofumi, N., & Ituro, I., (2012). Next-generation sequencing: Impact of exome sequencing in characterizing mendelian disorders. *Journal of Human Genetics, 57*, 621–632.

Blanca, E. H., Xiaofeng, J., Ruoxi, H., Ann, C. W., Lasky-Su, J. A., Barbara, J. K., John, Z., et al., (2012). Genome-wide association analysis in asthma subjects identifies Spats2l as a novel bronchodilator response gene. *PLoS Genetics,* 8.

Bleecker, E. R., Nelson, H. S., Kraft, M., Corren, J., Meyers, D. A., Yancey, S. W., Anderson, W. H., et al., (2010). Beta2-receptor polymorphisms in patients receiving salmeterol with or without fluticasone propionate. *Am J Respir Crit Care Med, 181*, 676–687.

Bleecker, E. R., Postma, D. S., Lawrance, R. M., Meyers, D. A., Ambrose, H. J., & Goldman, M., (2007). Effect of Adrb2 polymorphisms on response to longacting Beta2-agonist therapy: A pharmacogenetic analysis of two randomized studies. *Lancet, 370*, 2118–2125.

Brian, J. L., Kaninika, B., Helen, P. D., Roger, T., Donald, F. M., Simon, A. O., Colin, N. A. P., & Somnath, M., (2013). Tailored second-line therapy in asthmatic children with the Arg16 genotype. *Clinical Science, 124*, 521–528.

Bronson, D. R., Julie, O. C., Cécile, S., Leigha, A. S., June, A. P., Josephine, W. C., Callif-Daley, F., et al., (2012). Essential elements of genetic cancer risk assessment, counseling, and testing: Updated recommendations of the national society of genetic counselors. *Journal of Genetic Counseling, 21*, 151–161.

Chan, M. T., Leung, D. Y., Szefler, S. J., & Spahn, J. D., (1998). Difficult-to-control asthma: Clinical characteristics of steroid-insensitive asthma. *J. Allergy Clin. Immunol., 101*, 594–601.

Cheng, Y., Wang, G., Zhang, W., Fan, L., Chen, Y., & Zhou, H. H., (2013). Effect of CYP2c9 and SLCO1B1 polymorphisms on the pharmacokinetics and pharmacodynamics of nateglinide in healthy Chinese male volunteers. *Eur. J. Clin. Pharmacol., 69*, 407–413.

Cheung, V. G., Nowak, N., Jang, W., Kirsch, I. R., Zhao, S., Chen, X. N., Furey, T. S., et al., (2001). Integration of cytogenetic landmarks into the draft sequence of the human genome. *Nature, 409*, 953–958.

Choudhry, S., Ung, N., Avila, P. C., Ziv, E., Nazario, S., Casal, J., Torres, A., et al., (2005). Pharmacogenetic differences in response to albuterol between Puerto Ricans and Mexicans with asthma. *Am. J. Respir. Crit. Care Med., 171*, 563–570.

Chris, S., Bernhard, F., Roger, G., Flory, N., Derek, A. U., Steven, M., Christopher, A. R., et al., (2013). Fluticasone propionate pharmacogenetics: Cyp3a4* 22 polymorphism and pediatric asthma control. *The Journal of Pediatrics, 162*, 1222–1227. e2.

Clement, K., Pueyo, M. E., Vaxillaire, M., Rakotoambinina, B., Thuillier, F., Passa, P. H., Froguel, P. H., et al., (1996). Assessment of insulin sensitivity in glucokinase-deficient subjects. *Diabetologia, 39*, 82–90.

Crane, J., Pearce, N., Burgess, C., Jackson, R., & Beasley, R., (1995). End of New Zealand asthma epidemic. *Lancet, 345*(8955), 984, 985. doi: 10.1016/s0140-6736(95)90731-9.

Daniel, K. C. L., Graeme, P. C., Ian, P. H., John, J. L., & Brian, J. L., (2004). The arginine-16 B2-adrenoceptor polymorphism predisposes to bronchoprotective subsensitivity in patients treated with formoterol and salmeterol. *British Journal of Clinical Pharmacology, 57*, 68–75.

David, H. J., (2005). Gene-environment interactions in human diseases. *Nature Reviews. Genetics, 6*, 287–298.

Deborah, A. M., Eugene, R. B., John, W. H., & Stephen, T. H., (2014). Asthma genetics and personalized medicine. *The Lancet Respiratory Medicine, 2*, 405–415.

Diana, C., Karen, P., Richard, D. G., Castiglione-Gertsch, M., Carl-Magnus, R., Jurij, L., Martin, F. F., et al., (2003). Adjuvant endocrine therapy compared with no systemic therapy for elderly women with early breast cancer: 21-Year results of International Breast Cancer Study Group trial Iv. *Journal of Clinical Oncology, 21*, 4517–4523.

Douglas, F. E., Karen, A. P., Alison, M. D., Paul, D. P. P., Deborah, T., Dennis, G. B., Jeffery, P. S., et al., (2007). Genome-wide association study identifies novel breast cancer susceptibility loci. *Nature, 447*, 1087–1093.

Drazen, J. M., Yandava, C. N., Dube, L., Szczerback, N., Hippensteel, R., Pillari, A., Israel, E., et al., (1999). Pharmacogenetic association between Alox5 promoter genotype and the response to anti-asthma treatment. *Nat. Genet, 22*, 168–170.

Drysdale, C. M., McGraw, D. W., Stack, C. B., Stephens, J. C., Judson, R. S., Nandabalan, K., Arnold, K., et al., (2000). Complex promoter and coding region beta 2-adrenergic receptor haplotypes alter receptor expression and predict *in vivo* responsiveness. *Proc. Natl. Acad Sc.i U S A, 97*, 10483–10488.

Edward, H. R., Edith, A. P., John, B., Vera, J. S., Charles, E. G. Jr., Nancy, E. D., Tan-Chiu, E., et al., (2005). Trastuzumab plus adjuvant chemotherapy for operable HER2-positive breast cancer. *The New England Journal of Medicine, 353*, 1673–1684.

Elliot, I., Vernon, M. C., Jean, G. F., Homer, A. B., Reuben, C., Timothy, J. C., Aaron, D., et al., (2004). Use of regularly scheduled albuterol treatment in asthma: Genotype-stratified, randomized, placebo-controlled cross-over trial. *The Lancet, 364*, 1505–1512.

Étienne, L., Marie-Andrée, Y., Bettina, A. H., Gilles, O., LeBlanc, J., & Jacques, T., (1999). Influence of Cyp2d6 activity on the disposition and cardiovascular toxicity of the antidepressant agent venlafaxine in humans. *Pharmacogenetics, 9*, 435–443.

Eugene, R. B., Steven, W. Y., Leslie, A. B., Lisa, D. E., Michael, K., Wayne, H. A., & Paul, M. D., (2006). Salmeterol response is not affected by B2-adrenergic receptor genotype in subjects with persistent asthma. *Journal of Allergy and Clinical Immunology, 118*, 809–816.

Fernando, D. M., Penelope, E. G., Mauro, B., Susan, S., & Robert, E., (1997). Association between genetic polymorphisms of the Beta2-adrenoceptor and response to albuterol in children with and without a history of wheezing. *The Journal of Clinical Investigation, 100*, 3184–3188.

Flannick, J., Thorleifsson, G., Beer, N. L., Jacobs, S. B., Grarup, N., Burtt, N. P., Mahajan, A., et al., (2014). Loss-of-function mutations in Slc30a8 protect against type 2 diabetes. *Nat. Genet., 46*, 357–363.

Florence, R., Nicole, H., Eric, N., David, G., Raymond, L., & Philippe, A., (1997). APOE Genotyping and response to drug treatment in Alzheimer's disease. *The Lancet, 349*, 539.

Florez, J. C., Jablonski, K. A., Kahn, S. E., Franks, P. W., Dabelea, D., Hamman, R. F., Knowler, W. C., et al., (2007). Type 2 diabetes-associated missense polymorphisms Kcnj11 E23k and Abcc8 A1369s influence progression to diabetes and response to interventions in the diabetes prevention program. *Diabetes, 56*, 531–536.

Florez, J. C., Jablonski, K. A., Taylor, A., Mather, K., Horton, E., White, N. H., Barrett-Connor, E., et al., (2012). The C allele of Atm Rs11212617 does not associate with metformin response in the diabetes prevention program. *Diabetes Care, 35*, 1864–1867.

Francine, K., & Cambon-Thomsen, A., (2008). Tracing biological collections: Between books and clinical trials. *JAMA, 299*, 2316–2318.

Generoso, B., Fred, B., Thibaut, D., Heinz, H., Anne, J., Rupert, L., Denis, L., et al., (2010). The role of the pathologist in tissue banking: European consensus expert group report. *Virchows ARCHIV: An International Journal of Pathology, 456*, 449–454.

Geoffrey, S. G., & Huntington, F. W., (2009). Genomic and personalized medicine: Foundations and applications. *Translational Research, 154*, 277–287.

Ginsburg, G., & Willard, H., (2009). Genomics and personalized medicine: Foundations and applications. *Translational Research, 154*, 277–287.

Grainger, J., Woodman, K., Pearce, N., Crane, J., Burgess, C., Keane, A., & Beasley, R., (1991). Prescribed fenoterol and death from asthma in New Zealand, 1981-7: A further case-control study. *Thorax, 46*, 105–111.

Gregory, A. H., Ross, L., Richard, S. S., Kelan, G. T., Deborah, A. M., Stephen, P. P., Scott, T. W., & Eugene, R. B., (2009). The glucocorticoid receptor heterocomplex gene stip1 is associated with improved lung function in asthmatic subjects treated with inhaled corticosteroids. *Journal of Allergy and Clinical Immunology, 123*, 1376–1383. e7.

Guillaume, G., Yann, B., Gilles, V., & Sylviane, D., (2014). High-throughput sequencing of *Bacillus anthracis* in France: Investigating genome diversity and population structure using whole-genome SNP discovery. *BMC Genomics, 15*, 288.

Hara, K., Shojima, N., Hosoe, J., & Kadowaki, T., (2014). Genetic architecture of type 2 diabetes. *Biochem. Biophys. Res. Commun., 452*, 213–220.

Harold, S. N., Scott, T. W., Eugene, R. B., Steven, W. Y., Paul, M. D., & SMART Study Group, (2006). The salmeterol multicenter asthma research trial: A comparison of usual pharmacotherapy for asthma or usual pharmacotherapy plus salmeterol. *Chest, 129*, 15–26.

Hawkins, G. A., Tantisira, K., Meyers, D. A., Ampleford, E. J., Moore, W. C., Klanderman, B., Liggett, S. B., et al., (2006). Sequence, haplotype, and association analysis of ADRbeta2 in a multiethnic asthma case-control study. *Am. J. Respir. Crit. Care Med., 174*, 1101–1109.

Herbert, G., & Kurt, Z., (2007). Biobank governance: Trends and perspectives. *Pathobiology, 74*, 206–211.

Human Genome Sequencing, (2004). Finishing the euchromatic sequence of the human genome. *Nature, 431*, 931–945.

Ian, G., Paul, C., Wendi, B. J., Tanja, B., Michael, C., Suzanne, K., Sarah, F., et al., (2013). Whole-Genome sequencing of *Trypanosoma brucei* reveals introgression between subspecies that is associated with virulence. *MBio, 4*, e00197–13.

Inagaki, N., Gonoi, T., Th Clement, J. P., Namba, N., Inazawa, J., Gonzalez, G., Aguilar-Bryan, L., et al., (1995). Reconstitution of IKATP: An inward rectifier subunit plus the sulfonylurea receptor. *Science, 270*, 1166–1170.

International Human Genome Sequencing Consortium, (2001). Initial sequencing and analysis of the human genome. *Nature, 409*, 860–921.

Jan, A. K., Wouter, J. J., Aeilko, H. Z., Peter De, K., McPherson, R., Albert, V. G. B., Kong, I. L., & John, J. P. K., (1998). The role of a common variant of the cholesteryl ester transfer protein gene in the progression of coronary atherosclerosis. *New England Journal of Medicine, 338*, 86–93.

Jane, L. C., Sue, K. M., Lori, L. M. B., & Anne, L. P., (2014). Type 1 diabetes through the life span: A position statement of the American diabetes association. *Diabetes Care, 37*, 2034–2054.

Jed, E. R., Frédérique, M. B., Tomas, D., Catherine, J., & George, R. U., (2010). Personalized smoking cessation: Interactions between nicotine dose, dependence and quit-success genotype score. *Molecular Medicine (Cambridge, Mass.), 16*, 247–253.

Jeffrey, M. D., Edwin, K. S., & Tak, H. L., (2000). Heterogeneity of therapeutic responses in asthma. *British Medical Bulletin, 56*, 1054–1070.

Joanna, S. F., Mats, G. H., & Stefan, E., (2009). Changing perspectives in biobank research: From individual rights to concerns about public health regarding the return of results. *European Journal of Human Genetics, 17*, 1544–1549.

Johanne, T., & Pavel, H., (2013). Role of genomics on the path to personalized medicine. *Metabolism: Clinical and Experimental, 62*(Suppl 1), S2–S5.

John, A. T., Neil, M. W., Jason, D. C., Deborah, J. S., Kate, D., Vincent, P., Rebecca, B., et al., (2007). Robust associations of four new chromosome regions from genome-wide analyses of type 1 diabetes. *Nature Genetics, 39*, 857–864.

Judes, P., Marie-Claude, D., Remi, Q., Isabelle, A., Martin, F., Debmoi, L., Siu, H., et al., (1995). Apolipoprotein E4 allele as a predictor of cholinergic deficits and treatment outcome in alzheimer disease. *Proceedings of the National Academy of Sciences, 92*, 12260–12264.

Judith, M. V., Dirkje, S. P., Harm, M., Marcel, B., Gerard, H. K., & Herman, M., (2010). Arginase 1 and arginase 2 variations associate with asthma, asthma severity and B2 agonist and steroid response. *Pharmacogenetics and Genomics, 20*, 179–186.

Kalow, W. B. K. L., Tang, B. K., & and, L. E., (1998). Hypothesis: Comparisons of inter-and intra-individual variations can substitute for twin studies in drug research. *Pharmacogenetics, 8*, 283–290.

Katherine, A. D., Dara, G. T., Christopher, R. G., Joshua, M. G., Lindsey, A. R., Scott, H., Celeste, E., et al., (2014). A genome-wide association study of bronchodilator response in latinos implicates rare variants. *Journal of Allergy and Clinical Immunology, 133*, 370–378. e15.

Kathryn, A. P., David, L. V., Eyal, O., Jane, K. L., & Wolfgang, S., (2001). Potential role of pharmacogenomics in reducing adverse drug reactions: A systematic review. *JAMA, 286*, 2270–2279.

Kelan, G. T., Eun, S. H., Benjamin, A. R., Eric, S. S., Stephen, L. L., Brent, G. R., Stanford, L. P., et al., (2004). Tbx21: A functional variant predicts improvement in asthma with the use of inhaled corticosteroids. *Proceedings of the National Academy of Sciences, 101*, 18099–18104.

Kelan, G. T., Kersten, M. S., Augusto, A. L., Scott, T. W., & Stephen, B. L., (2005). Molecular properties and pharmacogenetics of a polymorphism of adenylyl cyclase type 9 in asthma: Interaction between β-agonist and corticosteroid pathways. *Human Molecular Genetics, 14*, 1671–1677.

Kelan, G. T., Lasky-Su, J., Michishige, H., Amy, M., Augusto, A. L., Blanca, E. H., Christoph, L., et al., (2011). Genomewide association between GLCCI1 and response to glucocorticoid therapy in asthma. *New England Journal of Medicine, 365*, 1173–1183.

Kelan, G. T., Stephen, L., Eric, S. S., Lyle, J. P., Ross, L., Edwin, K. S., Stephen, B. L., et al., (2004). Corticosteroid pharmacogenetics: Association of sequence variants in CRHR1 with improved lung function in asthmatics treated with inhaled corticosteroids. *Human Molecular Genetics, 13*, 1353–1359.

Kenneth, O., (2011). Personalized medicine: New genomics, old lessons. *Human Genetics, 130*, 3–14.

Kevin, F., Anna, B., William, Q., Jessica, S., Renta, H., Stuart, M., Satya, K., Rajeev, K., Klaus, C., & Kristina, Y., (2014). A subcutaneous, potent and durable RNAi platform targeting metabolic diseases, genes PCSK9, APOC3 and ANGPLT3. *Arteriosclerosis, Thrombosis, and Vascular Biology, 34*, A7-A7.

Kim, S. H., Ye, Y. M., Lee, H. Y., Sin, H. J., & Park, H. S., (2011). Combined pharmacogenetic effect of Adcy9 and Adrb2 gene polymorphisms on the bronchodilator response to inhaled combination therapy. *J. Clin. Pharm. Ther., 36*, 399–405.

Kirchheiner, J., Meineke, I., Müller, G., Bauer, S., Rohde, W., Meisel, C., Roots, I., & Brockmöller, J., (2004). Influence of Cyp2c9 and Cyp2d6 polymorphisms on the pharmacokinetics of nateglinide in genotyped healthy volunteers. *Clin. Pharmacokinet, 43*, 267–278.

Kleinberger, J. W., & Pollin, T. I., (2015). Personalized medicine in diabetes mellitus: Current opportunities and future prospects. *Ann. N Y Acad Sci., 1*, 45–56.

Klotsman, M., York, T. P., Pillai, S. G., Vargas-Irwin, C., Sharma, S. S., Van, D. O. E. J., & Anderson, W. H., (2007). Pharmacogenetics of the 5-lipoxygenase biosynthetic pathway and variable clinical response to montelukast. *Pharmacogenet Genomics, 17*, 189–196.

Lang, V. Y., Fatehi, M., & Light, P. E., (2012). Pharmacogenomic analysis of Atp-Sensitive potassium channels co-expressing the common type 2 diabetes risk variants E23k and S1369a. *Pharmacogenet Genomics, 22*, 206–214.

Laura, H. G., & Nicholas, J. S., (2018). Personalized medicine: Motivation, challenges, and progress. *Fertility and Sterility, 109*, 952–963.

Lazarou, J., Pomeranz, B. H., & Corey, P. N., (1998). Incidence of adverse drug reactions in hospitalized patients: A meta-analysis of prospective studies. *Jama, 279*(15), 1200–1205.

Lima, J. J., Thomason, D. B., Mohamed, M. H., Eberle, L. V., Self, T. H., & Johnson, J. A., (1999). Impact of genetic polymorphisms of the Beta2-adrenergic receptor on albuterol bronchodilator pharmacodynamics. *Clin. Pharmacol. Ther., 65*, 519–525.

Lima, J. J., Zhang, S., Grant, A., Shao, L., Tantisira, K. G., Allayee, H., Wang, J., et al., (2006). Influence of leukotriene pathway polymorphisms on response to montelukast in asthma. *Am. J. Respir. Crit. Care Med., 173*, 379–385.

Luba, F., Avital, G., Ofer, I., & Noam, S., (2013). Meeting summary: Ethical aspects of whole exome and whole genome sequencing studies (Wes/Wgs) in rare diseases, Tel Aviv, Israel. *Genetics Research, 95*, 53–56.

Maggie, S., Beverley, S., Sian, E., Rubio-Cabezas, O., & Andrew, T. H., (2009). A genetic diagnosis of HNF1A diabetes alters treatment and improves glycaemic control in the majority of insulin-treated patients. *Diabetic Medicine, 26*, 437–441.

Maja, D. J., Silvija, S., Miloš, M. J., Monika, M., Dragan, M., & Svetislav, N., (2008). A case of new mutation in maturity-onset diabetes of the young type 3 (mody 3) responsive to a low dose of sulphonylurea. *Diabetes Research and Clinical Practice, 81*, e1-e3.

Majid, N., Ondrej, Š., Johanne, T., Milan, P., Daniel, G., Theodore, A. K., Allen, W. C. Jr., & Pavel, H., (2012). Genetic mapping of habitual substance use, obesity-related traits, responses to mental and physical stress, and heart rate and blood pressure measurements reveals shared genes that are overrepresented in the neural synapse. *Hypertension Research: Official Journal of the Japanese Society of Hypertension, 35*, 585–591.

Marc, J. V. De. V., Yudong, D. H., Laura, J. V., V., Hongyue, D., Augustinus, A. M. H., Dorien, W. V., et al., (2002). A gene-expression signature as a predictor of survival in breast cancer. *New England Journal of Medicine, 347*, 1999–2009.

Maria, I., Emmanouil, P., Anna, T., Peristera, P., Athanasios, C., & Vangelis, G. M., (2012). G894t polymorphism of Enos gene is a predictor of response to a combination of inhaled corticosteroids with long-lasting B2-agonists in asthmatic children. *Pharmacogenomics, 13*, 1363–1372.

Maria, J. T., Blanca, M., Fernandez, J., Romano, A., De Weck, A., Aberer, W., Brockow, K., et al., (2003). Diagnosis of immediate allergic reactions to beta-lactam antibiotics. *Allergy, 58*, 961–972.

Mario, M., Vincenzo, B., Herbert, Y. M., Jeffrey, A. L., Serge, S., Fabio, M. M., Phil, C., et al., (1998). Serotonin subtype 2 receptor genes and clinical response to clozapine in schizophrenia patients. *Neuropsychopharmacology, 19*, 123–132.

Mark, T. S. C., Donald, Y. M. L., Stanley, J. S., & Joseph, D. S., (1998). Difficult-to-control asthma: Clinical characteristics of steroid-insensitive asthma. *Journal of Allergy and Clinical Immunology, 101*, 594–601.

Martine, J. Piccart-Gebhart, Marion, P., Brian Leyland-Jones, Aron, G., Michael, U., Ian, S., Luca, G., et al., (2005). Trastuzumab after adjuvant chemotherapy in Her2-positive breast cancer. *The New England Journal of Medicine, 353*, 1659–1672.

McLeod, J. F., (2004). Clinical pharmacokinetics of nateglinide: A rapidly-absorbed, short-acting insulinotropic agent. *Clin. Pharmacokinet, 43*, 97–120.

Md. Fakruddin, & Chowdhury, A., (2013). Pharmacogenomics-the promise of personalized medicine. *Bangladesh Journal of Medical Science, 12*, 346–356.

Mertens, G., (2010). Gene/environment interaction in atherosclerosis: An example of clinical medicine as seen from the evolutionary perspective. *International Journal of Hypertension*, 1–3.

Metzker, M. L., (2010). Applications of next-generation sequencing. *Nature Reviews. Genetics, 11*, 31–46.

Michael, E. W., Erik, L., Stephen, C. L., Robert, F. L. Jr., Homer, A. B., Aaron, D., John, V. F., et al., (2006). B-adrenergic receptor polymorphisms and response to salmeterol. *American Journal of Respiratory and Critical Care Medicine, 173*, 519–526.

Michael, E. W., Susan, J. K., Vernon, M. C., Eugene, B., Homer, A. B., William, J. C., Bill, T. A., et al., (2009). Effect of B2-adrenergic receptor polymorphism on response to longacting B2 agonist in asthma (large trial): A genotype-stratified, randomized, placebo-controlled, crossover trial. *The Lancet, 374*, 1754–1764.

Michael, J. B., Sarah, B. N., Abigail, W. B., Holly, K. T., Mary, J. E., Deborah, A. N., & Jay, S., (2011). Exome sequencing as a tool for mendelian disease gene discovery. *Nature Reviews Genetics, 12*, 745–755.

Michael, R. S., Peter, J. C., & Andrew, F. P., (2009). The cancer genome. *Nature, 458*, 719–724.

Miriam, F. M., Michael, K., Liming, L., Anna, L. D., David, S., Simon, H., Martin, D., et al., (2007). Genetic variants regulating Ormdl3 expression contribute to the risk of childhood asthma. *Nature, 448*, 470–473.

Mougey, E. B., Feng, H., Castro, M., Irvin, C. G., & Lima, J. J., (2009). Absorption of montelukast is transporter mediated: A common variant of Oatp2b1 is associated with reduced plasma concentrations and poor response. *Pharmacogenet Genomics, 19*, 129–138.

National Institutes of Health Consensus Development Panel, (2001). National institutes of health consensus development conference statement: Adjuvant therapy for breast cancer. *JNCI Monographs, 2001*, 5–15.

Ortega VE, Meyers DA, Bleecker ER. Asthma pharmacogenetics and the development of genetic profiles for personalized medicine. Pharmgenomics Pers Med. 2015 Jan 16;8:9-22. doi: 10.2147/PGPM.S52846. PMID: 25691813; PMCID: PMC4325626.

Ortega, V. E., Hawkins, G. A., Moore, W. C., Hastie, A. T., Ampleford, E. J., Busse, W. W., Castro, M., et al., (2014). Effect of rare variants in ADRB2 on risk of severe exacerbations and symptom control during longacting B agonist treatment in a multiethnic asthma population: A genetic study. *Lancet Respir Med., 2*, 204–213.

Oscar Diaz-Horta, Duygu, D., Joseph, F. IInd., Aslı, S., Michael, G., Nejat, M., Nikou, F., Mortaza, B., Filiz, B. C., & Ibis, M., (2012). Whole-exome sequencing efficiently detects rare mutations in autosomal recessive nonsyndromic hearing loss. *PloS One,* 7.

Pascal, B., Christian, W., David, Q. S., Tien-An, Y., Allan, W. W., & Markus, S., (2002). Decreased glibenclamide uptake in hepatocytes of hepatocyte nuclear factor-1α-deficient mice: A mechanism for hypersensitivity to sulfonylurea therapy in patients with maturity-onset diabetes of the young, type 3 (Mody3). *Diabetes, 51*, S343–S348.

Pearce, N., Burgess, C., Crane, J., & Beasley, R., (1997). Fenoterol, asthma deaths, and asthma severity. *Chest, 112*(4), 1148–1150. doi: 10.1378/chest.112.4.1148-b.

Pearce, N., Grainger, J., Atkinson, M., Crane, J., Burgess, C., Culling, C., Windom, H., & Beasley, R., (1990). Case-control study of prescribed fenoterol and death from asthma in New Zealand, 1977–1981. *Thorax, 45*, 170–175.

Pearson, E. R., Liddell, W. G., Shepherd, M., Corrall, R. J., & Hattersley, A. T., (2000). Sensitivity to sulphonylureas in patients with hepatocyte nuclear factor-1α gene mutations: Evidence for pharmacogenetics in diabetes. *Diabetic Medicine, 17*, 543–545.

Peters, S. P., Anthonisen, N., Castro, M., Holbrook, J. T., Irvin, C. G., Smith, L. J., & Wise, R. A., (2007). Randomized comparison of strategies for reducing treatment in mild persistent asthma. *N Engl. J. Med., 356*, 2027–2039.

Pierre, H., Elodie, C., Generoso, B., Fref, B., Thibaut, D., Heinz, H., Anne, J., et al., (2009). Pathology as the cornerstone of human tissue banking: European consensus expert group report. *Biopreservation and Biobanking, 7*, 157–160.

Pistoi, S., (2002). Mind the gap. *Scientific American, 286*(4), 25.

Qing, L. D., Brigitte, R. G., Gregory, A. H., Blanca, E. H., Eugene, R. B., Barbara, K., Charles, G. I., et al., (2011). Regulatory haplotypes in Arg1 are associated with altered bronchodilator response. *American Journal of Respiratory and Critical Care Medicine, 183*, 449–454.

Qing, L. D., Rose, D., Jessica Lasky-Su, Barbara, J. K., Amanda, B. P., Stephen, P. P., Charles, G. I., et al., (2013). A polymorphism in the thyroid hormone receptor gene is associated with bronchodilator response in asthmatics. *The Pharmacogenomics Journal, 13*, 130–136.

Ramsey, B. W., Davies, J., McElvaney, N. G., Tullis, E., Bell, S. C., Drevinek, P., Griese, M., et al., (2011). A CFTR potentiator in patients with cystic fibrosis and the G551d mutation. *N Engl. J. Med., 365*, 1663–1672.

Ramzi, A. A., & Katharine, R. O., (2014). Glucokinase mody and implications for treatment goals of common forms of diabetes. *Current Diabetes Reports, 14*, 559.

Randy, V. F., Carol, I. B., & Michael, P., (2010). Personalized medicine: Part 1: Evolution and development into theranostics. *P & T: A Peer-Reviewed Journal for Formulary Management, 35*, 560–576.

Ridha, J., Chawki, B., Kateri, B., André, T., Gustavo, T., Samarthji, L., David, B., Alain, L., Pierre, L., & Diane, F., (1999). T102c polymorphism in the 5ht2a gene and schizophrenia: Relation to phenotype and drug response variability. *Journal of Psychiatry and Neuroscience, 24*, 141.

Rigaud, A. S., Latchezar, T., Caputo, L., Guelfi, M. C., Latour, F., Couderc, R., Moulin, F., et al., (2000). The apolipoprotein E E4 allele and the response to tacrine therapy in Alzheimer's disease. *European Journal of Neurology, 7*, 255–258.

Robert, C. T., Carole, A. C., Valeria, F., Rury, R. H., & UK Prospective Diabetes Study Group, (1999). Glycemic control with diet, sulfonylurea, metformin, or insulin in patients with type 2 diabetes mellitus: Progressive requirement for multiple therapies (Ukpds 49). *JAMA, 281*, 2005–2012.

Robert, E. H., (2011). Biobanking: The foundation of personalized medicine. *Current Opinion in Oncology, 23*, 112–119.

Robert, S., Ghislain, R., Johan, R., Christian, D., Lishuang, S., David, S., Philippe, B., et al., (2007). A genome-wide association study identifies novel risk loci for type 2 diabetes. *Nature, 445*, 881–885.
Robin, T. D., Jeffrey, M. D., Peter, H. G., Chandra, N. Y., Robert, J. H., & Ian, T. G., (2000). Asthma exacerbations during long term β agonist use: Influence of B2 adrenoceptor polymorphism. *Thorax, 55*, 762–767.
Ronald, M. N., (2001). Clinical pharmacogenomics: Applications in pharmaceutical R&D. *Drug Discovery Today, 6*, 180–185.
Rosell, R., Bivona, T. G., & Karachaliou, N., (2013). Genetics and biomarkers in personalization of lung cancer treatment. *Lancet, 382*, 720–731.
Samir, N. K., James, H. D., and, W., N. H., (2010). Aacr-Fda-Nci cancer biomarkers collaborative consensus report: Advancing the use of biomarkers in cancer drug development. *Clinical Cancer Research, 16*, 3299–3318.
Sato, R., Watanabe, H., Genma, R., Takeuchi, M., Maekawa, M., & Nakamura, H., (2010). Abcc8 Polymorphism (Ser1369ala): Influence on severe hypoglycemia due to sulfonylureas. *Pharmacogenomics, 11*, 1743–1750.
Serena, S., Anne, U. J., Ramaiah, N., Cristen, J. W., Wei-Min, C., Lori, L. B., Haiqing, S., et al., (2008). Common variants in the GDF5-UQCC region are associated with variation in human height. *Nature Genetics, 40*, 198.
Sesti, G., Laratta, E., Cardellini, M., Andreozzi, F., Del, G, S., Irace, C., Gnasso, A., et al., (2006). The E23k variant of KCNJ11 encoding the pancreatic beta-cell adenosine 5'-triphosphate-sensitive potassium channel subunit Kir6.2 is associated with an increased risk of secondary failure to sulfonylurea in patients with type 2 diabetes. *J. Clin. Endocrinol. Metab., 91*, 2334–2339.
Shehadeh, N., Bakri, D., Njølstad, P. R., & Gershoni-Baruch, R., (2005). Clinical characteristics of mutation carriers in a large family with glucokinase diabetes (Mody2). *Diabetic Medicine, 22*, 994–998.
Shelley, R. S., Nicholas, S. B., Thomas, M. O., & Edwin, E. S., (2006). Meta-analysis: Effect of long-acting B-agonists on severe asthma exacerbations and asthma-related deaths. *Annals of Internal Medicine, 144*, 904–912.
Shields, B. M., Hicks, S., Shepherd, M. H., Colclough, K., Andrew, T. H., & Sian, E., (2010). Maturity-onset diabetes of the young (Mody): How many cases are we missing? *Diabetologia, 53*, 2504–2508.
Shweta, C., Ngim, U., Pedro, C. A., Elad, Z., Sylvette, N., Jesus, C., Alfonso, T., Jennifer, D. G., Keyan, S., & Rodriguez-Santana, J. R., (2005). Pharmacogenetic differences in response to albuterol between Puerto Ricans and Mexicans with asthma. *American Journal of Respiratory and Critical Care Medicine, 171*, 563–570.
Sian, E., Beards, F., Allen, L. I. S., Maggie, S., Ballantyne, E., Harvey, R., & Andrew, T. H., (2000). A high prevalence of glucokinase mutations in gestational diabetic subjects selected by clinical criteria. *Diabetologia, 43*, 250–253.
Simon, N. S., Andrei, M., Patrick, S., Thorunn, R., Julius, G., Sigurjon, A. G., Gisli, M., et al., (2007). Common variants on chromosomes 2q35 and 16q12 confer susceptibility to estrogen receptor–positive breast cancer. *Nature Genetics, 39*, 865–869.
Sorkness, C. A., Lemanske, R. F. Jr., Mauger, D. T., Boehmer, S. J., Chinchilli, V. M., Martinez, F. D., Strunk, R. C., et al., (2007). Long-term comparison of 3 controller

regimens for mild-moderate persistent childhood asthma: The pediatric asthma controller trial. *J Allergy Clin Immunol, 119*, 64–72.

Sørlie, T., Perou, C. M., Tibshirani, R., Aas, T., Geisler, S., Johnsen, H., Hastie, T., et al., (2001). Gene expression patterns of breast carcinomas distinguish tumor subclasses with clinical implications. *Proceedings of the National Academy of Sciences of the United States of America, 98*, 10869–10874.

Stephanie, S., Beth, A. P., & Deborah, A. D., (2005). Fragile X syndrome: Diagnostic and carrier testing. *Genetics in Medicine: Official Journal of the American College of Medical Genetics, 7*, 584–587.

Stephen, F. K., & Carol, J. S., (2011). Deep sequencing of patient genomes for disease diagnosis: When will it become routine? *Science Translational Medicine, 3*, 87ps23–87ps23.

Stockmann, C., Fassl, B., Gaedigk, R., Nkoy, F., Uchida, D. A., Monson, S., Reilly, C. A., et al., (2013). Fluticasone propionate pharmacogenetics: Cyp3a4*22 polymorphism and pediatric asthma control. *J. Pediatr., 162*, 1222–1227.

Stolley, P. D., (1972). Asthma mortality. Why the United States was spared an epidemic of deaths due to asthma. *Am. Rev. Respir. Dis., 105*, 883–890.

Sylvia, F. B., Dimitri, P., & Jorge, F., (2010). Epistasis of transcriptomes reveals synergism between transcriptional activators Hnf1α and Hnf4α. *PLoS Genetics,* 6.

Szefler, S. J., Phillips, B. R., Martinez, F. D., Chinchilli, V. M., Lemanske, R. F., Strunk, R. C., Zeiger, R. S., et al., (2005). Characterization of within-subject responses to fluticasone and montelukast in childhood asthma. *J. Allergy Clin. Immunol., 115*, 233–242.

Tantisira, K. G., Hwang, E. S., Raby, B. A., Silverman, E. S., Lake, S. L., Richter, B. G., Peng, S. L., et al., (2004). Tbx21: A functional variant predicts improvement in asthma with the use of inhaled corticosteroids. *Proc. Natl. Acad. Sci. U S A, 101*, 18099–18104.

Tantisira, K. G., Lake, S., Silverman, E. S., Palmer, L. J., Lazarus, R., Silverman, E. K., Liggett, S. B., et al., (2004). Corticosteroid pharmacogenetics: Association of sequence variants in crhr1 with improved lung function in asthmatics treated with inhaled corticosteroids. *Hum. Mol Genet, 13*, 1353–1359.

Tantisira, K. G., Lima, J., Sylvia, J., Klanderman, B., & Weiss, S. T., (2009). 5-lipoxygenase pharmacogenetics in asthma: Overlap with Cys-leukotriene receptor antagonist loci. *Pharmacogenet Genomics, 19*, 244–247.

Tantisira, K. G., Small, K. M., Litonjua, A. A., Weiss, S. T., & Liggett, S. B., (2005). Molecular properties and pharmacogenetics of a polymorphism of adenylyl cyclase type 9 in asthma: Interaction between beta-agonist and corticosteroid pathways. *Hum. Mol. Genet, 14*, 1671–1677.

Telleria, J. J., Blanco-Quiros, A., Varillas, D., Armentia, A., Fernandez-Carvajal, I., Jesus, A. M., & Diez, I., (2008). Alox5 promoter genotype and response to montelukast in moderate persistent asthma. *Respir Med., 102*, 857–861.

TG, National Heart HDL Working Group of the Exome Sequencing Project, Lung, and Blood Institute, (2014). Loss-of-function mutations in APOC3, triglycerides, and coronary disease. *New England Journal of Medicine, 371*, 22–31.

Toni, I. P., Coleen, M. D., Haiqing, S., Sandra, H. O., John, S., Richard, B. H., Wendy, P., et al., (2008). A null mutation in human Apoc3 confers a favorable plasma lipid profile and apparent cardioprotection. *Science, 322*, 1702–1705.

Van, L. N., Nijpels, G., Becker, M. L., Deshmukh, H., Zhou, K., Stricker, B. H., Uitterlinden, A. G., et al., (2012). A gene variant near atm is significantly associated with metformin treatment response in type 2 diabetes: A replication and meta-analysis of five cohorts. *Diabetologia, 55*, 1971–1977.

Venter, J. C., Adams, M. D., Myers, E. W., Li, P. W., Mural, R. J., Sutton, G. G., et al., (2001). The sequence of the human genome. *Science, 291*, 1304–1351.

Victor, E. O., Deborah. A. M., & Eugene, R. B., (2015). Asthma pharmacogenetics and the development of genetic profiles for personalized medicine. *Pharmacogenomics and Personalized Medicine, 8*, 9–22.

Win, C., Rick, F., John, H., & James, P., (1993). Serevent nationwide surveillance study: Comparison of salmeterol with salbutamol in asthmatic patients who require regular bronchodilator treatment. *BMJ, 306*, 1034–1037.

Ye, Y. M., Lee, H. Y., Kim, S. H., Jee, Y. K., Lee, S. K., Lee, S. H., & Park, H. S., (2009). Pharmacogenetic study of the effects of Nk2r G231e G> a and Tbx21 H33q C> G polymorphisms on asthma control with inhaled corticosteroid treatment. *Journal of Clinical Pharmacy and Therapeutics, 34*, 693–701.

Ye, Y. M., Lee, H. Y., Kim, S. H., Jee, Y. K., Lee, S. K., Lee, S. H., & Park, H. S., (2009). Pharmacogenetic study of the effects of Nk2r G231e G>a and Tbx21 H33q C>G polymorphisms on asthma control with inhaled corticosteroid treatment. *J. Clin. Pharm. Ther., 34*, 693–701.

Yixin, W., Jan, G. M. K., Yi, Z., Anieta, M. S., Maxime, P. L., Fei, Y., Dmitri, T., et al., (2005). Gene-expression profiles to predict distant metastasis of lymph-node-negative primary breast cancer. *The Lancet, 365*, 671–679.

Yuqiu, J., & Mu, W., (2010). Personalized medicine in oncology: Tailoring the right drug to the right patient. *Biomarkers in Medicine, 4*, 523–233.

Zhou, K., Bellenguez, C., Spencer, C. C., Bennett, A. J., Coleman, R. L., Tavendale, R., Hawley, S. A., et al., (2011). Common variants near atm are associated with glycemic response to metformin in type 2 diabetes. *Nat. Genet., 43*, 117–120.

CHAPTER 6

Stem Cell Technology and Regenerative Medicine: A Travel Back to Roots

BASHIR AHMAD SHEIKH,[1] AASHIQ SULTAN,[2]
BASHARAT AHMAD BHAT,[1] and MANZOOR AHMAD MIR[1]

[1]*Department of Bioresources, School of Biological Sciences, University of Kashmir, Srinagar, Jammu and Kashmir, India*

[2]*Department of Zoology, School of Biological Sciences, University of Kashmir, Srinagar, Jammu and Kashmir, India*

ABSTRACT

Stem cells are undifferentiated progenitor cells that can produce different cell phenotypes under specific biomolecular stimuli via the differentiation pathway. Stem cells and regenerative medicine are part of an ongoing research study and application process. Overall, increased interest in stem cell research is due to their specific biological nature and their potential to understand the process of aging and degenerative diseases in a better way. Stem cell biology, mainly stem cell-based therapy, has developed as an active field over the past decade. Significant research studies have led to critical breakthroughs in several fields through stem cell technology. Stem cell treatment has made significant strides in customization of therapy, tissue engineering, and disease mitigation in chronic and regenerative diseases. Adult stem cells (ASCs) are promising candidates for disease therapy because of their adaptability, minimal immunogenicity, and great anti-inflammatory capacity. Numerous disorders, including tissue and

Modern Biotechnology in Healthcare: Advances and Applications. Sheikh Umar Ahmad (Ed.)

organ damage, cardiovascular, and other functional organ failures, degenerative diseases, and cancer, provide a substantial challenge to the scientific community. Traditional therapies, such as pharmacological therapy and surgery, frequently have little effect on these disorders and fail to meet the growing medical needs. Recent advances in stem cell technology and its regenerative medicine have shown promising results and have given a good hint for their substantial application in disease treatments. Thus, stem cell-based regenerative medicine is expected to become the third treatment option after drug therapy and surgery. This chapter will explore how stem cell technology and regenerative medicine have contributed to disease treatment during the last decade. Moreover, we will also try to highlight the different opportunities and emerging challenges of stem cell-based therapies and regenerative medicine.

6.1 INTRODUCTION

Stem cells are undifferentiated or partially differentiated cells which retain their proliferative capacity throughout the lifetime of an animal (Dalerba, Diehn, Weissman, & Clarke, 2020; Fuchs & Segre, 2000). These cells lack tissue-specific differentiation markers and are present at both embryonic as well as at adult stages of life of an organism and are considered as earliest cell lineages of the body of an organism. Most adult animal cells are no longer capable of proliferation and are arrested at G_0. On their natural death due to aging, cells can be replaced by the proliferation of a subpopulation of less differentiated, self-renewing cells called stem cells (Cable et al., 2020; Dalerba et al., 2020; Gurley & Alvarado, 2019; Slack, 2008). These stem cells are found in most of an organism's adult tissues and are crucial for maintaining the majority of an organism's tissues and organs. However, differentiated cells remain arrested in the G_0 cell cycle stage but resume proliferation as needed to replenish cells lost during injury or death. The cell of this type includes fibroblasts dispersed in connective tissues where they secrete collagen; they are typically arrested in the G_0 stage but rapidly proliferate if needed to repair the damage resulting from cut or wound (Baksh, Song, & Tuan, 2004). The two essential properties of stem cells are self-renewal and potency. Self-renewal is the ability of a cell to go through one are more cycles of cell division to maintain the undifferentiated cell layer or cell mass. At the same time, potency is the capacity to differentiate into different specialized cell types (Ejtehadifar et al., 2015).

Stem cell biology has received significant interest during the last few decades. As a result, enhanced awareness of its properties and therapeutic potential for its application has been prioritized (Mohammadian et al., 2013). These cells are used in cell therapy and experimental research to treat human problems, including hematology, skin regeneration, and heart disease (Markoski, 2016). In the past, the work related to stem cells was performed by Friedenstein and colleagues in the 1960s. They identified, grew, and differentiated bone marrow (BM)-derived cells from guinea pigs into osteogenic cell lineages (Friedenstein, Chailakhyan, Latsinik, Panasyuk, & Keiliss-Borok, 1974). Their work led to a new perspective on the research related to stem cells. As a result, numerous further studies discovered that the BM contains fibroblast-like cells with congenic potential *in vitro*, capable of forming colonies (CFU-F) (Banas et al., 2007). For more than 60 years, transplantation of hematopoietic stem cells (HSC) approach has been one of the significant curative therapies for various genetic and hematological diseases and disorders (Ghimire, Weber, Mavin, Dickinson, & Holler, 2017). Till and McCulloch in 1963 described a peculiar single progenitor cell type in BM, having the potential to expand clonally and give rise to all hematopoietic cell lineages. This research work represented the first characterization of the HSCs (Becker, McCulloch, & Till, 1963; Volarevic et al., 2014). In addition to this, the recognition of mouse embryonic stem cells (ESCs) in 1981 proved revolutionary in the study of developmental biology (Volarevic et al., 2014). As stem cells can differentiate into specific cell types, it has become a possibility of a renewable source of replacement cells and tissues to treat diseases like leukemia, anemia, spinal cord injury (SCI), stroke, burns, heart diseases, diabetes, osteoarthritis, and rheumatoid arthritis (Lee et al., 2019; Makena, Ranjan, Thirumala, & Reddy, 2020; Tweed, 2019; Witkowski, Lasry, Carroll, & Aifantis, 2019; Zakrzewski, Dobrzyński, Szymonowicz, & Rybak, 2019). The sources of these stem cells can be ESC, somatic stem cells (SSCs) and induced pluripotent stem cells (iPSCs) (Bragança, Lopes, Mendes-Silva, & Santos, 2019). But human embryonic stem cell (hESC) research is ethically and politically controversial as it involves the mass destruction of human embryos like that of abortion. However, in modern times, we have learned a lot about stem cells and are obtained from sources like BM, umbilical cord blood, adipose tissue, amniotic fluid, and allografts (Acosta & Golub, 2016; McCoy & Sterngass, 2019).

6.2 CLASSIFICATION OF STEM CELLS

Based on differentiation capacity, stem cells are categorized into various types, comprising totipotent, pluripotent, multipotent, and unipotent (Amin et al., 2019). In addition to this, these cells are classified based on evolutionary stages. They include embryonic, fetal, infant, or umbilical cord blood and adult stem cells (ASCs) (Rajabzadeh, Fathi, & Farahzadi, 2019) shows an overview of stem cell classifications based on differentiation potency.

6.2.1 TOTIPOTENT CELLS

These cells can develop any cell of the developing organism, including embryonic and extra-embryonic (e.g., placenta) tissues, for example, 1–4 celled stage of the human zygote (Kalra & Tomar, 2014). During this stage, if cells get separated, they can develop into an entire embryo, including the tissues which support the developing embryo-like, in the natural case of monozygotic twins. Two cells of the single zygote naturally get separated in the mother's body and develop into two separate embryos, including their distinct supportive tissues of extra-embryonic membranes (Kalra & Tomar, 2014).

6.2.2 PLURIPOTENT

These cells can only make cells of the embryo proper, which includes all cells of three germ layers. Therefore, they can make any cell of the body, for example, the inner cell mass of an early embryo (Baker & Pera, 2018).

6.2.3 MULTIPOTENT

These cells can give rise to a few cells on differentiation; for example, HSCs of BM can give rise to different blood cell types (Cheng, Zheng, & Cheng, 2020).

6.2.4 UNIPOTENT

Unipotent stem cells have not received much attention in research compared with that totipotent and pluripotent stem cells. The unipotent stem cell is a cell that can give rise to cells with only one lineage differentiation (Giri, Alexander, Agrawal, Saraf, & Saraf, 2019). Muscle stem cells represent one of the instances of this type of cell (Giri et al., 2019). These cells have the lowest differentiation potential in adult tissues compared to other types of stem cells. More specifically, these cells do not have self-renewal potential. Furthermore, despite their limited differentiation potential, these cells depict their role in treating various disorders (Rajabzadeh et al., 2019).

6.2.5 EMBRYONIC STEM CELLS (ESCS)

Embryonic stem cells (ESCs) are self-renewing cells that originate in the blastocyst's inner cell mass and give birth to all cells during human development (Levenberg et al., 2003). In introduced embryos, these cells can develop into all types of body cells. Due to their unique capacity to generate all somatic cell lineages, these cells may be a good and promising source for cell transplantation and regenerative therapy (Lu, Li, Vida, & Honig, 2004). In other words, ESCs are pluripotent cells capable of differentiating to generate all of the body's specialized cell types (Lu et al., 2004). Additionally, ESCs capture the imagination due to their immortality and near-limitless growth potential. Due to ethical constraints on embryo collection and culture, these cells are employed in research less frequently (AE, 2004).

6.2.6 ADULT STEM CELLS (ASCS)/TISSUE-SPECIFIC STEM CELL

ASCs are undifferentiated cells and are usually located in various body tissues after embryonic development. These cells have the ability to multiplication by cell division and hence contribute significantly to the regeneration of damaged tissues (Mahla, 2016). Specific investigations have demonstrated that these cells can differentiate into cell types from multiple germ layers. For instance, BM stem cells originating from mesoderm can develop into cell lineages derived from mesoderm and endoderm, such as the gastrointestinal tract, liver, lung, and skin (Krause et al., 2001). Moreover, one more example of ASCs is neural stem cells (NSCs) derived from ectoderm. These cells can differentiate into other lineage such as endoderm and mesoderm (Clarke et al., 2000). The fact that ASCs have therapeutic potential has been proved by cell therapy and regenerative medicine (Mimeault, Hauke, & Batra, 2007).

6.2.7 INDUCED PLURIPOTENT STEM CELLS (IPSCS)

These cells are not present in the body but are created from body cells such as skin cells in a laboratory. Conversion of these somatic cells into iPSCs by gene reprogramming (Karagiannis et al., 2019). In 2006, Takahashi & Yamanaka were able to produce these engineered cells. To date, scientists are in a race to apply these cells in the advanced treatment of deadliest diseases for the substitution of traditional therapies such as drug therapy and surgery. These iPSCs cells have properties similar to embryonic stem cells (ESCs), which can develop into body cells, i.e., pluripotent cells (Karagiannis et al., 2019; Russo, 2017).

6.2.8 CANCER STEM CELLS (CSCS)

Cancer stem cells (CSCs) were primarily identified in acute myeloid diseases by John Dick in the late 1990s. These stem cells are cancerous and are located within tumors or hematological cancers. These cells possess the features of normal stem cells and have the ability to give rise to all other cell types found in the specific cancer sample (Yang et al., 2008). This hypothesis is supported by various evidence as well. In an adult living species, normal stem cells are responsible for repairing and regeneration of damaged and old tissues (Soltysova, Altanerova, & Altaner, 2005). Several research investigations have pointed out that the potential of a tumor to propagate and proliferate depends on the minute or small cellular subpopulation encompassing stem cell-like characteristics often named as CSCs (Li et al., 2007).

6.2.9 MESENCHYMAL STEM CELLS (MSCS)

The connective tissue (i.e., embryonic) consists of mesenchyme. All other hematopoietic and connective tissues originate from the close interaction of endoderm and ectoderm; however, MSCs do not differentiate into the hematopoietic cell (Fathi & Farahzadi, 2017). Alexander A. Maxi mow in 1924, used a histological detection to recognize a precursor cell within mesenchyme that ultimately develops and leads to various blood cell types (Sell, 2013). MSCs are a kind of cell that has the capacity for multilineage differentiation and self-renewal. They are found in various tissues and organs, including adipose tissue, skin, BM, peripheral blood, cord blood, fallopian tube, lungs, and liver (Ejtehadifar et al., 2015; Mohammadian et al., 2013). Today, stem cells are used for different applications, more importantly in human therapy, such as cell transplantation and cell engraftment. Types of stem cells, their source, uses, and limitations are enlisted in Table 6.1.

6.3 APPLICATION OF STEM CELL TECHNOLOGY AND REGENERATIVE MEDICINE IN DISEASE TREATMENT

Stem cell therapy is the novel process of implanting stem cells to promote the repair response of diseased, dysfunctional, and injured tissues. It is considered a novel substitution for traditional drug therapy and organ transplantation to eliminate these treatments' limitations or shortcomings (Chang, Wu, Harn, Lin, & Ding, 2018). As in the case of drug therapy, there are usually many side effects after treatment. In organ transplantation, it is hardly possible to get a donor. Even if we have the donor, there is unlikely to have a compatible match. In organ transplantation, there is always a chance of rejection from the body, so the person always relies on immunosuppressant drugs throughout life (Holzer, 2019). However, stem cell has shown promise in diseases like leukemia and anemia (BM stem cell transplantation), also in case of baldness and potential treatment in heart strokes (Segers & Lee, 2008), Alzheimer's disease (AD), diabetes, and Parkinson's disease (Duncan & Valenzuela, 2017; Zhu, Uezono, Yasui, & Nakashima, 2018). Stem cell therapy also has shown a lot of promise for wound healing and SCI, and their use can also be sensed in somebody with disabilities like blindness, deafness, and missing teeth (Price, 2020).

Regenerative Medicines are the use of body fluids, cell extracts, cultured stem cells, engineered regenerated tissues and organs for the treatment of severe injuries and chronic diseases to restore and establish back the normal functioning in the state where bodies own regenerative responses do not suffice (Fujita, Kadota, Araya, Ochiya, & Kuwano,

TABLE 6.1 Types of Stem Cells, Their Sources, Uses, and Limitations

SL. No.	Type	Source	Use	Limitation
1.	Embryonic stem cells (ESCs)	Inner cell mass from the blastocyst, an early stage of the embryo.	*Research purpose:* Study of cell commitment differentiation and maturation; drug testing, genomics, growth factor discovery. *Clinical purpose:* Tissue engineering, cell-based therapies.	Ethically exploitation of embryos is not permitted; ESCs can develop into uncontrollable cells if not correctly handled and rejection problems during therapy.
2.	Adult stem cells or somatic stem cells (SSCs)	Endodermal, e.g., gastrointestinal tract. Mesodermal, e.g., hematopoietic stem cells. Ectodermal, e.g., skin stems cells.	*Research purpose:* Drug testing, genomic study, growth factor discovery. *Clinical purpose:* Cell therapies.	These cells are very difficult to get isolated from a simple; difficult to be cultured in laboratory.
3.	Induced pluripotent stem cells (iPSCs)	Engineered in the laboratory from differentiated somatic cells by specific gene expression, e.g., Oct4, Sox2, Klf4, and c-Myc.	*Research purpose:* Disease modeling, drug development, organ, and tissue synthesis, cell type production. *Clinical purpose:* Tissue regeneration, most effective cell-based therapies.	In experimental trials, they have shown to form tumors in mice; mutations occur due to modifications of gene expression by growth factors.
4.	Perinatal stem cells (PSCs)	Cells in amniotic fluid and umbilical cord blood and tissues.	*Research purpose:* Disease modeling, drug discovery. *Clinical purpose:* Cell-based therapies.	–

2018). Regenerative medicine aims to amplify the body's natural healing abilities and encourage the growth of new body fluids, cells, and tissues in the desired body area. This latest and novel process is often used to treat sports injuries, skin burns, cartilage tissue regeneration, wound healing, etc. In the present scenario, donated tissues and organs cannot satisfy the transplantation needs of elderly and ailing populations, prompting the quest for other sources of tissue and organs. Stem cells are characterized by an unlimited capacity for cell division and the ability to transdifferentiate into different types of cells and have emerged as a primary source of regenerative medicine in recent years to repair tissues and organ anomalies caused by congenital disabilities, disease, and aging (Mahla, 2016). Stem cells serve as the basis for all of the body's tissues and organ systems and play a variety of roles in disease progression, development, and tissue repair. ESCs, tissue-specific progenitor stem cells (TSPSCs), mesenchymal stem cells (MSCs), umbilical cord stem cells (UCSCs), bone marrow stem cells (BMSCs), and iPSCs are the several types of stem cells used in regenerative medicine (Kotzabasaki, Sotiropoulos, & Sarimveis, 2020). Some challenging diseases such as AIDS (Figure 6.1) and their treatment by stem cell technology are described in Table 6.2.

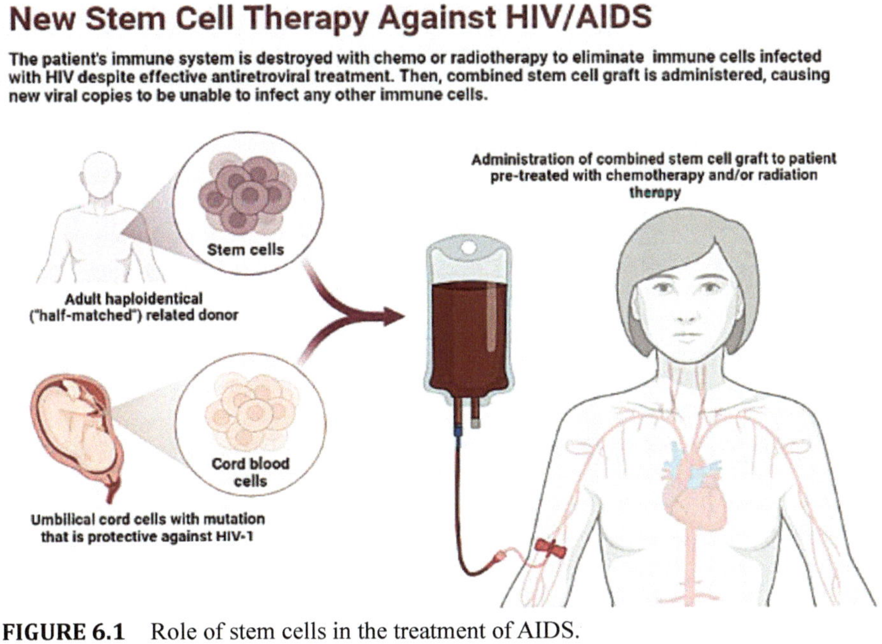

FIGURE 6.1 Role of stem cells in the treatment of AIDS.

TABLE 6.2 Some Challenging Diseases and Their Treatment by Stem Cell Technology

Disease	Factors Causing Disease	Mode of Stem Cells Application	Physiological and Mechanistic Aspects of Stem Cells Therapeutics	Improvements in Disease Signatures and Future Use	Stem Cells Used
Spinal cord injuries	Infection, cancer, accidents	ESCs transplantation to the affected site	Tissue homing is aided by ESCs and releases vasculogenic and neurogenic substances.	Regeneration of spinal tissue and improved balance and sensation	ESCs
Diabetes	Lifestyle, heart defects, and genetics	Transplantation of ESCs-derived PPCs	Progenitors (CD24+, CD133+, and CD49+) differentiate into β-cells, secrete insulin and express PDXL, GCK, and GLUT2	Improvement in glucose levels and obesity can be used for the treatment of T1DM and T2DM	ESCs
Anemia and blood cancer	Injury, genetics, and autoimmunity	Two-step infusion of lymphoid and myeloid	Immunity may be rebuilt by haploidentical BMSCs, which is a crucial step for minorities.	Treatment of aplastic anemia and hematological malignancies	BMSCs
Diabetes	Lifestyle and genetic factors	iPSCs-derived beta-cells transplantation	Skin to beta-cells reprogramming phase through cDE and cPF requires GPs	Treatment of T1DM and T2DM and insulin production	iPSCs
Neurodegenerative disorders	Accidents, age, trauma, and stroke	iGABA-1Ns and cortical spheroid transplantation	iGABA-1Ns secrete GABA; FOXIG cause ASD, spheroid mimics to brain	ASD, Alzheimer's, seizer, and obstinate epilepsies treatment	iPSCs
Cardiovascular disease	Diabetes, drugs, genetic factors, and lifestyle	ESCs-derived CMs and biomaterial coaxed ESCs	Cardiomyocytes express GCaMP3, secreting vasculogenic factors, and Tbox3 differentiates ESCs into SANPCs	Suppresses heart arrhythmias. CMs electrophysiologic ally integrates into the heart like a pacemaker	ESCs
Lung degeneration	Tuberculosis, cancer, and fibrosis	Biomaterial coaxed iPSCs transplantation	Miniature iPSCs lung resembles airways and alveoli, model drug testing	Regeneration of lung tissue	iPSCs
Congenital heart defects	Developmental errors	Transplantation of fibrin-coaxed AFSCs	The addition of VEFG to PEG-coaxed AFSCs promotes organogenesis	Regeneration of tissue repair for treatment of heart defects	UCSCs
AIDS	HIV1 infection	Transplantation of HIV1 resistant CD4+ cells	Anti-HIV1 CD4+ cells express HIV1 anti-RNA, which restricts HIV infection	Treatment of AIDS as an alternative to antiretroviral	BMSCs

Autologous, allogenic, or syngeneic stem cell transplantation can be used to induce tissue regeneration and immunological lysis of pathogen or cancer cells. To reduce the complications associated with host-versus-graft rejection, tissue typing of human leucocyte antigens (HLA) for tissue and organ transplantation and immunosuppressive agents are suggested. Stem cells exhibit the major histocompatibility complex (MHC) receptor at low levels. They secrete chemokines that attract endothelial and immune cells to the graft site, facilitating tissue tolerance (Morales-Molina et al., 2018). Current methods to stem cell regenerative medicine are based on tissue engineering technologies that integrate concepts of cell transplantation, material science, and microengineering to create organoids that may be utilized to restore the physiological function of damaged tissue and organs. Tissue engineering is a technique that creates nascent tissue on biodegradable three-dimensional scaffolds. The ideal scaffolds promote cell adhesion and ingrowth, match the mechanics of target tissue, promote angiogenesis and neovascularization for adequate tissue perfusion, and are non-immunogenic to the host, hence eliminating the need for systemic immune suppression (Rajabzadeh et al., 2019). In the upcoming section, we will discuss the approach of stem cell technology and regenerative medicine in treating a wide variety of diseases.

6.3.1 SPINAL CORD INJURIES (SCIS)

Spinal cord injury (SCI) is a devastating injury resulting in permanent neurological impairment. It occurs in young and adult people due to accidents, infections, and diseases (Mataliotakis & Tsirikos, 2016). Surgical spinal fixation and subsequent rehabilitation are the standard of care for acute SCI, and no neuroprotective and regenerative therapies capable of directly producing beneficial effects are available (Nagoshi & Okano, 2017). High-dose methylprednisolone has been used frequently. There is excellent room for stem cell treatment in this disease, as seen in treating rats and primates *in-vitro* experiments. Recent progress in stem cell research may pause such disease treatment by proper development of neural precursor cells (NPCs) and their proper therapy to the point of disease (Vincent, 2017). NPC cells can be developed by two means, one from the fetal tissues or ESCs and the second from the somatic mesodermal cell-like adipose cell (Radhakrishnan et al., 2019; Becker &

Mcdonald III, 2012) first demonstrated the efficacy of NPCs transplantation in SCI. They induced NPCs from mouse ESCs and transplanted them into rat spinal cord 9 days post-injury. The grafted cells differentiated into neurons, oligodendrocytes, and astrocytes in the damaged tissue. Transplanted cells showed mitotic neurogenesis and formed synapsis with host neurons in the injured spinal cord. The differentiated cells migrated along the rostro caudal axis from the lesion epicenter. After that, the analysis showed improved control and coordination of limbs due to motor function recovery (Venkatesh, Ghosh, Mullick, Manivasagam, & Sen, 2019).

6.3.2 DIABETES

Diabetes is defined as the loss of pancreatic B-cell bulk or function, which results in insulin insufficiency relative to the body's metabolic requirements. Historically, diabetic management has focused on medication, as demonstrated by insulin replacement therapy, to address the hormone's peripheral activities (Chen, Cohrs, Stertmann, Bozsak, & Speier, 2017). With an increased understanding of diabetes, new therapies to repair and restore failed B-cell activity are increasingly being examined as adjuncts to conventional diabetes treatment regimens. The emphasis is on exogenous pancreas/islets or artificial islets transplantation, improved proliferation and maturation of endogenous B-cells, prevention of B-cell loss, or accelerated regeneration of B-like cell populations from stem cell pools and non-B-cell sources (Sordi et al., 2017). There are two types of diabetes; Type 1 diabetes (T1D) is characterized by insulin insufficiency caused by autoimmune destruction of islet B-cells. Type 2 diabetes (T2D) is characterized by the insulin secretory capability gradually failing to meet peripheral insulin requirements. Exogenous insulin injections are essential for persons with T1D and advanced T2D (Gupta et al., 2018). Glucagon-like peptide-1 (GLP-1) analogs have been developed to stimulate insulin production while preserving B-cell function. Alternatively, DPP-4 inhibitors have been utilized to enhance endogenous insulin production in T2D by preventing GLP-1 inactivation. ESCs induce pluripotent cells' spontaneous differentiation potential via embryoid body formation, followed by selection for cells expressing B-cell/B-cell progenitor markers. It has been demonstrated that using insulin-producing cells generated from ESC offspring and chosen for B-cell-specific gene expression may correct hyperglycemia in diabetic mice (Meng, 2016). Additionally, stepwise differentiation techniques for each lineage have been adopted substantially from *in vitro* B-cell developmental blueprints. Thus, directed differentiation of ESCs resulted in the generation of insulin-producing cells; however, subsequent research cast doubt on the authenticity of generated progeny (Sackett et al., 2016).

6.3.3 ALZHEIMER'S DISEASE (AD)

Alzheimer's disease (AD) is a progressive neurodegenerative condition characterized by memory loss and cognitive impairment caused by synaptic dysfunction and an abnormal buildup of misfolded proteins (Castellazzi et al., 2019). Numerous fruitless attempts to produce novel chemicals or antibodies to intervene in the disease's etiology have been attempted. As a replacement or regeneration approach for AD, stem cell-based treatments offer fresh hope. AD is the most common kind of dementia, characterized by gradual memory loss and impairment of cognitive functioning. AD was first described in 1907 by Alois Alzheimer. Because AD is complex, it is difficult to pinpoint the precise pathophysiologic process. However, synaptic failure is the primary symptom of neuronal death in the cortex and hippocampus due to an abnormal buildup of neurofibrillary tangles and β-amyloid (Aβ) plaques (Kwak, Lee, Yang, & Park, 2018). AD is classified into two subtypes: familial and sporadic. Familial AD accounts for around 5% of cases and is caused by a unique autosomal genetic mutation in the amyloid precursor protein (APP), presenilin-1 (PSEN-1), or presenilin-2 (PSEN-2).

In comparison, the majority of occurrences of AD are sporadic. It often manifests itself late in life and is likely to be caused by interactions between a detailed genetic profile (including apolipoprotein [ApoE4]) and environmental variables such as cardiovascular disease, depression, and smoking (Stoychev et al., 2019). iPSCs were first developed from mouse fibroblasts in 2006. These cells are reprogramed into a state of pluripotency similar to that of ESCs. IPSCs are thought to be able to differentiate into various cells, including neurons and neurospheres. So neuronal subtypes can be generated and automated using iPSCs. For example, iPSC-derived glia could be used for research regarding inflammatory response in AD (Hunsberger et al., 2016). The workflow for iPSC-based cell therapy is shown in Figure 6.2.

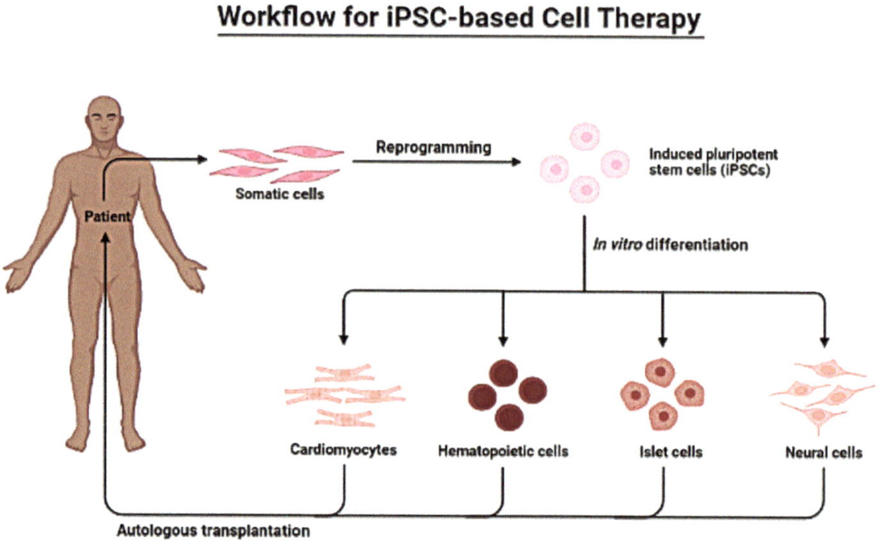

FIGURE 6.2 The workflow for iPSC-based cell therapy.

Additionally, MSCs have garnered considerable interest in the therapy of AD due to their outstanding accessibility, the relative simplicity of handling, substantial research, and a broad range of differentiating potential (including neuronal cells). Additionally, MSCs are beneficial as cell-based therapeutics since they may be injected intravenously, penetrate the blood-brain barrier, are low tumorigenic, and provoke a mild immune response (then do other cell-based therapies). Rather than being used to replace neurons, MSCs appeared to exert beneficial effects via their released factors, which promote the proliferation, differentiation, and survival of the neurogenic niche (Kwak et al., 2018). MSCs are obtained from a broad number of sources. MSCs generated from bone marrow (BM-MSCs) have been the subject of extensive research for an extended period. The immunomodulatory capacity of BM-MSCs is derived from the secretion of soluble factors such as IL-6, IL-10, TGF-β, and PGE2. They have been shown to impair the function of monocyte-derived dendritic cells and change the phenotypic of natural killer cells. Adipose tissue is an excellent source of MSCs. MSCs produced from adipose tissue (AT-MSCs) can differentiate into neuronal and astrocyte-like cells.

Additionally, AT-MSCs release a variety of neurotrophic factors that are required in the neurogenic niche. Finally, MSCs produced from umbilical cord blood (UCB-MSCs) can differentiate into neuron-like cells. These cells have been investigated in a mouse model of AD. One method of action that has been proposed is the activation of M2-like microglia (El-Omar et al., 2014).

6.3.4 APLASTIC ANEMIA (AA)

Aplastic anemia (AA) is a rare and potentially fatal BM failing illness characterized by pancytopenia and hypocellular BM. The majority of instances are caused by an autoimmune attack on BM stem and progenitor cells, which results in pancytopenia (Brodsky & Jones, 2005). While allogeneic BM transplantation may be curative in younger patients, most are ineligible for transplantation due to advanced age or a lack of a matched donor. With regular immunosuppressive therapy (IST), comparable long-term survival is possible in severe AA. However, one-quarter to one-third of patients treated with IST will not respond, and 30–40% of responders will relapse. These individuals may have chronic thrombocytopenia, require frequent platelet transfusions that are costly and inconvenient, and are at risk of developing more major bleeding issues. Thrombocytopenia is a significant cause of morbidity and mortality in patients with AA (Huber, Kumar, & Tefferi, 2003). Almost all patients with AA are thrombocytopenic upon presentation; platelet counts of 50,000 or 20,000/lL are used to diagnose moderate and severe AA. Thrombocytopenia is caused by decreased hematopoietic stem and progenitor cell numbers and function, which results in reduced megakaryocytopoiesis and insufficient mature platelet formation (Mansour et al., 2000).

Allogeneic hematopoietic stem cell transplantation (AHSCT) is the usual treatment case for severe patients, but there is always a risk of immune reaction from the recipient. Cultured hemopoietic stem cells can produce hematopoietic factors, mainly erythropoietic-stimulating agents (ESAs) and G-CSF and GM-CSF and IL-3 treat AA. Still, their infusion does not affect (Ogawa, 1993). The use of umbilical cord multipotent stromal cells and allogeneic HSCs in treating heavily transfused patients with severe AA have shown promise. It opens the door for the future use of umbilical cord cells for anemia disorders (Luan, Chen, Chen, Ding, & Ni, 2015).

6.3.5 LUNG DEGENERATION

Diseases such as TB, cancer, cystic fibrosis, alpha-1 antitrypsin (AAT), and smoking cause lung alveoli tissue degradation. Tissue degeneration can indicate reversible cell damage or necrosis, which indicates irreversible cell damage. Cell enlargement, cytoplasmic vacuolation, apparent perinuclear gaps, cytoplasmic bleb development, and cilia loss are all examples of reversible cell injury. Pyknosis, karyorrhexis or karyolysis, cell enlargement, and cell fragmentation are examples of permanent cell injury (Kumar, Abbas, Aster, & Deyrup, 2020).

Using stem cell technology, damaged pulmonary tissues can be regenerated. As previously stated for the use of mesodermal cells in the development of iPSCs for stem cell technology, here also the generation of disease-specific iPSCs from individuals with a variety of lung diseases, as there is a significant lack of conventional therapies or human model systems for several inherited or degenerative lung diseases (Cazzaniga, 2017). It has been hypothesized that pluripotent stem cells or differentiated offspring are promising options for reconstituting wounded lung tissues *in vivo* or simulating the etiology of lung disease *in vitro*. Therefore, scientists attempted to reprogram fibroblasts extracted from mice with lung disorders affecting the three major cell lineages of the adult lung: lung epithelium, lung endothelium, and lung interstitium (Karagiannis et al., 2019). CF or AAT deficiency-related emphysema (which affects the lung interstitium and epithelium)-affected mice were used to collect fibroblasts. The researchers focused on reprogramming dermal fibroblasts derived from 6-mm skin punch biopsies of mice with a hereditary CF mutation (homozygous DF508 mutant CFTR genotype). In one research work, researchers also obtained additional banked, frozen CF or AAT-deficient fibroblasts from the cell repositories. They used a single lentiviral "stem cell cassette" (STEMCCA) to encode all four reprogramming factors, Oct4, Sox2, Klf4, and c-Myc, in a single polycistronic vector. By combining all reprogramming transgenes in a single cassette, STEMCCA accomplished reprogramming of postnatal mouse fibroblasts with high efficiency and allowed the derivation of mouse iPSC containing a single viral integration of the hSTEMCCA-loxP vector allowed efficient generation of iPSC clones from all samples, regardless of the age of the individual from which the cells originated (Ahmed, 2016). Finally, to functionally assess pluripotency of the disease-specific iPSC cells generated with this method, researchers subdermally transplanted three representative iPSC cell lines into immunodeficient mice and found the cells gave rise to teratomas comprised of differentiated tissues characteristic of the three primary germ layers, ectoderm, mesoderm, and endoderm (Miranda, Fernandes, Diogo, & Cabral, 2020).

6.3.6 TOOTH REGENERATION

Various studies have been done on dental reconstruction with the help of MSCs. One of the studies has reported that dental pulp-derived stem cells (DPSCs) could promote periodontal regeneration in a canine model. Also, it was shown that canine DPSCs were successfully isolated and had a rapid proliferation and multi-lineage differentiation capacity (Khorsand et al., 2013).

6.3.7 WOUND HEALING

A chronic wound is one of the most frequent ailments and causes distress (Mansoub et al., 2018). MSCs generated from dental tissue are reliable donors of the cytokines and growth factors that promote wound healing. Several research findings indicated that stem cells produced from the horse's deciduous teeth might be a unique technique for wound care and could be used clinically to treat non-healing wounds (Srionrod, Bootcha, & Petchdee, 2016). However, additional study is required to understand the underlying mechanisms of beneficial growth factors that contribute to wound healing processes when using stem cells produced from deciduous teeth (Amirkhani et al., 2016). This early discovery suggests that stem cells made from deciduous teeth can improve wound healing in rabbit excisional wound models (Srionrod et al., 2016). In addition, one of the researchers experimented with a rat model and found that ADSCs provide a potentially suitable matrix for full-thickness wound healing (Lin et al., 2013).

6.4 CONCLUSION

The human body is strengthened by specialized cells collectively called stem cells, which can self-renew and develop into various cell types such as adipocytes, osteocytes, chondrocytes, and neurons. Additionally, to the qualities outlined above, these cells are readily separated, may be safely transplanted to wounded locations, and possess immune regulatory capabilities. Numerous *in vitro* and *in vivo* experiments using animal models have effectively proved the promise of multiple kinds of stem cells for treating various illnesses; nevertheless, clinical outcomes have been mixed. According to research, stem cells have a wide range of applications in treating disorders such as heart failure, wound healing, and tooth regeneration. Additionally, these cells are critical in the therapy of sub-branch neurodegenerative illnesses like Alzheimer's and Parkinson's disease.

ACKNOWLEDGMENT

This chapter was designed and initiated by MAM. It has been written by BAS, BAB, and AS and edited and compiled by MAM. The Authors are thankful to MAM for his assistance in the preparation of this manuscript.

CONFLICTS OF INTEREST

The authors have no conflicts of interest to declare.

ETHICAL STATEMENT

The authors are accountable for all aspects of the work in ensuring that questions related to the accuracy or integrity of any part of the work are appropriately investigated and resolved.

KEYWORDS

- **cancer stem cells**
- **embryonic stem cells**
- **induced pluripotent stem cells**
- **perinatal stem cells**
- **plasticity**
- **regenerative medicine**
- **stem cell therapy**

REFERENCES

Acosta, N. D., & Golub, S. H., (2016). The new federalism: State policies regarding embryonic stem cell research. *The Journal of Law, Medicine & Ethics, 44*(3), 419–436.

AE, H. R., (2004). Bishop embryonic stem cells. *Cell Prolif., 37*, 23–34.

Ahmed, M. J., (2016). *Directed Differentiation of Human Induced Pluripotent Stem Cells Into Endothelial Cells.* Boston University.

Amin, N., Tan, X., Ren, Q., Zhu, N., Botchway, B. O., Hu, Z., & Fang, M., (2019). Recent advances of induced pluripotent stem cells application in neurodegenerative diseases. *Progress in Neuro-Psychopharmacology and Biological Psychiatry, 95*, 109674.

Amirkhani, M. A., Shoae-Hassani, A., Soleimani, M., Hejazi, S., Ghalichi, L., & Nilforoushzadeh, M. A., (2016). Rejuvenation of facial skin and improvement in the dermal architecture by transplantation of autologous stromal vascular fraction: A clinical study. *BioImpacts: BI, 6*(3), 149.

Baker, C. L., & Pera, M. F., (2018). Capturing totipotent stem cells. *Cell Stem Cell, 22*(1), 25–34.

Baksh, D., Song, L., & Tuan, R. S., (2004). Adult mesenchymal stem cells: Characterization, differentiation, and application in cell and gene therapy. *Journal of Cellular and Molecular Medicine, 8*(3), 301–316.

Banas, A., Teratani, T., Yamamoto, Y., Tokuhara, M., Takeshita, F., Quinn, G., & Ochiya, T., (2007). Adipose tissue-derived mesenchymal stem cells as a source of human hepatocytes. *Hepatology, 46*(1), 219–228.

Becker, A. J., McCulloch, E. A., & Till, J. E., (1963). Cytological demonstration of the clonal nature of spleen colonies derived from transplanted mouse marrow cells. *Nature, 197*(4866), 452–454.

Becker, D., & Mcdonald, III. J. W., (2012). Approaches to repairing the damaged spinal cord: Overview. In: *Handbook of Clinical Neurology* (Vol. 109, pp. 445–461). Elsevier.

Bragança, J., Lopes, J. A., Mendes-Silva, L., & Santos, J. M. A., (2019). Induced pluripotent stem cells, a giant leap for mankind's therapeutic applications. *World Journal of Stem Cells, 11*(7), 421.

Brodsky, R. A., & Jones, R. J., (2005). Aplastic anemia. *The Lancet, 365*(9471), 1647–1656.

Cable, J., Fuchs, E., Weissman, I., Jasper, H., Glass, D., Rando, T., & Park, S., (2020). Adult stem cells and regenerative medicine—A symposium report. *Annals of the New York Academy of Sciences, 1462*(1), 27.

Castellazzi, M., Patergnani, S., Donadio, M., Giorgi, C., Bonora, M., Bosi, C., & Zuliani, G., (2019). Autophagy and mitophagy biomarkers are reduced in sera of patients with Alzheimer's disease and mild cognitive impairment. *Scientific Reports, 9*(1), 1–7.

Cazzaniga, A., (2017). *Space and Osteoporosis: How Gravity Affects Bone Microenvironment?* 29. ciclo, Anno Accademico 2016. [10.13130/cazzaniga-alessandra_phd2017-04-07]. Doctoral dissertation, University of Milan.

Chang, Y. H., Wu, K. C., Harn, H. J., Lin, S. Z., & Ding, D. C., (2018). Exosomes and stem cells in degenerative disease diagnosis and therapy. *Cell Transplantation, 27*(3), 349–363.

Chen, C., Cohrs, C. M., Stertmann, J., Bozsak, R., & Speier, S., (2017). Human beta cell mass and function in diabetes: Recent advances in knowledge and technologies to understand disease pathogenesis. *Molecular Metabolism, 6*(9), 943–957.

Cheng, H., Zheng, Z., & Cheng, T., (2020). New paradigms on hematopoietic stem cell differentiation. *Protein & Cell, 11*(1), 34–44.

Clarke, D. L., Johansson, C. B., Wilbertz, J., Veress, B., Nilsson, E., Karlström, H., & Frisen, J., (2000). Generalized potential of adult neural stem cells. *Science, 288*(5471), 1660–1663.

Dalerba, P., Diehn, M., Weissman, I. L., & Clarke, M. F., (2020). Stem cells, cell differentiation, and cancer. In: *Abeloff's Clinical Oncology* (pp. 97–107. e105). Elsevier.

Duncan, T., & Valenzuela, M., (2017). Alzheimer's disease, dementia, and stem cell therapy. *Stem Cell Research & Therapy, 8*(1), 111.

Ejtehadifar, M., Shamsasenjan, K., Movassaghpour, A., Akbarzadehlaleh, P., Dehdilani, N., Abbasi, P., & Saleh, M., (2015). The effect of hypoxia on mesenchymal stem cell biology. *Advanced Pharmaceutical Bulletin, 5*(2), 141.

El Omar, R., Beroud, J., Stoltz, J. F., Menu, P., Velot, E., & Decot, V., (2014). Umbilical cord mesenchymal stem cells: The new gold standard for mesenchymal stem cell-based therapies? *Tissue Engineering Part B: Reviews, 20*(5), 523–544.

Fathi, E., & Farahzadi, R., (2017). Enhancement of osteogenic differentiation of rat adipose tissue-derived mesenchymal stem cells by zinc sulphate under electromagnetic field via the PKA, ERK1/2 and Wnt/β-catenin signaling pathways. *PLoS One, 12*(3), e0173877.

Friedenstein, A. J., Chailakhyan, R. K., Latsinik, N. V., Panasyuk, A. F., & Keiliss-Borok, I. V., (1974). Stromal cells responsible for transferring the microenvironment of the hemopoietic tissues: Cloning *in vitro* and retransplantation *in vivo*. *Transplantation, 17*(4), 331–340.

Fuchs, E., & Segre, J. A., (2000). Stem cells: A new lease on life. *Cell, 100*(1), 143–155.

Fujita, Y., Kadota, T., Araya, J., Ochiya, T., & Kuwano, K., (2018). Clinical application of mesenchymal stem cell-derived extracellular vesicle-based therapeutics for inflammatory lung diseases. *Journal of Clinical Medicine, 7*(10), 355.

Ghimire, S., Weber, D., Mavin, E., Dickinson, A. M., & Holler, E., (2017). Pathophysiology of GvHD and other HSCT-related major complications. *Frontiers in Immunology, 8*, 79.

Giri, T. K., Alexander, A., Agrawal, M., Saraf, S., & Saraf, S., (2019). Current status of stem cell therapies in tissue repair and regeneration. *Current Stem Cell Research & Therapy, 14*(2), 117–126.

Gupta, R., Nguyen, D. C., Schaid, M. D., Lei, X., Balamurugan, A. N., Wong, G. W., & Bhatnagar, S., (2018). Complement 1q-like-3 protein inhibits insulin secretion from pancreatic β-cells via the cell adhesion g protein–coupled receptor BAI3. *Journal of Biological Chemistry, 293*(47), 18086–18098.

Gurley, K. A., & Alvarado, A. S., (2019). Stem cells in animal models of regeneration. *NISEB Journal, 12*(4).

Holzer, P. W., (2019). *Zero to One-Translational Advancements in the Field of Xenotransplantation.* Dartmouth College.

Huber, M. R., Kumar, S., & Tefferi, A., (2003). Treatment advances in adult immune thrombocytopenic purpura. *Annals of Hematology, 82*(12), 723–737.

Hunsberger, J. G., Rao, M., Kurtzberg, J., Bulte, J. W., Atala, A., LaFerla, F. M., & Schneider, L. S., (2016). Accelerating stem cell trials for Alzheimer's disease. *The Lancet Neurology, 15*(2), 219–230.

Kalra, K., & Tomar, P., (2014). Stem cell: Basics, classification and applications. *American Journal of Phytomedicine and Clinical Therapeutics, 2*(7), 919–930.

Karagiannis, P., Takahashi, K., Saito, M., Yoshida, Y., Okita, K., Watanabe, A., & Nakagawa, M., (2019). Induced pluripotent stem cells and their use in human models of disease and development. *Physiological Reviews, 99*(1), 79–114.

Khorsand, A., Eslaminejad, M. B., Arabsolghar, M., Paknejad, M., Ghaedi, B., Rokn, A. R., & Jahangir, S., (2013). Autologous dental pulp stem cells in regeneration of defect created in canine periodontal tissue. *Journal of Oral Implantology, 39*(4), 433–443.

Kotzabasaki, M. I., Sotiropoulos, I., & Sarimveis, H., (2020). QSAR modeling of the toxicity classification of superparamagnetic iron oxide nanoparticles (spions) in stem-cell monitoring applications: An integrated study from data curation to model development. *RSC Advances, 10*(9), 5385–5391.

Krause, D. S., Theise, N. D., Collector, M. I., Henegariu, O., Hwang, S., Gardner, R., & Sharkis, S. J., (2001). Multi-organ, multi-lineage engraftment by a single bone marrow-derived stem cell. *Cell, 105*(3), 369–377.

Kumar, V., Abbas, A. K., Aster, J. C., & Deyrup, A. T., (2020). *Robbins Essential Pathology E-Book*. Elsevier Health Sciences.

Kwak, K. A., Lee, S. P., Yang, J. Y., & Park, Y. S., (2018). Current perspectives regarding stem cell-based therapy for Alzheimer's disease. *Stem Cells International, 2018*.

Lee, M. W., Ryu, S., Kim, D. S., Lee, J. W., Sung, K. W., Koo, H. H., & Yoo, K. H., (2019). Mesenchymal stem cells in suppression or progression of hematologic malignancy: Current status and challenges. *Leukemia, 33*(3), 597–611.

Levenberg, S., Huang, N. F., Lavik, E., Rogers, A. B., Itskovitz-Eldor, J., & Langer, R., (2003). Differentiation of human embryonic stem cells on three-dimensional polymer scaffolds. *Proceedings of the National Academy of Sciences, 100*(22), 12741–12746.

Li, C., Heidt, D. G., Dalerba, P., Burant, C. F., Zhang, L., Adsay, V., & Simeone, D. M., (2007). Identification of pancreatic cancer stem cells. *Cancer Research, 67*(3), 1030–1037.

Lin, Y. C., Grahovac, T., Oh, S. J., Ieraci, M., Rubin, J. P., & Marra, K. G., (2013). Evaluation of a multi-layer adipose-derived stem cell sheet in a full-thickness wound healing model. *Acta Biomaterialia, 9*(2), 5243–5250.

Lu, S. J., Li, F., Vida, L., & Honig, G. R., (2004). Cd34+ cd38-hematopoietic precursors derived from human embryonic stem cells exhibit an embryonic gene expression pattern. *Blood, 103*(11), 4134–4141.

Luan, C., Chen, R., Chen, B., Ding, J., & Ni, M., (2015). Umbilical cord blood transplantation supplemented with the infusion of mesenchymal stem cell for an adolescent patient with severe aplastic anemia: A case report and review of literature. *Patient Preference and Adherence, 9*, 759.

Mahla, R. S., (2016). Stem cells applications in regenerative medicine and disease therapeutics. *International Journal of Cell Biology, 2016*.

Makena, M. R., Ranjan, A., Thirumala, V., & Reddy, A. P., (2020). Cancer stem cells: Road to therapeutic resistance and strategies to overcome resistance. *Biochimica et Biophysica Acta (BBA)-Molecular Basis of Disease, 1866*(4), 165339.

Mansoub, N. H., Gürdal, M., Karadadaş, E., Kabadayi, H., Vatansever, S., & Ercan, G., (2018). The role of PRP and adipose tissue-derived keratinocytes on burn wound healing in diabetic rats. *BioImpacts: BI, 8*(1), 5.

Mansour, A. M., Salti, H. I., Han, D. P., Khoury, A., Friedman, S. M., Salem, Z., & Saghir, N., (2000). Ocular findings in aplastic anemia. *Ophthalmologica, 214*(6), 399–402.

Markoski, M. M., (2016). Advances in the use of stem cells in veterinary medicine: From basic research to clinical practice. *Scientifica, 2016.*

Mataliotakis, G. I., & Tsirikos, A. I., (2016). Spinal cord trauma: Pathophysiology, classification of spinal cord injury syndromes, treatment principles and controversies. *Orthopaedics and Trauma, 30*(5), 440–449.

McCoy, E. L., & Sterngass, J., (2019). *Reproductive Technology: Indispensable or Problematic?* Cavendish Square Publishing, LLC.

Meng, Y., (2016). *The Role of the Mixed Lineage Leukemia Gene in Hematopoiesis and Leukemogenesis*. Doctoral dissertation, Imperial College London.

Mimeault, M., Hauke, R., & Batra, S. K., (2007). Stem cells: A revolution in therapeutics—Recent advances in stem cell biology and their therapeutic applications in regenerative medicine and cancer therapies. *Clinical Pharmacology & Therapeutics, 82*(3), 252–264.

Miranda, C. C., Fernandes, T. G., Diogo, M. M., & Cabral, J. M., (2020). Human pluripotent stem cells: Applications and challenges for regenerative medicine and disease modeling. *Current Applications of Pharmaceutical Biotechnology*, 189–224.

Mohammadian, M., Shamsasenjan, K., Nezhad, P. L., Talebi, M., Jahedi, M., Nickkhah, H., & Movassaghpour, A., (2013). Mesenchymal stem cells: New aspect in cell-based regenerative therapy. *Advanced Pharmaceutical Bulletin, 3*(2), 433.

Morales-Molina, Á., Gambera, S., Cejalvo, T., Moreno, R., Rodríguez-Milla, M. Á., Perisé-Barrios, A. J., & García-Castro, J., (2018). Antitumor virotherapy using syngeneic or allogeneic mesenchymal stem cell carriers induces systemic immune response and intratumoral leukocyte infiltration in mice. *Cancer Immunology, Immunotherapy, 67*(10), 1589–1602.

Nagoshi, N., & Okano, H., (2017). Applications of induced pluripotent stem cell technologies in spinal cord injury. *Journal of Neurochemistry, 141*(6), 848–860.

Ogawa, M., (1993). *Differentiation and Proliferation of Hematopoietic Stem Cells* (pp. 2844–2853).

Price, J., (2020). *The Future of Brain Repair: A Realist's Guide to Stem Cell Therapy*. MIT Press.

Radhakrishnan, S., Trentz, O. A., Reddy, M. S., Rela, M., Kandasamy, M., & Sellathamby, S., (2019). *In vitro* transdifferentiation of human adipose tissue-derived stem cells to neural lineage cells-a stage-specific incidence. *Adipocyte, 8*(1), 164–177.

Rajabzadeh, N., Fathi, E., & Farahzadi, R., (2019). Stem cell-based regenerative medicine. *Stem Cell Investigation, 6.*

Russo, J., (2017). *The Training of Cancer Researchers*. World Scientific.

Sackett, S., Brown, M., Tremmel, D., Ellis, T., Burlingham, W., & Odorico, J., (2016). Modulation of human allogeneic and syngeneic pluripotent stem cells and immunological implications for transplantation. *Transplantation Reviews, 30*(2), 61–70.

Segers, V. F., & Lee, R. T., (2008). Stem-cell therapy for cardiac disease. *Nature, 451*(7181), 937–942.

Sell, S., (2013). *Stem Cells Handbook*: Springer.

Slack, J., (2008). Origin of stem cells in organogenesis. *Science, 322*(5907), 1498–1501.

Soltysova, A., Altanerova, V., & Altaner, C., (2005). Cancer stem cells. *Neoplasma, 52*(6), 435.

Sordi, V., Pellegrini, S., Krampera, M., Marchetti, P., Pessina, A., Ciardelli, G., & Piemonti, L., (2017). Stem cells to restore insulin production and cure diabetes. *Nutrition, Metabolism and Cardiovascular Diseases, 27*(7), 583–600.

Srionrod, N., Bootcha, R., & Petchdee, S., (2016). Foal deciduous teeth stem cells enhance wound healing in rabbit wound model. *The Thai Journal of Veterinary Medicine, 46*(1), 155–161.

Stoychev, K. R., Stoimenova-Popova, M., Chumpalova, P., Ilieva, L., Swamad, M., & Kamburova-Martinova, Z., (2019). A clinical case of a patient carrying rare pathological PSEN1 gene mutation (l424v) demonstrates the phenotypic heterogenity of early onset familial ad. *Frontiers in Psychiatry, 10.*

Tweed, K. E., (2019). *Enhancing Stem Cell Delivery to the Trabecular Meshwork Using Magnetic Nanoparticles.* Doctoral thesis.

Venkatesh, K., Ghosh, S. K., Mullick, M., Manivasagam, G., & Sen, D., (2019). Spinal cord injury: Pathophysiology, treatment strategies, associated challenges, and future implications. *Cell and Tissue Research*, 1–27.

Vincent, P. H., (2017). *Specification and Potency of Human Neural Stem Cells for Clinical Transplantation.* Doctoral dissertation, Karolinska Institute (Sweden).

Volarevic, V., Bojic, S., Nurkovic, J., Volarevic, A., Ljujic, B., Arsenijevic, N., & Stojkovic, M., (2014). Stem cells as new agents for the treatment of infertility: Current and future perspectives and challenges. *BioMed Research International, 2014.*

Witkowski, M. T., Lasry, A., Carroll, W. L., & Aifantis, I., (2019). Immune-based therapies in acute leukemia. *Trends in Cancer, 5*(10), 604–618.

Yang, Z. F., Ho, D. W., Ng, M. N., Lau, C. K., Yu, W. C., Ngai, P., & Fan, S. T., (2008). Significance of CD90+ cancer stem cells in human liver cancer. *Cancer Cell, 13*(2), 153–166.

Zakrzewski, W., Dobrzyński, M., Szymonowicz, M., & Rybak, Z., (2019). Stem cells: Past, present, and future. *Stem Cell Research & Therapy, 10*(1), 1–22.

Zhu, Y., Uezono, N., Yasui, T., & Nakashima, K., (2018). Neural stem cell therapy aiming at better functional recovery after spinal cord injury. *Developmental Dynamics, 247*(1), 75–84.

CHAPTER 7

CRISPR-Cas Genome-Editing Tool: A Prospective Application to Study Human Diseases

AALIM MAQSOOD BHAT,[1,2] SHAHID HUSSAIN NAIKOO,[1,2*] NEELOFAR MAQSOOD BHAT,[3**] and SHEIKH TASDUQ ABDULLAH[1,2*]

[1]*Academy of Scientific and Innovative Research (AcSIR), Ghaziabad, Uttar Pradesh, India*

[2]*CSIR–Indian Institute of Integrative Medicine Canal Road, Jammu, Jammu and Kashmir, India*

[3]*Department of Botany, University of Kashmir, Jammu and Kashmir, India*

**These authors contributed equally.*

ABSTRACT

Clustered regularly interspaced short palindromic repeats (CRISPR) and CRISPR-associated protein (Cas) is a prokaryote-derived genome editing tool acquired from the defense system of prokaryotes. CRISPR-Cas has become an indispensable genome editing tool with many applications in biotechnology and medical research to treat various human diseases, including neurodegenerative disorders, cardiovascular diseases, and cancers. CRISPR-Cas no longer remained merely a gene-editing tool, but its applications diversified due to the reprogramming capacity of Cas9, including epigenetic editing, gene regulation, and chromatin engineering.

Modern Biotechnology in Healthcare: Advances and Applications. Sheikh Umar Ahmad (Ed.)

CRISPR-Cas is also used to develop animal models for studying and developing possible treatments for human genetic diseases. This study aims to overview the wide range of applications of CRISPR-Cas in various fields of biotechnology.

7.1 INTRODUCTION

Genome editing holds enormous value in the different areas of molecular biology, biotechnology, and in the field of medicines. DNA of eukaryotes consists of billions of nitrogenous bases, and the ability to alter these DNA bases at particular positions harbors tremendous significance. Discoveries of restriction enzymes fueled the recombinant DNA technology era and enabled researchers to manipulate the DNA at predetermined locations (Kelly & Smith, 1970; Smith & Wilcox, 1970; Danna & Nathans, 1971). Consequently, several targeted nucleases for genome editing emerged, which involves transcription activator-like effector nucleases (TALENs), meganucleases, and zinc-finger nucleases (ZFNs). However, these techniques were relatively low in target efficiencies and time-consuming, limiting their widespread usage (Chandrasegaran & Carroll, 2016). In this context, CRISPR has revolutionized the entire genome-editing field in terms of efficiency, flexibility, and more straightforward in use (Lino et al., 2018). This genome editing technique consists of endonuclease whose DNA cutting specificity is determined by a short guideRNA (gRNA). For decades, CRISPR had been known as a typical DNA repeat element in prokaryotes before it was discovered as an adaptive immune system in prokaryotes and subsequently developed as a robust genome-editing tool (Adli, 2018).

The Clustered Regularly Interspaced Short Palindromic Repeat DNA Sequences are responsible for immunity in about 40% of bacterial species, including *Escherichia coli*, Streptococcus pyogenes, Staphylococcus aureus, Streptococcus thermophiles, Streptococcus mutans, Staphylococcus epidermidis, Corynebacterium diphtheria, Neisseria meningitides, and in about 90% of archaeal species such as *Archaeoglobus fulgidus*, *Methanocaldococcus jannaschii*, *Pyrococcus furiosus*, *Haloferax mediterranei*, *Methanothermobacter thermoautotrophicum*, and *Sulfolobus solfataricus*. Immune system of prokaryotes is capable of responding and consequently eliminating the viral genetic elements by this gene-editing process (Horvath et al., 2008; Horvath & Barrangou, 2010; Jinek et al., 2012; Sangal, Fineran, and Hoskisson, 2013; Zhang et al., 2013; Ran et al., 2015).

CRISPR-Cas is a component of bacterial immune systems that can cut the DNA at predetermined locations with a pair of molecular scissors (Cas) directed by the programmable gRNA sequence. CRISPR locus is flanked by conserved protein-coding genes, which vary in their order and direction of orientation. Based on sequence similarity, about 93 *cas* genes are classified into 35 families, out of which 11 families encode the proteins which constitute *cas* core, including the families encoding Cas9 nuclease. It has been found that a complete CRISPR-Cas locus contains at least one gene from the *cas* core (Hille & Charpentier, 2016).

Cas nucleases are endonucleases that binds to DNA and create double-stranded breaks. At the time of viral infection, the bacterium utilizes Cas nucleases to cut off a fragment of viral DNA known as protospacer which is reserved in the genome of the bacterium with the fragments from the other viruses that have earlier infected the bacterium. These viral spacer pieces are positioned in the middle of repeated palindromic sequences resulting in the formation of spacers and palindromic repeat arrangements from where the CRISPR has derived its name (Figure 7.1). CRISPR-Cas system comprises of two indispensable components: CRISPR-associated (Cas) endonuclease, which binds to DNA and creates DSBs, and a gRNA sequence, which directs target of Cas nucleases (Richter, Chang, and Fineran, 2012).

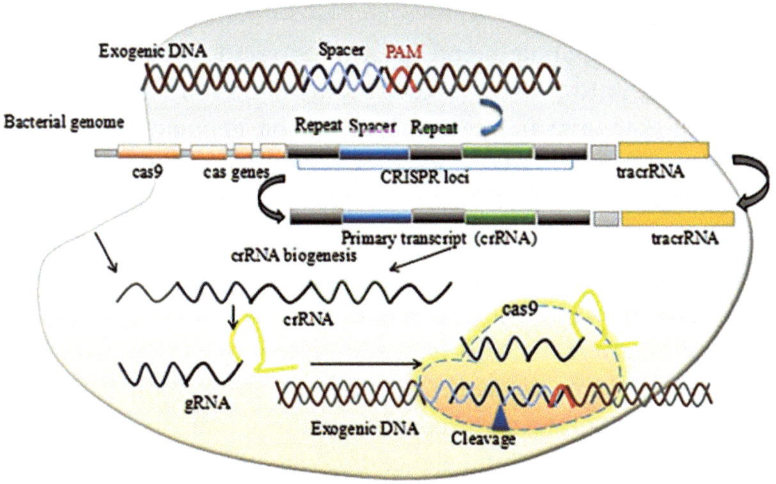

FIGURE 7.1 Molecular mechanism of CRISPR/Cas9.

Source: Reprinted from Khan et al. (2016). © 2016 Khan et al. https://creativecommons.org/licenses/by/4.0/

Reinfection of bacteria with the same virus is recognized and subsequently destroyed by Cas9. Activity of Cas9 relies upon CRISPR RNA (crRNA) and a trans-activating CRISPR RNA (tracrRNA) (Chylinski, Le Rhun, & Charpentier, 2013). During the course of the original infection, the genetic element which was stored in CRISPR locus serves as crRNA which is complementary to the viral spacer, while tracrRNA acts as a scaffold; complex of these pair of RNAs is known as guideRNA (gRNA) directing molecular scissors Cas9 to introduce a cut. Prior to the introduction of breaks, Cas9 scans the foreign DNA for protospacer adjacent motif (PAM), located downstream of the target site (Gleditzsch et al., 2019). After recognizing PAM, endonuclease Cas9 scans upstream region where if it detects the target pointed by the gRNA will introduce a double-strand break (DSB) inactivating the virus since viruses do not possess DNA repairing mechanisms (Chylinski, Le Rhun, & Charpentier, 2013; Gleditzsch et al., 2019).

7.2 GENOME EDITING UTILIZING CRISPR-CAS9

The complex of gRNA can be specifically engineered into single chimeric guideRNA (sgRNA) directing Cas9 to the predetermined locations to introduce DSBs offering a simple and cost-effective method of genome editing. Depending upon the presence of correct PAM sequence, the only required thing is to provide a new gRNA and then Cas9 can introduce breaks at a compass of targeted sites in the genome of any organism (Dong, Gou, & Lian, 2022). As it was originally detected in bacterial defense systems, wherein the DNA of any invading bacteriophages (virus) is broken down into fragments rendering them inactive due to lack of DNA repairing machinery (Rath et al., 2015). Once the mechanism underlying its DNA cleavage activity was determined, it was swiftly developed as genome editing tool. This genome editing tool (CRISPR-Cas9) takes the advantage of DNA repair pathways after the introduction of DSBs. At the cellular level, several DNA repairing pathways are operational among them only a few are used in gene editing, including non-homologous end joining (NHEJ) and homology-directed repair (HDR). NHEJ is commonly used in gene inactivation, while HDR is used to insert new fragments of genetic material (Xue & Greene, 2021).

7.3 MOLECULAR MECHANISMS OF CRISPR GENOME EDITING

CRISPR is a robust genome editing technology with a wide range of applications. Some of them are described in subsections.

7.3.1 CRISPR GENE KNOCKOUT

Since Cas9 creates DSBs and the most likely DNA repair mechanism by which these DSBs are repaired is through NHEJ. However, as known NHEJ is error-prone, and usually ends up the insertion or deletion mutations (indels) in the region of the gene being repaired. Frameshift mutation commences if the insertion or deletion mutations (indels) takes place within the coding region of the gene, leading to the non-functionality of the gene commonly called as gene knockout (KO). Gene KOs are the backbone of many research areas such as pathway analysis, functional genomics, screening, disease modeling and drug discovery. Targeting various regions of the gene by using multiple gRNAs certifies high efficiency gene knock outs (Ishibashi et al., 2020).

7.3.2 CRISPR KNOCK-IN

DSBs created by Cas9 can also be repaired by HDR and this repair mechanism offers a window of opportunity to insert a new piece of DNA or whole gene. This process of insertion is known as gene knock-in. In order to achieve gene knock-in, a DNA template known as donor template must consist of a sequence of gene of interest flanked by homology regions that must match on either side of the DNA break. This donor template along with other editing machinery of Cas9 and sgRNA is delivered to cells to achieve desired results. Gene knock in has achieved key position in biotechnology, including gene therapies to correct genetic mutations, diseases modeling, and recombinant protein production (Banan, 2020).

Gene knock-ins are more challenging in comparison to the gene KOs, because HDR is very less occurring pathway as compared to NHEJ, as it occurs only in certain phases of cell cycle. Moreover, HDR-mediated knock-in efficiencies are very less. In order to overcome such issues people started experimental optimization, synchronization of cell cycle and various treatments that either disable NHEJ or boost HDR (Song et al., 2016).

7.3.3 CRISPR ACTIVATION (CRISPRa) AND CRISPR INTERFERENCE (CRISPRi)

CRISPR-Cas9 can be employed not only to insert or delete the genes, but it can also be used to regulate gene expression as well which is known as CRISPR activation (CRISPRa) and CRISPR interference (CRISPRi). Upregulation and down regulation of gene expression can be performed by using CRISPRa and CRISPRi, respectively, by using an engineered variant of Cas9 called as catalytically dead Cas9 (dCas9), therefore cannot cut the DNA. Integrating dCas9 with transcriptional effectors can modulate the expression of the target gene. Unification of dCas9 with transcriptional effectors simply leads to the delivery of transcriptional effectors to the gene of interest and modulate their expression as dCas9 can't cleave the DNA. Modulation of gene expression has tremendous applications in biotechnology, developmental biology, functional genomics, infectious diseases, and screening for identification of genetic elements that imparts resistance to various drugs. Since CRISPR-Cas9 is highly efficient, therefore screening studies for the discoveries of drugs has become simple and time-efficient. Screening by CRISPR technique involves the formation of a huge sgRNA library, targeting various genes to analyze phenotypic effects of these wide range of edits delineating the relationship between genotype and phenotype. One is able to identify the genes that are actually involved in disease pathogenesis by performing knock out studies on normal cell line. Similarly, one can identify the drug targets in a diseased/cancerous cell line using the same methodology. CRISPR screening is more accurate with comparatively less off targets than RNAi screens of drug discovery (Kampmann, 2018).

7.3.4 CRISPR-CAS BASE EDITING AND PRIME EDITING

Recently, CRISPR methods have been adopted as base and prime editing, both of which are based on the same principle and both of these methods do not cleave the target DNA. Instead, base editing involves the use of either catalytically dCas9 or a nickase Cas9 (nCas9) While the dCas9 is catalytically dead therefore cannot cleave DNA, nCas9 is capable of producing either nick or single-strand breaks (Figure 7.2). Therefore, a fusion of either of the two to the DNA-modifying enzyme can modify nucleotides at specific points (Anzalone, Koblan, & Liu, 2020). However, base editing cannot be performed to alter every possible nucleotide this led to the discovery of prime editing.

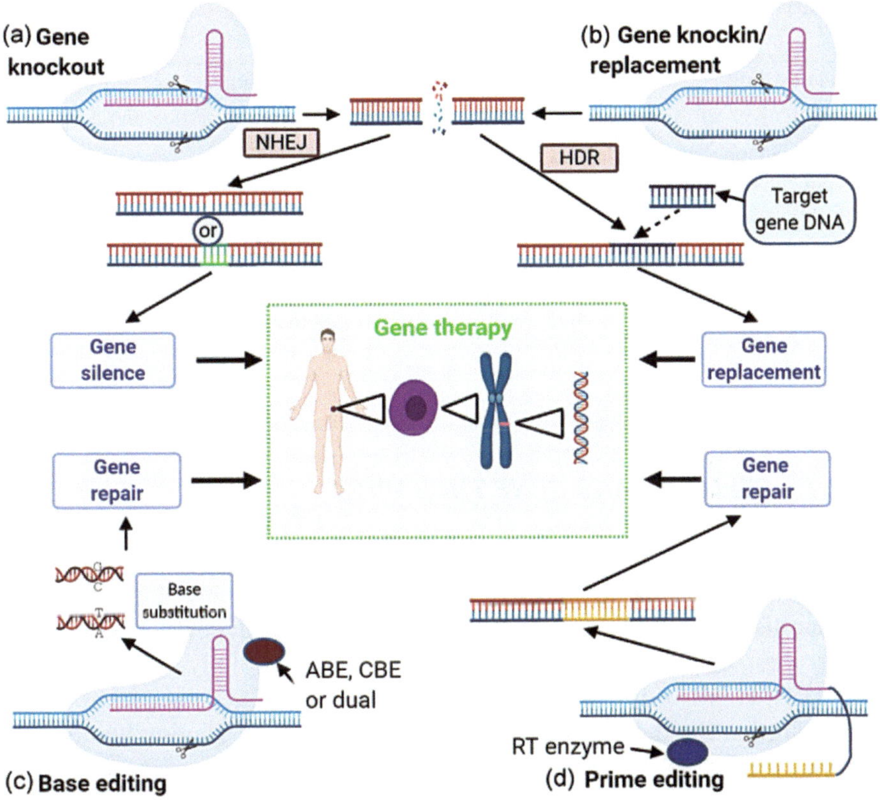

FIGURE 7.2 CRISPR/Cas9-based gene therapy. (a) Gene knock-out; (b) gene knock-in/replacement; (c) base editing; and (d) prime editing – can be employed to correct and/or replace the disease-related genes in human genome.

Source: Reprinted with permission from Zhang, 2020. © 2020 John Wiley and Sons.

Prime base editing involves the fusion of designed reverse transcriptase with nickase nCas9 and prime editing guideRNA (pegRNA). This prime editing gRNA pegRNA has two parts which performs two different functions. One guides it to the desired region while other harbors desired substitution mutations to be incorporated after the repair of single-stranded break that has been produced (Nelson et al., 2022). After the alteration of one strand by the prime editor, the complementary strand can also be modified by an additional gRNA and nCas9 that will induce a nick followed by its repairing with earlier edited strand as template. It has been predicted that prime editing can treat about 89% of genetic mutations in humans (Kantor, McClements, & MacLaren, 2020).

Base Editing	Prime Editing
• It Involves substitution mutations. • It is reliable, predictable, and efficient genetic outcomes. • It does not require template-based HDR. • It avoids undesirable double-stranded DNA breaks and oversized genomic rearrangements.	• It Involves insertion and deletion mutations. • It has low error rates. • It can perform transition and transverse mutations. • It can insert up to 45 bps and can delete up to 80 bps.

7.4 APPLICATIONS

7.4.1 CELLULAR AND GENE THERAPIES

CRISPR has enormous capability to transform to the field of medicine, it has ability to treat wide range of diseases such as cancer, blood disorders, ocular disorders, and neurodegenerative diseases (Liu et al., 2021) (Figure 7.3). First cell therapy utilizing CRISPR methodology was performed to treat patients with sickle cell diseases restoring the fetal hemoglobin, thus eliminating the requirement of a functional copy of adult hemoglobin (Park & Bao, 2021). Recently, in 2021, an important trail of CRISPR was performed to treat transthyretin amyloidosis, a neurodegenerative disease, displayed very encouraging results (Gillmore et al., 2021). Not only has the field of medicine, CRISPR also transformed the field of pediatrics. CRISPR has been used to treat pediatric cancer by the generation of chimeric antigen receptor (CAR) T cells, a kind of immunotherapy utilized in treatment of cancer. Lymphocytes T-cells are engineered after their extraction from a person to express CAR before injecting back to the patients. These T cells expressing CARs helps in identifying and eliminating the specific type of cancer efficiently from which a patient suffers (Razeghian et al., 2021).

CRISPR technique is still in its early phases of clinical trials, and this technique has the capability to treat a wide range of genetic conditions in the near future, including ovarian and breast cancer associated with BRCA mutations, cystic fibrosis, beta-thalassemia, Alzheimer's, and Huntington's disease (Hirakawa et al., 2020).

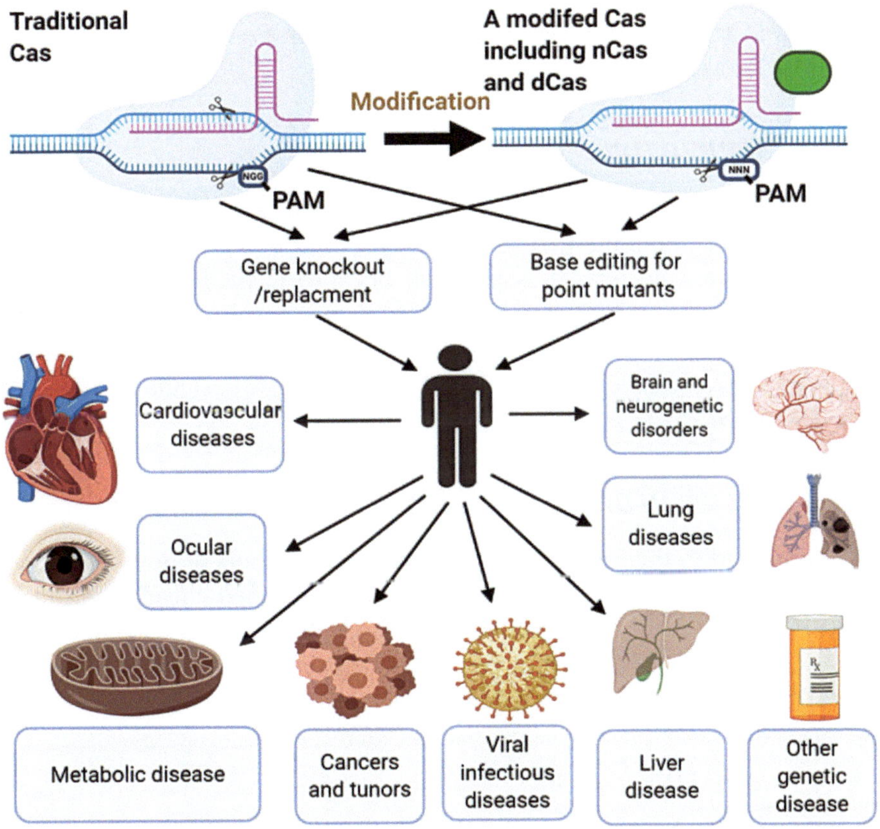

FIGURE 7.3 CRISPR-Cas genome editing tool has a huge capability to treat various genetic diseases, including cardiovascular diseases, brain and neurodegenerative disorders, respiratory diseases, hepatic diseases, and malignancies, as well as metabolic syndromes and virulent infectious diseases.

Source: Reprinted with permission from Zhang, 2020. © 2020 John Wiley and Sons.

7.4.2 DIAGNOSTICS

CRISPR can be used not only as a therapeutic tool but also as a diagnostic tool as well. For example, during COVID-19 pandemic (Dara and Talebzadeh, no date), CRISPR was used as a diagnostic tool such as CRISPR SARS-CoV-2 SHERLOCK™ was permitted for emergency use authorized by federal authorities. Another CRISPR-based COVID-19 kit DETECTOR was developed by Mammoth Biosciences, which uses Cas9's scan function to search and detect genetic material of Coronavirus (Joung et al., 2020). Utilizing the search function of Cas9, many other diagnostic tools have been engineered to identify other infectious and genetic diseases.

7.4.3 AGRICULTURE

CRISPR-Cas has also widespread application in agriculture as well and scientists are claiming that CRISPR-modified foods will be available within the next decade. It is possible because CRISPR can be used to develop drought and disease resistance plants (Joshi, Bharat, & Mishra, 2020). It has the capability to increase the shelf-life of perishable foods, thereby increasing the access to healthy foods at low cost and reducing the food wastage.

7.4.4 CRISPR IN DISEASE MODELS AND ANIMAL MODELS

As compared to the traditional gene targeting methods of development of animal models, the CRISPR-Cas system has provided cheaper and faster development of animal models with a high degree of precision. It also has potential to provide more precise disease models. Some of them are described in subsections.

7.4.4.1 PRIMARY CELLS

Human primary cell lines are very important for disease modeling as they produce more reliable data as compared to immortalized cell line. Primary cells edited by the CRISPR technology can be used for cell therapy and disease modeling. Genetic manipulation of primary cells poses significant challenges; however, scientists have developed methods to increase CRISPR editing efficiency in primary human cells, which involves use of chemically modified sgRNAs, synchronizing the cell cycle, generating ribonucleoprotein (RNP) complex for delivery, and cell-specific optimization of delivery methods which leads to the production of many cell therapies, including T cell immunotherapies (Mangeot et al., 2019).

7.4.4.2 STEM CELLS AND ORGANOIDS

CRISPR system has revolutionized the world of stem cell research. As known stem cells are pluripotent in nature, i.e., they can differentiate into any type of cell, making them extremely beneficial in the field of medical research. The emergence of induced pluripotent stem cells (iPSCs) was played a vital role because they are adult cells that can reprogramed to become pluripotent.

Human iPSCs are very difficult to grow and reprogram, because they are quite resistant to genetic reprogramming, however CRISPR has made it easier and also delivery efficiency is much greater as compared to earlier gene-editing tools and techniques used to manipulate iPSCs (Ben Jehuda, Shemer, & Binah, 2018).

Three-dimensional models of human organs that are developed in the lab are known as organoids and the CRISPR-Cas9 edited iPSCs can be used to produce such organoids. Organoids are more readable to study human diseases because they provide complexity related to the real human organs, thus providing a good model to analyze response to treatments and mimic the progression of diseases more accurately than cell cultures which are grown as monolayers or suspensions (Driehuis & Clevers, 2017). Recently, brain organoids are generated using CRISPR edited iPSCs to study the evolution of the human brain (McTague et al., 2021).

7.4.4.3 ANIMAL MODELS

Development of animal models that mimic human diseases is imperative in the studies of various diseases but do have some drawbacks, including a lack of translational capacity. CRISPR-edited transgenic animal models are closer to mimic human diseases and are comparatively easier and more efficient than earlier genome manipulation technologies (Shrock & Güell, 2017). It has the capacity to edit at any desired location and has also the ability to generate multiple edits within the same organism. Humanized knock-ins in animals are generated by the use of CRISPR technology which involves deletion of a particular region of the gene or whole gene followed by replacing it with a human version. Humanized knock-ins are key in understanding as well as treating human diseases (Artegiani et al., 2020). Recently a significant breakthrough has been achieved through gene-edited models of Duchenne Muscular Dystrophy in treating this disease, giving rise to the development of a possible gene therapy (Choi & Koo, 2021). CRISPR has also been used to generate a wide variety of neurodegenerative animal models as well (Yang et al., 2016).

7.5 CONCLUSION

The development of genome editing tool CRISPR-Cas has revolutionized the whole field of genomic research from last few years, enabling the advancement in basic genomic research for a wide range of applications. Also, CRISPR-Cas-based technologies are used to develop both *in vitro* and *in vivo* diseases models. CRISPR-Cas is now regarded as an umbrella term and is believed to be a promising approach to treat various human diseases. Continuous progression of CRISPR-based technologies not only revolutionizes the medical research but also offers various therapeutic opportunities to treat various diseases. No doubt CRISPR technology has proven a game-changer for genome engineering, but it's just the beginning, more possibilities and valuable benefits of this technology will be seen in the near future.

KEYWORDS

- **animal model**
- **clustered regularly interspaced short palindromic repeats**
- **CRISPR-Cas**
- **gene therapy**
- **genetic disorders**
- **genome editing**
- **protospacer adjacent motif**
- **transcription activator-like effector nucleases**

REFERENCES

Adli, M., (2018). The CRISPR tool kit for genome editing and beyond. *Nature Communications, 9*(1), 1911. doi: 10.1038/s41467-018-04252-2.

Anzalone, A. V., Koblan, L. W., & Liu, D. R., (2020). Genome editing with CRISPR–Cas nucleases, base editors, transposases and prime editors. *Nature Biotechnology, 38*(7), 824–844. doi: 10.1038/s41587-020-0561-9.

Artegiani, B., et al., (2020). Fast and efficient generation of knock-in human organoids using homology-independent CRISPR–Cas9 precision genome editing. *Nature Cell Biology, 22*(3), 321–331. doi: 10.1038/s41556-020-0472-5.

Banan, M., (2020). Recent advances in CRISPR/Cas9-mediated knock-ins in mammalian cells. *Journal of Biotechnology, 308*, 1–9. doi: 10.1016/j.jbiotec.2019.11.010.

Ben, J. R., Shemer, Y., & Binah, O., (2018). Genome editing in induced pluripotent stem cells using CRISPR/Cas9. *Stem Cell Reviews and Reports, 14*(3), 323–336. doi: 10.1007/s12015-018-9811-3.

Chandrasegaran, S., & Carroll, D., (2016). Origins of programmable nucleases for genome engineering. *Journal of Molecular Biology, 428*(5), 963–989. doi: 10.1016/j.jmb.2015.10.014.

Choi, E., & Koo, T., (2021). CRISPR technologies for the treatment of Duchenne muscular dystrophy. *Molecular Therapy, 29*(11), 3179–3191. doi: 10.1016/j.ymthe.2021.04.002.

Chylinski, K., Le Rhun, A., & Charpentier, E., (2013). The tracrRNA and Cas9 families of type II CRISPR-Cas immunity systems. *RNA Biology, 10*(5), 726–737. doi: 10.4161/rna.24321.

Danna, K., & Nathans, D., (1971). Specific cleavage of simian virus 40 DNA by restriction endonuclease of *Hemophilus influenzae*. *Proceedings of the National Academy of Sciences of the United States of America, 68*(12), 2913–2917. doi: 10.1073/pnas.68.12.2913.

Dara, M., & Talebzadeh, M., (no date). CRISPR/Cas as a potential diagnosis technique for COVID-19. *Avicenna Journal of Medical Biotechnology, 12*(3), 201, 202. Available at: Http://www.ncbi.nlm.nih.gov/pubmed/32695284.

Dong, C., Gou, Y., & Lian, J., (2022). SgRNA engineering for improved genome editing and expanded functional assays. *Current Opinion in Biotechnology, 75*, 102697. doi: 10.1016/j.copbio.2022.102697.

Driehuis, E., & Clevers, H., (2017). CRISPR/Cas 9 genome editing and its applications in organoids. *American Journal of Physiology-Gastrointestinal and Liver Physiology, 312*(3), G257–G265. doi: 10.1152/ajpgi.00410.2016.

Gillmore, J. D., et al., (2021). CRISPR-Cas9 *in vivo* gene editing for transthyretin amyloidosis. *The New England Journal of Medicine, 385*(6), 493–502. doi: 10.1056/NEJMoa2107454.

Gleditzsch, D., et al., (2019). PAM identification by CRISPR-Cas effector complexes: Diversified mechanisms and structures. *RNA Biology, 16*(4), 504–517. doi: 10.1080/15476286.2018.1504546.

Hille, F., & Charpentier, E., (2016). CRISPR-Cas: Biology, mechanisms and relevance. *Philosophical Transactions of the Royal Society B: Biological Sciences, 371*(1707), 20150496. doi: 10.1098/rstb.2015.0496.

Hirakawa, M. P., et al., (2020). Gene editing and CRISPR in the clinic: Current and future perspectives. *Bioscience Reports*, *40*(4). doi: 10.1042/BSR20200127.

Horvath, P., & Barrangou, R., (2010). CRISPR/Cas, the immune system of bacteria and archaea. *Science (New York, N.Y.)*, *327*(5962), 167–170. doi: 10.1126/science.1179555.

Horvath, P., et al., (2008). Diversity, activity, and evolution of CRISPR loci in *Streptococcus thermophilus*. *Journal of Bacteriology*, *190*(4), 1401–1412. doi: 10.1128/JB.01415-07.

Ishibashi, A., et al., (2020). A simple method using CRISPR-Cas9 to knock out genes in murine cancerous cell lines. *Scientific Reports*, *10*(1), 22345. doi: 10.1038/s41598-020-79303-0.

Jinek, M., et al., (2012). A programmable dual-RNA-guided DNA endonuclease in adaptive bacterial immunity. *Science (New York, N.Y.)*, *337*(6096), 816–821. doi: 10.1126/science.1225829.

Joshi, R. K., Bharat, S. S., & Mishra, R., (2020). Engineering drought tolerance in plants through CRISPR/Cas genome editing. *3 Biotech*, *10*(9), 400. doi: 10.1007/s13205-020-02390-3.

Joung, J., et al., (2020). *Point-of-Care Testing for COVID-19 Using SHERLOCK Diagnostics*. medRxiv: The Preprint Server for Health Sciences [Preprint]. doi: 10.1101/2020.05.04.20091231.

Kampmann, M., (2018). CRISPRi and CRISPRa screens in mammalian cells for precision biology and medicine. *ACS Chemical Biology*, *13*(2), 406–416. doi: 10.1021/acschembio.7b00657.

Kantor, A., McClements, M., & MacLaren, R., (2020). CRISPR-Cas9 DNA base-editing and prime-editing. *International Journal of Molecular Sciences*, *21*(17), 6240. doi: 10.3390/ijms21176240.

Kelly, T. J., & Smith, H. O., (1970). A restriction enzyme from *Hemophilus influenzae*. II. *Journal of Molecular Biology*, *51*(2), 393–409. doi: 10.1016/0022-2836(70)90150-6.

Khan, F. A., et al., (2016). CRISPR/Cas9 therapeutics: A cure for cancer and other genetic diseases. *Oncotarget*, *7*(32), 52541–52552. doi: 10.18632/oncotarget.9646.

Lino, C. A., et al., (2018). Delivering CRISPR: A review of the challenges and approaches. *Drug Delivery*, *25*(1), 1234–1257. doi: 10.1080/10717544.2018.1474964.

Liu, W., et al., (2021). Applications and challenges of CRISPR-Cas gene-editing to disease treatment in clinics. *Precision Clinical Medicine*, *4*(3), 179–191. doi: 10.1093/pcmedi/pbab014.

Mangeot, P. E., et al., (2019). Genome editing in primary cells and *in vivo* using viral-derived nanoblades loaded with Cas9-sgRNA ribonucleoproteins. *Nature Communications*, *10*(1), 45. doi: 10.1038/s41467-018-07845-z.

McTague, A., et al., (2021). Genome editing in iPSC-based neural systems: From disease models to future therapeutic strategies. *Frontiers in Genome Editing*, 3. doi: 10.3389/fgeed.2021.630600.

Nelson, J. W., et al., (2022). Engineered pegRNAs improve prime editing efficiency. *Nature Biotechnology*, *40*(3), 402–410. doi: 10.1038/s41587-021-01039-7.

Park, S. H., & Bao, G., (2021). CRISPR/Cas9 gene editing for curing sickle cell disease. *Transfusion and Apheresis Science: Official Journal of the World Apheresis Association: Official Journal of the European Society for Haemapheresis*, *60*(1), 103060. doi: 10.1016/j.transci.2021.103060.

Ran, F. A., et al., (2015). *In vivo* genome editing using *Staphylococcus aureus* Cas9. *Nature, 520*(7546), 186–191. doi: 10.1038/nature14299.

Rath, D., et al., (2015). The CRISPR-Cas immune system: Biology, mechanisms and applications. *Biochimie, 117*, 119–128. doi: 10.1016/j.biochi.2015.03.025.

Razeghian, E., et al., (2021). A deep insight into CRISPR/Cas9 application in CAR-T cell-based tumor immunotherapies. *Stem Cell Research & Therapy, 12*(1), 428. doi: 10.1186/s13287-021-02510-7.

Richter, C., Chang, J. T., & Fineran, P. C., (2012). Function and regulation of clustered regularly interspaced short palindromic repeats (CRISPR) / CRISPR associated (Cas) systems. *Viruses, 4*(10), 2291–2311. doi: 10.3390/v4102291.

Sangal, V., Fineran, P. C., & Hoskisson, P. A., (2013). Novel configurations of type I and II CRISPR-Cas systems in *Corynebacterium diphtheriae. Microbiology (Reading, England), 159*(Pt 10), 2118–2126. doi: 10.1099/mic.0.070235-0.

Shrock, E., & Güell, M., (2017). CRISPR in animals and animal models. *Progress in Molecular Biology and Translational Science, 152*, 95–114. doi: 10.1016/bs.pmbts.2017.07.010.

Smith, H. O., & Wilcox, K. W., (1970). A restriction enzyme from *Hemophilus influenzae*. I. Purification and general properties. *Journal of Molecular Biology, 51*(2), 379–391. doi: 10.1016/0022-2836(70)90149-x.

Song, J., et al., (2016). RS-1 enhances CRISPR/Cas9- and TALEN-mediated knock-in efficiency. *Nature Communications, 7*(1), 10548. doi: 10.1038/ncomms10548.

Xue, C., & Greene, E. C., (2021). DNA repair pathway choices in CRISPR-Cas9-mediated genome editing. *Trends in Genetics: TIG, 37*(7), 639–656. doi: 10.1016/j.tig.2021.02.008.

Yang, W., et al., (2016). CRISPR/Cas9: Implications for modeling and therapy of neurodegenerative diseases. *Frontiers in Molecular Neuroscience, 9*. doi: 10.3389/fnmol.2016.00030.

Zhang, B., (2020). CRISPR/Cas gene therapy. *Journal of Cellular Physiology, 236*(4), 2459–2481. doi: 10.1002/jcp.30064.

Zhang, Y., et al., (2013). Processing-independent CRISPR RNAs limit natural transformation in *Neisseria meningitidis. Molecular Cell, 50*(4), 488–503. doi: 10.1016/j.molcel.2013.05.001.

CHAPTER 8

Emerging Immunotherapies for Healthcare Systems

BHAWANA SINGH and VIMAL VERMA

Department of Medicine, Institute of Medical Sciences, Banaras Hindu University, Varanasi, Uttar Pradesh, India

ABSTRACT

The human immune system is equipped with the ability to fight against invading pathogens. It has been an exciting area to target this ability of the immune system for improving human health. Recent technological advances have greatly accelerated the research toward the development of targeted therapeutics. The steadfast progress for the development of immunotherapies has paved the way for better understanding and fine-tuning of the immune system for fighting against different diseases. Considering the important role of T-cells in driving immunological responses, their differentiation and functional subsets immunotherapies are designed with an aim to release the brakes on the immune system. Immunotherapy uses a wide variety of immunomodulators that augment immune responses using various strategies ranging from altering the cultured cells to monoclonal antibodies and autologous immune enhancement therapies. This chapter aims to provide a comprehensive outlook on the pre-clinical and clinical findings for the development of different immunotherapeutic approaches for the management of different diseases.

Modern Biotechnology in Healthcare: Advances and Applications. Sheikh Umar Ahmad (Ed.)

8.1 INTRODUCTION

The host immune system is the key determinant of the infection outcome (i.e., disease progression/regression), so harnessing the immune responses for fighting against diseases remains the hottest areas of biomedical research. Immunotherapy is a treatment approach that harnesses the host's own immune system to combat infections or diseases. It harnesses innate and adaptive arms of immunity for fighting the disease and/or enhance immune resistance against pathogens.

The history of modern immunotherapy began back in the late 19th century, when Dr. William B Coley first attempted to harness the host immune system to fight against cancer, when most of the cancer cases underwent remission and developed erysipelas. Dr. Coley attempted to treat these subjects by injecting a mixture of live but inactivated *Streptococcus pyrogens* and *Serratia marcescens* that yielded durable and complete remission in the sarcoma, lymphoma, and testicular carcinoma cases. Unfortunately, the mechanism behind the action of Coley's toxin was not deciphered and potential risks associated with injection of pathogenic bacteria instigated the oncologists to opt for surgery and radiotherapy as the standard treatment option in early 20th century (Decker & Safdar, 2009). The use of attenuated bacteria for treating malignancies re-emerged in 1976, when BCG (Bacillus-Calmette-Guerin) was successfully used for preventing recurrence of bladder cancer (Morales et al., 1976) and has been used till date (Morales et al., 2017). In the same year, IL-2 was identified, that was used for metastatic cancer cases for enhancing the T-cell population, that eventually gained Food and Drug Administration (FDA) approval in 1991 for treatment of metastatic kidney cancer and for treating metastatic melanoma (in 1998) (Rosenberg et al., 1985). In 1970, Kohler and Milstein used hybridoma technology for developing monoclonal antibodies and later, research on antibody-based therapies led to the development of rituximab that targeted the immature B-cells (Rudnicka et al., 2013). In 1997, FDA approved monoclonal antibody-based therapy for the treatment of non-Hodgkin's lymphoma. In 2010, FDA approved sipuleucel-T, as the first vaccine for prostate cancer. Another breakthrough in immunotherapy came after the discovery of immune checkpoint molecules, in 2011 and 2014, when Ipilimumab (targeting CTLA-4) and Nivolumab (targeting PD-1), respectively, were approved for treating melanoma cases. Further, promising results have been obtained with the use of MPDL3280A (anti-PD-L1 antibody) in melanoma, lung cancer and bladder cancer (Rosenberg et al., 2016; Patel et al., 2017).

In recent years, immunotherapy has become the hottest areas of biomedical research for treating several diseases. With the aim of manipulating host immune responses to fight diseases, immunotherapy is also referred to as the biological therapy, due to its ability to use substances called biological response modifiers (BRM). Human body usually produces small amount of BRM in response to invading pathogens however, in laboratory conditions these substances are produced in large quantities for therapeutic purpose. These BRMs or immunomodulators have fewer side effects and lesser potent in terms of creating drug resistance in disease causing microbes (Masihi, 2001).

Based on the mechanism of action immunotherapies can be broadly categorized into two classes: passive immunotherapy and active immunotherapy. Passive immunotherapy (also called adoptive immunotherapy) employs donated or laboratory-made immune cells and/or recombinant molecules (monoclonal antibodies) for fighting against diseases. This approach can be specific (monoclonal antibody therapy) or non-specific (for example, injection of patient's cultured T-cells, after *in vitro* expansion and or activation). Active immunotherapy directly triggers host immune system/immune response to fight infections. It is again divided into specific (generates cell-mediated and antibody responses) and non-specific (elicits cytokine and other cell signaling molecules) arms. Overall, immunotherapies can be agent based or cell based that aims to trigger the host immune responses by targeting different virulence factors and brings about immunological manipulations for therapeutic benefits.

8.2 TYPES OF IMMUNOTHERAPEUTIC AGENTS

As discussed in the above sections, there are two types of immunotherapeutic approaches – the specific and non-specific. In this section, we discuss the targeted therapeutic approaches for stimulating immune responses. Over the decades, there have been diverse therapies that block the biochemical pathways and have been successfully used for fighting against several diseases. This chapter aims to give a comprehensive outlook on different approaches that have revolutionized the biomedical sector.

8.2.1 SPECIFIC IMMUNOTHERAPIES

Treatment for several diseases includes the use of specific immunotherapies that aims to target particular immune cell population and/or molecule for immunomodulation. Recent specific immunotherapies are focused on modulating the signaling pathways/molecules as PD-1, CTLA-4, etc., for altering the host immune responsiveness.

8.2.1.1 mRNA VACCINES

Nucleic acid therapeutics has been one of the best innovative alternatives to the conventional vaccine approaches, which was successfully used in animal model (in 1990) (Wolff et al., 1990). Later in 1992, documented the effect of vasopressin encoding mRNA injection elicited the physiological responses (Jirikowski et al., 1992). But the instability of mRNA, inefficient *in vivo* delivery and high innate immunogenicity restricted subsequent research for mRNA-based therapeutics. Recent technological advances have again brought the mRNA-based therapeutic strategies to the limelight for vaccinology. RNA-based vaccines remains potentially advantageous as compared to the DNA-based vaccination approaches in terms of its safety (non-infectious), it can be easily degraded during normal cellular processes and easy modification options provide flexibility in regulating its half-life during *in vivo* administration (Kariko et al., 2008; Kauffman et al., 2016; Thess et al., 2015). Furthermore, the easy manufacturing and production of mRNA vaccines makes this field exciting avenue to explore for different diseases. Therefore, RNA vaccines fulfill different criterion for an ideal vaccine so, vast array of clinical trials are undergoing for translation of these trials for clinical success.

Broadly speaking, there are two forms of RNA vaccines available against infectious agents: self-amplifying (replicating) vaccines and non-replicating vaccines. Self-amplifying vaccines (SAV) are alpha virus genome-based formulation, where genes for RNA replication machinery remain unaffected while the genes for structural proteins are replaced with the desired antigenic fragment. SAVs facilitates the production of large amount of antigen from small vaccine dosage, that eventually elicits stronger T- and B-cells mediated immune responses against diverse range of viral infections (Fleeton et al., 2001; Geall et al., 2012; Bogers et al.,

2015) and bacterial infections (Maruggi et al., 2017). Further, SAVs create their own adjuvants (double stranded RNA structures) that improve the vaccine efficacy; however, greater size and higher immunogenicity restricts its repeated usage. Conversely, non-replicating mRNA vaccines can further be divided into two sub-classes, depending on their mode of delivery: *ex vivo* loading of DCs or direct injection method.

Ex vivo DCs loading method remains the choice for generating cell-mediated immune responses for cancer, HIV (electroporation of autologous DCs with mRNA for HIV antigens). The intervention was found to successfully elicit CD4 and CD8 T-cells responses in HIV cases; however, it didn't confer any clinical advantage to the cases (Routy et al., 2010; Allard et al., 2012). Similarly, DC-based vaccination for cytomegalovirus (CMV) in stem cell recipients induced cell-mediated immune responses (Van Craenenbroeck et al., 2015).

Recently, non-replicating mRNA-based directly injectable vaccines are emerging as simple and easy administrable option in resource poor settings. This vaccine format was first demonstrated against influenza virus infection in 2012 (Petsch et al., 2012). Thus, mRNA vaccines in its uncomplexed form were sufficiently immunogenic for eliciting protective responses.

Similarly, protamine-based RNActive platform encoding for rabies virus induced protective cell- and antibody-mediated immune responses in animal model (Schnee et al., 2016). The prophylactic vaccine was also tested in humans where needle-free delivery method induced elicited variable levels of functional antibody responses against the virus, with reasonable tolerability profile (Alberer et al., 2017).

Different formats of mRNA vaccines are available that use lipid- or polymer-based delivery platforms. In mice model, cationic 1,2-dioleoyloxy-3-trimethylammoniumpropane (DOTAP) and dioleoylphosphatidylethanolamine (DOPE) lipid-complexed mRNA vaccines against HIV are known inducers of antigen-specific CD4+ and CD8+ T-cell responses (Pollard et al., 2013). Similarly, PEI-complexed mRNA vaccine (against HIV) and lipid complexed vaccine (against influenza) successfully induced T-cell activation and responses (Kranz et al., 2016; Zhao et al., 2016).

Nucleoside-modified mRNA vaccines have been the new innovation; however, limited data on its efficacy has been reported. First data on the successful use of the vaccine was demonstrated for Zika virus in mice and macaques (Pardi et al., 2017). A subsequent study reported for the efficacy of vaccine in terms of moderate immune responses however, the booster dose elicited protective immune responses in vaccinated group (Richner et al., 2017). Another trial against influenza virus, reported for protective efficacy of low-dose LNP-complexed mRNA vaccine. The successful pre-clinical findings paved the way for human trials where the vaccine was potently immunogenic and elicited protective antibody responses (Bahl et al., 2017). Presently, several vaccines are entering clinical trials and are expected to deal with a diversity of antigens for fighting against infectious diseases.

In context with the cancer vaccine, mRNA-based vaccines aim to target tumor or its associated antigens. Mostly, cancer vaccines are therapeutic rather than prophylactic and the idea of RNA-based cancer vaccines emerged about two decades ago (Boczkowski et al., 1996). DCs based mRNA vaccines were potent inducers of protective responses that can be further enhanced with the use of mRNA-encoded adjuvants. Several reports about the use of costimulatory molecules (CD83, OX40, and 4-1BB) significantly enhanced the immunostimulatory potential of DCs based vaccines (Dannull et al., 2005; Aerts-Toegaert et al., 2007; De Keersmaecker et al., 2011). Similarly, mRNA encoding for pro-inflammatory cytokines and trafficking molecules have also been successful in inducing anti-tumor T-cells and NK cells responses (Bontkes et al., 2008; Dorrie et al., 2008). Furthermore, a cocktail of mRNA-encoded adjuvants (named TRiMix, includes CD70, CD40L, and TLR4) has also shown promising results in fighting cancer in several pre-clinical studies (Bonehill et al., 2008; Van Lint et al., 2014). The cocktail serves as a potent inducer of DC activation and triggers TH1 type immune responses that eventually cause tumor regression in different cancer cases (Wilgenhof et al., 2013; Wilgenhof et al., 2016; Batich et al., 2017). Therefore, there has been tremendous progress in writing the success story of RNA vaccines however, this is not the end since, dozens of clinical trials for different diseases are still ongoing and potency of RNA vaccines against several infectious pathogens are still under exploration.

8.2.1.2 DNA VACCINES

DNA vaccination approach although first used in the early 90s, has recently been evolving as a smart strategy for the development of immunotherapeutics, with the aim to stimulate the humoral as well as the cellular immune responses. In animal models, DNA vaccines have been successfully used to prevent or treat some infectious diseases, cancer, autoimmunity, and allergies (Ulmer et al., 1996).

DNA vaccines are aimed to deliver one or several antigen encoding genes (under the control of appropriate promoter) with the aim to elicit antigen-specific immune responses. With the direct injection of genetic material, a relatively smaller number of cells incorporates the genetic material in them that eventually leads to inappropriate expression of the introduced gene that has important immunological consequences. The inappropriate gene expression forms the key for immune activation since most of the cells do not present the protein to the T-cells that eventually triggers the CTL responses by CD8+ T-cells (if presented to the immune cells with the MHC-I) that inhibits viruses. Conversely, CD4+ T-cells are activated (if the peptide fractions are presented in context with MHC-II by the antigen presenting cells). CD4+ T-cells (Th1 type) stimulate the B-cells and the antibody production (Encke et al., 1999). These vaccines have significantly gained attention in recent years due to their stability, improved safety, and feasibility for large scale manufacturing. Furthermore, strategies aiming at improving their immunogenicity and antigen presentation by employing innovative delivery platforms, adjuvants, immune-stimulatory molecules, or blockade of immune checkpoint molecules are still being underway.

The success of DNA vaccines has been seen in the clinical settings, where several trials are ongoing. Clinical trial targeting the breast cancer has validated the safety and efficacy of DNA plasmid encoding for Her-2 gene, where remittent cases (stage II, III, and IV) upon vaccination with DNA along with cytokine adjuvant and GM-CSF yielded humoral and cell-mediated immune responses. It is well known that the success of DNA vaccine regimen usually depends on the preventing the disease recurrence, this formulation was successfully known to elicit immunological memory (Disis et al., 2004).

A veterinary DNA vaccine for West Nile virus safety in horses has been already approved. Some previous studies show that intramuscular injection of 'naked' plasmids produces immunogens that would elicit immune responses against influenza virus antigens in mice (Ulmer et al., 1993). Plasmid DNAs which express the influenza virus hemagglutinin

(HA) glycoproteins to increase protective immunity against has also been investigated (Fynan et al., 1993).

HA, the primary viral antigen in influenza, is commonly used in DNA vaccines against influenza virus. Other viral proteins are being used infrequently, usually in combination with HA, but sequences encoding PB1, M2, NP, and M1 have also been used in DNA vaccines (Patel et al., 2012; Lim et al., 2013; Lee et al., 2018). Plasmids encoding for NP and M2 have been shown to minimize the viral load and increase survival in BALB/c mice (Ulmer et al., 1993; Tompkins et al., 2007). Fusion protein antigen (combined use of can be used to expand the range of host responses (Lee et al., 2018). In mice, plasmid-encoded chimeric proteins of H1N1 HA and the conserved M2-ectodomain increase cross-reactivity of antibody response compared to when using HA alone viruses in DNA vaccine (Park et al., 2011). The USDA approved conditionally, the first commercial DNA vaccine against H5N1 in chickens. Effective plasmid delivery and the use of suitable adjuvants are still major obstacles to overcome before influenza DNA vaccines can be used in humans.

In SARS-CoV-2 infection antibodies are formed against the P and S protein. The N protein is covering the viral genome as well as it helps in the release of virus particles from cells. The S (S1, S2, and S2') protein, on the other hand, plays an important role in pathogenesis through binding to the host cell via its receptor-binding domain and thus inducing infection (Zhang et al., 2022). All DNA vaccines currently being under clinical trials have used the S protein as their antigen. One of them is "Study of COVID-19 DNA Vaccine (AG0301-COVID-19)" (ClinicalTrials.gov Identifier: NCT04463472). In this study they used two experimental groups: low dose group and high dose group. In both cases they recruited 15 people aged between 20 to 65 years and expected date of completion of the project is July 31, 2021 (ClinicalTrials.gov) (Table 8.1).

8.3 STEM CELL THERAPY

Stem cells have the capacity of proliferation, migration, and differentiation, making them a perfect candidate for potential therapeutic applications in tissue regeneration and repair. This self-renewable capacity is also the characteristics of tumor cell, but the difference is that stem cell divides in a very controlled manner while tumor cell divide in an uncontrolled

TABLE 8.1 . Recent COVID-19 Nucleic Acid Vaccine Clinical Trials

Estimated Study Start Date	Estimated Study Completion Date	Project Title	Clinical Trials. gov Identifier	Phase	Subject (Age)	Number of Subjects	Study Location
June 17, 2020	June 17, 2022	Safety and Immunogenicity Study of GX-19, a COVID-19 Preventive DNA Vaccine in Healthy Adults	NCT04445389	Phase I/II	18 to 50	210	Korea
December 30, 2020	March 30, 2022	Safety and Immunogenicity Study of GX-19N, a COVID-19 Preventive DNA Vaccine in Healthy Adults	NCT04715997	Phase I/II	18 to 55	170	Korea
February 25, 2021	June 2022	Safety and Immunogenicity of COVID-eVax, a Candidate Plasmid DNA Vaccine for COVID-19, in Healthy Adult Volunteers	NCT04788459	Phase I/II	18 to 65	160	Italy
December 30, 2020	May 2022	CORVax12: SARS-CoV-2 Spike (S) Protein Plasmid DNA Vaccine Trial for COVID-19 (SARS-CoV-2)	NCT04627675	Phase I	18 and older	36	United States
March 2021	August 2022	A Clinical Trial of a Plasmid DNA Vaccine for COVID-19 [Covigenix VAX-001] in Adults	NCT04591184	Phase I	18 to 84	72	Canada
August 31, 2020	September 30, 2021	Study of COVID-19 DNA Vaccine (AG0302-COVID-19)	NCT04527081	Phase I/II	20 to 65	30	Japan
June 29, 2020	July 31, 2021	Study of COVID-19 DNA Vaccine (AG0301-COVID-19)	NCT04463472	Phase I/II	20 to 65	30	Japan
November 23, 2020	March 31, 2022	Phase II/III Study of COVID-19 DNA Vaccine (AG0302-COVID-19)	NCT04655625	Phase II/III	18 and older	500	Japan
February 15, 2021	December 31, 2022	The Safety and Immunogenicity of a DNA-based Vaccine (COVIGEN) in Healthy Volunteers	NCT04742842	Phase I	18 to 75	150	Australia

Source: Assess at ClinicalTrials.gov as of March 22, 2021.

manner. Using this property stem cells are used in therapy called stem cell therapy in which stem cells used to cure, prevent a disease or disorder (Mahla, 2016). The most common problems that can benefit from this type of treatment are strokes (Liu, 2013), diabetes (Shahjalal et al., 2018), macular degenerations (Sun et al., 2017) and osteoarthritis (Gangji et al., 2011).

Extensive research in stem cell provides clue about its usage for treating cancer patients due to its characteristics like self-renewal, modulatory effects on other cells, directional migration, and self-differentiation, which can be used as drug targeting, immune cell modulation, and therapeutic carriers. Embryonic stem cells (ESCs) can give rise to any form of cell except those in the placenta. We can make pluripotent cells directly from mouse embryonic or adult fibroblast cultures, which we call induced pluripotent stem (iPSCs) cells when cultured in defined factors (Oct3/4, Sox2, c-Myc, and Klf4) (Takahashi & Yamanaka, 2006). Both hESCs and iPSCs have recently been used to induce effector T- and NK cells, and also to produce anti-cancer vaccines (Matsushita et al., 2014; Kooreman et al., 2018). The second type of stem cell is adult stem cells (ASCs) is a group of three types of cell mesenchymal stem cells (MSCs), neural stem cells (NSCs) and hematopoietic stem cells (HSCs) used in cancer treatment (Table 8.2). Another Third type of stem cell is called as the CSCs (cancer stem cells) which is produces by epigenetic mutations in natural stem cells or progenitor/precursor cells and play important role in growth of cancerous cell, its metastasis, recurrence, and treatment resistance (Chang, 2016). Therefore, targeting CSCs could be a good option for cancer treatment.

In tumors, patients need intravenous infusion of autologous or allogeneic HSCs. These HSCs are thought to have a homing mechanism that allows them to migrate quickly into established stem cell niches in the bone marrow (BM) where interaction of the CXCR4 receptor (present on stem cells) and the gradient of SDF-1 (secreted by cells lining of BM stem cell niches) taken place (Juarez et al., 2012). Some other examples of molecular signaling of HSC homing process are Ceramide-1 phosphate and sphingosine-1-phosphate (Seitz et al., 2005; Massberg & von Andrian, 2009), calcium sensing receptor (Adams et al., 2006), proton sensing GPCR (Okajima, 2013). HSCs migrate across the bloodstream and through the endothelial vasculature to various organs and in the BM niche via interaction between endothelial (through LFA-1, VLA-4/5,

CD44) and secretion of matrix degradable enzyme MMP-2/9 (Lapidot et al., 2005) (Table 8.2).

TABLE 8.2 Different Types of Stem Cells and Their Therapeutic Utility for Cancer

Cell Type	Location	Role	Use in Cancer
Hematopoietic stem cells (HSCs)	Bone marrow	Can form all mature blood cells	Multiple myeloma, leukemia, and other cancers
Mesenchymal stem cells (MSCs)	Bone marrow, adipose tissue, and fetal tissue (umbilical cord)	Tissue repair as well as regeneration	Liver, lung cancer, pancreatic tumors, and other cancers
Neural stem cells (NSCs)	Central nervous system	Generate new neurons and glial cells	Metastatic breast, lung, and prostate cancers and other cancers

A large number of registered clinical trials using MSCs to treat solid tumors are currently underway. The world's first gastrointestinal tumor clinical trial (phase I/II) using genetically engineered MSCs is successfully completed in humans (TREATME1) (Niess et al., 2015). There are other trials have also been completed and many more clinical trials currently underway (Table 8.3).

MSCs are extensively studied stem cell which is widely being used to treat autoimmune disorders, involvement in hematopoietic stem cell transplantation (HSCT), and treatment of infections and their complications, including septic shock and acute respiratory distress syndrome. BM-MSCs may protect against infectious challenges by activation of the host immune responses and modulating proinflammatory host immune response (Auletta et al., 2012).

Stem cell therapy has recently been developed to treat parasitic infections. *In vivo*, MSCs have been shown to strengthen the liver damage caused by *S. japonicum* infection, and this effect is improved when MSCs are combined with praziquantel. These results indicate that treating *S. japonicum*-induced liver injury and fibrosis with MSCs is a novel therapeutic strategy (Xu et al., 2012). In mice infected with *Plasmodium berghei*, massive recruitment of MSCs has been observed in secondary lymphoid organs, and transplantation of these cells into naive mice provides host resistance against malaria infection (Thakur et al., 2013). BM mononuclear cells have been shown to reduce inflammation

TABLE 8.3 Some Current Ongoing Stem Cell Therapy Clinical Trials in Cancer

Estimated Study Start Date	Estimated Study Completion Date	Project Title	Clinical Trials. gov Identifier	Phase	Subject (Age)	Number of Subjects	Study Location
August 18, 2018	October 22, 2021	Mesenchymal Stem Cell Therapy for Liver Cirrhosis: A Phase I/II Study	NCT03626090	Phase I/II	21 to 69	20	Singapore
January 11, 2007	April 24, 2013 (completed)	Adoptive Cell Therapy for B-Cell Malignancies After Allogeneic Hematopoietic Stem Cell Transplantation with Costimulated, Tumor-Derived Lymphocytes	NCT01445132	Phase I	18 to 75	11	United States
April, 2017	July 2023	Efficacy of Doxycycline on Metakaryote Cell Death in Patients with Resectable Pancreatic Cancer	NCT02775695	Phase II	18 years and old	12	United States
March 5, 2019	September 1, 2025	Targeted Stem Cells Expressing TRAIL as a Therapy for Lung Cancer	NCT03298763	Phase I/II	18 and old	46	United Kingdom
January 16, 2019	July 31, 2024	A Phase 1 Study of IDH1 Inhibition Using Ivosidenib as Maintenance Therapy for IDH1-mutant Myeloid Neoplasms Following Allogeneic Stem Cell Transplantation	NCT03564821	Phase I	18 and old	22	United States
February, 2018	November, 2021	High-dose Chemotherapy with Autologous Hematopoietic Stem Cell Transplantation as Adjuvant Treatment for Triple Negative Breast Cancer Patients Without Complete Pathological Response to Neoadjuvant Chemotherapy	NCT02670109	Phase II	18 years to 60	20	Mexico
November 2005	December 2022	Hematopoietic Cell Transplantation in the Treatment of Infant Leukemia and Myelodysplastic Syndrome	NCT00357565	Phase II	Up to 3 Years	20	United States

Source: Assess at ClinicalTrials.gov as of March 22, 2021.

and fibrosis in mice infected with *Trypanosoma cruzi* in mouse models of Chagas disease (Soares et al., 2004).

Targeting MSCs or NO appears to be an effective therapeutic intervention for the development of new prevention strategies for *Mycobacterium tuberculosis* (TB). Combining standard anti-TB chemotherapy with autologous MSC transplantation could eventually improve anti-TB treatment efficacy in MDR-TB patients. Studies have shown that MSCs are vulnerable to infection by viruses from the herpes family such as Epstein-Barr virus, HSV-1, HSV-2, and CMV and when MSCs are infected with herpes viruses, they lose their ability to function. The highly pathogenic avian influenza A/H5N1 infection is permissive to human MSCs, and infection can cause apoptosis and loss of immunomodulatory function of MSCs. MSCs secrete extracellular vesicles (ECVs) that have anti-inflammatory and anti-influenza properties hence it can be used as a therapeutic agent (Khatri et al., 2018). In mice, Treatment with MSCs reduces inflammation and mortality associated with the Japanese encephalitis virus, which is the most common cause of viral encephalitis in Asia (Bian et al., 2017) (Table 8.4).

8.4 REGULATORY T-CELL THERAPY

Tregs are a type of T cell that helps to regulate immune responses. The main function of Treg is to suppress a variety of pathological and physiological immune responses in the host, allowing immune homeostasis to be maintained (Sakaguchi, 2000). Despite the fact that Tregs are characterized as T cells that suppress immune responses, it has been established that regulatory T cell populations are diverse; some are induced in response to infectious challenge, while others are considered natural regulators (Belkaid, 2007). These inducible Tregs cells produces inhibitory cytokines such as IL-10 and (TGF)-β. Tregs are CD4+CD25+ T cells that make up about 10% of total peripheral CD4+ T cells in both mice and humans (Lee et al., 2011). Foxp3 transcription factor is highly expressed in the CD4+CD25+ population (Bacchetta et al., 2007), which is required for Treg cell differentiation, function, and programming of suppressor T cell function (Fontenot et al., 2003; Haque et al., 2011). Impaired regulatory T-cell function is linked to autoimmune, infectious, and allergic diseases in humans (Taams et al., 2006). Taking all together, these cells play a vital role in the maintenance of both innate and acquired immune responses.

TABLE 8.4 Some Current Ongoing Stem Cell Therapy Clinical Trials in Different Infectious Diseases

Estimated Study Start Date	Estimated Study Completion Date	Project Title	Clinical Trials. gov Identifier	Phase	Subject Age (Year)	Number of Subjects	Study Location
April, 2020	September 2020 (completed)	Clinical Use of Stem Cells for the Treatment of COVID-19	NCT04392778	Phase I/II	40 to 60	30	Turkey
June 1, 2020	June 30, 2021	Efficacy of Intravenous Infusions of Stem Cells in the Treatment of COVID-19 Patients	NCT04437823	Phase II	30 to 70	20	Pakistan
January 4, 2019	October 2021	Antigen Specific Adoptive T Cell Therapy for Adenovirus Infection After Hematopoietic Stem Cell Transplantation	NCT03378102	Phase 1	3 Months and old	20	United States
October 2016	December 2022	CMV Specific T Cell Therapy After Allogeneic Stem Cell Transplantation	NCT03067155	Phase II	16 to 75	20	Belgium
January 15, 2020	June 30, 2022	Escalation Antifungal Prophylaxis for Patients Undergoing Allogeneic Hematopoietic Stem Cell Transplantation	NCT04273178	Phase II	16 to 65	175	China
May 21, 2017	December 30, 2020	Effectivity of Mesenchymal Stem Cell on Vertebral Bone Defect Due to *Mycobacterium tuberculosis* Infection	NCT04493918	Phase II	15 to 64	20	Indonesia

Source: Assess at ClinicalTrials.gov as of March 22, 2021.

In lung, pancreatic, breast, liver, and skin cancer patients having high numbers of CD4+ CD25+ Tregs in the blood or in the tumor itself (Liyanage et al., 2002; Woo et al., 2002; Wolf et al., 2003; Ormandy et al., 2005). These Treg cells stop NK cell mediated cytotoxicity and also the production of IFN-γ from CD4+ and CD8+ T cells. Treg cells are found abundantly in the tumor microenvironment, where they suppress antitumor immune responses, promoting tumor growth and progression. Infiltrating lymphocytes from Hodgkin lymphoma contain both CD4+ CD25+ and IL-10-producing Tregs, which suppress mitogen- and antigen-specific peripheral blood mononuclear cell responses (Marshall et al., 2004). Chemokines are a type of secreted chemotactic protein that regulates leukocyte migration by binding to G protein-coupled receptors on target cells. They are involved in the recruitment of inflammatory cells as well as immunomodulatory cells, such as Treg cells, for antitumor responses, and play an important role in selective homing of neoplastic B and T cells (Nagarsheth et al., 2017).

The immune response to infection is a delicate balancing act between the successful induction of proinflammatory responses and the anti-inflammatory responses needed to limit tissue damage. Tregs (CD4+ CD25+ Treg and IL-10-secreting Treg) undoubtedly play a key role in maintaining this balance during infection, with outcomes ranging from highly harmful to the host to highly beneficial to both the host and the pathogen (O'Garra et al., 2004). Studies have shown that in comparison to uninfected controls, infected patients with *Helicobacter pylori* have more CD4+ CD25+ FOXP3+ T cells in the stomach and duodenal mucosa (Lundgren et al., 2005). There are some more examples of infectious dieses in which Treg cells might contribute to chronic infection (*Helicobacter hepaticus, Listeria monocytogenes, Leishmania major, Schistosoma masoni,* herpes simplex virus, cytomegalovirus) (Belkaid & Rouse, 2005).

Targeting Treg cells can be used as the main therapeutic approach against both cancer and infectious diseases. In mice, Treg cell depletion can elicit highly effective antitumor immune responses in preclinical cancer models (Onda et al., 2019). CD25 can be blocked or used to deplete Treg cells depending on the interaction of CD25-specific antibodies with the Fc receptor (Huss et al., 2016). An alternative strategy, use of a recombinant IL-2 diphtheria toxin fusion protein (ONTAK) significantly decreases Treg cells in peripheral blood (Cheung et al., 2019). Treg cells in tumors may have a unique chemokine receptor expression pattern that can be used for therapeutic purposes. CCR4 is highly expressed on Treg cells and other CD4+ T cell subsets, is essential for Treg cell migration, and its ligands are expressed by a variety of TME cells (Faget et al., 2011). Further comparisons of Treg cells from malignant and normal tissues showed that CCR8 is a more specific marker for tumor-resident Treg cells than CCR4 and has a much more limited expression pattern (Magnuson et al., 2018). Similarly, CCR4 antagonists could be used to improve cellular and humoral immune responses to antigens from *Mycobacterium, Haematopinus suis*, HBV, and *Plasmodium.* High cell surface expression of molecules like PD-1, ICOS, TIGIT, LAG3, TIM3, TNFR2, and 4-1BB could be used to antibody development for Treg-specific targeting (Bour-Jordan & Bluestone, 2009). Inhibition of the PD1PDL1/PDL2 pathway increases the IFN reaction to *M. tuberculosis in vitro* (Stephen-Victor et al., 2015) (Table 8.5).

8.4 CHECKPOINT INHIBITORS

This type of immunotherapy blocks the proteins that serve as the brake for the immune system for fighting against the invading pathogens or tumors. During chronic infections and cancer, the immune system is manipulated to stop and/or slow down its functions so that pathogen/tumor can establish a niche for its survival. Checkpoint inhibitors are designed to release the brakes on the immune system and enhance the existing immune responses to promote the elimination of cancer cells and/or pathogens. In 2011, FDA approved the first immune checkpoint inhibitor therapy for treating cancer – the CTLA4-blocking antibody ipilimumab. Many other checkpoint inhibitors are currently under clinical trial or in different phases of pre-clinical trials alone and in combination with other treatment regimens. These molecules have revolutionized medical oncology field and the discovery of their therapeutic utility won the Nobel prize for Medicine for Prof. J. Allison and T. Honjo in 2018. Till 2020, immune checkpoint inhibitors are amongst the most successful immunotherapies developed so far.

Impairment of the T-cells functions remains the hallmark for several diseases; CTLA-4, PD-1, and other immune checkpoint molecules are known to regulate the magnitude of T-cell responses. During malaria infection, checkpoint molecules are upregulated on T-cells that induces CTL dysfunctions. PD-1 deficient mice have shown to exhibit improved

TABLE 8.5 Some Current Ongoing Regulatory T-Cell Clinical Trials in Different Cancers and Infectious Diseases

Estimated Study Start Date	Estimated Study Completion Date	Project Title	Clinical Trials. gov Identifier	Phase	Subject Age (Year)	Number of Subjects	Study Location
July 1, 2020	December 31, 2022	PD-L1-Expressing Regulatory T Cells in Localized Prostate Cancer Patients Undergoing Iodine-125 Permanent Brachytherapy	NCT04369508	–	30 years to 90	20	Beijing
February 22, 2018	February 2022	Allogeneic Immunotherapy for Hematological Malignancies by Selective Depletion of Regulatory T Cells (DLI-Boost)	NCT03236129	Phase III	Child, adult, older adult	–	France
July 8, 2019	May 31, 2024	A Phase 1 Trial of CD25/Treg-depleted DLI Plus Ipilimumab for Myeloid Disease Relapse After Matched-HCT	NCT03912064	Phase I	18 years and old	25	United States
May 5, 2020	April 2021	Effector and Regulatory T-Cell Receptor Repertoire Analyzes in Patients Affected by COVID-19 (CovRep)	NCT04379466	–	18 years to 75	60	France
March 2021	September 2022	Safety of T-Regulatory Cell Therapy in Subjects With COVID-19 Induced Acute Respiratory Distress Syndrome	NCT04737161	Phase I	18 years to 75	20	United States

Source: Assess at ClinicalTrials.gov as of March 22, 2021.

parasitic clearance and monoclonal antibody-based therapy targeting PD-1 accelerated the reduction of parasitemia (Butler et al., 2011). In contrast, anti-CTLA4 and anti-PD-L1 antibody are known to predispose the *Plasmodium berghei* infected mice to fatal cerebral malaria (Hafalla et al., 2012). This indicated for the varying tissue-specific effects of checkpoint inhibition therapy. A similar increase of the checkpoint molecules expression has been observed in visceral leishmaniasis (VL), where these molecules abrogate the long-term parasite control by blocking the CTL functions. PD-L1 blockade is known to improve the CTL survival and functions by inducing the parasite clearance from the spleen. Similar effects were observed with anti-CTLA-4 therapy, which markedly reduced the parasitic burdens in the liver and spleen in mice model (Joshi et al., 2009; Singh et al., 2019). Cutaneous form of leishmaniasis is also known to upregulate the expression of PD-1, which impairs the IFN-γ production, proliferation of TH1 cells while, its blockade is associated with reversal of the adverse immunological outcomes (Mou et al., 2013).

Anti-PD-1 therapy in macaques, is known to enhance the antigen-specific CTL responses, viral clearance, improved survival and attenuated the excessive immune activation (Dyavar Shetty et al., 2012; Mylvaganam et al., 2018). The therapy was known to enhance the viral control after the therapy was discontinued. The use of checkpoint therapy for treating HIV-infected malignancy cases reported for its safety without interfering efficacy of the antiviral therapy (Chang et al., 2018). In case of hepatitis B infection (HBV), there is a strong association between the polymorphism in checkpoint molecules has been found to be associated with the viral clearance and clinical outcomes (Chihab et al., 2018). Checkpoint molecules are usually upregulated during acute HBV infection and declines once the viral load is cleared (Zhang et al., 2008). Conversely, during chronic infection, checkpoint molecules are upregulated, monoclonal antibody-based blockade of these molecules substantially improves the antibody production, CTL proliferation and their functions (Fisicaro et al., 2010; Schurich et al., 2011; Salimzadeh et al., 2018). Similar to HBV, hepatitis C virus (HCV) clearance is also regulated by the polymorphisms in the checkpoint molecules that eventually determines the clinical outcomes (Xiao et al., 2015; Sepahi et al., 2017). High levels of PD-1 have been observed on the CTLs that possessed exhausted T-cells phenotype and were not able to clear the infection, blockade of PD-1 was found to relieve the intrahepatic T-cell exhaustion (Golden-Mason et al., 2007; Nakamoto et al., 2008).

Similarly, blockade of CTLA-4 was found to significantly improve the infection outcomes (Sangro et al., 2013). Checkpoint molecules are also known to be induced during several acute respiratory viral infections; however, it does not always result in T-cell exhaustion. Administration of anti-PD-1 monoclonal antibody to influenza infected mice, significantly improved the CTL number and virus clearance (Rutigliano et al., 2014).

In bacterial infection, during gastrointestinal *H. pylori* infection, monoclonal antibody therapy targeting CTLA-4 has been known to reduce gastric inflammation without reducing the bacterial load (Watanabe et al., 2004; Lee et al., 2010). Conversely, PD-1 deficient mice have shown potential improvement in the survival rates (reduced mortality) and bacterial clearance abilities (Yao et al., 2009). PD-1 expression is also known to be associated with reduced proliferative abilities of T-cells, poor T-cell infiltrations and impaired effector functions which can be reversed by blocking PD-1/PD-L1 (Day et al., 2018). Anti-CTLA-4 based therapy has also been efficacious in improving the T-cells functions without affecting the granuloma formation or IFN-γ expression (Kirman et al., 1999). Checkpoint molecules (PD-1 and CTLA-4) have also been important determinants of lepromatous immunological responses. These are upregulated on cutaneous T-cells in lepromatous and tubercular leprosy (Palermo Mde et al., 2012). Since these infections are poorly responsive to antimicrobial therapies; therefore, novel therapeutic strategies like checkpoint blockade can serve to open ways for disease management.

Conventional anti-fungal therapies have been used for fungal infections, however, histoplasmosis (caused by *Histoplasma capsulatum*) affecting the advanced AIDS cases and SOT exhibited marked enhancement in the expression of PD-1 (Lazar-Molnar et al., 2008). Genetic deficiency of PD-1 in mice model was found to significantly improve the survival rates by inducing fungal clearance, reducing airspace edema and necrosis. The mortalities due to invasive candidiasis is significantly higher (Pappas et al., 2018), anti-CTLA-4, anti-PD1 and anti-PD-L1 monoclonal antibody therapies have been shown to markedly improve the survival rates. The checkpoint molecule-based therapy upregulates the MHC-II expression by APCs and IFN-γ production (by T-cells) that eventually skews the immune response to TH1 type immunity (Chang et al., 2013; Roussey et al., 2017). Unfortunately, excessive TH1 type immune responses are associated with the development of immune reconstitution inflammatory syndrome (IRIS) (Balasko & Keynan, 2019) therefore, future studies warrant the understanding of the impact of therapy and development of adverse events. Overall early pre-clinical and clinical success of checkpoint inhibitor therapy in different diseases has paved the path for adapting the checkpoint-based therapy for management of different pathologies.

8.5 AUTOLOGOUS IMMUNE ENHANCEMENT THERAPY (AIET)

Autologous immune enhancement therapy (AIET) is the type of immunotherapy where the cells from the patients' blood are isolated, cultured, and processed to activate them until their resistance to fight against the disease/cancer is enhanced. The cultured cells are then added back in the patient to immune system to clear off the infection/cancer. The cells, antibodies, and organs of the immune system defend the body against invading micro-organisms and cancer. AIET was first used for the treatment of cancer, in the late 1980s, when Rosenberg et al. reported that the immunotherapy lowered the tumor regression rates (2.6–3.3%) in metastatic cancer cases (Rosenberg, 1984). AIET along with chemotherapy/radiotherapy was considered as one of the biggest revolutions in the field of cancer management. Till date, different strategies of autologous and allogenic immune cells, NK cells, CTLs, DCs, and genetically manipulated autologous as well as allogenic immune cells have been widely used in the immunotherapeutics. The current technology of AIET uses the autologous NK cells and activated T-cells for treating cancers. Though the treatment concept emerged in 1980s, but it was started to be used since early 1990s, after several successful clinical trials in gastric, lung, ovarian, and liver cancers (Egawa, 2004). When this cell-based immunotherapy is combined with the conventional therapeutic approaches, it showed marked improvement in the treatment efficacy by 20–30% and it was also efficacious for treating relapse cases (Egawa, 2004; Manjunath et al., 2012).

Adoptive transfer of tumor infiltrating T-cells (TILs) cultured in the presence of recombinant IL-2 induced the lysis of autologous tumors, improved the survival rates, and regressed the tumors in a variety of cancers (1483950). TILs are also known to induce cell-mediated immune responses by immune activation reactions in tumor cases (Rosenberg, 1991). AIET with IL-2 and lymphokine activated killer (LAK) cells has been used for treating the malignancies, when used in combination with chemotherapy/radiotherapy improved the survival in lung carcinoma cases

(Kimura & Yamaguchi, 1997). Another piece of evidence, reported for the adoptive transfer of TILs after chemotherapy as an advantageous approach for the complete cure of the ovarian cancer cases (Fujita et al., 1995). Adoptive immunotherapy has also been shown to be safe and potential treatment option for reducing the post-surgical recurrence and improving the recurrence free surgical outcomes in hepatocellular carcinoma cases (Takayama et al., 2000). Similarly, AIET with tumor-associated lymphocytes in combination with chemotherapy (cisplatin/5-fluorouracil) effectively prolonged the survival in stage IV gastric cancer cases (Kono et al., 2002). AIET has mostly been used for cancer management, it has also been reported in infectious diseases however, in majority of clinical situations, the host immune responses take the charge and resolve the infection, and therefore, AIET is generally not the preferred treatment choice. Though AIET cannot be the preferred option for viral disease management however, there are clinical scenarios (solid organ and BM transplantation) where complete ablation/suppression of host T-cell immunity is needed for preventing the allograft rejection. This increases the risk for reactivation of persistent viruses such as cytomegalovirus (CMV) and Epstein Barr virus (EBV), which can have life threatening consequences in patients' (Barnes & Stallard, 2001; Sivaraman & Lye, 2001). AIET has shown potential applications for controlling viral diseases, EBV-specific T-cells have shown an important role in the resolution of infection and long-term viral control. Adoptive transfer of T-cells in immunosuppressed patients restored the protective immune functions and polyclonal T-cells populations which eventually was proved beneficial for treating virus-induced lymphoproliferative syndromes (Reddehase et al., 1987; Riddell et al., 1992; Papadopoulos et al., 1994).

CD25 immunotoxin-based depletion of T-cells *ex vivo* and allo-depleted haplodeficient donor cells has served to provide momentum to the anti-viral immune responses in the recipients (Andre-Schmutz et al., 2002; Amrolia et al., 2006). AIET trial of infusion of adenoviral specific CD4+ and CD8+ T-cells in adenovirus infected children after hemopoietic stem cell therapy led to sustained decline in the viral loads and eventually resolution of infection (Feuchtinger et al., 2006). Therefore, AIET based therapy has now found widespread application in different diseases and opens the ways for application of similar strategies for other infectious diseases.

8.5.1 NON-SPECIFIC IMMUNOTHERAPY

Cytokines are essential regulators of the immune system that alters the properties of cells to determine the infection outcomes. These are produced by leukocytes and act other leukocytes; therefore, they are also referred to as the interleukins (ILs). Interferons are another class of signaling molecules which are released in response to invading pathogens and help in their elimination from the body. Interferons are named after their ability to interfere with the viral replication and fight against the pathogens (Nora et al., 1975). Cytokines have been developed as the protein therapies as a means to boost host immune responses and circumvent infections.

8.5.2 INTERFERONS

Cytokines are known modifiers of cellular proliferation and differentiation. The synergy and antagonism between different cytokines remain the key determinant of the immunological outcomes. The story of cytokines dates back to last century when interferons were discovered. Interferons are categorized into three classes: α-interferons, β-interferons, and γ-interferons. Interferon-α (IFN-α) and β are class I, i.e., virus-induced interferons, which shares several structural and biological properties and compete for binding to the same receptor. IFN-α are produced by leukocytes in response to viruses or double stranded RNA. It was one of the earliest cytokines, known to exert anti-tumor activity that was further explored in several pre-clinical and clinical trials during the 1980s and 1990s (Gresser & Bourali, 1970). There are two recombinant forms of human IFN-α that are approved by FDA (in 1995) for licensed use in clinics: IFN-α2a (Roferon-A or Sylatron) and IFN-α2b (Intron-A). IFN-α is the most commonly used interferons for the treatment of more than 14 different types of cancers, including hematological malignancies and solid tumors (as melanoma, Kaposi's sarcoma, etc.) (Pfeffer et al., 1998). Before the discovery of imatinib, IFN-α was the standard therapy for chronic myeloid leukemia (CML). Later research focused on improving the efficacy of imatinib by combining with pegylated-IFN-α, that yielded significant improvement in the treatment of the cancer cases (Simonsson et al., 2011; Johnson-Ansah et al., 2013; Talpaz et al., 2013).

Clinical trial for IFN-α2a significantly improved the survival rates of high-risk melanoma patients and was found to be beneficial as compared to ganglioside vaccine and it was also found effective for treating inoperable melanomas (Livingston et al., 1994; Creagan et al., 1995). The existing cancer treatment are associated with a high risk of relapses and mortality that accentuated the need for adjunctive therapy. High dose of IFN-α2b based therapy has also been tested in several clinical trials for improving the survival rates and promoting the relapse free survival in cancer cases (Livingston et al., 1994; Kirkwood et al., 1996). The high-dose treatment was associated with significant toxicity; therefore, low-dose interferon therapy was tested as an adjunctive option for treating high-risk melanoma cases. Unfortunately, the trial reported for no significant improvements in the relapse free survival rates of the melanoma cases (Pehamberger et al., 1998). Recombinant IFN-α has also been used for viral diseases; in HCV treatment, pegylated IFN-α was introduced in the clinics in 2001, that required a single weekly injection rather than thrice a week. Pegylated IFN-α was used for some time as the standard therapy for chronic HCV infection in combination with ribavirin (Zeuzem et al., 2003; Aghemo et al., 2010). Unfortunately, severe side effects associated with the therapy brought up other interferons as albumin IFN-α, IFN-λ and consensus IFN (CIFN) as potential options.

Similar to HCV, pegylated IFN-α2a was shown to be less successful for HBV with development of relapses in 5–10% of cases. Since the side effects can easily be managed, therefore, despite of the potential side effects, it remains the standard of care for the HBV infections (Huang et al., 2013; Wang et al., 2014).

Albumin IFN-α based phase III trial raised concerns regarding its safety which led to its failure to gain FDA approval (Balan et al., 2006; Zeuzem et al., 2010). CIFN is a recombinant protein that contains the common amino acids from different type I interferons. It has shown higher anti-viral ability than the conventional IFNs thus, it is required in low dose that overcomes the potential side effects associated with higher doses. Though useful in low doses, its short half-life, rapid absorption, and fast elimination from urine (similar to IFN-α) called upon for daily injections, however, successful phase III trials won it approval for preferred treatment option (Sjogren et al., 2007; Bacon et al., 2009).

Sub-cutaneous injections of IFN-α2b have also been used in combination with the standard lopinavir for management of novel coronavirus infection (COVID-19) that shortened the hospitalization duration and accelerated the viral clearance and recovery in the cases (Pereda et al., 2020).

IFN-β is produced by fibroblasts and epithelial cells. It shares approximately 25% amino acid sequence homology with IFN-α. It is mostly used for the treatment of multiple sclerosis and as an anti-viral therapy (Ahn et al., 2009; Inoue et al., 2009). The efficacy of IFN-β alone or in combination with standard therapeutic regimen has been reported in HCV infection and for patients who are poorly responsive to IFN-α2b/ribavirin treatment alone (Ishikawa et al., 2012). Another study reported for the efficacy of two weeks of IFN-β/ribavirin treatment followed by pegylated IFN-α/ribavirin treatment that significantly reduced the viral loads as compared to the cases who were treated with pegylated IFN-α/ribavirin alone. Furthermore, IFN-β is known to lower the incidence of treatment-associated adverse events as compared to IFN-α alone (23173698). Immunotherapy aimed to block the type I interferon on inflammatory cells is in the early phase of development, likewise, type I IFN blocking antibody has been used in systemic lupus erythematosus (SLE) cases (Yao et al., 2009). Thus, IFN-α and IFN-β are prototypic immunotherapies that has remarkably improved the clinical outcomes in cancer cases due to their ability to control diverse cellular events. Recent research has elucidated their role as therapeutics however, further insight into their role in regulating the immunological outcomes is warranted.

The second class of cytokine, IFN-γ is structurally and functionally distinct from other interferons. It is produced by the T-cells activation, in response to mitogenic stimulation or non-specifically in response to IL-2. IFN-γ is one of the critical cytokine that determines the success of immunotherapy, in response to several immune checkpoint therapies IFN-γ and signature genes (HLA-DR, IDO1, STAT1, CXCL10, CD274, CXCL9) are the read-outs for the successful therapeutic outcomes in different cancers (Higgs et al., 2018; Karachaliou et al., 2018; Ni & Lu, 2018). The combination of checkpoint therapy with IFN-γ has shown to improve the efficacy of cancer immunotherapy. The combination of nivolumab or pembrolizumab along with IFN-γ (NCT02614456 and NCT 03063632) is still ongoing (Mojic et al., 2017).

IFN-γ has direct anti-tumor effects due to its anti-angiogenic, pro-apoptotic, and immune stimulatory abilities (Pujade-Lauraine et al., 1996; Wall et al., 2003; Marth et al., 2006). It has been used in combination with chemotherapy to induce apoptosis in ovarian cancer cells and it has shown to improve the response rates (from 56% to 68%) and prolong the progression free survival rates in cases (Windbichler et al., 2000). It has also been under trial for its use as an adjuvant for vaccine and chemotherapies (NCT00428272, NCT00499772, NCT00824733, NCT00004016).

In a targeted clinical trial using IFN-γ in the form of aerosol for tuberculosis cases resulted in a marked decline in the mycobacterial burden in the sputum, increased body mass index (BMI) and reduced cavitary lesion sizes. IFN-γ based adjunctive therapy has shown to activate the signal transduction and gene expression in alveolar macrophages and improve the disease related clinical outcomes (Condos et al., 1997; Suarez-Mendez et al., 2004). In chronic HBV infection, IFN-γ1b alone has no impact on viral infection but exerts an immunomodulatory effect (Lau et al., 1991). Similar findings were reported for HCV infection (Muir et al., 2006), but the pre-treatment with IFN-γ prior to IFN-α treatment significantly enhanced immunological responsiveness in HCV cases which could be attributed to the interferon mediated enhanced viral clearance (Saez-Royuela et al., 1991; Katayama et al., 2001). Adeno-IFN-γ (non-replicating adenovirus with IFN-γ cDNA insert) has been tested in cutaneous lymphoma cases, after the success of phase I trial, the drug has now been tested under phase II trial (Dummer et al., 2004; Dummer, 2005).

Anti-IFN-γ based monoclonal antibody treatment (Fontolizumab and AMG 811) has also been used for management of multiple sclerosis (Skurkovich et al., 2001; Skurkovich & Skurkovich, 2003), inflammation associated with Crohn's disease (Hommes et al., 2006), rheumatoid arthritis (Sigidin et al., 2001) and lupus (Machold & Smolen, 1990; Theofilopoulos et al., 2001).

Usually, interferon therapy is an immunoregulatory strategy that affects a vast variety of immune cells, and their clinical application is yet to be explored for the management of different diseases.

8.5.3 INTERLEUKINS (ILS)

Recent advances in understanding the immune responses have encouraged newer strategies to come over the horizon for fighting against different diseases. Immunity against invading micro-organisms can be enhanced by inducing the expression of cytokines that stimulates the immune activation and differentiation of immune cells to fight against invading microorganisms or suppressing the negative regulatory cytokines. Different recombinant interleukins (ILs) have been undergoing trials while others are under development for potentiating the immune responses by improving the antigen priming and orchestrating the pharmacological properties of the chemotherapy for better immunological outcomes.

IL-7 is an important cytokine in reinforcing the immune system, which can restore normal T-cells population that wanes naturally over the period of time. It can be used in patients who have undergone thymic involution (thymus becomes less efficient in T-cell generation), chemotherapy, HIV cases and BM transplantation. IL-7 is known trigger for the expansion of different T-cells and increases overall T-cell repertoire diversity (Sportes et al., 2008; Parrish et al., 2009; Levy et al., 2012). Unlike, naïve and memory cells, regulatory T-cells, effector memory cells and senescent T-cells minimally expand in response to the therapy emphasizing for the fact that IL-7 enhances the T-cells reactivity against chronic infections (Rosenberg et al., 2006). Increasing dose of recombinant IL-7 is well-tolerated, with non-significant adverse events associated with the therapy in the clinical trial (Rosenberg et al., 2006; Sportes et al., 2008, 2010). It has been tested in cancers, idiopathic lymphopenia, chronic viral infections, immunodeficiencies, and/or BM transplantation. It has also been tested as adjuvant for improving the vaccine, adoptive cell transfer therapies and chemotherapeutic efficacy (Mackall et al., 2011).

IL-2 is another crucial cytokine that has been widely used for different diseases including chronic myelogenous leukemia (CML) cases after HSCT, where low dose of IL-2 with Treg infusion significantly improved the clinical effects of HSCT (Pulster et al., 2009). IL-2 based therapy for the management of HCV related vasculitis was not associated with any adverse events and it significantly improved the vasculitis and attenuated the signatures associated with inflammation and oxidative stress (Saadoun et al., 2011). Administration of IL-2 has also raised promising results for the treating several forms of cancers (Chavez et al., 2009), graft-versus-host disease (GVHD) (Koreth et al., 2011; Kennedy-Nasser et al., 2014), SLE (Mizui & Tsokos, 2016; Humrich & Riemekasten, 2019) and type 1 diabetes (T1D) (Long et al., 2013; Dwyer et al., 2016).

IL-15, produced by myeloid cells has immune potentiating effects and has shown anti-tumor activity by inducing the expansion of NK cells and CD8+ T-cells in head and neck cancer, renal cell carcinoma (RCC) and melanomas (Miller et al., 2018) however, the toxicity associated with the treatment led to its limited its use as a therapeutic agent. It has also been thought of as a possible alternative for rheumatic disorders and Kawasaki disease (McInnes & Gracie, 2004).

IL-21, relatively newcomer in the field of cytokine has also shown potential anti-tumor effects in early trials in humans. It has been tested across different cancer lines including RCC, melanoma, and breast cancer, adenocarcinoma, etc. Similar to the IL-15, it mediates the antitumor effects via NK cells (mediated by antibody dependent cell mediated cytotoxicity and cytokine production) and CD8+ T-cells. The effects were further enhanced when it was used in combination with monoclonal antibody therapy (as Rituxan) (Bhave & Carson, 2009).

IL-1β is another cytokine that has a crucial role to play in inflammatory disorders. Recombinant IL-1 receptor antagonist (Anakinra) has been an FDA approved drug for treating rheumatoid arthritis. It has been used for NOMID cases (where blockade of receptor helped in restoring hearing and vision) thus prevents organ specific damage and disability (Goldbach-Mansky, 2009).

Ustekinumab, anti-IL-12 monoclonal antibody has been used for inhibiting inflammatory cell infiltration and epidermal hyperplasia in psoriasis cases and has recently been approved by FDA. Apart from chronic psoriasis, it has also been used for treating the psoriatic arthritis and Crohn's disease (20074279). Systemic IL-12 administration in cancer cases led to systemic toxicity and remained fatal in some cases, thus, ruled its use in cancer treatment (Brunda et al., 1993; Jenks, 1996); however, recent use of nanogold delivery system has been used for improving the therapeutic index (Gasparri et al., 2019).

These are just to name a few; however, different cytokine immuno-therapies have been used for treating different diseases and several other therapies are under pre-clinical and clinical trials.

8.5.4 RISKS OR SIDE EFFECTS ASSOCIATED WITH IMMUNOTHERAPY

Immunotherapy has been the best choice for certain diseases as cancer; however, potential risks associated with the therapy limits its usage. The time duration of manifestation of side effects varies but mostly the patients' complaints of potential problems in first weeks to months.

In case of checkpoint-based therapy, the overactive T-cells are the cause for the side effects seen as tiredness, dry and itchy skin, diarrhea, breathlessness, or dry cough due to lung inflammation. Usually, the side effects depend on the type of immunotherapy; top side effect associated with immunotherapy is inflammation and fatigue. Additionally, immunotherapies are associated with development of flu like symptoms including fever, chills, weakness, dizziness, nausea, sinus congestion, low/high blood pressure and fatigue. Conversely, pain, swelling, redness, itchiness, rash, diarrhea, heart palpitation, stuffy head and weight gain due to deposition of extra fluids in the different parts of the body can be observed in many treated cases. However, the symptoms subsides and/or ease up after the first treatment dose. The potential side effects associated with immunotherapy can also manifest after its usage in combination with other drugs leading to serious pathological outcomes including diabetes and inflammatory arthritis. Some immunotherapies have been to be associated with severe or even fatal allergic and inflammatory reactions however, the occurrence is rare. Serious outcomes can be controlled by using steroid and other treatment options that potentially helps in inhibiting the signals on cells that causes damage (Michot et al., 2016; Spain et al., 2016). Furthermore, potential side effects accentuate the need for research to avoid these issues.

8.5.5 MANAGEMENT OF SIDE EFFECTS OF IMMUNOTHERAPY

Early reporting of the potential side effects remains the key for its management, if left untreated the symptoms may get worse and even be life threatening in some cases. With proper treatment the side effects wanes in one to three weeks. However, it is important to get monitored for any symptoms of unwellness before self-medication. Usually, the side effects are characterized by the grading system on the scale of 1–4 that greatly facilitates its management, depending on the severity. Mild to severe side effects are graded in the range of 2 to 4, that can be managed with the steroids such as prednisone. The management of severe side effects (with a grading score of 3–4) relies on immediate hospitalization of the patient for intravenous steroid treatment/other medicines and immediately terminating the immunotherapy till the resolution/control of the side effects. Conversely, in case of the persistence of severe side effects, it is usually recommended to permanently terminate the immunotherapy. It is assumed that the previous immunotherapy doses can "train the immune system" to recognize the cancer cells and fight against them. Despite the risks associated with severe side effects, most of the patients on immunotherapy experiences only mild side effects. Guidelines are available for managing the immunotherapy associated adverse events. This includes the laboratory tests for adrenocorticotrophic hormone (ACTH), cortisol, thyroid profiles, and testosterone in males to identify endocrine dysregulations. If hypophysitis is suspected, with headache and/or visual symptoms, then magnetic resonance imaging (MRI) scans are recommended. If adrenal anomalies (dehydration, hypotension) are suspected, then administration of stress-dose steroids and hormone replacement therapy is recommended in an inpatient setting along with a delay/discontinuation of therapy (Weber et al., 2016).

Dehydration, hypotension, and/or shock out of proportion to the current illness, administer stress-dose steroids in an inpatient setting, delay or discontinue treatment with anti-PD-1 therapy, and consider hormone-replacement therapy. Hypothyroidism can also be preceded by the signs and symptoms of hyperthyroidism over a period of days to weeks, which mimics the clinical presentation of Hashimoto's thyroiditis.

Guidelines are available on how to appropriately manage endocrine AEs following immune checkpoint inhibition. Laboratory testing for adrenocorticotropic hormone (ACTH), cortisol, T4, thyroid-stimulating hormone, and testosterone in males, may help identify endocrine dysfunction. If hypophysitis is suspected, especially with a headache or visual symptoms, consider a MRI scan of the brain with pituitary cuts and visual field testing. Treatment with immune checkpoint inhibition may continue with low-grade endocrine toxicity, although patients should be closely monitored.

8.6 CONCLUSION

The main challenge faced in the field of immunotherapy remains to be the low rates of participation for clinical trials, only 3% of cancer cases willingly participate while 40% of trials fails to achieve minimum number of enrollments for their trials. Secondly, the high cost (approx. 1 million per patient per year) of the treatment restricts its application in underdeveloped nations. Thirdly, the therapy is successful for a certain number of cases while others fail to respond to the treatment; overall, the success rates of participation remain to be 15–20%. Unfortunately, it has still been an enigma to dissect who will respond to the therapy and who won't respond. Thus, these challenges call upon for much-needed effort for successfully establishing immunotherapy as a first-line treatment option.

KEYWORDS

- **adoptive cell transfer**
- **autologous immune enhancement therapy**
- **biological response modifiers**
- **cytomegalovirus**
- **hemagglutinin**
- **immunotherapy**
- **monoclonal antibody**

REFERENCES

Adams, G. B., Chabner, K. T., Alley, I. R., Olson, D. P., Szczepiorkowski, Z. M., Poznansky, M. C., Kos, C. H., et al., (2006). Stem cell engraftment at the endosteal niche is specified by the calcium-sensing receptor. *Nature, 439*(7076), 599–603.

Aerts-Toegaert, C., Heirman, C., Tuyaerts, S., Corthals, J., Aerts, J. L., Bonehill, A., Thielemans, K., & Breckpot, K., (2007). CD83 expression on dendritic cells and T cells: Correlation with effective immune responses. *Eur. J. Immunol., 37*(3), 686–695.

Aghemo, A., Rumi, M. G., & Colombo, M., (2010). Pegylated interferons alpha2a and alpha2b in the treatment of chronic hepatitis C. *Nat. Rev. Gastroenterol Hepatol., 7*(9), 485–494.

Ahn, S. H., Lee, H. W., Kim, Y. S., Kim, J. K., Han, K. H., Chon, C. Y., & Moon, Y. M., (2009). Recombinant interferon-Beta-1alpha plus ribavirin for the treatment of chronic HCV infection: A prospective, randomized, comparative pilot study. *Gut Liver, 3*(1), 20–25.

Alberer, M., Gnad-Vogt, U., Hong, H. S., Mehr, K. T., Backert, L., Finak, G., Gottardo, R., et al., (2017). Safety and immunogenicity of a mRNA rabies vaccine in healthy adults: An open-label, non-randomized, prospective, first-in-human phase 1 clinical trial. *Lancet, 390*(10101), 1511–1520.

Allard, S. D., De Keersmaecker, B., De Goede, A. L., Verschuren, E. J., Koetsveld, J., Reedijk, M. L., Wylock, C., et al., (2012). A phase I/IIa immunotherapy trial of

HIV-1-infected patients with Tat, Rev and Nef expressing dendritic cells followed by treatment interruption. *Clin. Immunol., 142*(3), 252–268.

Amrolia, P. J., Muccioli-Casadei, G., Huls, H., Adams, S., Durett, A., Gee, A., Yvon, E., et al., (2006). Adoptive immunotherapy with allodepleted donor T-cells improves immune reconstitution after haploidentical stem cell transplantation. *Blood, 108*(6), 1797–1808.

Andre-Schmutz, I., Le Deist, F., Hacein-Bey-Abina, S., Vitetta, E., Schindler, J., Chedeville, G., Vilmer, E., et al., (2002). Immune reconstitution without graft-versus-host disease after haemopoietic stem-cell transplantation: A phase 1/2 study. *Lancet, 360*(9327), 130–137.

Auletta, J. J., Deans, R. J., & Bartholomew, A. M., (2012). Emerging roles for multipotent, bone marrow-derived stromal cells in host defense. *Blood, 119*(8), 1801–1809.

Bacchetta, R., Gambineri, E., & Roncarolo, M. G., (2007). Role of regulatory T cells and FOXP3 in human diseases. *J. Allergy Clin. Immunol., 120*(2), 227–235; quiz 236–227.

Bacon, B. R., Shiffman, M. L., Mendes, F., Ghalib, R., Hassanein, T., Morelli, G., Joshi, S., et al., (2009). Retreating chronic hepatitis C with daily interferon alfacon-1/ribavirin after nonresponse to pegylated interferon/ribavirin: DIRECT results. *Hepatology, 49*(6), 1838–1846.

Bahl, K., Senn, J. J., Yuzhakov, O., Bulychev, A., Brito, L. A., Hassett, K. J., Laska, M. E., et al., (2017). Preclinical and clinical demonstration of immunogenicity by mRNA vaccines against H10N8 and H7N9 influenza viruses. *Mol. Ther., 25*(6), 1316–1327.

Balan, V., Nelson, D. R., Sulkowski, M. S., Everson, G. T., Lambiase, L. R., Wiesner, R. H., Dickson, R. C., et al., (2006). A Phase I/II study evaluating escalating doses of recombinant human albumin-interferon-alpha fusion protein in chronic hepatitis C patients who have failed previous interferon-alpha-based therapy. *Antivir. Ther., 11*(1), 35–45.

Balasko, A., & Keynan, Y., (2019). Shedding light on IRIS: From pathophysiology to treatment of cryptococcal meningitis and immune reconstitution inflammatory syndrome in HIV-infected individuals. *HIV Med., 20*(1), 1–10.

Barnes, R. A., & Stallard, N., (2001). Severe infections after bone marrow transplantation. *Curr. Opin. Crit. Care 7*(5), 362–366.

Batich, K. A., Reap, E. A., Archer, G. E., Sanchez-Perez, L., Nair, S. K., Schmittling, R. J., Norberg, P., et al., (2017). Long-term survival in glioblastoma with cytomegalovirus pp65-targeted vaccination. *Clin. Cancer Res., 23*(8), 1898–1909.

Belkaid, Y., & Rouse, B. T., (2005). Natural regulatory T cells in infectious disease. *Nat. Immunol., 6*(4), 353–360.

Belkaid, Y., (2007). Regulatory T cells and infection: A dangerous necessity. *Nat. Rev. Immunol., 7*(11), 875–888.

Bhave, N. S., & Carson, W. E. 3rd., (2009). Immune modulation with interleukin-21. *Ann. N Y Acad. Sci., 11, 82,* 39–46.

Bian, P., Ye, C., Zheng, X., Yang, J., Ye, W., Wang, Y., Zhou, Y., et al., (2017). Mesenchymal stem cells alleviate Japanese encephalitis virus-induced neuroinflammation and mortality. *Stem Cell Res. Ther., 8*(1), 38.

Boczkowski, D., Nair, S. K., Snyder, D., & Gilboa, E., (1996). Dendritic cells pulsed with RNA are potent antigen-presenting cells *in vitro* and *in vivo. J. Exp. Med., 184*(2), 465–472.

Bogers, W. M., Oostermeijer, H., Mooij, P., Koopman, G., Verschoor, E. J., Davis, D., Ulmer, J. B., et al., (2015). Potent immune responses in rhesus macaques induced by nonviral delivery of a self-amplifying RNA vaccine expressing HIV type 1 envelope with a cationic nanoemulsion. *J. Infect. Dis., 211*(6), 947–955.

Bonehill, A., Tuyaerts, S., Van, N. A. M., Heirman, C., Bos, T. J., Fostier, K., Neyns, B., & Thielemans, K., (2008). Enhancing the T-cell stimulatory capacity of human dendritic cells by co-electroporation with CD40L, CD70 and constitutively active TLR4 encoding mRNA. *Mol. Ther., 16*(6), 1170–1180.

Bontkes, H. J., Kramer, D., Ruizendaal, J. J., Meijer, C. J., & Hooijberg, E., (2008). Tumor associated antigen and interleukin-12 mRNA transfected dendritic cells enhance effector function of natural killer cells and antigen specific T-cells. *Clin. Immunol., 127*(3), 375–384.

Bour-Jordan, H., & Bluestone, J. A., (2009). Regulating the regulators: Costimulatory signals control the homeostasis and function of regulatory T cells. *Immunol. Rev., 229*(1), 41–66.

Brunda, M. J., Luistro, L., Warrier, R. R., Wright, R. B., Hubbard, B. R., Murphy, M., Wolf, S. F., & Gately, M. K., (1993). Antitumor and antimetastatic activity of interleukin 12 against murine tumors. *J. Exp. Med., 178*(4), 1223–1230.

Butler, N. S., Moebius, J., Pewe, L. L., Traore, B., Doumbo, O. K., Tygrett, L. T., Waldschmidt, T. J., et al., (2011). Therapeutic blockade of PD-L1 and LAG-3 rapidly clears established blood-stage plasmodium infection. *Nat. Immunol., 13*(2), 188–195.

Chang, E., Sabichi, A. L., Kramer, J. R., Hartman, C., Royse, K. E., White, D. L., Patel, N. R., et al., (2018). Nivolumab treatment for cancers in the HIV-infected population. *J. Immunother., 41*(8), 379–383.

Chang, J. C., (2016). Cancer stem cells: Role in tumor growth, recurrence, metastasis, and treatment resistance. *Medicine (Baltimore), 95*(1 Suppl 1), S20–S25.

Chang, K. C., Burnham, C. A., Compton, S. M., Rasche, D. P., Mazuski, R. J., McDonough, J. S., Unsinger, J., et al., (2013). Blockade of the negative co-stimulatory molecules PD-1 and CTLA-4 improves survival in primary and secondary fungal sepsis. *Crit. Care, 17*(3), R85.

Chavez, A. R., Buchser, W., Basse, P. H., Liang, X., Appleman, L. J., Maranchie, J. K., Zeh, H., De Vera, M. E., & Lotze, M. T., (2009). Pharmacologic administration of interleukin-2. *Ann. N Y Acad. Sci., 11, 82*, 14–27.

Cheung, L. S., Fu, J., Kumar, P., Kumar, A., Urbanowski, M. E., Ihms, E. A., Parveen, S., Bullen, C. K., Patrick, G. J., Harrison, R., Murphy, J. R., Pardoll, D. M., & Bishai, W. R., (2019). Second-generation IL-2 receptor-targeted diphtheria fusion toxin exhibits antitumor activity and synergy with anti-PD-1 in melanoma. *Proc. Natl. Acad. Sci. USA, 116*(8), 3100–3105.

Chihab, H., Jadid, F. Z., Foka, P., Zaidane, I., El Fihry, R., Georgopoulou, U., Marchio, A., et al., (2018). Programmed cell death-1 3'-untranslated region polymorphism is associated with spontaneous clearance of hepatitis B virus infection. *J. Med. Virol., 90*(11), 1730–1738.

Condos, R., Rom, W. N., & Schluger, N. W., (1997). Treatment of multidrug-resistant pulmonary tuberculosis with interferon-gamma via aerosol. *Lancet, 349*(9064), 1513–1515.

Creagan, E. T., Dalton, R. J., Ahmann, D. L., Jung, S. H., Morton, R. F., Langdon, R. M. Jr., Kugler, J., & Rodrigue, L. J., (1995). Randomized, surgical adjuvant clinical trial of recombinant interferon alfa-2a in selected patients with malignant melanoma. *J. Clin. Oncol., 13*(11), 2776–2783.

Dannull, J., Nair, S., Su, Z., Boczkowski, D., DeBeck, C., Yang, B., Gilboa, E., & Vieweg, J., (2005). Enhancing the immunostimulatory function of dendritic cells by transfection with mRNA encoding OX40 ligand. *Blood, 105*(8), 3206–3213.

Day, C. L., Abrahams, D. A., Bunjun, R., Stone, L., De Kock, M., Walzl, G., Wilkinson, R. J., et al., (2018). PD-1 expression on mycobacterium tuberculosis-specific CD4 T cells is associated with bacterial load in human tuberculosis. *Front Immunol., 9*, 1995.

De Keersmaecker, B., Heirman, C., Corthals, J., Empsen, C., Van, G. L. A., Allard, S. D., Pen, J., et al., (2011). The combination of 4–1BBL and CD40L strongly enhances the capacity of dendritic cells to stimulate HIV-specific T cell responses. *J. Leukoc. Biol., 89*(6), 989–999.

Decker, W. K., & Safdar, A., (2009). Bioimmunoadjuvants for the treatment of neoplastic and infectious disease: Coley's legacy revisited. *Cytokine Growth Factor Rev., 20*(4), 271–281.

Disis, M. L., Schiffman, K., Guthrie, K., Salazar, L. G., Knutson, K. L., Goodell, V., Dela, R. C., & Cheever, M. A., (2004). Effect of dose on immune response in patients vaccinated with an her-2/neu intracellular domain protein--based vaccine. *J. Clin. Oncol., 22*(10), 1916–1925.

Dorrie, J., Schaft, N., Muller, I., Wellner, V., Schunder, T., Hanig, J., Oostingh, G. J., et al., (2008). Introduction of functional chimeric E/L-selectin by RNA electroporation to target dendritic cells from blood to lymph nodes. *Cancer Immunol. Immunother., 57*(4), 467–477.

Dummer, R., (2005). Emerging drugs in cutaneous T-cell lymphomas. *Expert Opin. Emerg Drugs, 10*(2), 381–392.

Dummer, R., Hassel, J. C., Fellenberg, F., Eichmuller, S., Maier, T., Slos, P., Acres, B., et al., (2004). Adenovirus-mediated intralesional interferon-gamma gene transfer induces tumor regressions in cutaneous lymphomas. *Blood, 104*(6), 1631–1638.

Dwyer, C. J., Ward, N. C., Pugliese, A., & Malek, T. R., (2016). Promoting immune regulation in type 1 diabetes using low-dose interleukin-2. *Curr. Diab. Rep., 16*(6), 46.

Dyavar, S. R., Velu, V., Titanji, K., Bosinger, S. E., Freeman, G. J., Silvestri, G., & Amara, R. R., (2012). PD-1 blockade during chronic SIV infection reduces hyperimmune activation and microbial translocation in rhesus macaques. *J. Clin. Invest., 122*(5), 1712–1716.

Egawa, K., (2004). Immuno-cell therapy of cancer in Japan. *Anticancer Res., 24*(5C), 3321–3326.

Encke, J., Zu Putlitz, J., & Wands, J. R., (1999). DNA vaccines. *Intervirology, 42*(2, 3), 117–124.

Faget, J., Biota, C., Bachelot, T., Gobert, M., Treilleux, I., Goutagny, N., Durand, I., et al., (2011). Early detection of tumor cells by innate immune cells leads to T(reg) recruitment through CCL22 production by tumor cells. *Cancer Res., 71*(19), 6143–6152.

Feuchtinger, T., Matthes-Martin, S., Richard, C., Lion, T., Fuhrer, M., Hamprecht, K., Handgretinger, R., et al., (2006). Safe adoptive transfer of virus-specific T-cell immunity for the treatment of systemic adenovirus infection after allogeneic stem cell transplantation. *Br. J. Haematol., 134*(1), 64–76.

Fisicaro, P., Valdatta, C., Massari, M., Loggi, E., Biasini, E., Sacchelli, L., Cavallo, M. C., et al., (2010). Antiviral intrahepatic T-cell responses can be restored by blocking programmed death-1 pathway in chronic hepatitis B. *Gastroenterology, 138*(2), 682–693, 693 e681–684.

Fleeton, M. N., Chen, M., Berglund, P., Rhodes, G., Parker, S. E., Murphy, M., Atkins, G. J., & Liljestrom, P., (2001). Self-replicative RNA vaccines elicit protection against influenza A virus, respiratory syncytial virus, and a tickborne encephalitis virus. *J. Infect. Dis., 183*(9), 1395–1398.

Fontenot, J. D., Gavin, M. A., & Rudensky, A. Y., (2003). Foxp3 programs the development and function of CD4+CD25+ regulatory T cells. *Nat. Immunol., 4*(4), 330–336.

Fujita, K., Ikarashi, H., Takakuwa, K., Kodama, S., Tokunaga, A., Takahashi, T., & Tanaka, K., (1995). Prolonged disease-free period in patients with advanced epithelial ovarian cancer after adoptive transfer of tumor-infiltrating lymphocytes. *Clin. Cancer Res., 1*(5), 501–507.

Fynan, E. F., Webster, R. G., Fuller, D. H., Haynes, J. R., Santoro, J. C., & Robinson, H. L., (1993). DNA vaccines: Protective immunizations by parenteral, mucosal, and gene-gun inoculations. *Proc. Natl. Acad. Sci. U S A, 90*(24), 11478–11482.

Gangji, V., De Maertelaer, V., & Hauzeur, J. P., (2011). Autologous bone marrow cell implantation in the treatment of non-traumatic osteonecrosis of the femoral head: Five-year follow-up of a prospective controlled study. *Bone, 49*(5), 1005–1009.

Gasparri, A. M., Sacchi, A., Basso, V., Cortesi, F., Freschi, M., Rrapaj, E., Bellone, M., et al., (2019). Boosting interleukin-12 antitumor activity and synergism with immunotherapy by targeted delivery with isoDGR-tagged nanogold. *Small, 15*(45): e1903462.

Geall, A. J., Verma, A., Otten, G. R., Shaw, C. A., Hekele, A., Banerjee, K., Cu, Y., et al., (2012). Nonviral delivery of self-amplifying RNA vaccines. *Proc. Natl. Acad. Sci. U S A, 109*(36), 14604–14609.

Goldbach-Mansky, R., (2009). Blocking interleukin-1 in rheumatic diseases. *Ann. N Y Acad. Sci., 11, 82,* 111–123.

Golden-Mason, L., Palmer, B., Klarquist, J., Mengshol, J. A., Castelblanco, N., & Rosen, H. R., (2007). Upregulation of PD-1 expression on circulating and intrahepatic hepatitis C virus-specific CD8+ T cells associated with reversible immune dysfunction. *J. Virol., 81*(17), 9249–9258.

Gresser, I., & Bourali, C., (1970). Antitumor effects of interferon preparations in mice. *J. Natl. Cancer Inst., 45*(2), 365–376.

Hafalla, J. C., Claser, C., Couper, K. N., Grau, G. E., Renia, L., De Souza, J. B., & Riley, E. M., (2012). The CTLA-4 and PD-1/PD-L1 inhibitory pathways independently regulate host resistance to plasmodium-induced acute immune pathology. *PLoS Pathog., 8*(2), e1002504.

Haque, R., Lei, F., Xiong, X., & Song, J., (2011). The regulation of FoxP3-expressing regulatory T cells. *Endocr. Metab. Immune. Disord. Drug Targets, 11*(4), 334–346.

Higgs, B. W., Morehouse, C. A., Streicher, K., Brohawn, P. Z., Pilataxi, F., Gupta, A., & Ranade, K., (2018). Interferon gamma messenger RNA signature in tumor biopsies predicts outcomes in patients with non-small cell lung carcinoma or urothelial cancer treated with durvalumab. *Clin. Cancer Res., 24*(16), 3857–3866.

Hommes, D. W., Mikhajlova, T. L., Stoinov, S., Stimac, D., Vucelic, B., Lonovics, J., Zakuciova, M., et al., (2006). Fontolizumab, a humanized anti-interferon gamma antibody, demonstrates safety and clinical activity in patients with moderate to severe Crohn's disease. *Gut, 55*(8), 1131–1137.

Huang, R., Hao, Y., Zhang, J., & Wu, C., (2013). Interferon-alpha plus adefovir combination therapy versus interferon-alpha monotherapy for chronic hepatitis B treatment: A meta-analysis. *Hepatol. Res., 43*(10), 1040–1051.

Humrich, J. Y., & Riemekasten, G., (2019). Low-dose interleukin-2 therapy for the treatment of systemic lupus erythematosus. *Curr. Opin. Rheumatol., 31*(2), 208–212.

Huss, D. J., Pellerin, A. F., Collette, B. P., Kannan, A. K., Peng, L., Datta, A., Wipke, B. T., & Fontenot, J. D., (2016). Anti-CD25 monoclonal antibody Fc variants differentially impact regulatory T cells and immune homeostasis. *Immunology, 148*(3), 276–286.

Inoue, K., Watanabe, T., Yamada, M., Yoshikumi, H., Ogawa, O., & Yoshiba, M., (2009). Efficacy of interferon beta combined with cyclosporine induction and intensified therapy for retreatment of chronic hepatitis C. *Transplant Proc., 41*(1), 246–249.

Ishikawa, T., Kubota, T., Abe, H., Nagashima, A., Hirose, K., Togashi, T., Seki, K., et al., (2012). Efficacy of the regimen using twice-daily beta-interferon followed by the standard of care for chronic hepatitis C genotype 1b with high viral load. *Hepatol Res., 42*(9), 864–869.

Jenks, S., (1996). After initial setback, IL-12 regaining popularity. *J. Natl. Cancer Inst., 88*(9), 576–577.

Jirikowski, G. F., Sanna, P. P., Maciejewski-Lenoir, D., & Bloom, F. E., (1992). Reversal of diabetes insipidus in Brattleboro rats: Intrahypothalamic injection of vasopressin mRNA. *Science, 255*(5047), 996–998.

Johnson-Ansah, H., Guilhot, J., Rousselot, P., Rea, D., Legros, L., Rigal-Huguet, F., Nicolini, F. E., et al., (2013). Tolerability and efficacy of pegylated interferon-alpha-2a in combination with imatinib for patients with chronic-phase chronic myeloid leukemia. *Cancer, 119*(24), 4284–4289.

Joshi, T., Rodriguez, S., Perovic, V., Cockburn, I. A., & Stager, S., (2009). B7-H1 blockade increases survival of dysfunctional CD8(+) T cells and confers protection against *Leishmania donovani* infections. *PLoS Pathog., 5*(5), e1000431.

Juarez, J. G., Harun, N., Thien, M., Welschinger, R., Baraz, R., Pena, A. D., Pitson, S. M., et al., (2012). Sphingosine-1-phosphate facilitates trafficking of hematopoietic stem cells and their mobilization by CXCR4 antagonists in mice. *Blood, 119*(3), 707–716.

Karachaliou, N., Gonzalez-Cao, M., Crespo, G., Drozdowskyj, A., Aldeguer, E., Gimenez-Capitan, A., Teixido, C., et al., (2018). Interferon gamma, an important marker of response to immune checkpoint blockade in non-small cell lung cancer and melanoma patients. *Ther. Adv. Med. Oncol., 10,* 1758834017749748.

Kariko, K., Muramatsu, H., Welsh, F. A., Ludwig, J., Kato, H., Akira, S., & Weissman, D., (2008). Incorporation of pseudouridine into mRNA yields superior nonimmunogenic vector with increased translational capacity and biological stability. *Mol. Ther., 16*(11), 1833–1840.

Katayama, K., Kasahara, A., Sasaki, Y., Kashiwagi, T., Naito, M., Masuzawa, M., Katoh, M., et al., (2001). Immunological response to interferon-gamma priming prior to interferon-alpha treatment in refractory chronic hepatitis C in relation to viral clearance. *J. Viral Hepat., 8*(3), 180–185.

Kauffman, K. J., Webber, M. J., & Anderson, D. G., (2016). Materials for non-viral intracellular delivery of messenger RNA therapeutics. *J. Control Release, 240*, 227–234.

Kennedy-Nasser, A. A., Ku, S., Castillo-Caro, P., Hazrat, Y., Wu, M. F., Liu, H., Melenhorst, J., et al., (2014). Ultra low-dose IL-2 for GVHD prophylaxis after allogeneic hematopoietic stem cell transplantation mediates expansion of regulatory T cells without diminishing antiviral and antileukemic activity. *Clin. Cancer Res., 20*(8), 2215–2225.

Khatri, M., Richardson, L. A., & Meulia, T., (2018). Mesenchymal stem cell-derived extracellular vesicles attenuate influenza virus-induced acute lung injury in a pig model. *Stem. Cell Res. Ther., 9*(1), 17.

Kimura, H., & Yamaguchi, Y., (1997). A phase III randomized study of interleukin-2 lymphokine-activated killer cell immunotherapy combined with chemotherapy or radiotherapy after curative or noncurative resection of primary lung carcinoma. *Cancer, 80*(1), 42–49.

Kirkwood, J. M., Strawderman, M. H., Ernstoff, M. S., Smith, T. J., Borden, E. C., & Blum, R. H., (1996). Interferon alfa-2b adjuvant therapy of high-risk resected cutaneous melanoma: The eastern cooperative oncology group trial EST 1684. *J. Clin. Oncol., 14*(1), 7–17.

Kirman, J., McCoy, K., Hook, S., Prout, M., Delahunt, B., Orme, I., Frank, A., & Le Gros, G., (1999). CTLA-4 blockade enhances the immune response induced by mycobacterial infection but does not lead to increased protection. *Infect. Immun., 67*(8), 3786–3792.

Kono, K., Takahashi, A., Ichihara, F., Amemiya, H., Iizuka, H., Fujii, H., Sekikawa, T., & Matsumoto, Y., (2002). Prognostic significance of adoptive immunotherapy with tumor-associated lymphocytes in patients with advanced gastric cancer: A randomized trial. *Clin. Cancer Res., 8*(6), 1767–1771.

Kooreman, N. G., Kim, Y., De Almeida, P. E., Termglinchan, V., Diecke, S., Shao, N. Y., Wei, T. T., et al., (2018). Autologous iPSC-based vaccines elicit anti-tumor responses *In Vivo. Cell Stem Cell, 22*(4), 501–513 e507.

Koreth, J., Matsuoka, K., Kim, H. T., McDonough, S. M., Bindra, B., Alyea, E. P. 3rd., Armand, P., et al., (2011). Interleukin-2 and regulatory T cells in graft-versus-host disease. *N Engl. J. Med., 365*(22), 2055–2066.

Kranz, L. M., Diken, M., Haas, H., Kreiter, S., Loquai, C., Reuter, K. C., et al., (2016). Systemic RNA delivery to dendritic cells exploits antiviral defense for cancer immunotherapy. *Nature, 534*(7607), 396–401.

Lapidot, T., Dar, A., & Kollet, O., (2005). How do stem cells find their way home? *Blood, 106*(6), 1901–1910.

Lau, J. Y., Lai, C. L., Wu, P. C., Chung, H. T., Lok, A. S., & Lin, H. J., (1991). A randomized controlled trial of recombinant interferon-gamma in Chinese patients with chronic hepatitis B virus infection. *J. Med. Virol., 34*(3), 184–187.

Lazar-Molnar, E., Gacser, A., Freeman, G. J., Almo, S. C., Nathenson, S. G., & Nosanchuk, J. D., (2008). The PD-1/PD-L costimulatory pathway critically affects host resistance to the pathogenic fungus *Histoplasma capsulatum*. *Proc. Natl. Acad. Sci. U S A., 105*(7), 2658–2663.

Lee, J. C., Hayman, E., Pegram, H. J., Santos, E., Heller, G., Sadelain, M., & Brentjens, R., (2011). *In vivo* inhibition of human CD19-targeted effector T cells by natural T regulatory cells in a xenotransplant murine model of B cell malignancy. *Cancer Res., 71*(8), 2871–2881.

Lee, L. Y. Y., Izzard, L., & Hurt, A. C., (2018). A review of DNA vaccines against influenza. *Front Immunol., 9*, 1568.
Lee, S. J., O'Donnell, H., & McSorley, S. J., (2010). B7-H1 (programmed cell death ligand 1) is required for the development of multifunctional Th1 cells and immunity to primary, but not secondary, Salmonella infection. *J. Immunol., 185*(4), 2442–2449.
Levy, Y., Sereti, I., Tambussi, G., Routy, J. P., Lelievre, J. D., Delfraissy, J. F., Molina, J. M., et al., (2012). Effects of recombinant human interleukin 7 on T-cell recovery and thymic output in HIV-infected patients receiving antiretroviral therapy: Results of a phase I/IIa randomized, placebo-controlled, multicenter study. *Clin. Infect. Dis., 55*(2), 291–300.
Lim, K. L., Jazayeri, S. D., Yeap, S. K., Mohamed, A. N. B., Bejo, M. H., Ideris, A., & Omar, A. R., (2013). Antibody and T cell responses induced in chickens immunized with avian influenza virus N1 and NP DNA vaccine with chicken IL-15 and IL-18. *Res. Vet. Sci., 95*(3), 1224–1234.
Liu, J., (2013). Induced pluripotent stem cell-derived neural stem cells: New hope for stroke? *Stem Cell Res. Ther., 4*(5), 115.
Livingston, P. O., Wong, G. Y., Adluri, S., Tao, Y., Padavan, M., Parente, R., Hanlon, C., Calves, M. J., Helling, F., Ritter, G., et al., (1994). Improved survival in stage III melanoma patients with GM2 antibodies: A randomized trial of adjuvant vaccination with GM2 ganglioside. *J. Clin. Oncol., 12*(5), 1036–1044.
Liyanage, U. K., Moore, T. T., Joo, H. G., Tanaka, Y., Herrmann, V., Doherty, G., Drebin, J. A., et al., (2002). Prevalence of regulatory T cells is increased in peripheral blood and tumor microenvironment of patients with pancreas or breast adenocarcinoma. *J. Immunol., 169*(5), 2756–2761.
Long, S. A., Buckner, J. H., & Greenbaum, C. J., (2013). IL-2 therapy in type 1 diabetes: Trials and tribulations. *Clin. Immunol., 149*(3), 324–331.
Lundgren, A., Stromberg, E., Sjoling, A., Lindholm, C., Enarsson, K., Edebo, A., Johnsson, E., et al., (2005). Mucosal FOXP3-expressing CD4+ CD25high regulatory T cells in helicobacter pylori-infected patients. *Infect. Immun., 73*(1), 523–531.
Machold, K. P., & Smolen, J. S., (1990). Interferon-gamma induced exacerbation of systemic lupus erythematosus. *J. Rheumatol., 17*(6), 831–832.
Mackall, C. L., Fry, T. J., & Gress, R. E., (2011). Harnessing the biology of IL-7 for therapeutic application. *Nat. Rev. Immunol., 11*(5), 330–342.
Magnuson, A. M., Kiner, E., Ergun, A., Park, J. S., Asinovski, N., Ortiz-Lopez, A., Kilcoyne, A., et al., (2018). Identification and validation of a tumor-infiltrating Treg transcriptional signature conserved across species and tumor types. *Proc. Natl. Acad. Sci. U S A, 115*(45), E10672–E10681.
Mahla, R. S., (2016). Stem cells applications in regenerative medicine and disease therapeutics. *International Journal of Cell Biology*, 1–24, 6940283.
Manjunath, S. R., Ramanan, G., Dedeepiya, V. D., Terunuma, H., Deng, X., Baskar, S., Senthilkumar, R., et al., (2012). Autologous immune enhancement therapy in recurrent ovarian cancer with metastases: A case report. *Case Rep. Oncol., 5*(1), 114–118.
Marshall, N. A., Christie, L. E., Munro, L. R., Culligan, D. J., Johnston, P. W., Barker, R. N., & Vickers, M. A., (2004). Immunosuppressive regulatory T cells are abundant in the reactive lymphocytes of Hodgkin lymphoma. *Blood, 103*(5), 1755–1762.
Marth, C., Windbichler, G. H., Hausmaninger, H., Petru, E., Estermann, K., Pelzer, A., & Mueller-Holzner, E., (2006). Interferon-gamma in combination with carboplatin and

paclitaxel as a safe and effective first-line treatment option for advanced ovarian cancer: Results of a phase I/II study. *Int. J. Gynecol Cancer, 16*(4), 1522–1528.
Maruggi, G., Chiarot, E., Giovani, C., Buccato, S., Bonacci, S., Frigimelica, E., Margarit, I., et al., (2017). Immunogenicity and protective efficacy induced by self-amplifying mRNA vaccines encoding bacterial antigens. *Vaccine, 35*(2), 361–368.
Masihi, K. N., (2001). Fighting infection using immunomodulatory agents. *Expert. Opin. Biol. Ther., 1*(4), 641–653.
Massberg, S., & Von, A. U. H., (2009). Novel trafficking routes for hematopoietic stem and progenitor cells. *Ann. N Y Acad Sci., 11, 76*, 87–93.
Matsushita, N., Kobayashi, H., Aruga, A., & Yamamoto, M., (2014). Establishment of induced pluripotent stem cells from adipose tissue-derived stem cells for dendritic cell-based cancer vaccines. *Gan To Kagaku Ryoho., 41*(4), 467–470.
McInnes, I. B., & Gracie, J. A., (2004). Interleukin-15: A new cytokine target for the treatment of inflammatory diseases. *Curr. Opin. Pharmacol., 4*(4), 392–397.
Michot, J. M., Bigenwald, C., Champiat, S., Collins, M., Carbonnel, F., Postel-Vinay, S., Berdelou, A., et al., (2016). Immune-related adverse events with immune checkpoint blockade: A comprehensive review. *Eur. J. Cancer, 54*, 139–148.
Miller, J. S., Morishima, C., McNeel, D. G., Patel, M. R., Kohrt, H. E. K., Thompson, J. A., Sondel, P. M., et al., (2018). A first-in-human phase I study of subcutaneous outpatient recombinant human IL15 (rhIL15) in adults with advanced solid tumors. *Clin. Cancer Res., 24*(7), 1525–1535.
Mizui, M., & Tsokos, G. C., (2016). Low-dose IL-2 in the treatment of lupus. *Curr. Rheumatol. Rep., 18*(11), 68.
Mojic, M., Takeda, K., & Hayakawa, Y., (2017). The dark side of IFN-gamma: Its role in promoting cancer immunoevasion. *Int. J. Mol. Sci., 19*(1).
Morales, A., Eidinger, D., & Bruce, A. W., (1976). Intracavitary bacillus Calmette-Guerin in the treatment of superficial bladder tumors. *J. Urol., 116*(2), 180–183.
Morales, A., Eidinger, D., & Bruce, A. W., (2017). Intracavitary bacillus Calmette-Guerin in the treatment of superficial bladder tumors. *J. Urol., 197*(2S), S142–S145.
Mou, Z., Muleme, H. M., Liu, D., Jia, P., Okwor, I. B., Kuriakose, S. M., Beverley, S. M., & Uzonna, J. E., (2013). Parasite-derived arginase influences secondary anti-leishmania immunity by regulating programmed cell death-1-mediated CD4+ T cell exhaustion. *J. Immunol., 190*(7), 3380–3389.
Muir, A. J., Sylvestre, P. B., & Rockey, D. C., (2006). Interferon gamma-1b for the treatment of fibrosis in chronic hepatitis C infection. *J. Viral Hepat., 13*(5), 322–328.
Mylvaganam, G. H., Chea, L. S., Tharp, G. K., Hicks, S., Velu, V., Iyer, S. S., Deleage, C., et al., (2018). Combination anti-PD-1 and antiretroviral therapy provides therapeutic benefit against SIV. *JCI Insight, 3*(18).
Nagarsheth, N., Wicha, M. S., & Zou, W., (2017). Chemokines in the cancer microenvironment and their relevance in cancer immunotherapy. *Nat. Rev. Immunol., 17*(9), 559–572.
Nakamoto, N., Kaplan, D. E., Coleclough, J., Li, Y., Valiga, M. E., Kaminski, M., Shaked, A., et al., (2008). Functional restoration of HCV-specific CD8 T cells by PD-1 blockade is defined by PD-1 expression and compartmentalization. *Gastroenterology, 134*(7), 1927–1937, 1937 e1921–1922.

Ni, L., & Lu, J., (2018). Interferon gamma in cancer immunotherapy. *Cancer Med., 7*(9), 4509–4516.

Niess, H., Von, E. J. C., Thomas, M. N., Michl, M., Angele, M. K., Huss, R., Gunther, C., et al., (2015). Treatment of advanced gastrointestinal tumors with genetically modified autologous mesenchymal stromal cells (TREAT-ME1): Study protocol of a phase I/II clinical trial. *BMC Cancer, 15*, 237.

Nora, T., Laszlo, L., & Laszlo, V., (1975). Pulmonary embolism in patients with internal diseases. *Orv. Hetil., 116*(17), 970–972.

O'Garra, A., Vieira, P. L., Vieira, P., & Goldfeld, A. E., (2004). IL-10-producing and naturally occurring CD4+ Tregs: Limiting collateral damage. *J. Clin. Invest., 114*(10), 1372–1378.

Okajima, F., (2013). Regulation of inflammation by extracellular acidification and proton-sensing GPCRs. *Cell Signal, 25*(11), 2263–2271.

Onda, M., Kobayashi, K., & Pastan, I., (2019). Depletion of regulatory T cells in tumors with an anti-CD25 immunotoxin induces CD8 T cell-mediated systemic antitumor immunity. *Proc. Natl. Acad. Sci. U S A., 116*(10), 4575–4582.

Ormandy, L. A., Hillemann, T., Wedemeyer, H., Manns, M. P., Greten, T. F., & Korangy, F., (2005). Increased populations of regulatory T cells in peripheral blood of patients with hepatocellular carcinoma. *Cancer Res., 65*(6), 2457–2464.

Palermo Mde, L., Trindade, M. A., Duarte, A. J., Cacere, C. R., & Benard, G., (2012). Differential expression of the costimulatory molecules CD86, CD28, CD152 and PD-1 correlates with the host-parasite outcome in leprosy. *Mem. Inst. Oswaldo Cruz, 107*(Suppl 1), 167–173.

Papadopoulos, E. B., Ladanyi, M., Emanuel, D., Mackinnon, S., Boulad, F., Carabasi, M. H., Castro-Malaspina, H., Childs, B. H., Gillio, A. P., Small, T. N., et al., (1994). Infusions of donor leukocytes to treat Epstein-Barr virus-associated lymphoproliferative disorders after allogeneic bone marrow transplantation. *N Engl. J. Med., 330*(17), 1185–1191.

Pappas, P. G., Lionakis, M. S., Arendrup, M. C., Ostrosky-Zeichner, L., & Kullberg, B. J., (2018). Invasive candidiasis. *Nat. Rev. Dis. Primers, 4,* 18026.

Pardi, N., Hogan, M. J., Pelc, R. S., Muramatsu, H., Andersen, H., DeMaso, C. R., Dowd, K. A., et al., (2017). Zika virus protection by a single low-dose nucleoside-modified mRNA vaccination. *Nature, 543*(7644), 248–251.

Park, K. S., Seo, Y. B., Lee, J. Y., Im, S. J., Seo, S. H., Song, M. S., Choi, Y. K., & Sung, Y. C., (2011). Complete protection against a H5N2 avian influenza virus by a DNA vaccine expressing a fusion protein of H1N1 HA and M2e. *Vaccine, 29*(33), 5481–5487.

Parrish, Y. K., Baez, I., Milford, T. A., Benitez, A., Galloway, N., Rogerio, J. W., Sahakian, E., et al., (2009). IL-7 Dependence in human B lymphopoiesis increases during progression of ontogeny from cord blood to bone marrow. *J. Immunol., 182*(7), 4255–4266.

Patel, A., Gray, M., Li, Y., Kobasa, D., Yao, X., & Kobinger, G. P., (2012). Co-administration of certain DNA vaccine combinations expressing different H5N1 influenza virus antigens can be beneficial or detrimental to immune protection. *Vaccine, 30*(3), 626–636.

Patel, R., Bock, M., Polotti, C. F., & Elsamra, S., (2017). Pharmacokinetic drug evaluation of atezolizumab for the treatment of locally advanced or metastatic urothelial carcinoma. *Expert Opin. Drug Metab. Toxicol., 13*(2), 225–232.

Pehamberger, H., Soyer, H. P., Steiner, A., Kofler, R., Binder, M., Mischer, P., Pachinger, W., et al., (1998). Adjuvant interferon alfa-2a treatment in resected primary stage II cutaneous melanoma. Austrian malignant melanoma cooperative group. *J. Clin. Oncol., 16*(4), 1425–1429.

Pereda, R., Gonzalez, D., Rivero, H. B., Rivero, J. C., Perez, A., Lopez, L. D. R., Mezquia, N., et al., (2020). Therapeutic effectiveness of interferon Alpha 2b treatment for COVID-19 patient recovery. *J. Interferon Cytokine Res., 40*(12), 578–588.

Petsch, B., Schnee, M., Vogel, A. B., Lange, E., Hoffmann, B., Voss, D., Schlake, T., et al., (2012). Protective efficacy of *in vitro* synthesized, specific mRNA vaccines against influenza A virus infection. *Nat. Biotechnol., 30*(12), 1210–1216.

Pfeffer, L. M., Dinarello, C. A., Herberman, R. B., Williams, B. R., Borden, E. C., Bordens, R., Walter, M. R., et al., (1998). Biological properties of recombinant alpha-interferons: 40th anniversary of the discovery of interferons. *Cancer Res., 58*(12), 2489–2499.

Pollard, C., Rejman, J., De Haes, W., Verrier, B., Van, G. E., Naessens, T., De Smedt, S., et al., (2013). Type I IFN counteracts the induction of antigen-specific immune responses by lipid-based delivery of mRNA vaccines. *Mol. Ther., 21*(1), 251–259.

Pujade-Lauraine, E., Guastalla, J. P., Colombo, N., Devillier, P., Francois, E., Fumoleau, P., Monnier, A., et al., (1996). Intraperitoneal recombinant interferon gamma in ovarian cancer patients with residual disease at second-look laparotomy. *J. Clin. Oncol., 14*(2), 343–350.

Pulster, E. L., Smalling, K. L., Zolman, E., Schwacke, L., & Maruya, K. A., (2009). Persistent organochlorine pollutants and toxaphene congener profiles in bottlenose dolphins (*Tursiops truncatus)* frequenting the turtle/Brunswick river estuary (TBRE) in coastal Georgia, USA. *Environ. Toxicol. Chem., 28*(7), 1390–1399.

Reddehase, M. J., Mutter, W., Munch, K., Buhring, H. J., & Koszinowski, U. H., (1987). CD8-positive T lymphocytes specific for murine cytomegalovirus immediate-early antigens mediate protective immunity. *J. Virol, 61*(10), 3102–3108.

Richner, J. M., Himansu, S., Dowd, K. A., Butler, S. L., Salazar, V., Fox, J. M., Julander, J. G., et al., (2017). Modified mRNA vaccines protect against zika virus infection. *Cell, 168*(6), 1114–1125 e1110.

Riddell, S. R., Watanabe, K. S., Goodrich, J. M., Li, C. R., Agha, M. E., & Greenberg, P. D., (1992). Restoration of viral immunity in immunodeficient humans by the adoptive transfer of T cell clones. *Science, 257*(5067), 238–241.

Rosenberg, J. E., Hoffman-Censits, J., Powles, T., Van Der, H. M. S., Balar, A. V., Necchi, A., Dawson, N., et al., (2016). Atezolizumab in patients with locally advanced and metastatic urothelial carcinoma who have progressed following treatment with platinum-based chemotherapy: A single-arm, multicentre, phase 2 trial. *Lancet, 387*(10031), 1909–1920.

Rosenberg, S. A., (1984). Adoptive immunotherapy of cancer: Accomplishments and prospects. *Cancer Treat Rep., 68*(1), 233–255.

Rosenberg, S. A., (1991). Immunotherapy and gene therapy of cancer. *Cancer Res., 51*(18 Suppl), 5074s-5079s.

Rosenberg, S. A., Lotze, M. T., Muul, L. M., Leitman, S., Chang, A. E., Ettinghausen, S. E., Matory, Y. L., Skibber, J. M., Shiloni, E., Vetto, J. T., et al., (1985). Observations on the systemic administration of autologous lymphokine-activated killer cells and recombinant interleukin-2 to patients with metastatic cancer. *N Engl. J. Med., 313*(23), 1485–1492.

Rosenberg, S. A., Sportes, C., Ahmadzadeh, M., Fry, T. J., Ngo, L. T., Schwarz, S. L., Stetler-Stevenson, M., et al., (2006). IL-7 administration to humans leads to expansion of CD8+ and CD4+ cells but a relative decrease of CD4+ T-regulatory cells. *J. Immunother., 29*(3), 313–319.

Roussey, J. A., Viglianti, S. P., Teitz-Tennenbaum, S., Olszewski, M. A., & Osterholzer, J. J., (2017). Anti-PD-1 antibody treatment promotes clearance of persistent cryptococcal lung infection in mice. *J. Immunol., 199*(10), 3535–3546.

Routy, J. P., Boulassel, M. R., Yassine-Diab, B., Nicolette, C., Healey, D., Jain, R., Landry, C., et al., (2010). Immunologic activity and safety of autologous HIV RNA-electroporated dendritic cells in HIV-1 infected patients receiving antiretroviral therapy. *Clin. Immunol., 134*(2), 140–147.

Rudnicka, D., Oszmiana, A., Finch, D. K., Strickland, I., Schofield, D. J., Lowe, D. C., Sleeman, M. A., & Davis, D. M., (2013). Rituximab causes a polarization of B cells that augments its therapeutic function in NK-cell-mediated antibody-dependent cellular cytotoxicity. *Blood, 121*(23), 4694–4702.

Rutigliano, J. A., Sharma, S., Morris, M. Y., Oguin, T. H. 3rd., McClaren, J. L., Doherty, P. C., & Thomas, P. G., (2014). Highly pathological influenza A virus infection is associated with augmented expression of PD-1 by functionally compromised virus-specific CD8+ T cells. *J. Virol., 88*(3), 1636–1651.

Saadoun, D., Rosenzwajg, M., Joly, F., Six, A., Carrat, F., Thibault, V., Sene, D., et al., (2011). Regulatory T-cell responses to low-dose interleukin-2 in HCV-induced vasculitis. *N Engl. J. Med., 365*(22), 2067–2077.

Saez-Royuela, F., Porres, J. C., Moreno, A., Castillo, I., Martinez, G., Galiana, F., & Carreno, V., (1991). High doses of recombinant alpha-interferon or gamma-interferon for chronic hepatitis C: A randomized, controlled trial. *Hepatology, 13*(2), 327–331.

Sakaguchi, S., (2000). Regulatory T cells: Key controllers of immunologic self-tolerance. *Cell, 101*(5), 455–458.

Salimzadeh, L., Le Bert, N., Dutertre, C. A., Gill, U. S., Newell, E. W., Frey, C., Hung, M., et al., (2018). PD-1 blockade partially recovers dysfunctional virus-specific B cells in chronic hepatitis B infection. *J. Clin. Invest., 128*(10), 4573–4587.

Sangro, B., Gomez-Martin, C., De La Mata, M., Inarrairaegui, M., Garralda, E., Barrera, P., Riezu-Boj, J. I., et al., (2013). A clinical trial of CTLA-4 blockade with tremelimumab in patients with hepatocellular carcinoma and chronic hepatitis C. *J. Hepatol., 59*(1), 81–88.

Schnee, M., Vogel, A. B., Voss, D., Petsch, B., Baumhof, P., Kramps, T., & Stitz, L., (2016). An mRNA vaccine encoding rabies virus glycoprotein induces protection against lethal infection in mice and correlates of protection in adult and newborn pigs. *PLoS Negl. Trop. Dis., 10*(6), e0004746.

Schurich, A., Khanna, P., Lopes, A. R., Han, K. J., Peppa, D., Micco, L., Nebbia, G., et al., (2011). Role of the coinhibitory receptor cytotoxic T lymphocyte antigen-4 on apoptosis-prone CD8 T cells in persistent hepatitis B virus infection. *Hepatology, 53*(5), 1494–1503.

Seitz, G., Boehmler, A. M., Kanz, L., & Mohle, R., (2005). The role of sphingosine 1-phosphate receptors in the trafficking of hematopoietic progenitor cells. *Ann. N Y Acad. Sci., 10, 44*, 84–89.

Sepahi, S., Pasdar, A., Gerayli, S., Rostami, S., Gholoobi, A., & Meshkat, Z., (2017). CTLA-4 gene haplotypes and the risk of chronic hepatitis C infection; a case control study. *Rep. Biochem. Mol. Biol., 6*(1), 51–58.

Shahjalal, H. M., Abdal, D. A., Lim, K. M., Jeon, T. I., & Cho, S. G., (2018). Generation of pancreatic beta cells for treatment of diabetes: Advances and challenges. *Stem Cell Res. Ther., 9*(1), 355.

Sigidin, Y. A., Loukina, G. V., Skurkovich, B., & Skurkovich, S., (2001). Randomized, double-blind trial of anti-interferon-gamma antibodies in rheumatoid arthritis. *Scand J. Rheumatol., 30*(4), 203–207.

Simonsson, B., Gedde-Dahl, T., Markevarn, B., Remes, K., Stentoft, J., Almqvist, A., Bjoreman, M., et al., (2011). Combination of pegylated IFN-alpha2b with imatinib increases molecular response rates in patients with low- or intermediate-risk chronic myeloid leukemia. *Blood, 118*(12), 3228–3235.

Singh, B., Bhushan, C. S., Kumar, R., Singh, S. S., Ng, S., Amante, F., De Labastida, R. F., et al., (2019). A molecular signature for CD8(+) T cells from visceral leishmaniasis patients. Parasite *Immunol., 41*(11), e12669.

Sivaraman, P., & Lye, W. C., (2001). Epstein-Barr virus-associated T-cell lymphoma in solid organ transplant recipients. *Biomed Pharmacother., 55*(7), 366–368.

Sjogren, M. H., Sjogren, Jr. R., Lyons, M. F., Ryan, M., Santoro, J., Smith, C., Reddy, K. R., et al., (2007). Antiviral response of HCV genotype 1 to consensus interferon and ribavirin versus pegylated interferon and ribavirin. *Dig. Dis. Sci., 52*(6), 1540–1547.

Skurkovich, B., & Skurkovich, S., (2003). Anti-interferon-gamma antibodies in the treatment of autoimmune diseases. *Curr. Opin. Mol. Ther., 5*(1), 52–57.

Skurkovich, S., Boiko, A., Beliaeva, I., Buglak, A., Alekseeva, T., Smirnova, N., Kulakova, O., et al., (2001). Randomized study of antibodies to IFN-gamma and TNF-alpha in secondary progressive multiple sclerosis. *Mult. Scler., 7*(5), 277–284.

Soares, M. B., Lima, R. S., Rocha, L. L., Takyia, C. M., Pontes-de-Carvalho, L., De Carvalho, A. C., & Ribeiro-dos-Santos, R., (2004). Transplanted bone marrow cells repair heart tissue and reduce myocarditis in chronic chagasic mice. *Am. J. Pathol., 164*(2), 441–447.

Spain, L., Diem, S., & Larkin, J., (2016). Management of toxicities of immune checkpoint inhibitors. *Cancer Treat Rev., 44*, 51–60.

Sportes, C., Babb, R. R., Krumlauf, M. C., Hakim, F. T., Steinberg, S. M., Chow, C. K., Brown, M. R., et al., (2010). Phase I study of recombinant human interleukin-7 administration in subjects with refractory malignancy. *Clin. Cancer Res., 16*(2), 727–735.

Sportes, C., Hakim, F. T., Memon, S. A., Zhang, H., Chua, K. S., Brown, M. R., Fleisher, T. A., et al., (2008). Administration of rhIL-7 in humans increases *in vivo* TCR repertoire diversity by preferential expansion of naive T cell subsets. *J. Exp. Med., 205*(7), 1701–1714.

Stephen-Victor, E., Saha, C., Sharma, M., Holla, S., Balaji, K. N., Kaveri, S. V., & Bayry, J., (2015). Inhibition of programmed death 1 ligand 1 on dendritic cells enhances mycobacterium-mediated interferon gamma (IFN-gamma) production without modulating the frequencies of IFN-gamma-producing CD4+ T cells. *J. Infect. Dis., 211*(6), 1027–1029.

Suarez-Mendez, R., Garcia-Garcia, I., Fernandez-Olivera, N., Valdes-Quintana, M., Milanes-Virelles, M. T., Carbonell, D., Machado-Molina, D., et al., (2004). Adjuvant interferon gamma in patients with drug-resistant pulmonary tuberculosis: A pilot study. *BMC Infect. Dis., 4*, 44.

Sun, S., Li, Z., Glencer, P., Cai, B., Zhang, X., Yang, J., & Li, X., (2017). Bringing the age-related macular degeneration high-risk allele age-related maculopathy susceptibility 2 into focus with stem cell technology. *Stem Cell Res. Ther., 8*(1), 135.

Taams, L. S., Palmer, D. B., Akbar, A. N., Robinson, D. S., Brown, Z., & Hawrylowicz, C. M., (2006). Regulatory T cells in human disease and their potential for therapeutic manipulation. *Immunology, 118*(1), 1–9.

Takahashi, K., & Yamanaka, S., (2006). Induction of pluripotent stem cells from mouse embryonic and adult fibroblast cultures by defined factors. *Cell, 126*(4), 663–676.

Takayama, T., Sekine, T., Makuuchi, M., Yamasaki, S., Kosuge, T., Yamamoto, J., Shimada, K., et al., (2000). Adoptive immunotherapy to lower postsurgical recurrence rates of hepatocellular carcinoma: A randomized trial. *Lancet, 356*(9232), 802–807.

Talpaz, M., Hehlmann, R., Quintas-Cardama, A., Mercer, J., & Cortes, J., (2013). Re-emergence of interferon-alpha in the treatment of chronic myeloid leukemia. *Leukemia, 27*(4), 803–812.

Thakur, R. S., Tousif, S., Awasthi, V., Sanyal, A., Atul, P. K., Punia, P., & Das, J., (2013). Mesenchymal stem cells play an important role in host protective immune responses against malaria by modulating regulatory T cells. *Eur. J. Immunol., 43*(8), 2070–2077.

Theofilopoulos, A. N., Koundouris, S., Kono, D. H., & Lawson, B. R., (2001). The role of IFN-gamma in systemic lupus erythematosus: A challenge to the Th1/Th2 paradigm in autoimmunity. *Arthritis Res., 3*(3), 136–141.

Thess, A., Grund, S., Mui, B. L., Hope, M. J., Baumhof, P., Fotin-Mleczek, M., & Schlake, T., (2015). Sequence-engineered mRNA without chemical nucleoside modifications enables an effective protein therapy in large animals. *Mol. Ther., 23*(9), 1456–1464.

Tompkins, S. M., Zhao, Z. S., Lo, C. Y., Misplon, J. A., Liu, T., Ye, Z., Hogan, R. J., et al., (2007). Matrix protein 2 vaccination and protection against influenza viruses, including subtype H5N1. *Emerg. Infect. Dis., 13*(3), 426–435.

Ulmer, J. B., Donnelly, J. J., Parker, S. E., Rhodes, G. H., Felgner, P. L., Dwarki, V. J., Gromkowski, S. H., Deck, R. R., DeWitt, C. M., Friedman, A., et al., (1993). Heterologous protection against influenza by injection of DNA encoding a viral protein. *Science, 259*(5102), 1745–1749.

Ulmer, J. B., Sadoff, J. C., & Liu, M. A., (1996). DNA vaccines. *Curr. Opin. Immunol., 8*(4), 531–536.

Van, C. A. H., Smits, E. L., Anguille, S., Van De, V. A., Stein, B., Braeckman, T., Van, C. K., et al., (2015). Induction of cytomegalovirus-specific T cell responses in healthy volunteers and allogeneic stem cell recipients using vaccination with messenger RNA-transfected dendritic cells. *Transplantation, 99*(1), 120–127.

Van, L. S., Wilgenhof, S., Heirman, C., Corthals, J., Breckpot, K., Bonehill, A., Neyns, B., & Thielemans, K., (2014). Optimized dendritic cell-based immunotherapy for melanoma: The TriMix-formula. *Cancer Immunol. Immunother., 63*(9), 959–967.

Wall, L., Burke, F., Smyth, J. F., & Balkwill, F., (2003). The anti-proliferative activity of interferon-gamma on ovarian cancer: *In vitro* and *in vivo*. *Gynecol. Oncol., 88*(1 Pt 2), S149–151.

Wang, L., Zou, Z. Q., Liu, C. X., & Liu, X. Z., (2014). Immunotherapeutic interventions in chronic hepatitis B virus infection: A review. *J. Immunol. Methods, 407*, 1–8.

Watanabe, K., Murakami, K., Sato, R., Okimoto, T., Maeda, K., Nasu, M., Nishizono, A., & Fujioka, T., (2004). CTLA-4 blockade inhibits induction of helicobacter pylori-associated gastritis in mice. *Clin. Exp. Immunol., 135*(1), 29–34.

Weber, J. S., Postow, M., Lao, C. D., & Schadendorf, D., (2016). Management of adverse events following treatment with anti-programmed death-1 agents. *Oncologist, 21*(10), 1230–1240.

Wilgenhof, S., Corthals, J., Heirman, C., Van, B. N., Lucas, S., Kvistborg, P., Thielemans, K., & Neyns, B., (2016). Phase II study of autologous monocyte-derived mRNA electroporated dendritic cells (TriMixDC-MEL) plus ipilimumab in patients with pretreated advanced melanoma. *J. Clin. Oncol., 34*(12), 1330–1338.

Wilgenhof, S., Van, N. A. M. T., Benteyn, D., Corthals, J., Aerts, C., Heirman, C., Van, R. I., et al., (2013). A phase IB study on intravenous synthetic mRNA electroporated dendritic cell immunotherapy in pretreated advanced melanoma patients. *Ann. Oncol., 24*(10), 2686–2693.

Windbichler, G. H., Hausmaninger, H., Stummvoll, W., Graf, A. H., Kainz, C., Lahodny, J., Denison, U., et al., (2000). Interferon-gamma in the first-line therapy of ovarian cancer: A randomized phase III trial. *Br. J. Cancer, 82*(6), 1138–1144.

Wolf, A. M., Wolf, D., Steurer, M., Gastl, G., Gunsilius, E., & Grubeck-Loebenstein, B., (2003). Increase of regulatory T cells in the peripheral blood of cancer patients. *Clin. Cancer Res., 9*(2), 606–612.

Wolff, J. A., Malone, R. W., Williams, P., Chong, W., Acsadi, G., Jani, A., & Felgner, P. L., (1990). Direct gene transfer into mouse muscle *in vivo*. *Science, 247*(4949 Pt 1), 1465–1468.

Woo, E. Y., Yeh, H., Chu, C. S., Schlienger, K., Carroll, R. G., Riley, J. L., Kaiser, L. R., & June, C. H., (2002). Cutting edge: Regulatory T cells from lung cancer patients directly inhibit autologous T cell proliferation. *J. Immunol., 168*(9), 4272–4276.

Xiao, W., Zhang, Q., Deng, X. Z., Jiang, L. F., Zhu, D. Y., Pei, J. P., Xu, M. L., et al., (2015). Genetic variations of IL-28B and PD-1 are in association with the susceptibility and outcomes of HCV infection in Southeast China. *Infect. Genet. Evol., 32*, 89–96.

Xu, H., Qian, H., Zhu, W., Zhang, X., Yan, Y., Mao, F., Wang, M., et al., (2012). Mesenchymal stem cells relieve fibrosis of *Schistosoma japonicum*-induced mouse liver injury. *Exp. Biol. Med. (Maywood), 237*(5), 585–592.

Yao, S., Wang, S., Zhu, Y., Luo, L., Zhu, G., Flies, S., Xu, H., et al., (2009). PD-1 on dendritic cells impedes innate immunity against bacterial infection. *Blood, 113*(23), 5811–5818.

Yao, Y., Richman, L., Higgs, B. W., Morehouse, C. A., De Los Reyes, M., Brohawn, P., Zhang, J., et al., (2009). Neutralization of interferon-alpha/beta-inducible genes and downstream effect in a phase I trial of an anti-interferon-alpha monoclonal antibody in systemic lupus erythematosus. *Arthritis Rheum, 60*(6), 1785–1796.

Zeuzem, S., Sulkowski, M. S., Lawitz, E. J., Rustgi, V. K., Rodriguez-Torres, M., Bacon, B. R., Grigorescu, M., et al., (2010). Albinterferon Alfa-2b was not inferior to pegylated interferon-alpha in a randomized trial of patients with chronic hepatitis C virus genotype 1. *Gastroenterology, 139*(4), 1257–1266.

Zeuzem, S., Welsch, C., & Herrmann, E., (2003). Pharmacokinetics of peginterferons. *Semin Liver Dis., 23*(Suppl 1), 23–28.

Zhang, M., Wang, H., Foster, E. R., Nikolov, Z. L., Fernando, S. D., & King, M. D., (2022). Binding behavior of spike protein and receptor binding domain of the SARS-CoV-2 virus at different environmental conditions. *Scientific Reports, 12*(1), 1–13.

Zhang, Z., Zhang, J. Y., Wherry, E. J., Jin, B., Xu, B., Zou, Z. S., Zhang, S. Y., et al., (2008). Dynamic programmed death 1 expression by virus-specific CD8 T cells correlates with the outcome of acute hepatitis B. *Gastroenterology, 134*(7), 1938–1949, 1949 e1931–1933.

Zhao, M., Li, M., Zhang, Z., Gong, T., & Sun, X., (2016). Induction of HIV-1 gag specific immune responses by cationic micelles mediated delivery of gag mRNA. *Drug Deliv., 23*(7), 2596–2607.

Index

B

G

H

I

K

L

M

Q

R

T

U

V